TRACER HYDROLOGY 97

PROCEEDINGS OF THE 7TH INTERNATIONAL SYMPOSIUM ON WATER TRACING
PORTOROŽ/SLOVENIA/26-31 MAY 1997

Tracer Hydrology 97

Edited by
A. KRANJC
Karst Research Institute ZRC SAZU, Postojna, Slovenia

A.A.BALKEMA/ROTTERDAM/BROOKFIELD/1997

7th
INTERNATIONAL SYMPOSIUM
ON WATER TRACING

The texts of the various papers in this volume were set individually by typists under the supervision of each of the authors concerned.

Published by
A.A. Balkema, P.O. Box 1675, 3000 BR Rotterdam, Netherlands (Fax: +31.10.413.5947)
A.A. Balkema Publishers, Old Post Road, Brookfield, VT 05036-9704, USA (Fax: 802.276.3837)

ISBN 90 5410 875 4
© 1997 A.A. Balkema, Rotterdam
Printed in the Netherlands

Tracer Hydrology 97, Kranjc (ed.)© 1997 Balkema, Rotterdam, ISBN 90 5410 875 4

Table of contents

Preface   XI

Organization   XIII

Methods

Movement of bacteriophage and fluorescent tracers through underground river sediments   3
M.Bricelj & M.Mišič

About the influence of the injection mode on tracer test results   11
S.Brouyère & C.Rentier

Laboratory tracer experiments in carbonate porous media from Slovenia   19
B.Čenčur Curk, M.Obal, J.Kogovšek & M.Veselič

In situ experimental evaluation of aquifer-tracers interaction: First results   27
A.Di Fazio & M.Benedini

The use of bacteriophages for multi-tracing in a lowland karst aquifer in western Ireland   33
D.Drew, N.Doerfliger & K.Formentin

Determination of bacteriophage migration and survival potential in karstic groundwaters   39
using batch agitated experiments and mineral colloidal particles
K.Formentin, P.Rossi, M.Aragno & I.Müller

Performance and persistence of Indium-EDTA as activable tracer in field experiments   47
E.Gaspar, R.D.Gaspar & I.Paunica

Activation analysis in investigation of hydrokarst systems   51
E.Gaspar, S.P.Stanescu & R.D.Gaspar

New fluorescent tracers   55
S.Hadi, Ch.Leibundgut, K.Friedrich & P.Maloszewski

Can ^{222}Rn be used as a partitioning tracer to detect mineral oil contaminations?   63
D.Hunkeler, E.Hoehn, P.Höhener & J.Zeyer

A contribution to toxicity of fluorescent tracers   69
Ch.Leibundgut & S.Hadi

Sorption of Uranine and Eosine on an aquifer material containing high organic carbon   77
C.Mikulla, F.Einsiedl, Ch.Schlumprecht & S.Wohnlich

^{222}Rn concentration variations owing to dynamic condition changes induced in the waters 85
of the Apulian karst aquifer (Southern Italy)
T.Tadolini, M.Spizzico & D.Sciannamblo

Use of optical brighteners in applied hydrogeology 91
A.Uggeri & B.Vigna

Fluorescent polystyrene microspheres as tracers of colloidal and particulate materials: 99
Examples of their use and developments in analytical technique
R.S.Ward, I.Harrison, R.U.Leader & A.T.Williams

The importance of tracer technology at combined borehole investigations 105
H.Zojer

Surface waters

Natural tracers – Indicators of the origin of the water of the Vrana Lake on Cres Island, 113
Croatia
B.Biondić, S.Kapelj & S.Mesić

Field tracer tests for the simulation of pollutant dispersion in the Doller river (France) 121
O.François, P.Calmels & F.Merheb

Restoration of tritium inflow dynamics in regulated river systems 127
D.I.Gudkov & M.I.Kuz'menko

Transport of Caesium radionuclides and Strontium-90 in the Dnieper water reservoirs 135
within 10 year period after the Chernobyl nuclear accident
A.E.Kaglyan, V.G.Klenus, M.A.Fomovsky, Yu.B.Nabivanets, O.I.Nasvit & V.V.Belyaev

'In situ' determination of the dilution efficiency of submarine sewage outfall's 143
with the help of fluorescent dye tracers
J.Roldão, J.Pecly & L.Leal

Unsaturated zone

Tracer investigations in the unsaturated zone under different cultivation types 153
in a mountainous catchment area
T.Harum, J.Fank & W.Stichler

Behaviour and comparison of dissolved silica and oxygen-18 as natural tracers during 161
snowmelt
A.C.Hildebrand, M.Lindenlaub & Ch.Leibundgut

Water tracing tests in vadose zone 167
J.Kogovšek

Conceptual runoff model for small catchments in the crystalline border mountains of Styria, 173
as developed from isotopic investigations of single hydrological events
D.Rank & W.Papesch

Investigation of preferential flow in the unsaturated zone using artificial tracers 181
S.Uhlenbrook & Ch.Leibundgut

In-situ tracer investigations on the water balance of an alternative surface covering system   189
on a sanitary landfill
R.Zischak & H.Hötzl

Aquifer

Karstic sources in Malatya Province, east of Turkey   199
I.Atalay

A negative dye tracing in the Grintovec massif (Kamnik Alps)   203
P.Audra

Ground water flow in the fresh water lens of northern Guam   205
W.L.Barner

Experiences in monitoring Timavo River (Classical Karst)   213
F.Cucchi, F.Giorgetti, E.Marinetti & A.Kranjc

Tracer tests in the Joèu karstic system (Aran Valley, Central Pyrenees, NE Spain)   219
A.Freixes, M.Monterde & J.Ramoneda

An example of geological investigation in a karst area in Lower Austria   227
G.Gangl

Karst water tracing in some of the speleological features (caves and pits) in Dinaric karst area   229
in Croatia
M.Garašić

Isotopic and hydrochemical tracing for a Cambrian-Ordovician carbonate aquifer system   237
of the semi-arid Datong area, China
Gu Wei-Zu, Ye Gui-Jun, Lin Zeng-Ping, Chang Guang-Ye, Fu Rong-An, Fei Guang-Chan,
Jing Zhi-mno & Zheng Ping-Sheng

Analysis of fracture-induced water mixing by natural tracers at a granitic pluton   245
(El Berrocal, Spain)
J.Guimerà, L.Martínez & P.Gómez

Properties of underground water flow in karst area near Lunan in Yunnan Province, China   255
J.Kogovšek, H.Liu & M.Petrič

Preliminary results of the submarine outfall survey near Piran (northern Adriatic sea)   263
V.Malačič & A.Vukovič

Separation of groundwater-flow components in a karstified aquifer using environmental   269
tracers
R.Nativ, G.Günay, L.Tezcan, H.Hötzl, B.Reichert & K.Solomon

Drainage basin boundaries of major karst springs in Croatia determined   273
by means of groundwater tracing in their hinterland
A.Pavičić & D.Ivičić

The Muschelkalk karst in Southwest Germany   279
T.Simon

Development of tracer techniques for natural and artificial recharge to confined
and unconfined aquifers, India   287
B.S.Sukhija, D.V.Reddy & P.Nagabhushanam

Water balance investigations in the Bohinj region   295
N.Trišič, M.Bat, J.Polajnar & J.Pristov

The use of tracer tests in UK aquifers   299
A.T.Williams

Contamination transport and protection

Groundwater exploration and contaminant migration testing in a confined karst aquifer   305
of the Swabian Jura (Germany)
H.Behrens, W.Drost, M.Wolf, J.P.Orth & G.Merkl

Use of artificial and natural tracers for the estimation of urban groundwater contamination   313
by chemical grout injections
M.Eiswirth, R.Ohlenbusch & K.Schnell

Agriculture – Potential polluter of waters in karst region in Slovenia   321
B.Matičič

Tracer tests applied to nitrate transfer and denitrification studies in a shaly aquifer   327
(Coët-dan basin, Brittany, France)
H.Pauwels, W.Kloppmann, J.-C.Foucher, P.Lachassagne & A.Martelat

Tracer experiments on the site of NPP Jaslovské Bohunice – The Slovak Republic   331
J.Plško, M.Kostolanský, T.Kovács, J.Benko & J.Hulla

Microbiologic activities in karst aquifers with matrix porosity and consequences   339
for ground water protection in the Franconian Alb, Germany
K.-P.Seiler & A.Hartmann

Simulation of pollutant-immission using geoelectric mapping of the migration   347
of an artificially infiltrated salt tracer
R.Supper, W.F.H.Kollmann & A.Kuvaev

Environmental isotope studies and tracer testing to determine capture and protection zones   353
in the Zechstein-Karst on the northern verge of the Thuringian Forest
C.Treskatis & K.Hartsch

Hydrogeological investigations at the mineral springs of Stuttgart (Muschelkalk karst,   361
South West Germany) – New results
W.Ufrecht

The use of groundwater tracers for assessment of protection zones around water supply   369
boreholes – A case study
R.S.Ward, A.T.Williams & D.S.Chadha

Development of a tracer test in a flooded uranium mine using Lycopodium clavatum   377
C.Wolkersdorfer, I.Trebušak & N.Feldtner

Aquifer parameters and modelling

A deuterium-calibrated compartment model of transient flow in a regional aquifer system   389
M.E.Campana, W.R.Sadler, N.L.Ingraham & R.L.Jacobson

2D and 3D groundwater simulations to interpret tracer test results in heterogeneous   397
geological contexts
A.Dassargues, S.Brouyère & G.Carabin

Interpretation of tracer tests in a granitic formation (El Berrocal site, Spain)   405
J.Guimerà, J.Carrera, I.Benet, M.Saaltink, L.Vives, M.García-Gutiérrez & M.d'Alessandro

Geohydraulic parameters in hard rocks of SW-Germany determined by tracer tests   415
A.E.Jakowski & G.Ebhardt

Dual-tracer transport experiments in a heterogeneous porous aquifer: Retardation   423
measurements at different scales and non-parametric numerical stochastic transport
modelling
T.Ptak

Some aspects of the functioning of *careos* determined by tracer experiments: 431
Example of La Alpujarra (Spain)
A.Pulido-Bosch, Y.Ben Sbih & A.Vallejos

Differentiation of flow components in a karst aquifer using the $\delta^{18}O$ signature   435
M.Sauter

Simulation of cave hydrology using a conventional computer spreadsheet   443
J.D.Wilcock

Author index   449

Tracer Hydrology 97, Kranjc (ed.) © 1997 Balkema, Rotterdam, ISBN 90 5410 875 4

Preface

Five years ago a book entitled 'Tracer Hydrology' was published containing papers presented at the 6th Symposium on Water Tracing (SWT) which was held in the German town of Karlsruhe. In these five years many new discoveries have been made relating to karst hydrology and methodology, and the technology of water tracing – from new techniques of tracer determination to new tracers and improved knowledge of their properties and use. The impact of tracers on the environment and human beings is also much better known and it can assist in planning research using tracers.

Obviously this is not the only matter. Water tracing techniques, as with so many other techniques, tightly connect theoretical and laboratory experiments with practice and applied water tracing tests, but they are also important as a method of solving various questions relating to hydrology and hydrogeology.

I would like to say that the present book contains the reports of a number of researchers over several years. Among the numerous authors from all over the world (from the United States of America to China, from Great Britain to Brazil) there are contributions from both the theoretical and practical points of view, made by researchers working either in laboratories or in the field using water tracing methods to study new underground water connections, the velocity of underground flows and the spreading of pollution along streams.

Although this book is strictly professional it nevertheless covers a very broad domain of interest. A reader may find out which new tracers have appeared, how they may be used in different spheres and also the problems which can be solved by water tracing methods. One can discover much detail about a water tracing test and such information may be of direct use in future work; one can get acquainted with regional reviews of water tracings and their results. The most recent approaches to solving these questions, such as computer simulation in particular, are not neglected either.

This book is the proceedings of a Symposium, but it is an invaluable manual for all sorts of questions related to water tracing but at the same time it is a work that provides an overview of how these branches of hydrology, hydrogeology and karst hydrology have developed in the last five years, what scientific achievements have been accomplished, and the patterns of development for the future. Unfortunately, due to limited space, the short time available for preparation and for a variety of other reasons not all the contributions could be included.

Andrej Kranjc
Organizing Secretary

Tracer Hydrology 97, Kranjc (ed.) © *1997 Balkema, Rotterdam, ISBN 90 5410 875 4*

Organization

Organized by

International Association of Tracer Hydrology (ATH)
and
Karst Research Institute ZRC SAZU, Postojna

Under the auspices of

The Minister of Science and Technology of the Republic of Slovenia
and
The Minister of Environment and Physical Planning of the Republic of Slovenia

Promoted by

International Association of Hydrogeologists (IAH)
 – Karst Commission
International Association of Hydrological Sciences (IAHS)
 – International Committee on Tracers
International Atomic Energy Agency (IAEA)
Slovenian National Commission for the UNESCO
National Committee of Slovenia for the IHP UNESCO

Financially supported by

Ministry of Science and Technology of the Republic of Slovenia
Ministry of Environment and Physical Planning of the Republic of Slovenia

General sponsor

HIT Hotels Casinos Tourism

Scientific Committee

Dr Peter Habič (Chairman)
Postojna, Slovenia
Prof. Dr Mitja Brilly
Faculty of Civil Engineering and Geodesy,
University of Ljubljana, Slovenia
Dr Klaus Fröhlich
International Atomic Energy Agency,
Vienna, Austria
Prof. Dr Heinz Hötzl
Department of Applied Geology,
University of Karlsruhe, Germany
Prof. Dr Werner Käss
Umkirch, Germany
Dr Andrej Kranjc
Karst Research Institute ZRC SAZU,
Postojna, Slovenia
Prof. Dr Christian Leibundgut
Institute of Hydrology, Albert Ludwig
University of Freiburg, Germany

Prof. Dr Heribert Moser
Munich, Germany
Dr Dieter Rank
Bundesforschungs- und Prüfzentrum Arsenal,
Geotechnisches Institut, Vienna, Austria
Dr Barbara Reichert
Department of Applied Geology,
University of Karlsruhe, Germany
Prof. Dr Klaus-Peter Seiler
GSF-Forschungszentrum für Umwelt und
Gesundheit, Institut für Hydrologie,
Oberschleißheim, Germany
Prof. Dr Miran Veselič
Institute of Mining, Geotechnology and
Environment, Ljubljana, Slovenia
Prof. Dr Hans Zojer
Joanneum Research, Institut für Hydrogeologie
und Geothermie, Graz, Austria

Honorary Committee

Dr Ciril Baškovič
Ministry of Science and Technology
of the Republic of Slovenia
Prof. Dr France Bernik
President of the Slovenian Academy
of Sciences and Arts
Mitja Bricelj
Nature Protection Administration
of the Republic of Slovenia
Prof. Dr Matija Drovenik
Secretary General of the Slovenian Academy
of Sciences and Arts
Franko Fičur
Mayor of Piran
Dr Miloš Komac
Ministry of Science and Technology
of the Republic of Slovenia

Lucijan Korva
Nature Protection Administration
of the Republic of Slovenia
Albin Krapež
Nature Protection Administration
of the Republic of Slovenia
Dr Oto Luthar
Scientific Research Centre of the Slovenian
Academy of Sciences and Arts
Ada Štravs-Brus
Nature Protection Administration
of the Republic of Slovenia
Peter Volasko
Ministry of Science and Technology
of the Republic of Slovenia
Prof. Dr Peter Tancig
President of the Slovenian National Commission
for the UNESCO

Organizing Committee

Andrej Kranjc (Chairman)
Janja Kogovšek (Scientific Secretary)
Andrej Juren
Martin Knez

Metka Petrič
Andrej Mihevc
Tadej Slabe
Stanka Šebela

Methods

Tracer Hydrology 97, Kranjc (ed.) © 1997 Balkema, Rotterdam, ISBN 90 5410 875 4

Movement of bacteriophage and fluorescent tracers through underground river sediments

M. Bricelj
National Institute of Biology, Ljubljana, Slovenia

M. Mišič
Institute of Geology, Geotechnics and Geophysics, Ljubljana, Slovenia

ABSTRACT: Underground river sediment cores were taken from the river Unec in Slovene central karst region. The sediment comprised 27% clay minerals, of which constituent parts were muscovite, illite, chlorite, montmorillonite and their mixed layer counterparts. Disturbed and undisturbed cores were used in continuos flow conditions to assess the velocity of movement and recovered quantity of salmonella phage P22H5, coliphage T7 and fluorescent dies uranine and rodamine B through the sediment. Only uranine (recovery-up to 43.1%) and phage P22H5 (recovery-up to 2.6%) appeared through disturbed sediments in measurable quantities, while rhodamine B and coliphage T7 appeared only in traces through some columns. Coliphage T7 was not detected in outflow of ten undisturbed cores but rhodamine B appeared only through some of them with recovery values between 0.009 to 3.8%. Recovery values of uranine and salmonella phage after 42 days were between 14.5 to 62.5 and 0.045 to 74.9 %, respectively. Corresponding velocity was 0.48 to 0.69 cm day_{-1} for rhodamine B, 1.7 to 7.4 cm day^{-1} for uranine and 11.4 to 40 cm day^{-1} for salmonella phage P22H5.

1 INTRODUCTION

In several tracing experiments in the karst regions of Slovenia (HABIČ, et al., 1990), Austria (BEHRENS et al., 1992) and Greece (BRICELJ et al.,1986) salmonella phage tracer P22H5 was used. The recovery values of phage tracer were low, considering those the of fluorescent tracers that were used in almost the same environmental conditions. It is well documented, that different phages adsorb very quickly to the surfaces of clay minerals (CARLSON et al., 1968; STOTZKY et al., 1981; The degree of reversibility or irreversibility of attachment varies for different phages and also for different fluorescent dyes used as tracers. For this reason salmonella phage P22H5 and coliphage T7 and fluorescent dyes rhodamine B and uranine were used in mixing experiments with clay minerals (BRICELJ, 1994) and in column experiments with disturbed and undisturbed sediments of the underground river, that are presented in this contribution.

2 MATERIAL AND METHODS

2.1 Bacterial and bacteriophage strains

The *Salmonella typhimurium* TL747(LT2) w.t. strain and the salmonella phage P22H5 were obtained from dr. Miklavž Grabnar, Department of Molecular Biology, Biotechnical Faculty, University of Ljubljana. The *Salmonella typhimurium*, nal^{+} is a mutant of *Salmonella typhimurium* TL747(LT2) w.t. by nitric acid mutagenesis. The coliphage T7 w.t. was obtained from Kinnunen. The *Escherichia coli* B strain, host for coliphage T7, was also obtained from dr. Grabnar.

2.2 Bacterial media

Brain-Heart Infusion was from , yeast extract, triptose bouillon was from bioMérieux, physiological solution was 0.85% NaCl; 50 x E medium was composed from citric acid 100 g, $MgSO_4.7H_2O$ 10 g, K_2HPO_4. anh. 500 g, $NaHNH_4PO_4.4H_2O$ 175 g dissolved in bidestilled water of 670 ml.

Table 1: Quantity of clay suspension, total number of phages P22H5 and T7, and fluorescent tracers rhodamine B and uranine in 1 m long and 30 mm in diameter.

col.. No.	clay ml	P22H5 pfu	T7 pfu	Rhod. B µg	Uranine µg
3	30	(250 ml) 3.0×10^{10}	(250 ml) 2.0×10^{9}	(250 ml) 42.5	not added
4	30	(165 ml) 3.3×10^{10}	(165 ml) 1.1×10^{9}	(165 ml) 28.1	(165 ml) 8.3
6	60	(237 ml) 8.1×10^{10}	(237 ml) 8.1×10^{8}	not added	(237 ml) 11.9
7	60	(238 ml) 2.9×10^{10}	(238 ml) 1.9×10^{9}	(238 ml) 40.5	not added
8	60	(250 ml) 5.0×10^{10}	(250 ml) 1.7×10^{9}	(250 ml) 42.5	(250 ml) 12.5

Table 2: Quantity of clay suspension, total number of phages P22H5 and T7, and fluorescent tracers rhodamine B and uranine in 1 m to 3 m long and 30 mm in diameter.

col. No.	h c/s. cm	P22H5 pfu	T7 pfu	rhod. B mg	uranine mg
1	300 15	1,600 ml 3.7×10^{11}	1,600 ml 6.1×10^{9}	1,600ml 208	1,600ml 302
2	300 10	1,600 ml 3.7×10^{11}	1,600 ml 6.1×10^{9}	1,600ml 208	1,600ml 302
3	300 5	1,600 ml 3.7×10^{11}	1,600 ml 6.1×10^{9}	1,600ml 208	1,600ml 302
4	200 15	1,100 ml 2.5×10^{11}	1,100 ml 4.2×10^{9}	1,100ml 143	1,100ml 220
5	200 10	1,100 ml 2.5×10^{11}	1,100 ml 4.2×10^{9}	1,100ml 143	1,100ml 220
6	200 5	1,100 ml 2.5×10^{11}	1,100 ml 4.2×10^{9}	1,100ml 143	1,100ml 220
7	100 15	490 ml 1.1×10^{11}	490 ml 1.9×10^{9}	490 ml 63.7	490 ml 98
8	100 10	510 ml 1.2×10^{11}	510 ml 1.9×10^{9}	510 ml 66.3	510 ml 102
9	100 5	500 ml 1.2×10^{11}	500 ml 1.9×10^{9}	500 ml 65	500 ml 100

Legend: c/s   -height of column and sediment

Table 3: Movement of phage P22H5 and fluore--scent dyes rhodamine B and uranine through the disturbed sediment cores from the sands of the Planina cave.

column prop.			P22H5 % of rec. quant.			FD* % of rec.quant.	
col. No.	sed. q.	th.fl ml/d	7 d.	28 d.	140d	rho. B	ura.
3	30	11.6	1.1	2.5	2.6	32.7	n.a.
4	30	3.1	0.01	0.20	0.22	n.a.	43.1
6	60	5.6	0.25	0.38	0.38	0.0	29.9
7	60	5.6	0.03	0.04	0.09	0.0	n.a.
8	60	6.1	0.00	0.26	0.26	0.0	12.9

Legend:
n.a.   - fluorescent dye was not added
*   - % of recovered quant. of fluorescent dyes was calculated for the period of 98 days

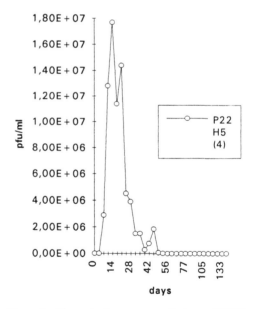

Figure 1: Through flow curve of phage P22H5 through column No. 4 filled with disturbed sediment.

2.3. Propagation of bacteriophages

Both bacteriophage strains were propagated in 10 l flasks filled with 7 l of growing medium composed of Triptose bullion 26 g, glucose 1 g, yeast extract 5 g, 50 x E medium 25 ml, Vitamins PLIBEX 0.5 ml, anti foam agent 0.05 ml and Aminosteril L Forte 50 ml per 1 liter of the growth solution. Growth solution was held at 37 $^{\circ}$C and vigorously aerated with the vacuum pump. During propagation the neutralisation of growth medium was achieved with the addition of 2 N sodium hydroxide. At the end of propagation period, resting bacteria were killed with 25 ml chloroform per 1 liter of growth solution. Phage suspensions were titrated for viable phage according to the agar layer method of Adams (1966), using host bacteria as the indicator strains.

2.4. Sampling and analysis of clay minerals

Clay minerals from different locations were taken with PVC tube with the diameter of 30 mm. Samples were taken to the depth of 20 cm. In the laboratory sediment was shed into 1 liter plastic flask and overflowed with 1 l of bidestilate water. The sample was shaken on the reciprocal shaker for 24 hours at the room temperature. Coarse particles (> 10 μm) were separated with gravitational settlement. Fraction of particles with dimensions 10-5 μm and 5-2 μm were separated with centrifugalization. Preparation of clay minerals for the X-ray difraction was done by the method described by Mišič (1992).

2.5. Column experiments with bacteriophage and fluorescent tracers

Disturbed cores

Clay minerals for studying movement of the tracer through cores were taken from the sandy bank of the Unec river at the entrance to the Planinska cave. The sediment was shed into plastic bottle and overflowed with 500 ml of filtered river water (Wattmann GF/C glass filter) and shaken for 24 hours, then the sediment was poured into 1 m long PVC tube with diameter of 30 mm. Columns were flushed for two weeks with the filtered river water. Afterwards fluorescent dyes and bacteriophages were introduced to pass through the sediment. The movement of tracers was followed for four month. The data of clay quantity, and the quantity of introduced tracers are given in Table 1.

Undisturbed cores

Clay minerals for studying movement of tracer through the undisturbed cores were also taken from the sandy bank of the river Unec at the entrance to the Planinska cave. On the surface of 1 m2, 1 meter long PVC tubes (φ = 30 mm) were pushed directly into the sediment to the depth of 5, 10 and 15 cm, so that the natural structures of micro and macropores in the sediments weren't at risk. In the laboratory, the plugs with glass wool were added to the lower part of the tube. The tubes were flushed with filtered river water (Wattmann GF/C glass filter) for two weeks and afterwards the tracers were added. The movement of added tracers was being controlled for three months. The data of clay quantity and the quantity of introduced tracers are given in Table 2.

The third experiment was done with 11 columns with 10 cm of undisturbed sediment from the same location. All the columns were 2m long and after the two week period of flushing 1.000 ml of bacteriophage P22H5 suspension with total titre of 2.6×10^{11} pfu were poured to each column. Sampling was performed for 8 days in the 3 hour period. The sample quantity was measured by its weighting.

3 RESULTS

3.1 Movement of tracers through disturbed sediment cores

In the laboratory experiments of disturbed sediments, the homogenised sediment settled down in the tube, the coarse particle at the bottom and fine sediment at the sediment-water surface. In all columns both batcteriophages P22H5 and T7 were added, but fluorescent dyes were added as a single tracer or both to the same column.

Bacteriophage P22H5 appeared in the outflow of all five columns in the first sample which was a collection of three day sampling period. The recovered quantity of the phage tracer were very low. The highest quantity of 2.6% of initial added tracer was obtained in the outflow of column No. 4, which was the only column where also zero values were obtained before the end of the experiment.

Coliphage T7 appeared through some columns only in traces; through the column No. 4 only with 5 pfu ml^{-1} after 46 days and through column No. 6

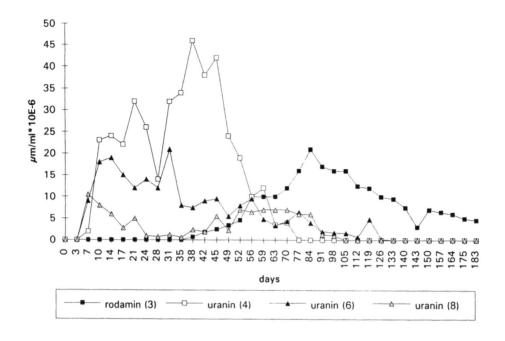

Figure 2: Through flow curve of fluorescent dyes rhodamine B and uranine through the columns with disturbed sediments. The column numbers are added to the name of dye.

Table 4: Movement of phage tracer P22H5 and fluorescent dyes rhodamine B and uranine through the sediment cores with undisturbed macro and micropores taken from sands at Planinska cave.

column properties				% of rec. quant. of phage P22H5				% of rec. quant. of uranine			
column No.	h.of col. cm	h. of sed. cm	th. flow ml/day	8 days	28 days	42 days	98 days	8 days	28 days	42 days	98 days
1	300	15	15.0	3.7	4.0	6.1	15.9	25.4	37.7	43.5	n.m.
2	300	10	10.9	0.026	0.032	0.045	4.3	1.5	9.3	14.5	n.m.
3	300	5	13.0	1.2	1.2	1.5	6.7	4.1	15.2	20.6	n.m.
4	200	15	30.0	52.0	62.0	74.9	89.2	11.7	39.6	55.7	n.m.
5	200	10	24.9	52.0	58.0	66.0	91.0	10.1	42.2	56.4	n.m.
6	200	5	12.6	23.8	24.5	27.5	35.2	5.9	21.3	28.7	n.m.
7	100	15	20.3	15.0	26.1	40.7	48.5	13.1	60.4	94.1	n.m.
8	100	10	9.6	5.2	5.2	2.7	13.3	2.9	25.2	62.5	n.m.
9	100	5	15.8	10.9	12.1	12.9	15.6	15.6	55.3	93.0	n.m.

Legend: n.m. - not measured

6

with 41 pfu ml^{-1} after 56 days. Fluorescent dye rhodamine B appeared after 42 days in the samples of column No. 3. The recovered quantity of the dye was 18.4% after 98 days and 32.7% after 183 days and there was still traces of dye when experiment was brought to an end.. Through the columns No. 4 and 7 rhodamine B did not appear in the time of the experiment. The fluorescent dye uranine appeared through all the columns to which it had been added and in all three columns the negative samples were also detected before the end of the experiment (see Figure 2). The recovered quantities of the uranine tracer are presented in Table 3.

3.2. Movement of the tracers through undisturbed cores

In the second series of experiments the sediments cores were taken directly with the columns used in the experiments, so that the macro and micropore structures of the sediment were not disturbed. Salmonella phage P22H5, as well as coliphage T7 and fluorescent dyes rhodamine B and uranine were added to all columns. The total quantity of phage tracer varied between 1.2×10^{11} - 3.7×10^{11} pfu for phage P22H5 and between 1.9×10^9 - 6.1×10^9 pfu for phage T7. Quantities of added rhodamine B were between 63.7 in 208 mg ml^{-1}, and those of uranine between 98 in 302 mg ml^{-1}.

The fluorescent tracer rhodamine B did not appear in the samples of water in any column at all collected during the experiment proceeded 98 days. The bacteriophage P22H5 appeared through all the columns and the samples were also positive at the end of the experiment. In comparison with the disturbed sediment cores the recovery values of salmonella phage tracer were very high, as a result 4.3 and 91.1 % for various columns in 98 days period (see Table 4). The recovered quantity of phage tracer correlates with the water through flow with regression coefficient 0.92 (P=0.05).

Despite the relatively high quantity of rhodamine B added to the columns, we detected the dye only in the outflow samples of four columns. From the column No. 9 the dye appeared already on the 9th day and recovered quantity of the dye for 28 days period was 3.8 %. The appearance of tracer in the column No. 4 occurred on the 20th day, from column No. 5 on the 22nd day and from the column No. 6 on the 25th day. The corresponding recovery values were very low (0.34%; 0.08% and 0.009%).

The recovered quantities of uranine, that appeared in the samples of all columns already on the 2nd day (column No. 2 on the 4th day) were relatively high. On the 28th day after the injection of tracer we got 9.3 to 60.4% of injected tracer. Correlation analysis did not reveal any correlation between the recovered quantity of uranine tracer and the average through flow through the columns. There was no correlation either with the hydrostatic pressure and the quantity of sediment in the core.

Because bacteriophage P22H5 appeared very quickly in the experiments with disturbed and undisturbed sediments in samples of outgoing water in the third experiment the same quantity of bacteriophage tracer (2.6×10^{11} pfu) was used with 11 columns with the same quantity of undisturbed sediment (10 cm). Bacteriophage tracer appeared very quickly; from the column No. 5 after 3 hours, from the other columns after 6 and 9 hours, and from the column No. 11 only after 21 hours. The recovered quantities were, in spite of the same quantity of sediment, very different, between 0.0002 and 7.3 % (see Table 5). Correlation between the average through flow of water and recovered quantity of tracer was evident (($r = 0.86$; P=0.05).

4 CONCLUSIONS

Column experiments for detection of movement of bacteria, viruses and bacteriophages (HUYSMAN et al., 1993; POWELSON et al., 1993) were done with various types of soil in saturated and unsaturated conditions. Additional conservative tracers were chlorine ion (ALHAJJAR et al., 1988), ^{36}Cl (LINDQVIST et al., 1992), heavy water (TAN, et al., 1992), potassium and sodium bromide (POWELSON s sod., 1993). There were no data of simultaneous use of uranine and rhodamine B tracers with micro-organisms in column experiments. In our experiments of movement of bacteriophage tracers P22H5 and T7 and fluorescent tracers uranine and rhodamine B through disturbed and undisturbed cores of sands' sediment, we can notice the correlation between the retention of tracer in columns and the largeness of adsorbtion of tracers on clay minerals in mixing experiments (BRICELJ, 1994). In the experiments with disturbed cores we can conclude that retention of uranine is correlated with the quantity of sediment. High recovery values of rhodamine B and phage P22H5 in column No. 3 correlate with the high through flow. With the high through flow the adsorbtion of viruses and phages is rather low (POWELSON and al., 1993). High retention of rhodamine B tracer in the columns with 60 ml sediment can be connected with high adsorbtion and migration into the micropores, t.i. with chromatographic effect (HARVEY et al., 1989). Higher adsorbtion of rhodamine B to the sediment is also evident from the data of first appearance of tracer which was much longer for rhodamnine B (42 days) than for uranine (11 days).

The difference in the movement of phages P22H5 and T7 through the sediments with disturbed structure is evident and probably depends on the surface characteristics of both phages, t.i. hydrophobic and hydrophilic structures of the bacteriophage head and tail parts. Bales and co-workers (1991) found out in column experiments with phages MS-2 and PRD-1 that hydrophobic effect is also rather important in the adsorbtion of even relatively hydrophylic biocolloids.

The comparison of recovered quantities of bacteriophage P22H5 in the experiments with disturbed and undisturbed sediments, shows the recovered quantities greater that the sediments with the undisturbed structure. It is evident that sediments and soil samples with undisturbed structure retain the systems of macro and micropores unaffected, which enables the molecules, bacteria, viruses and phages to pass the strata quicklier (POWELSON et al., 1993). The greater deactivation of phage T7 could be contributed to the temperature (YATES et al., 1992), micro-organisms (GANNON et al., 1991), but reversible adsorbtion of QEASS type or irreversible adsorbtion is also possible (GRANT et al., 1993).

Velocity of fluorescent tracer uranine movement in undisturbed cores varied between 1.7 - 7.4 cm day^{-1} and for rhodamine B between 0.48 - 0.69 cm day^{-1} and almost the same was the velocity of T7 phage, t.i. 0.43 - 0.48 cm day^{-1} that we could determine from the scarce appearance of phage T7 in some columns. The difference between the velocity of both tracers could be contributed to the nature of dye. Uranine is anionic dye and adsorbtion to the negatively charged clay particles is lower than of the cationic, positively charged dye rhodamine B. From almost the same values of velocity for rhodamine B and coliphage T7, we could conclude that the negatively charged parts predominate on the surface structures.

Velocity of movement of P22H5 tracer was determined in the experiment with 11 columns filed with 10 cm of sediment with undisturbed structure.

Although the sediments were withdrawn from the sands of a very closed area, we can conclude that the macropores and micropores were differently arranged in individual columns and also the clay mineral composition should be different, because no correlation was detected between the first appearance of tracer and the velocity of through flow.

Calculated velocity of phage P22H5 was between 9.2×10^{-3} - 1.3×10^{-4} cm sec^{-1}. This range of velocity is from 2.1 - 692 times lower than the velocity of movement of polio virus type 1 and echovirus type 1 through the columns filled with sieved soil in the experiments of Wang and co-workers (WANG et al., 1981). Higher velocities of movement of polio and echovirus are comprehensible as the soil contains 90% of sand and only 3 to 4% of clay minerals. In our sediments clay content was 27% that means much bigger absorbtion surface for the retention of phage tracer.

REFERENCES

ADAMS, M.H., 1959. Bacteriophages. *Methods of Study of Bacterial Viruses*. Interscience Publishers, a division of John Wiley & Sons Inc., New York. 443 - 522.

ALHAJJAR, B. J., STRAMER, S. L., CLIVER, D. O. & HARKIN, J. M., 1988. Transport Modelling of Biological Tracers from Septic Systems. *Water Research*, 22(7): 907 - 915.

BALES, R. C., HINKLE, S. R., KROEGER, T. W. & STOCKING, K. in dr., 1991. Bacteriophage Adsorption during Transport through Porous Media: Chemical Perturbations and Reversibility. *Environmental Science & Technology*, 25(12): 2088 - 2095.

BEHRENS, H., BENISCHE, R., BRICELJ, M., HARUM, T., KASS, W., KOSI, G., LEDITZKY, H.P., LEIBUNDGUT, Ch., MAŁOSZEWSKI P., MAURIN, V., RAJNER, V., RANK., D., REICHERT, B., STADLER, H., STICHLER, W., TRIMBORN, P., ZOJER, H. & ZUPAN, M., 1992. Investigations with Natural and Artificial Tracers in the Karst Aquifer of the Lurbach System (Peggau-Tanneben-Semriach, Austria).*Steiriche Beitrage zur Hydrology*, 43: 9 - 158.

BRICELJ, M., KOSI, G. & VRHOVŠEK, D.,

1986. Tracing with Salmonella-phage P22H5. *Steirische beitrage zur Hydrogeologie*, **37/38**: 269 - 271.

BRICELJ, M., 1994. Underground Water Tracing with the Phages of *Salmonella typhimurium*, Dissertation thesis, Biotechnical Faculty, University of Ljubljana., 113 pp.

CARLSON, G. F. Jr., WOODWARD, F. E., WENTWORTH D. F. & SPROUL O. J., 1968. Virus Inactivation on Clay Particles in Natural Waters. *Journal of Water Pollution Control Federation*, **40**(2/2): R89 - R106.

GANNON, J: T., MANILAL, V. B. & ALEXANDER, M., 1991. Relationship between Cell Surface Properties and Transport of Bacteria through Soil. *Applied and Environmental Microbiology*, **57**(1). 190 - 193.

GRANT, S. B., LIST, E. J. & LINDSTROM, M. E., 1993. Kinetic Analysis of Virus Adsorption and Inactivation in Batch Experiments. *Water Resources Research*, **29**(7): 2067 - 2085.

HABIČ, P., KOGOVŠEK, J., BRICELJ M. & ZUPAN M., 1990. Dobličica springs and their wider karst background. *Acta Carsologica*, **XIX**: 5 - 100.

HARVEY, R. W., LEAH, H. G., SMITH, R. L. & LeBLANC D. R., 1989. Transport of Micro-spheres and Indigenous Bacteria through a Sandy Aquifer: Results of Natural- and Forced-Gradient Tracer Experiments. *Environmental Science & Technology*, **23**(1): 51 - 56.

HUYSMAN, F. & VERSTRAETE, W., 1993. Water-Facilitated Transport of Bacteria in Unsaturated Soil Columns: Influence of Inoculation and Irrigation Methods. *Soil Biology & Biochemistry*, **25**(1): 91 - 97.

LINDQVIST, R. & ENFIELD, C. G., 1992. Cell Density and Non-Equilibrium Sorption Effect on Bacterial Dispersal in Groundwater Microcosmos. *Microbial Ecology*, **24**: 25 - 41.

MIŠIČ, M. 1992. Glineni minerali z zmesno strukturo tipa illit/montmorillonit. *Magistersko delo*. Fakulteta za naravoslovje in tehnologijo, Oddelek za montanistiko. 73 pp.

POWELSON, D. K., GERBA C. P. & YAHYA M. T., 1993. Virus Transport and Removal in Wastewater during Aquifer Recharge. *Water Research*, **27**(4): 583 - 590.

STOTZKY, G., SCHIFFENBAUER, M., LIPSON, S. M. & YU, B. H., 1981. Surface Interactions Between Viruses and Clay Minerals and Microbes: Mechanisms and Implications. In: *Viruses and Wastewater Treatment*, ed. by M. Goddard & M. Butler. Pergamon Press, Oxford, pp. 199 - 204.

TAN, Y., BOND, W. J. & GRIFFIN D. M., 1992. Transport of Bacteria during Unsteady Unsaturated Soil Water Flow. *Soil Science Society of America Journal*, **56**(5): 1331 - 1340.

WANG, D-S., GERBA, C. P. & LANCE J. C., 1981. Effect of Soil Permeability on Virus Removal Through Soil Columns. *Applied and Environmental Microbiology*, **42**(1): 83 - 88.

YATES, M. V. & OUYANG, T., 1992. VIRTUS, a Model of Virus Transport in Unsaturated Soils. *Applied and Environmental Microbiology*, **58**(5): 1609 - 1616.

Tracer Hydrology 97, Kranjc (ed.) © 1997 Balkema, Rotterdam, ISBN 90 5410 875 4

About the influence of the injection mode on tracer test results

S. Brouyère
Laboratoires de Géologie de l'Ingénieur, d'Hydrogéologie et de Prospection Géophysique (L.G.I.H.), University of Liège & National Fund for Scientific Research of Belgium

C. Rentier
Laboratoires de Géologie de l'Ingénieur, d'Hydrogéologie et de Prospection Géophysique (L.G.I.H.), University of Liège, Belgium

ABSTRACT: One of the most common techniques used to evaluate the aquifer transport properties is the well-known two-well injection-withdrawal tracer test. A radially converging flow field is created by a pumping well. A pulse of tracer-labelled water is introduced into the groundwater through the injection well. The time evolution of the tracer concentration is monitored at the pumping well. The resulting breakthrough curve can be interpreted with analytical or numerical solutions to evaluate the transport properties. Most often, the injection is supposed to be instantaneous. Such an assumption allows to interpret the breakthrough curve as the impulse response of the aquifer-piezometer system.

As mentioned by many authors, the way the injection is conducted can differ dramatically from the usual Dirac-type pulse of tracer, due to the complex interaction between the piezometer equipment and the aquifer material. This can have a critical influence on the shape of the breakthrough curve leading at the extreme to a double-peak. The interpretation of such a result disregarding the way the injection was conducted could prove erroneous.

Different tracer tests performed in the alluvial plain of the river Meuse in Belgium have shown double-peaked breakthrough curves. One of the tracer tests was selected to be replicated under different injection conditions. The breakthrough curves did not present two peaks anymore.

After some theoretical and field considerations, the tracer test results are analysed and criticised. It is shown how important it is to conduct further investigation in order to gain a better understanding of the influence of the injection mode on the breakthrough curve.

1 INTRODUCTION

One of the most common techniques used to evaluate the aquifer transport properties is the well-known two-well injection-withdrawal tracer test. A radially converging flow field is created by a pumping well. A pulse of tracer-labelled water is introduced into the groundwater through the injection well. The time evolution of the tracer concentration is monitored at the pumping well, giving the breakthrough curve, which can be used to evaluate the aquifer transport properties.

The main advantage of this technique is that, theoretically, the whole injected mass of tracer is recovered at the pumping well. This allows to interpret the results in a quantitative way by comparing the actual breakthrough curves with theoretical ones obtained with analytical or numerical solutions.

Usually, these tests are performed by injecting the tracer at a known level in a piezometer. A chase fluid accompanies and follows the injection to force the tracer to penetrate in the aquifer. A good advice is to use a volume of chase water at least equal to the volume of the piezometer, to make sure that the tracer actually reaches the aquifer. This chase fluid is often performed by imposing a water flow rate at two points surrounding the position of the tracer injection. This is done to determine the level at which the tracer is injected into the aquifer and to prevent the tracer to reach the lowest, often blind, part of the piezometer.

Two injection modes are often conducted. The most usual is an instantaneous injection. In practice, the duration of the tracer injection is very short compared to the total duration of the tracer test. The best way to do this is to use a high chase fluid of short duration. Lepiller and Mondain (1986) give a review of the notions of system theory applied to hydrogeological studies. The aquifer can be viewed as a system receiving an input (the tracer injection) and giving an output (the tracer breakthrough curve). If the injection is of the Dirac type, the output can be considered as the impulse response of the system and interpreted with such theories. Sometimes, a continuous injection is performed. It actually consists in a long time step injection, which is often very difficult to control. This injection mode is better suited when the experimenter is mainly concerned with the retardation mechanisms affecting the tracer transport. Indeed, this enhances the breakthrough curve tailing, separating it from the concentration noise of the aquifer.

It is usually difficult to get a precise idea of the actual input function of the tracer in the aquifer. The concentrations measured directly in the injection piezometer are not necessarily equal to the concentrations in the aquifer in the vicinity of the piezometer. A borehole-mixing effect is frequently observed. Moreover, the interactions between the piezometer and the aquifer are often more complex than usually supposed. Ackerer et al. (1982) present an experimental device to study that interaction. Figure 1 shows how a piezometer modifies the groundwater flow at its vicinity. The streamlines skirt round the piezometer. This confirms that the concentrations measured in the piezometer can differ from the concentrations in the aquifer at the same location, the piezometer acting more or less as a dead-water zone. To get more representative concentrations, a pumping is necessary to force an exchange of water between the piezometer and the aquifer. Locally, as the pumping can modify the flow field and the experimental conditions, it is not always a good solution.

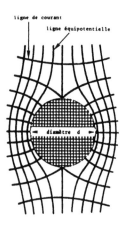

Figure 1. Modification of the groundwater flow near a piezometer (from Ackerer et al., 1982)

There are different methods to monitor the concentration evolution at the injection well. Porel (1988) measures the residual tracer concentrations at different levels in the injection piezometers with a manual multi-sampler. For saline tracers (Li^+,...), it is possible to use a conductivity probe to evaluate the concentration history in the piezometer. Meus (1993) used such a method to monitor the input function of Li^+ and KNO_3 for tracer tests conducted in a chalk aquifer. Derouane (1994) followed the conductivity (for saline tracers) and took manual samples in different injection piezometers during and after the injections of different tracers (Li^+, I^-, uranine, naphtionate and rhodamine WT). This allowed him to conclude that most of the tracer had left the piezometer after one day (more than 99% of the total injected mass). But he did not use the input concentration profile to simulate the tracer tests. To be really efficient, a multilevel conductivity measurement should be made, which implies the use of a very sophisticated probe.

The tracer nature is usually not taken into account. Laboratory experiments have often been conducted to study the influence of the tracer nature (Sabatini and Austin, 1991; Shiau et al., 1992), but field experiments are still missing.

The interpretation of the tracer test results is often done by taking advantage of analytical solutions (Sauty and Kinzelbach, 1987; Moench, 1995) or numerical methods (Biver and Meus, 1992). At this stage, other numerical or conceptual problems can be encountered. These are the subject of a companion paper (Dassargues et al., this issue).

2 INFLUENCE OF THE TRACER INJECTION AND CHASE FLUID

As mentioned before, impulse injections are often considered to be of the Dirac type (or square type) when interpreting the results. Moreover, the volume of water used to inject and to chase the tracer is supposed to have a negligible influence on the flow field in the vicinity of the piezometer. In practical terms, the way the injection is performed can have a dramatic influence on the shape of the breakthrough curve. Apart from the duration of the injection, the main factor influencing the result is the chase fluid that accompanies the tracer injection. Guvanasen and Guvanasen (1986) have developed an approximate semi-analytical solution for tracer injection tests in a confined aquifer with a radially converging flow field and a finite volume of tracer and chase fluid. They show how, according to the porous media properties, the way the tracer injection and the chase fluid are made can modify the shape of the breakthrough curve.

To obtain the solution, the tracer transport is conceptually divided in two parts: (1) the tracer injection and chase is considered first, (2) the effective transport of the tracer to the pumping well is calculated. During the injection, the dispersion mechanism is supposed to be negligible compared to the advection mechanism, inducing a piston-like tracer transport in all directions around the injection well. Near the injection piezometer, the combination of the groundwater velocity induced by the pumping well and the velocity due to the injection causes the initial distribution of the tracer to be non-circular and ringlike (figure 2). This initial tracer distribution is evaluated through a tracking along the pathlines. Then, the resulting tracer geometry is approximated by a series of contiguous rectangles of uniform concentration. The radial form of the advection-dispersion equation is applied to all these rectangles to calculate the resulting breakthrough curve at the pumping well.

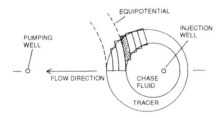

Figure 2 Initial tracer distribution in the aquifer (adapted from Guvanasen and Guvanasen, 1986)

Four non-dimensional parameters are used to determine the shape of the tracer ring at the end of the injection stage.

$$g1 = (Q_{Tr} \cdot T_{Tr} + Q_{ch} \cdot T_{ch}) / (\pi \, b \theta \, R_0{}^2)$$
$$g2 = Q_{ch} \cdot T_{ch} / (Q_{Tr} \cdot T_{Tr} + Q_{ch} \cdot T_{ch})$$
$$g3 = Q_{Tr} \cdot R_0 / Q_p \cdot r_{inj}$$
$$g4 = Q_{ch} / Q_{Tr}$$

where Q_{tr} and Q_{ch} are the tracer and chase fluid injection rates, T_{tr} and T_{ch} the time period of tracer and chase fluid injection, b is the saturated thickness of the aquifer, θ is the effective porosity of the aquifer, R_0 is the distance between the injection piezometer and the pumping well, r_{inj} is the injection piezometer radius and Q_p is the pumping rate. Parameter $g1$ compares the total injection volume to the effective volume of water in the aquifer in a circle

of radius R_0, $g2$ compares the chase fluid volume to the total injection volume, $g3$ compares the velocities due to the tracer injection to the velocities near the injection well due to the pumping Q_p, $g4$ compares the chase fluid rate and the tracer injection rate. Figure 3 shows the tracer ring shape according to these different adimensional parameters. Guvanasen and Guvanasen (1986) compare on the basis of a few simple examples their results to an analytical solution considering a Dirac-type tracer injection. The differences can be dramatic, showing the important effect of the volume of injection and chase water on the shape of the breakthrough curves. At the extreme, the breakthrough curve can be double-peaked. The interpretation of such a result, disregarding the way the injection is conducted, could prove erroneous.

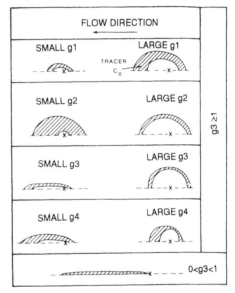

Figure 3. Effects of geometrical parameters on the tracer plume shape (adapted from Guvanasen and Guvanasen, 1986)

Different tracer tests performed in the alluvial plain of the river Meuse in Belgium have shown double-peaked breakthrough curves. One of the tracer test was replicated under different injection conditions and did not present two peaks.

3 CASE STUDIES

Both studied sites are situated in the gravel aquifer located in the alluvial plain of the river Meuse in Belgium. These sites consist in water catchment areas. Hydrogeological studies and groundwater modelling were conducted in order to define protection zones around the production wells (Derouane, 1994; Comeaga et al., 1995;). Site 1 is located at Vivegnis, near the city of Liege, on the left bank of the river Meuse. Site 2 is situated at Dinant, on the right bank of the river (figure 4).

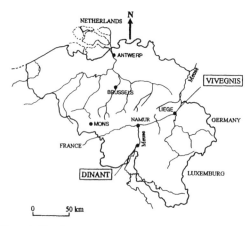

Figure 4. Situation map

Site 1

The geology of this site is composed of an upper carboniferous bed-rock consisting in shales and sandstones. This constitutes the impervious basis of the aquifer. Loose materials deposited by the river Meuse on that substratum are composed of gravel bodies imbedded in old channels filled with sandy to silty or clayey sediments. At the studied site, the vertical succession shows, from bottom to top, aquifer well-sorted gravels, sandy to clayey gravels and surface deposits. The total thickness of the main gravel layer is about 7 meters. There are four pumping wells extracting about 8000 m³/d. For the needs of the study, ten piezometers have been drilled, completing the set of boreholes already existing in the area. The aquifer is unconfined on nearly the whole of the studied area.

Seven tracer tests have been performed on that site, with saline tracers (Li$^+$, I$^-$) and fluorescent tracers (uranine, naphtionate and rhodamine WT). One of these tests, performed with Li$^+$, presented a double-peaked breakthrough curve (figure 5). This test was the replication, with a higher chase fluid, of a previous one conducted with I$^-$. This first injection was not performed under 'good' experimental conditions: due to a density effect the tracer remained in the blind deepest part of the injection piezometer. A breakthrough curve could be identified at the pumping well only when a further bottom

chase fluid was performed in the piezometer, four days after the injection of I$^-$.

If one compares the two results, the first arrivals are almost the same. The peak of Li$^+$ arrives later than the peak of I$^-$ but this difference can be attributed to different injection conditions. Since the experimental conditions are not perfectly controlled, the comparison between the two tracer tests can only be made on a qualitative level. As two peaks are not identifiable on the iodide breakthrough curve, the hypothesis of a transport of the tracer along two different paths can be rejected and the influence of the injection mode has to be considered.

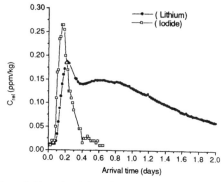

Figure 5. Tracer breakthrough curves at Vivegnis

Site 2

The geology of site 2 is composed of a devonian and carboniferous bed-rock made of sandstones, limestones and shales. The bed-rock cannot be considered as the impervious basis of the aquifer. The different studies conducted on the site have set into evidence a incoming water flux from the bed-rock to the gravel aquifer. The loose materials have the same structure and succession as in Vivegnis. Due to the presence of sandy to clayey gravel over the well-sorted gravel, the aquifer can be considered as semi-confined. There are two pumping wells (P1 and P2) extracting a mean water flux of 1500 m³/d from the gravel aquifer. For the purpose of the study, seven piezometers have been drilled in the gravel aquifer and in the bed-rock.

In a first phase, three tracer tests were performed. The injections were made at piezometers situated in the vicinity (20 to 30 meters) of the pumping well P1. These tests are summarised in table 1 (together with the tests conducted at Vivegnis). The breakthrough curves (normalised according to the injected mass of tracer) are shown in figure 6.

14

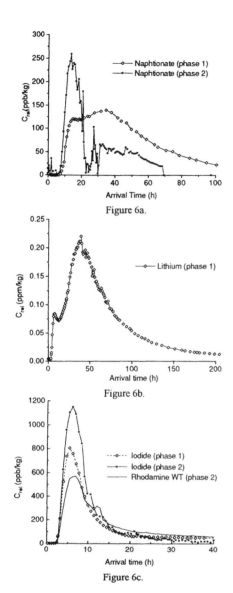

Figure 6a.

Figure 6b.

Figure 6c.

Figure 6. Tracer breakthrough curves at Dinant

Naphtionate and lithium present a double-peaked breakthrough curve (figure 6a and 6b, phase 1). Iodide presents a unimodal breakthrough curve with a very long and flat tailing (figure 6c, phase 1). First, the double-peaked tracer tests were interpreted as the result of a tracer transport along two different paths. These could be two gravel layers with different permeabilities or a gravel layer and a preferential fissured path in the bed-rock. One year after, the naphtionate and iodide tracer tests were performed under similar experimental conditions. Rhodamine WT was added to I⁻ to check the

influence of the tracer nature. The extracted flow rate at P1 was nearly unchanged. The main difference was in the injection mode. The chase fluid (which was suspected to be the main cause of the double-peaked results) was dramatically diminished. The breakthrough curves observed during the second phase are superimposed to the corresponding curves during the first phase in figure 6a and 6c. When the chase fluid is lower, both peaks are enhanced. Furthermore, there is no more double peak visible on the naphtionate breakthrough curve. The iodide breakthrough curve shows a rapidly decreasing tailing. This was not the case during the first phase where the longer tailing was due to a higher dispersion of the tracer in the aquifer during the high chase fluid.

So, the first interpretation considering the double-peaked shape of the breakthrough curve has to be rejected. The influence of the chase fluid seems to provide a better explanation. If one considers only the tracer transfer aspects, the influence seems to be minimal since the first arrival and modal times are nearly the same. But if one is concerned with the retardation effects, one has to be very careful about the way the injection is conducted.

Figure 6a also shows the influence of the tracer nature on the result. Rhodamine WT was injected with I⁻. The breakthrough curve clearly shows that the peak of rhodamine WT appears later than the peak of I⁻. This is probably due to a greater sorption of rhodamine WT on shales. This is a lesser problem than the previous one mentioned since it is easily possible to conduct an experiment (even in laboratory) to demonstrate the sorption of rhodamine WT.

4 INTERPRETATION OF THE RESULTS

Table 2 shows the four adimensional parameters defined by Guvanasen and Guvanasen (1986), calculated for both tracer tests. In fact, we are not exactly working on the assumptions of their paper since the aquifer should be confined. We can suppose that we are not far from that hypothesis since the gravels are overlaid by sandy to silty gravels which make them semi-confined. The transport parameter values are those obtained through the simulations of the tracer tests with finite elements (Derouane, 1994) and finite differences (Rentier, 1996).

As shown in table 2, parameters g2 and g4 are nearly the same for the two phases conducted on site 2. The corresponding parameters g1 and g3 are shorter in phase 2 than in phase 1 (approximately 1/3 and 1/2). Remembering that g1 is a measure of the total chase volume and g3 a comparison between the

Table 1. Tracer tests results

Site	Qp (m³/h)	Tracer	d (m)	I.M.(kg)	Vtr (l)	Ttr (min.)	Vch (l)	Tch(min.)	Vmax (m/h)	Vmod (m/h)	R (%)
Vivegnis	120	iodide	26.85	2	---	---	---	---	~16.11	6.78	14.8
		lithium	26.85	3.48	70	5	2100	120	11.93	5.1	88
Anseremme		lithium	18.63	3.25	170	13	1145	30	6.1	2.07/0.48	87
phase 1	57.9	iodide	19.78	7.65	288	6	2016	72	9.13	3.65	61
		naphtionate	31.12	5	272	5	1720	45	5.31	2.09/0.93	49
phase 2	53.6	iodide	19.78	1.53	75	3	241	17	7.657	3.25	52
		naphtionate	31.12	0.5	77	3	234	12	4.5	2.28	24

Qp : pumping rate
d: distance between pumping well and piezometer
I.M. : injected mass
R : Recovery factor

Vtr : tracer injection volume
Ttr : tracer injection duration
Vch : chase volume
Tch : chase duration

Table 2. Geometrical parameters for the different tracer test according to Guvanasen and Guvanasen (1986)

Site	Qp (m³/h)	Tracer	d (m)	r_{inj} (m)	b(m)	θ (%)	g1	g2	g3	g4
Vivegnis	120	iodide	26.85	---	---	---	---	---	---	---
		lithium	26.85	0.08	7.0	5.8	0.00236	0.968	2.352	1.250
Anseremme		lithium	18.63	0.058	6.0	9.0	0.00200	0.871	4.390	2.920
phase 1	57.9	iodide	19.78	0.058	6.0	6.5	0.00480	0.875	17.110	0.583
		naphtionate	31.12	0.058	6.0	9.0	0.00120	0.863	30.510	0.703
phase 2	53.6	iodide	19.78	0.058	6.0	6.5	0.00066	0.763	9.630	0.567
		naphtionate	31.12	0.058	6.0	9.0	0.00019	0.750	17.880	0.760

r_{inj} : injection piezometer radius
b : aquifer thickness
θ : effective porosity
g1-g4 : geometrical parameters defined by Guvanasen and Guvanasen (1986)

velocities due to the pumping in the vicinity of the piezometer and the velocity due to the injection, it is easy to understand the reasons why the tracer tests have been disturbed by the injection mode. In the first phase, the chase fluid volume was too big, which modified the velocity field near the injection piezometer, inducing a ring-like shape of tracer. The double peak is due to that shape. The first peak is the result of the part of the ring which was 'pushed' in the direction of the pumping well by the chase fluid, the second peak is the result of the part of the ring which was 'pushed' upstream with regard to the main flow direction and was, therefore, delayed. Even the tracer test performed with I⁻ was disturbed by the injection. The results of the first phase show a lower peak intensity and a very long and flat tailing, which is probably the consequence of a high artificial dispersion of the tracer in the medium due to the chase fluid.

5 CONCLUSIONS

According to the presented results, it is obvious that tracer test experimenters should be very careful with the way they conduct the injections, particularly when the tracer test distances are short (10 to 30m). Of course it is interesting to chase the injected tracers with a volume of water added after the injection, but this chase should not be too strong to avoid a high deformation of the initial tracer cloud in the aquifer. If the tracing distances are longer (50 to 100 meters), the influence of the injection is probably mitigated by the hydrodynamic dispersion in the media.

As far as we know, the case studies presented here are the first mentioned in the literature. But this is perhaps because other tracer tests have not been interpreted with the knowledge of the possible occurrence of such problems.

In the future, two kinds of experiments should be undertaken. It would be interesting to model such results with numerical methods and to check if it is actually possible to reproduce the influence of the chase. To do so, the difficulty is that one is faced with numerical problems: oscillations, dispersion and mass conservation. Representing the injection remains tremendously difficult. In the field, it should be very interesting to conduct a research on a simple experimental site composed of one or two injection piezometers (aligned if possible) and a pumping well. Various injection modes should be checked to gain a better understanding of what actually occurs and different tracers should be used to compare their respective response.

REFERENCES

Ackerer P., P.Muntzer, L.Zilliox. 1982. Le piézomètre, une discontinuité dans un aquifère alluvial. Expérimentation d'un dispositif pour acquérir des données représentatives du milieu perturbé. Colloque B.R.G.M. en hommage à G.Castany : les milieux discontinus en hydrogéologie, Doc 45: 1-8.

Biver P. and P.Meus. 1992. The use of tracer tests to identify and quantify the processes in an heterogeneous aquifer. Tracer Hydrology. Hötzl & Werner (eds). Balkema. Rotterdam. pp. 415-421.

Comeaga Th., S.Brouyère, A.Dassargues, A.Monjoie. 1995. Prise d'eau Prieuré à Anseremme. Essais de traçage et modélisation dans le cadre de l'étude des zones de prévention. L.G.I.H. Report SWDE/958. Unpublished.

Dassargues A., S.Brouyère, G.Carabin. 1997. 2D and 3D groundwater simulations to interpret tracer tests results in heterogeneous geological contexts. This issue.

Derouane J. 1994. Etude hydrogéologique du site de captage de Vivegnis (Plaine alluviale de la Meuse). Détermination des zones de protection. Travail de fin d'études en vue de l'obtention du grade d'ingénieur géologue. Université de Liège. Faculté des Sciences Appliquées. Belgium. 172p.

Guvanasen V. and V.M.Guvanasen. 1987. An approximate solution for tracer injection tests in a confined aquifer with radially converging flow field and finite volume of tracer and chase fluid. Water Resour. Res. 23(8): 1607-1619.

Lepiller M., P.H.Mondain. 1986. Les traçages artificiels en hydrogéologie karstique. Mise en oeuvre et interprétation. Hydrogéologie. 1: 33-52.

Meus P. 1993. Hydrogéologie d'un aquifère karstique dans les calcaires carbonifères (Néblon-Anthisnes, Belgique). Apport des traçages à la connaissance des milieux fissurés et karstiques. Thèse de doctorat. Faculté des Sciences Appliquées, Université de Liège, Belgium. 323p.

Moench A.F. 1995. Convergent radial dispersion in a double-porosity aquifer with fracture skin: Analytical solution and application to a field experiment in fractured chalk. Water Resour. Res. 31(8): 1823-1835.

Rentier C. 1996. Etude hydrogéologique du site de captage de Dinant-Anseremme. Essais de traçage et modélisation du transport de polluant pour la détermination des zones de protection. Travail de fin d'études en vue de l'obtention du grade d'ingénieur géologue. Université de Liège. Faculté des Sciences Appliquées. Belgium. 160p.

Porel G. 1988. Transfert de soluté en aquifère crayeux. Causes de modifications des résultats de traçages. Thèse de doctorat. Université des sciences et techniques de Lille-Flandres-Artois, France. 327p.

Sabatini D.A. and T.A.Austin. 1991. Characteristics of rhodamine WT and fluorescein as adsorbing ground-water tracers. Groundwater 29(3): 341-349.

Sauty J.P. and W.Kinzelbach. 1987. CATTI. User's manual. International ground water modeling center.

Shiau B., D.A.Sabatini, J.H.Harwell. 1992. Sorption of rhodamine WT as affected by molecular properties. Tracer Hydrology. Hötzl & Werner (eds). Balkema. Rotterdam. pp. 57-64.

Tracer Hydrology 97, Kranjc (ed.) © 1997 Balkema, Rotterdam, ISBN 90 5410 875 4

Laboratory tracer experiments in carbonate porous media from Slovenia

Barbara Čenčur Curk
IRGO-Institute for Mining, Geotechnology and Environment & PGD, Ljubljana, Slovenia

Marija Obal & Miran Veselič
IRGO-Institute for Mining, Geotechnology and Environment, Ljubljana, Slovenia

Janja Kogovšek
Karst Research Institute ZRC SAZU, Postojna, Slovenia

ABSTRACT: More than 40 % of Slovenia is covered with carbonate rocks, urbanisation is proliferating thus increasing pollution. Previous facts result in intensified research of solute transport in fractured and karstified rocks. Laboratory experiments were made to study behaviour (sorption) of tracers in porous media as a preliminary study for in-situ tracer experiments in fractured rocks and final modelling of mass transport.

Experiments were carried out in the laboratory of Institute for Mining, Geotechnology and Environment in three permeameters which were connected in a series. Each permeameter was 1,5 m long and permitted individual sampling. Huge reservoirs containing fluid fed the flow which was propelled through gravitation. Constant flow was maintained through uniform fluid levels in the supply and outlet reservoir. Salt and uranine were utilised as tracers. Two different concentrations of salt and uranine water solutions were used. Three porous materials of different grain size distribution from the nearby limestone quarry were used as porous media with 99,3% concentration of $CaCO_3$. The grain size distribution of applied material was in range between 0,160-4 mm and 4-8 mm, whereas part of the material was a mixture of both ranges with an even distribution.

During the tests various measurements of parameters like flow, pressure, temperature, pH, conductivity, oxygen content, alkalinity, acidity and uranine content were constantly performed. The experiment with salt water solution confirmed that there was no sorption. Conductivity of uranine solution increased after beginning of the experiment and then stabilised at a value 20% higher than the initial value. Experiments with uranine water solution indicated very fast sorption and desorption. Based on this fact this process was assumed to be connected only to the grain surface. Because of detection problems, the results so far can not be determined. Hence, additional batch-tests are planned to confirm the sorption process.

1. EXPERIMENTAL DESIGN

During the laboratory test we examined tracers which were used later in terrain experiments on porous karst limestone (two in-situ polygons in slovenian karst region: Unška Koliševka and Sinji Vrh).

We were particularly interested in behaviour of the tracer during the flow through the karst rock - especially the sorption of both tracers used: NaCl and uranine, respectively. The first tracer does not react with a rock (*Di Fazio, 1996*) and the second one belongs to the group of fluorescent dyes that are prone to sorption (*Käss 1992*).

The main goals of the experiments were to determine the sorption of the uranine on the limestone sample from Razdrto quarry and to determine the effects of water flow through unsaturated media (moistening of dry limestone samples) on uranine sorption on the limestone. Additional tasks were to determine the influence of grain size (porosity) and retardation time on uranine sorption on limestone sample.

The experiment schedule is shown in table 1.

2. EXPERIMENTAL SYSTEM

2.1 Description of experiment system

Experiments were carried out at the Institute for Mining, Geotechnology and Environment.

Table 1 Permeameter experiment schedule: experimental design

	MOISTENING		TRACER SOLUTION		TRACER SOLUTION	GRANULOMETRY
1	sample+DM H₂O	→	NaCl+DM H₂O (MF)	→	uranine(HC)+DM H₂O (HF)	sample 1
2	sample+DM H₂O	→	uranine(HC)+DM H₂O (MF)			sample 1
3	dry sample	→	uranine(LC)+DM H₂O (LF)			sample 1, 2 and 3
4	dry sample	→	NaCl+ DM H₂O (LF)	→	uranine(LC)+DM H₂O (LF)	sample 2 and 3
5	dry sample	→	uranine(HC)+DM H₂O (LF)			sample 2
6	dry sample	→	uranine(HC)+DM H₂O (LF)	→	demineralised H₂O	sample 3
7	dry sample	→	uranine(HC)+natural H₂O(LF)			sample 2

DM=demineralised H_2O, HC=high concentration (about $1mg/m^3$), LC= low concentration (about $0,1mg/m^3$), LF=low flow (23,7-34,3 l/h), MF=mid-flow (41,1; 55,8 l/h), HF=high flow (110,4 l/h).

Experimental system consisted of three equal permeameters, each 1,5 m long with a diameter of 15 cm which were sequentially interconnected (figure 1). At the end of each third of this series, that is at the end of each single permeameter, fluid sampling was possible. The flow has been restored naturally. Lineal and constant flow has been enabled by large 100 l collecting reservoirs (figure 1) containing fluid which maintained constant hydraulic pre-determined head through pouring of liquid.

Fig. 1   Experimental system

2.2 Characterisation of material

Permeameters have been filled with limestone granulate of various grain sizes. Samples from quarry Razdrto with technical signatures of 0/4 and 4/8 have been used. Because of an excessive amount of experiment samples necessary, samples could not have been taken directly on terrain, where the field tracing experiments were carried out.

The fraction under 0,160 mm has been removed from the sample 0/4 (from 0 to 4 mm) through rinsing of particles and the first sample of granulation between 0,160 - 4 mm was thus obtained. Granulate 4/8 (4-8 mm) has been taken as a sample 2 and a sample 3 was obtained through the combination of both samples in ratio 1:1 (0,160 - 8 mm). Figure 2 depicts granulometric analysis by sifting all three samples.

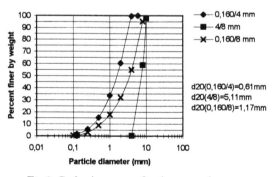

Fig.2 Grain-size curves for three samples

Limestone granulate has been taken from quarry Razdrto which is also situated on karst terrain. Sample from the quarry is close to pure limestone and contains 99,3 % of $CaCO_3$. The rock from the first location (Unška Koliševka) of in-situ test polygon is pure limestone containing 99,4 % $CaCO_3$ while on the second (Sinji Vrh) in-situ polygon next to the limestone of 99,3 - 99,6 $CaCO_3$ there also exists dolomitic limestone (35 % $CaCO_3$ and 65 % $MgCO_3$).

2.3 Properties of the flow through the rock samples

During the experiment we wanted to study the laminar flow through the rock sample so the Reynolds value $\left(Re = \dfrac{v * d_e}{v} \right)$ had to be lower than 10 (*Bear, 1972*).

Permeability coefficient k of single rock samples was calculated using empirical equation USBR (*Veselič, 1984*) which takes into account effective grain diameter (d_{20} in mm) from granulometric curves (figure 2):

$$k_{USBR} = 0,36 * (d_{20})^{2,3} * 10^{-3} \left[\frac{m}{s} \right].$$

For each sample the highest hydraulic head Δh (table 2) was calculated, where the flow remained laminar:

$$\Delta h = \frac{Re* v(m^2 / s) * l(m)}{k(m / s) * d_e * 10^{-3}(m)} ,$$

where Re = 10,
$v_{(20^\circ C)} = 1,003*10^{-6} m^2/s$ (*de Marsily, 1981*),
$l = 1,5 \times 3$ m and
$d_e = d_{20}$ in mm.

So we could have optional hydraulic head in the first (4,08 m) and the third sample (0,5 m). In the second sample the hydraulic head was only 0,1 m (flow has been hindered on 30 l/h.

Table 2 The uppermost hydraulic heads (Δh) for each limestone sample

sample	d_{20} (mm)	k_{USBR} (m/s)	Δh (m)
1 0,160-4 mm	0,61	$1,2*10^{-4}$	616,6
2 4-8 mm	5,11	$1,5*10^{-2}$	0,59
3 0,160-8 mm	1,17	$5,2*10^{-4}$	74,2

2.4 Applied tracers

NaCl and uranine have been used as tracers.

NaCl is a salt that is very easily dissolved in water (358,5 g/l at $20^0 C$) and dissociates on Na^+ and Cl^- ion. Cations are subject to ionic interchange regarding the ion concentration reduction. During this process the other ions bond due to established balance of charge - creators of hardness (*Käss 1992*). Since anions do not react with the rock, their concentration better describes the flow through the rock (*Di Fazio, 1996*). During the experiment NaCl was used as a comparison of sorption with uranine and we have measured only conductivity.

Fluorescent dye uranine is a sodium fluorescein which has a structural formula $C_{20}H_{10}O_5Na_2$. Fluorescein anion form is at present the best known fluorescent substance and has thus been the most useful tracer for an extended period of time.

Fluorescent dyes have various sorption properties on rocks. Generally the cation dye form is more subject to sorption than anionic form. Uranine cation has also sorption properties so alkaline media and anionic uranine form is requested. At the same time the intensity of fluorescence is higher in alkaline media (*Käss, 1992*).

In practice pseudo ideal substance is being used for determination of retardation factor - in most cases it is uranine (*Käss, 1992*).

During the experiment NaCl concentration of 10^5 mg/m^3 and uranine concentrations 0,1 and 0,01 mg/m^3 were used. Demineralised water was used as a media in order to provide similar circumstances compared to those in the nature, where rain leaks through the rock. In one experiment the natural karst water from Malni spring was also used (system 4).

3. EXPERIMENTAL PROCESS

During the experiment permanent measuring of flow (flow meters) and pressure (Hg manometers) in permeameters was carried out. Gained samples were measured with laboratory instrument - temperature, pH, conductivity and oxygen content. Acidity and alkalinity have been determined at our Institute using the ASTM method. Content of uranine in samples was determined at the Institute For Karst Research, ZRC SAZU.

14 experiments were carried out (table 3) on experimental system described above and one on smaller experimental system (glassy columns 0,5 m high with diameter of 0,055 m).

Experiment systems are schematically shown in table 3: the first column depicts sample grain size, the second and sixth column system signatures. The third column describes used tracer and used media (demineralised water and natural karst water from Malni spring) and the fourth column represents given concentration of tracer. The succession of experimental phases is shown in table 1.

From the system 3 a constant flow sized from 15 to 35 l/h was sustained; except at small system 4 (table 1, No.7), where the flow was calculated for small system (equivalent for other system).

Each experiment performed on the same system with signatures a and b, was first carried out with NaCl, then followed by uranine, and ended up with a study of interactions between tracers.

All experiments were performed with permanent in-flow of fresh solution until the balance was restored. During the last experiment (system 10) a clear demineralised water was added after a determined time and after circulating a constant tracer concentration.

Batch tests were carried out in large 100 l reservoirs to provide enough volume of samples. Experiments were carried out to determine commutation of content of Ca^{2+} and HCO_3^- ions, uranine, acidity, alkalinity, and other parameters (particularly pH and conductivity). Demineralised water and natural karst water from Malni spring was used as a media. For the limestone samples only the granulation of 0,160-8mm (third sample) was considered.

4. RESULTS

4.1 Results of tests with NaCl solution

For the NaCl experiments the conductivity measurements were predominant (systems 1a, 5a and 8a). The conductivity has raised (figure 3, left) at the beginning of the experiment because of the dissolution of the limestone and stabilised at a value a little bit higher than value of the initial NaCl solution when limestone dissolving equilibrium has been attained.

4.2 Results of tests with uranine solution

The results of experiments with small uranine concentration (round 0,1 mg/m^3) were inconclusive due to various errors (analytical, problems with cleaning of the permeameters) and because of some results that were higher than initial value (figure 4).

At experiments with higher uranine concentration (round 1 mg/m^3) such difficulties did not appear. The uranine concentration remained more or less constant during the experiment (figure 5). A fall of concentration (time about 24 and 48 hours) was due to higher retardation time (velocity drops during the night time).

The last experiment (system 10) was not performed like other experiments (system 1-9) - after moistening with uranine solution the initial solution was replaced with pure demineralised water. The uranine solution in the permeameters was quickly replaced with pure water, that shows rapid reduction of uranine concentration (figure 6).

In batch test the uranine concentration slightly oscillates around initial uranine concentration (figure 7) similar to the experiments in permeameters.

Table 3 Experiment schedule on three limestone samples (sample 1: 0,160-4 mm, sample 2: 4-8 mm and sample 3: 0,160-8 mm and with two tracers (NaCl and Uranine) in permeameters.

	GRAIN SIZE	SYSTEM	TRACER	CONC. mg/m^3	Q l/h	SYSTEM	TRACER	CONC. mg/m^3	Q l/h
1	0,160/4 mm	S1a	NaCl +demin.H_2O	$1*10^5$	55,8	S1b	uranine+demin.H_2O	0,952	110,4
2	0,160/4 mm	S2	uranine+demin. H_2O	0,952	41,1	-	-	-	-
3	0,160/4 mm	S3	uranine+demin. H_2O	0,095	28,2	-	-	-	-
4	4/8 mm	S4****	uranine+natural water	0,952	4,0	-	-	-	-
5	4/8 mm	S5a	NaCl +demin. H_2O	$0,98*10^5$	26,0	S5b	uranine+demin.H_2O	0,095	27,2
6	4/8 mm	S6	uranine+demin. H_2O	0,094	27,6	-	-	-	-
7	4/8 mm	S7	uranine+demin. H_2O	0,941	26,9	-	-	-	-
8	0,160/8 mm	S8a	NaCl +demin. H_2O	$0,98*10^5$	34,3	S8b	uranine+demin.H_2O	0,093	24,1
9	0,160/8 mm	S9	uranine+demin. H_2O	0,095	23,7	-	-	-	-
10	0,160/8 mm	S10a	uranine+demin. H_2O	0,938	23,8	S10b	demineralised H_2O	pure	22,7

*** small system

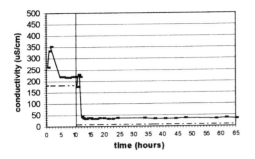

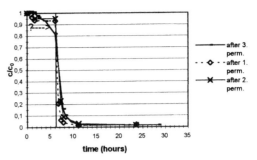

Fig 3 The conductivity of NaCl solution samples (left of the vertical line) after the 3rd permeameter (full line) and initial concentration ($0,98*10^5$ mg/m^3) of NaCl solution in demineralised water (hatched line) - system 5a. Right from the vertical line: the conductivity of the uranine solution (full line) and initial uranine solution in demineralised water (hatched line) - system 5b.

Fig. 6 On the left side of vertical line: uranine breakthrough curve for system 10 (0,160/8mm, uranine concentration: 0,938 mg/m^3). Right side: initial uranine concentration was replaced with pure demineralised water.

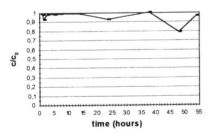

Fig. 4 Breakthrough curve for uranine for system 6 (4/8mm, uranine concentration: 0,094 mg/m^3).

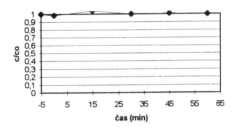

Fig. 7 Breakthrough curve of the uranine solution in natural karst water in batch test (limestone sample 0,160-8mm, uranine concentration 0,96 mg/m^3).

4.3 Experimental remarks

pH

A pH value of NaCl and uranine solution in demineralised water stabilised soon at a value of 9,5 (figure 8), regardless of concentration, sort of the tracer and grain size of the limestone sample.

Demineralised water was used for the experiment, for the purpose of better approximation of the natural conditions, where rain water, with conductivity of about 10 µS/cm and pH 4-5, infiltrated into the ground. Demineralised water from the IRGO institute demineralisator had also a low pH value: from 5,5 to 6,5. The water was not desaerated due to the large amount of water needed for these experiments (600 - 1000 litres for each experiment). Water with pH value approximately 6 ought to have in equilibrium H$_2$CO$_3$ radicals (figure 9) and dissolved limestone.

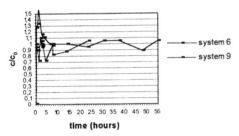

Fig. 5 Uranine breakthrough curve for system 7 (4/8mm, uranine concentration 0,941 mg/m^3). A fall of concentration (time about 24 and 48 hours) due to diminishing velocity(higher retardation time).

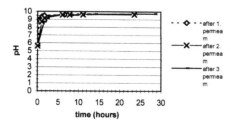

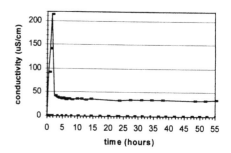

Fig. 8 pH value of the uranine solution (in demineralised H_2O) samples after 1st, 2nd and 3rd permeameter (system 10; 0,938 mg/m^3 uranine solution).

Fig. 10 The conductivity of the uranine solution samples after third permeameter (full line) and initial uranine solution (0,094 mg/m^3) in demineralised water (hatched line) - system 6.

Demineralised water mineralised in permeameters with limestone samples and pH value rose to 9,5. According to our estimation, these effects did not influence results of above experiments.

In samples with more mineralised natural karst water, an approximate pH value 8 was attained (system 4).

and stabilised in all experiments with uranine at a certain value of 35 - 40 µS/cm.

Net alkalinity

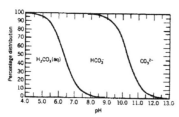

Fig. 9 Percentages of total dissolved carbon dioxide species in solution as a function of pH (25°C, 1 atm), (from *Montgomery, 1985*)

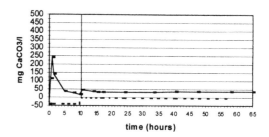

Fig. 11 To the left of vertical line: net alkalinity of NaCl solution samples (full line) and initial NaCl solution (hatched line) - system 5a. To the right side: net alkalinity of uranine solution samples (full line) and initial uranine solution (hatched line) system 5b.

O2

Oxygen content was within the limits from 3 to 6 mg/l in the samples of salt or uranine solution and throughout experiments. The samples were exposed to the oxygen immediately after sampling, since they were taken from the pipe (after 1st and 2nd permeameter) or tube (after 3rd permeameter) into a Erlenmeyer, which was not closed. Measurements were performed immediately after sampling and values were significant for the oxygen content.

Conductivity

At the experiments with uranine solution in demineralised water conductivity raised at the moistening (figure 10: first three points) of material in permeameters (dissolving of the limestone sample)

The value of net alkalinity raised (figure 11) like conductivity at moistening (three points between 100 and 250 mg CaCO$_3$) from the negative value of initial sample (NaCl or uranine solution in demineralised water) to stabilised value at 10 - 40 mg CaCO$_3$/l (limestone dissolving equilibrium).

Batch tests

In the batch tests the pH value (9,5) of uranine solution in demineralised water was confirmed. The conductivity value stabilised similar to the experiments in permeameters at value of 35-40 µS/cm. Increase of Ca^{2+} (figure 12) concentration (from 0,2 to 7 mg/l) was attained in the moment of contact between carbonate and uranine solution in demineralised water (limestone dissolving reaction). The Ca^{2+} and HCO_3^- concentration stabilised until the end of experiment, so a conclusion was drawn that the equilibrium of the carbonate dissolving reaction was reached: $CaCO_3+H_2O+CO_2 \rightarrow Ca^{2+}+2HCO_3^-$.

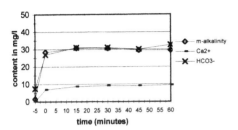

Fig. 12 Ca^{2+} and HCO_3^- concentration and alkalinity of the uranine solution samples in the contact with limestone (0,160-8mm).

In batch tests the natural karst water from the spring Malni was also used. In this case the pH value of uranine solution samples was about 8. In this case the Ca^{2+} and HCO_3^- contents slightly decreased after contact with the limestone and then remained constant until the end of the experiment.

5. CONCLUSIONS

A final conclusion could not be drawn for experiments with low uranine concentration (about 0,1 mg/m3) since results were inconsistent (the sum of errors: analytical, sampling, cleaning the system and sampling material). In system 5a a perceptible uranine content was obtained in samples not containing uranine.

Experiments with high uranine concentration (about 1 mg/m^3) gave better results. The sorption of uranine was negligible. Regarding precision of experiments the sorption depended on active grain surface and contact time of uranine solution and limestone sample from Razdrto quarry, which confirm diminishing of uranine concentration in the times of low velocity (larger contact time solution-sample).

The rate of specific sorption in the experiment system 10 (sample 3) was possible to estimate from flow, duration of the experiment and diminishing of uranine concentration. Considering 5,5% diminishing of uranine concentration 0,938 mg/m^3, flow 23,8 l/h, duration time 4,5 hours and the specific grain surface

387 m^2 the rate of specific sorption was $3,2*10^{-6}$ mg/m^{2}h. This applied to the limestone from Razdrto quarry and there was a simplification to the sphere grains. Considering the Heywood factor 1 - 2 the sorption extent was in range from $1,6*10^{-6}$ to $3,2*10^{-6}$ mg/m^{2}h. Exact calculus demands additional analysis of sphericity.

This sorption result was only an evaluation, because it was calculated only from one experiment system (system 10). In other experiments the results were either inconclusive or there was no diminishing of uranine concentration through constant flow (system 7). Only a repetition of experiments could confirm this conclusion.

When moistening limestone with uranine solution in demineralised water the sample bound water caused initial reduced retardation of uranine concentration drop because of negligible uranine sorption. In experiments with low concentration this effect could not be reliably established.

By decreasing the flow through limestone sample in permeameters the residence time increased and diffusion and sorption processes became perceptible (diminishing of uranine concentration in system 7, figure 9), also in the case where the specific grain surface (sample 2) was smaller than in experiment system 10. Yet no rate of specific sorption determination was attempted due to the possible masking effect of diffusion (within the retention porosity volume).

6. ACKNOWLEDGEMENT

Financial support by the Ministry of Science and Technology of the Republic of Slovenia, under the research grant No. J1-7172-0210-96 is gratefully acknowledged.

7. REFERENCES

Bear, J. 1972: *Dynamics of fluids in porous media.-* Dover Publications, 764p., New York.

Di Fazio, A. 1996: *Una valutazione sperimentale dell'interazione tra il mezzo poroso ed i traccianti usati per misure di parametri idrogeologici.-* Mem.Soc.Geol.It. 56 (1996): in Techware, Tecnology for Water Resources, Cosenza, Italy.

Käss, W. 1992: *Geohydrologische Markierungstechnik.-* Lehrbuch der Hydrogeologie, Band 9, 519p., Gebrüder Bornträger, Berlin-Stuttgart, Germany.

Marsily, G. de 1986: *Quantitative Hydrogeology.* Academic Press, 440p., San Diego, California.

Montgomery, J.M. 1985: *Water Treatment Principles and Design.*- John Wiley & Sons, 675p., New York.

Veselič, M. 1984: *Hidrogeologija.* Lectures, Univ. of Ljubljana, Slovenia.

Tracer Hydrology 97, Kranjc (ed.) © 1997 Balkema, Rotterdam, ISBN 90 5410 875 4

In situ experimental evaluation of aquifer-tracers interaction: First results

A. Di Fazio & M. Benedini
CNR, Water Research Institute, Bari, Italy

ABSTRACT: After intensive experiments at the laboratory scale, in situ tests have been carried out in a carbonatic aquifer to estimate the linear effective velocity in uniform flow. The two-well test technique has been adopted for this purpose. Two typical tracers, NaCl (conservative) and Rhodamine WT (non conservative), were injected simultaneously for comparative evaluations. The results show different breakthrough curves of concentration at sampling well for the two tracers. The travel time for NaCl is shorter than for Rhodamine WT. This experiment confirms the necessity to know the retardation of a specific tracer used, in conjunction with the aquifer characteristics, in order to interpret correctly the results of tracer tests.

1 INTRODUCTION

The tracers are characterised according to their capability of interacting with the filtering medium; this interaction shows a "retardation" of the advancing front of the tracer, in respect to the flow in which it is injected. It is necessary to quantify such a retardation in order to correct the data obtained by means of the raw breakthrough curves of the tracer.

Solute transport in groundwater has been exhaustively studied in its physical, chemical and biological aspects and in its relevant mathematical interpretation. The results of such studies have allowed experts to develop reliable methods to quantify the interaction ("sorption") between the solid and liquid phases of the aquifer that causes the retardation of the solute front in the water flow in which it is, sometimes accidentally, injected.

Tracers used in hydrogeology research belong to a solute class that can suffer from interaction phenomena with the solid matrix: this means that the hydrogeological and hydrodispersive parameters are often misinterpreted.

Starting from the general transport equation, the definition of the "retardation factor" (R) for solutes can be obtained (Freeze & Cherry, 1979; Bauer, 1991):

$$R = \frac{V_l}{V_c} = 1 + \frac{\rho}{\eta_e}\frac{\partial S}{\partial C} \qquad (1)$$

where

V_l : average linear velocity of groundwater [L/T];

V_c : velocity of the C/Co=0,5 point on the concentration profile of the retarded solute [L/T];

η_e : effective porosity [L^3/L^3];

ρ : bulk mass density of porous medium [M/L^3];

S : solute concentration in the solid phase [M/M];

C : solute concentration in the liquid phase [M/L^3];

$\partial S/\partial C$: partitioning of the solute between the solution and the solid phases.

Eq. (1) shows that R can be obtained from data given by batch procedures ($\partial S/\partial C$) or as a velocity ratio (V_l /V_c) given in-flow conditions (Di Fazio & Vurro, 1994).

The batch procedure consists of a simple laboratory development of the adsorption isotherms through standardised procedures (Roy et Al. 1986), allowing the definition of $\partial S/\partial C$ ratio to be defined; it represents the tracer's partition and/or distribution between the solution and the solid matrix. It is expressed by an equilibrium isotherm of the interactive phenomenon at the liquid/solid interface in a graphic representation. The adsorbed solute mass per unit of dry solid matrix mass (S) is expressed as a function of tracer concentration in solution (C) when the solute adsorption reactions are again in equilibrium.

If the interaction at the liquid/solid interface is mainly due to the substrate adsorption on the solid phase, the phenomena, measured through suitable laboratory experiments, can be expressed by the Freudlich equation:

$$S = K_d C^{1/n} \tag{2}$$

where K_d (L^3/M) and "$1/n$" depend on the type of solute and the nature of the porous medium. By knowing these two terms, the value S can be obtained, which thermodynamically represents the saturation capacity of solute in the considered soil.

The flow condition determination are carry out both in laboratory by column procedure and in situ by tracer tests. The column procedure ("permeameter") is based on the solute seepage, in a given water flow condition, through a bed made of the solid matrix examined; the solute concentration is then recorded at the column output; the ratio between water flow velocity and solute output velocity will give the retardation value. If the tracer solution is the mixture of two tracers, a conservative one (NaCl) and an interacting one (Rhodamine WT) (Smart & Laidlaw, 1977), the comparison between the two breakthrough curves gives useful elements to quantify the retardation in that specific experimental condition. In situ test is a classic tracer test for direct measure of hydrogeological parameters.

The laboratory tests carried out both through the determination of equilibrium isotherms and by means of column tests have confirmed the interaction between Rhodamine WT (Rh.WT) and the carbonatic porous media (sand from limestone with 98,16% of $CaCO_3$). At the same time a retard R = 6,5 - 8,5 (Di Fazio & Benedini, 1995; Di Fazio, 1996) has been determined, with a range probably due to different contact time. This test showed also that the relation between Rh.WT sorption and the carbonatic matrix is not linear.

This paper reports the results of a first tracer test carried out for the determination of groundwater velocity by means of the two-well technique at the "CNR-IRSA well-field", in a carbonatic and carsified aquifer. Also in this case a comparison will be made between the breakthrough curves of NaCl and Rh.WT simultaneously injected. The in situ test seem to confirm the R value obtained in laboratory.

2 STRUCTURAL AND HYDROGEOLOGICAL CHARACTERISTICS OF THE "CNR-IRSA WELL FIELD"

The "CNR-IRSA well field" is located in the vicinity of Bari, South Italy, Apulia region, at about 3 km from the Adriatic sea coast. It consists of five wells drilled into the carbonatic subsoil at an elevation of 15 m a.s.l.

The field has a central well of 300 mm diameter and four lateral wells of 142 mm diameter, located in 2 orthogonal directions and at a distance of 10 m from the central one. The wells are named following the initial of the cardinal point to which they are oriented (fig. 1).

The morphology of the area is characterised by subsequent planes that debase going toward the coast with scarps, not more than 10 m high. These are linked to the terraced platform structure derived from tectonic faults.

The geology of Murgia region is characterised by a thick rocky body, originated by the process of stone formation from sea sediments accumulated during the Mesozoic period, in an environment essentially *epioceanic*. After the final emerging of the carbonate *Apulo-Garganic* platform, the natural tectonic evolution of sediments caused the actual morphology and rock's composition. The layers sequence observed when the wells were drilling, from the bottom to upward, was the following: Cretaceous dolomite (with a thickness of 21 m), Cretaceous limestone (26 m) and Pleistocene calcarenite (5 m). (see fig. 2)

The groundwater flows under a pressure gradient ranging from 0.1% to 0.5%. The wells specific discharges measured in this area were lower than 100 l/s*m. The water salinity in the wells of this coastal aquifer was about 2-5 g/l.

All the wells are lined with a PVC pipe, which is carefully holed for the length corresponding to the whole thickness of the saturated aquifer (from -15 m to - 51m), with a total opening greater than the 20% of the lateral surface.

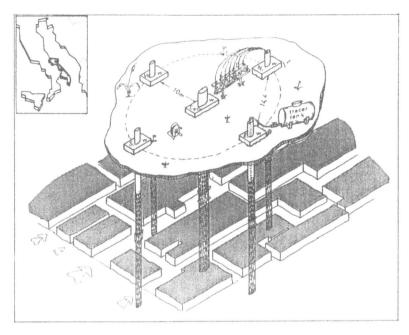

Figure 1: Schematic arrangement device in the fractured limestone of the well field CNR-IRSA \located in Bari, Apulia region (South Italy).

The wells are also equipped with a stir-free sampler (Di Fazio, 1984 ;   ) with captors at different depths under the water table and at a distance of 5 m each other (fig. 3). This sampler provides a simultaneous water sampling at different levels and in almost continuous manner.

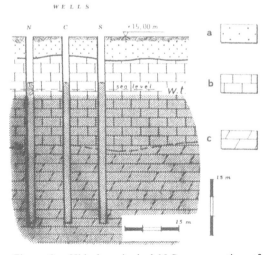

Figure 2 - Hidrologeological N-S cross section of well field (a) - Calcarenite; (b) - Limestone; (c) - Dolomite.

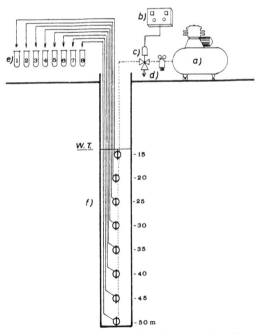

Figure 3 - Arrangement and working principle of the sampler. a)- air compressor; b)- timer; c)- three-wai solenoid valve; d)- discharge; e)- samples; f)- samplers.

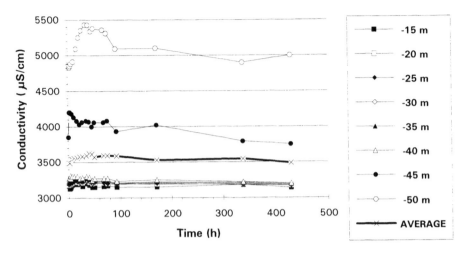

Figure 4: Breakthrogh curves of NaCl at different depth in the North well.

3 TRACER TEST

The first aim of the tracer test has been the direct measurement of the average linear velocity under non-uniform laminar flow conditions, with a piezometric gradient close to 0.5%; the second one has been an attempt to simulate the tracer's migration in the fractured aquifer.

Only two wells have been used for this purpose, approximately aligned with overall direction of ground-water flow: the upstream E-well that was used for injection and the N-well for observation.

In analogy with the laboratory tests and for comparative evaluations, the tracer was a mixture of 600 l (corresponding to the well's volume) of 40 g/l NaCl and 0.02 g/l Rhodamine WT; the analysis of the prepared mixture gave 60,000 μS/cm of electric conductivity and 825,000 Fluorimetric Units (U.F.).

The tracers mixture was poured for 8 minutes into the E-well using a semi-rigid pipe siphon with its outlet at 0.5 m under water table, following an experienced procedure of tracer injection. In fact, such a procedure allows the high density to determine the tracer mixing during its drop into the water well; the negative effects of this slug have the end in about 50 minute.

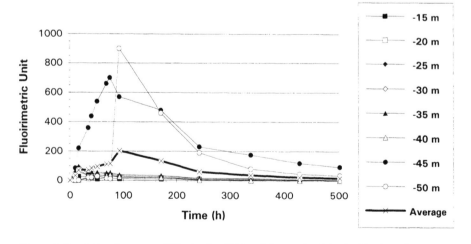

Figure 5: Breakthrogh curves of Rhodamine WT at different depth in the North well.

After the tracer injection, all the wells have been monitored almost continuously in time. Piezometric heads, salinity and tracers concentration were recorded at several depths under the water table; the average breakthrough curves (BTC) were obtained also by means of an average value of the vertical profiles of tracer concentration.

This paper deals only with the tracer's travel from E-well to N-well for retard considerations and does not consider other possibilities of result interpretation (bore dilution, record of tracer spreading in the other wells, etc.).

Figg. 4 and 5 show the breakthrough curves respectively of NaCl and Rh.WT recorded at different depths of N-well; it is also possible to notice the different groundwater salinity at the different depths in the same well and the layers having higher hydraulic conductivity.

4 CONCLUSIONS

The availability of experimental data and the efficiency of both the field and laboratory facilities have confirmed the validity of the procedure adopted for the determination of fundamental aspects of groundwater hydrology as the retardation and the solute-matrix interaction. For the specific case investigated, a high discrepancy for NaCl and Rh.WT traverl time has been noticed, which does not permit a complete and quantitative definition of the retard on all the depth of water well. Besides to a possible incertitude of the tracer injection techniques, such a drawback can be due to the natural NaCl groundwater salinity as a consequence of sea water intrusion.

The necessity of new tracer tests is therefore evident, with other injection techniques and other conservative tracers.

ACKNOWLEDGMENTS to Mr. Giovanni Evaristo for field operation.

REFERENCES

Bauer, H. 1991. Simple derivation of the Retardation equation and application to preferential flow and macrodispersion. *Ground Water*. 29(1):41-46.

Freeze R.A. & Cherry, J.A. 1979. *Groundwater*. Prantice Hall Inc., Englewood Cliffs, N.J.

Di Fazio, A.1984. Stir-free sampling, World Water, Aug.: 29.

Di Fazio, A. & Vurro, M. 1994. Experimental tests using Rhodamine WT as tracer. *Advances in Water Resources*. 17: 375-378.

Di Fazio, A.1994. Una valutazione dell'interazione tra la matrice solida ed i traccianti usati per misure di parametri idrogeologici". *Procedings of the 77° Congr. Naz. S.G.I. Bari 24-28/9. Mem. Soc. Geol. It. 51*: 321-333.

Di Fazio, A. & Benedini, M., 1995. Sulla determinazione sperimentale del fattore di ritardo in traccianti utilizzati dagli idrogeologi. *Idrotecnica*. 6:323-328

Roy, W.R.; I.G. Krapac, S.F.I.Chov And R.A. Griffen. 1986.Batch type adsorption procedures for estimating soil attenuation of chemicals - *Tecnical report prepared for U.S. EPA, Office of solid wast and emergency respons*. Washington, D.C. 20460.

Smart, P.L. & Laidlaw, I.M.S. 1977. An evaluation of some fluorescent dyes for water tracing. *Water Resour. Res.* 13: 15-33.

Tracer Hydrology 97, Kranjc (ed.) © 1997 Balkema, Rotterdam, ISBN 90 5410 875 4

The use of bacteriophages for multi-tracing in a lowland karst aquifer in western Ireland

David Drew
Department of Geography, Trinity College Dublin, Ireland

Nathalie Doerfliger
Center of Hydrogeology, Institute of Geology, University of Neuchâtel, Switzerland

Kitty Formentin
Laboratory of Microbiology, University of Neuchâtel, Switzerland

Abstract
Bacteriophage tracers were used successfully in a lowland karst aquifer in western Ireland. The particular hydrological conditions rendered the use of other tracers only partially successful. Data concerning the nature of the karst groundwater systems were also obtained.

1. The Area

The most southerly part of County Galway in western Ireland consists of a north south range of low hills of Devonian sandstones (Slieve Aughty) from which a series of rivers drain westwards on to the adjacent Carboniferous limestone strata. The limestone forms a lowland, less than 30m in elevation and extending northwards and westwards to the Atlantic Ocean in Galway Bay (Figure 1).

The rivers generated on the Devonian rocks sink underground at or close to the geological contact, though the presence of locally thick deposits of till distorts the behaviour of the allogenic streams in some instances. The rivers appear to flow in large (<25m diameter conduits) at depths of 15-25m below present sea level. In the east of the area in particular the rivers several times reappear and flow on the surface before sinking once more. Other autogenic rivers are present on the limestone in the south of the area and these also sink underground (Drew D.P. 1984; Coxon C. and Drew D.P. 1986.

The outlets for the karst waters are two groups of large springs in the inter-tidal zone of Galway Bay (Figure 1). Mean water flux through the karst aquifer is estimated at c. 4-5m^{3}s^{-1} with high winter total flows exceeding 15m^{3}s^{-1}. There are also several lakes, both perennial;l and seasonal (Coxon 1987) on the limestone lowland.

It is assumed, though evidence is lacking, that the system of major karst conduits represents fragments of an ancient system of karst drainage graded to a lower base level than that of the present day. It also seems likely that glacial erosion and deposition has destroyed and infilled some of the karst conduits and that the present hydrogeological system is in part using the original drainage system and in part developing new, predominantly epikarstic drainage routes, Drew D.P. and Daly D. (1993).

2. Choice of Tracers

The area described above has, in recent winters been subject to severe and prolonged winter flooding lasting up to three months. The fundamental cause of the flooding is that there has been a succession exceptionally prolonged rainy periods in recent winters but other factors such as changes in afforested areas and changes in agricultural practices in the area may also have contributed to the flooding.

A major study to investigate the causes of this flooding and to propose alleviative measures was commissioned by the Irish Government (Anon. 1995; Daly 1992). An essential part of this study is the development of a conceptual and ultimately a numerical model of the functioning of groundwater and surface water within the area. Results obtained from sophisticated water tracing has been an integral part of the data input to the process.

Basic groundwater flow connections in the area have been established used fluorescent tracers. However, the use of such tracers has limitations in the context of the area and of the data output required. Flow routes and rates in the area are very stage sensitive, with flow directions reversing in some instances in response to changes in head gradient. Thus multi-tracing was essential, using comparable tracers at a large number of inputs simultaneously. Also, some of the sinking waters reappear for a short distance on one or more occasion at the surface and in some instances the water enters lakes (the largest having a mean area of 0.5km^2 and containing a mean volume of water exceeding one million cubic metres (Figure 1). Fluorescent tracers could not be used in quantities that would colour the water in these lakes and indeed, any dye present would probably be decolorised by exposure to sunlight and by adsorption on the organic suspended sediments (derived from upland peat).

3. Bacteriophage Tracing

It was therefore decided to carry out a multi-trace using 5 strains of Bacteriophage. The use of bacteriophages as

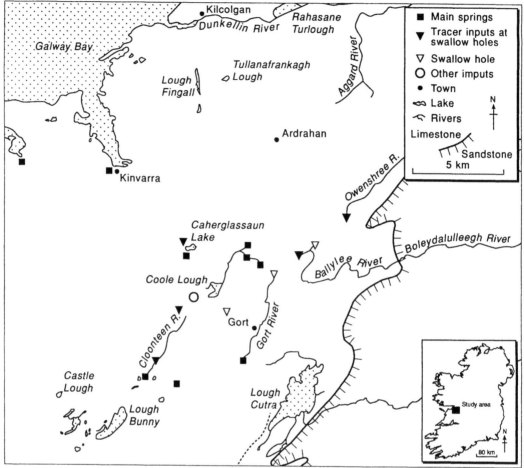

Figure 1 The location of the study area and the major hydrological features

artificial tracers for surface and subsurface waters dates back to 1972 (Winpenny et al.). In the 1980's the Microbiology Laboratory associated with the Centre of Hydrogeology of the University of Neuchatel had accomplished several encouraging experiments (Aragno and Müller 1982). Detailed research on the use of bacteriophages as biological tracers has been undertaken by Rossi (1994).

These biological tracers have the following characteristics:
[i] They are not harmful either to man, animals or plants, not even affecting any bacteria
other than that of their host . Standard ecological toxicity tests have been conducted on
Daphnia and trout fry and all results were completely negative (Rossi, 1994).
[ii] They do not pollute an aquifer as a maximum of some tens of grams of biologic material is introduced and it is biodegradable. Particular care is taken to chose the types

of bacteriophages/host bacteria systems that cannot already be found naturally in the
aquifer.
[iii] It is also possible to differentiate and count a mixture of phages in a single sample.

The structure of the bacteriophages induces an interaction with their environment. In the common range of pH of karstic groundwater, they are strongly negatively charged. They can then be adsorbed on colloids in the water. The adsorption on these particles does not inactivate all bacteriophages (Moore et al. 1975). On the contrary, the adsorption can confer some advantages as they can be protected from a rapid inactivation, due to the (turbulent flows, torrential or laminar flows).

Grant et al. (1993) define three possible types of adsorption behaviour, depending on the phage and on the kind of particle with which it interacts. This behaviour ranges from a feeble interaction, with no influence on the survival of the phage, to a very strong adsorption, result-

Table 1 Summary of Characteristics of phages used in Galway tracer test

Phage	Host Bacterium	Amount Used (litres)	Phage Numbers (pfu)**	Input
T7	*Escherichia coli B (ATCC* 11303)*	19.5	*5.02 E+13*	Lake sink
Psf2	*Pseudomonas fluorescens (ATCC 27663)*	15	*3.92 E+13*	River sink
H6/1	*H6 Bacterial strains not yet registered*	15	*3.92 E + 13*	Borehole
H40/1	*H40 Bacterial strains not yet registered*	10	*1.47 E+14*	River sink
H4/4	*H4 Bacterial strains not yet registered*	10	*2.0 E + 14*	River sink

(*ATCC = American Type Collection
**pfu = plaque forming units is equivalent to the virus or bacteriophage per ml.)

ing in a considerable loss of virulence. For example phage T7 is adsorbed strongly by clay minerals on their positive charged sites; it reacts strongly with the mineral colloids. A great number of phages are inactivated quickly after an initial reaction stage, but nevertheless this reaction in its turn produces a strong protection for the phages Phages Psf2, H40/1, H4/4 and H6/1 do not react practically to colloids minerals in laboratory experiments and therefore they are the best potential tracers.

The bacteriophages used in the multi-tracing test were as shown in Table 1. The phage T7 belongs to the collection of the Laboratory of Microbiology of the University of Neuchatel. It has the host-bacteria Escherichia *coli* and has a small size. It is member of the Podoviridae family characterised by a polyhedral head and a small non contractile tail (Ackermann and DuBow 1987).

The bacteriophage Psf2 belongs to the Siphoroviridae family of the type B1. Its head is polyhedral (icosahedron) and its tail flexible and non contractile. Its was isolated from the waste water of a water treatment plant.

The bacteriophages H40/1, H6/1 and H4/4 are marine bacteriophages; they were isolated during marine explorations in the Atlantic by K. Moebus (Moeubus K.,

Hermann F. 1987; Moebus K., Nattkemper H. 1989; Moebus K. 1989; Moebus K. 1992). H40/1 belongs to the Sophorviridae family. It has an icosahedral head and a flexible noncontractile tail. H6/1 is an isosahedron without a tail and is a member of the Leviviridae family.

These phages were chosen for this multi-trace, firstly they are not present in the studied karst environment, except in very low concentrations and only sporadically for T7. The analysis of some blank water samples gave negative results. Then, as described above, these phages do not react with colloids in the water and finally they had been used many times in various karst aquifers in the Swiss Alps and in the Swiss Jura and provided good results, qualitatively as well as quantitatively. (Jeannin, Wildberger and Rossi 1995; Doerfliger 1996)

4. Results and Interpretation

All five phage types were recovered in varying amounts. Recovery rates were clearly largely determine by the dilutions, the rates of flow and the input and sampling site characteristics and thus the relative efficacy of the individual phages as tracers in this area cannot be assessed. The actual results of the tracing is mainly of local interest only and is not the concern of this paper.

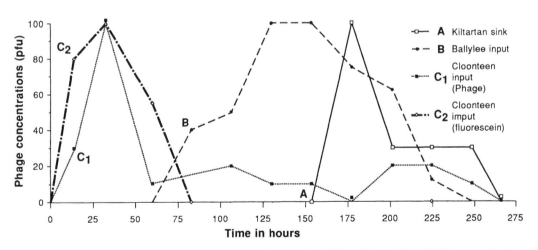

Figure 2 Breakthrough curves for phage at the main springs (A) flow via lakes (B) Some flow via lakes (C1) Flow bypassing lakes, and comparison with a fluorescein trace (C2) from the same input as C1.

Two findings from the tests are given below.

Figure 2 shows graphically a comparison of the breakthrough curves for tracers at one of the main coastal springs. All inputs were stream sinks. Tracer from (A) was all routed through the main lake, tracer from (B) showed divergent flow with some water going through the lake and some bypassing the lake to flow directly to the springs. It can be seen that (unlike fluorescent tracer used earlier) the phage crossed the lake and re-entered the karst groundwater system lagged by some 50 hours behind water flowing directly to the springs. Line (C) is for a phage input at the same sink as fluorescein (line D) and shows the greater level of resolution obtained in the phage breakthrough curve as compared to the dye breakthrough curve. On all graphs the tracer concentrations are expressed as percentages of peak concentration.

An important aim of the tracing was to determine the character of the karst flow system, in particular whether it was unimodal, bimodal or trimodal in terms of permeability type Conduit, epikarstic and deeper distributed flows were the categories thought likely to exist. Sampling sites included coastal springs, shallow (epikarstic?) springs and wells, lakes, deep boreholes. Figure 3 shows aggregated breakthrough curves for deep borehole sampling sites (3 sites), shallow/epikarstic sites (2 sites),major conduit sites (4 sites) and coastal springs (2 sites). The values are in absolute phage/ml concentrations.

The graphs are interpreted as showing three discrete flow systems in the aquifer.

[i] A major conduit flow system carrying a large proportion of the flow (high tracer concentration) and having a rapid follow-through and a rapid falloff to low concentrations of tracer.

[ii] A shallow flow system (probably in the 2-5m deep epikarstic layer, showing rapid initial follow-through but also a prolonged tail indicating considerable storage.

[iii] A deeper flow system, independent of the major conduits with peak tracer concentrations lagged 50-60 hours behind the other flow systems. This may be a less well developed conduit or a partly distributed flow system.

The breakthrough curve at the springs most closely corresponds to that within the major conduits.

5. Conclusions

The use of phages compared very favourably with results obtained from tests using fluorescent tracers allowing a much greater level of sensitivity and being usable where dyes could not work. A major aim of the tracing was to characterise the karst flow systems and in this respect it was also useful. Further developments in phage tracing are detailed by Rossi (this volume).

Acknowledgements

The authors wish to thank Professor Emer Colleran and Ms Clodagh Jordan of the Department of Microbiology, University College Galway for making available the facilities of their laboratory and for much practical help and advice. Also Dr Matthew Stout (Trinity College Dublin) for cartography and layout of the paper.

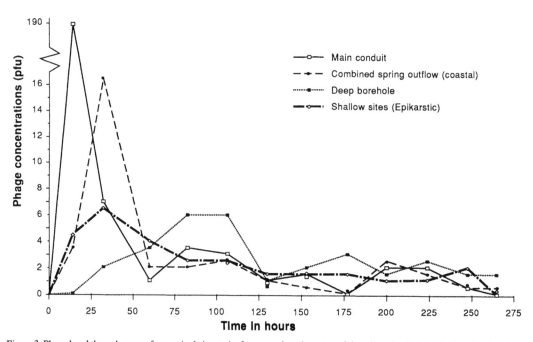

Figure 3 Phage breakthrough curves from a single input site from samples taken at conduit, epikarstic, deep borehole and spring sites.

REFERENCES

Anon. 1995 *An investigation of the flooding problems in the Gort-Ardrahan area of south Galway,* Preliminary report to Office of Public Works Dublin by Southern Water McDowells Ltd, Crawley, England, 62 pp

Aragno, M., Müller, I. 1982. *Premières expériences de traçages des eaux souterraines dans le karst neuchatelois à l'aide de bactériophages.* Bull. Centre Hydrogéol., Neuchatel,

Coxon C. 1987a The spatial distribution of turloughs. *Irish Geography* 20: 11-23

Coxon C. and Drew D.P. 1986 Groundwater flow in the lowland limestone aquifer of eastern Co. Galway and eastern Co. Mayo, western Ireland. In Paterson K & Sweeting M. (Eds), *New Directions in Karst*, 259-80, Norwich.

Daly D. 1992 *A report on the flooding in the Gort - Ardrahan area* Geological Survey of Ireland, 209pp

Doerfliger N., 1996: *Advances in karst groundwater protection strategy using tracer tests analysis and multiattribute vulnerability mapping (EPIK method).* PhD. thesis, University of Neuchâtel, 225 pp.

Drew D.P. 1984 The effect of human activity on a lowland karst aquifer. (In) Burger A., (Ed) *Hydrogeology of Karstic Terrains*: Case Histories. Int. Assoc. Hydrogeologists, Hanover, pp 195-99.

Drew D.P. and Daly D. 1993 *Groundwater and karstification in mid-Galway, south Mayo and North Clare.* Geological Survey of Ireland Report Series RS 93/3. Dublin, 86 pp.

Grant, S.B., List, E.J., Lindstrom, M.E. 1993. Kinetic analysis of virus adsorption and inactivation in batch experiments. *Water Res. Research* 29: 2067-2085.

Jeannin P-Y., Wildberger A, Rossi P., 1995: Multitracing-Versuche 1992 und 1993 im Karstgebiet der Silberen (Muotatal und Klöntal, Zentralschweiz). *Beiträge zur Hydrogeologie*, 46; 43-88.

Moebus, K. 1992. Laboratory investigations on the survival of marine bacteriophages in raw and treated seawater. *Helgolander. Meersunters.* 46: 251-273

Moebus, K., Nattkemper, H. 1989. Bacteriophages sensitivity patterns among bacteria isolated from marine waters. Helgolander. Meersunters. 34: 375-385

Moore, B.E., Sagik, B.P., Malina, J.F. 1975. Viral association with suspended solids. *Water Research* 9: 197-203

Moore, R.S., Taylor, D.H., Sturman, L.S., Reddy, M.M., Fuhs, G.W. 1981. Poliovirus adsorption by 34 minerals and soils. *Appl. Environ. Microbiol.* 42: 963-975

Moore, R.S., Taylor, D.H., Reddy, M.M.M., Sturman, L.S. 1982. Adsorption of reovirus by minerals and soils. *Appl. Environ. Microbiol.* 44: 852-859

Rossi, P. 1994. *Advances in biological tracer techniques for hydrology and hydrogeology using bacteriophages: optimization of the methods and investigation of the behaviour of the bacterial viruses in surface waters and in porous and fractured aquifers.* Ph. D. thesis, University of Neuchâtel, 200 pp.

Wimpenny, J.W.T., Cotton, N., Statham, M. 1972. Microbes as tracers of water *Water Research* 6: 731-739.

Tracer Hydrology 97, Kranjc (ed.) © 1997 Balkema, Rotterdam, ISBN 90 5410 875 4

Determination of bacteriophage migration and survival potential in karstic groundwaters using batch agitated experiments and mineral colloidal particles

K. Formentin, P. Rossi & M. Aragno
Laboratoire de Microbiologie, University of Neuchâtel, Switzerland

I. Müller
CHYN, University of Neuchâtel, Switzerland

ABSTRACT: This paper presents the results of a new method for the determination of rates of inactivation and of adsorption to colloidal clay particles (CCP) by bacteriophages. These rates are used to determine the suitability of bacteriophages as biological tracers. Inactivation rates were determined by agitating a vessel containing the phages in a standard liquid medium. Adsorption rates were obtained by adding a small concentration (0.030mg/ml) of Attapulgite CCP to the medium. In the presence of this small amount of CCP, the behaviors of six phages (T7, f1, MS2, Psf2, H6/1 and H40/1) change radically from the one in water only. Each bacteriophage reacts differently to the presence of CCP. For three phages (T7, H40/1 and H6/1) the presence of CCP offers an effective protection from the physical inactivation due to agitation of the liquid medium. However, reactions to presence of CCP range from a rapid inactivation by a massive adsorption (f1) to a dramatic enhancement of the virulence (MS2). These data are in good correspondence with the breakthrough results of tracing experiments performed in two karstic aquifers in the Swiss Jura. They show that it is possible to predict the breakthrough of a phage according to its absorption rates. But these experiments also demonstrate that it is difficult to generalize the behavior of a particular phage. Consequently, it may not be valid to use randomly selected bacteriophages to model the migration of vertebrate pathogenic viruses.

1 INTRODUCTION

The use of bacteriophages as biological water tracers answers the growing needs for environmentally harmless investigation tools. These instruments are necessary, not only to acquire a better understanding of natural systems, but also to estimate the impact of human activities on these systems. The choice of potentially interesting bacteriophages is made usually through tracing experiments. These time-consuming trials give qualitative results (i.e. good or poor breakthrough) that allow a subjective selection of the best phages. Furthermore, the results are poorly correlated to environmental factors such as chemical, mineral or geological parameters. One way of addressing the problem of the selection of the phages is to experimentally determine the interactions that occur between phages and material originating from soils and aquifers. This can be done by migration experiments in columns (Bales et al., 1993; Lance and Gerba, 1984) or using small experimental sites (Bales et al., 1989; Rossi et al., 1994). Laboratory inactivation-adsorption experiments have the advantage of allowing a small scale work in controlled and standardized conditions. The protocol generally used has been described by Moore et al. (1981). But these assays, which generally last several hours or even days, are too slow to follow the rapid inactivation and adsorption reactions of phages. Moreover, the Freudlich or Langmuir equilibrium adsorption models used for the computing of the results give very little information about the changes within the phages population, since they cannot tell us anything about the variations of their various sub-fractions. All one obtains is an overview of the reaction between the free and virulent (FV) fraction of phages and the substrate. This type of experiment cannot be used to follow the evolution of the inactivated fraction (I) or of the adsorbed and virulent (AV) one. As Payment et al. (1988) have shown already, a large proportion of viruses (77%) and coliphages (about 65%) are associated in natural environments with particles suspended in water, though they retain their virulence. Therefore, if one only takes into account the FV sub-fraction of viruses, the kind of analysis described above will result in an incomplete picture of the behavior of viruses in water.

The goal of this work is to present a new batch inactivation-adsorption method designed for the selection of the bacteriophages used in water tracing experiments. The inactivation experiments are performed in standard buffered medium, under constant agitation. The adsorption experiments are performed under the same conditions, in the presence of Attapulgite colloidal clay particles (CCP).

The use of CCP within the water medium allows us to describe efficiently the adsorption kinetics of the bacteriophages with a simultaneous description of all their sub-fractions. Using these results, the potential of a specific phage as a biological water tracer is presented and correlated to the results of two different water tracing experiments in the Swiss Jura (karst).

2 MATERIAL AND METHODS

2.1 Inactivation kinetics
A 500ml glass vessel treated with 2ml of Repel-Silane (Sigma) was used for all experiments in order to minimize the interaction between the viruses and the glass walls. The glass container was then carefully rinsed with distilled water and 200 ml of Synthetic freshwater (SF), prepared after Moore et al. (1982) and buffered with 25mM Tris-HCl buffer (pH=7.4). The pH was then precisely adjusted to pH=7.4. 2 to $4*10^5$ bacteriophages from an unpurified stock (10^{-5} dilution) were added to the flask, which was briefly stirred manually and deposited into an agitated water bath (180 mvts min^{-1}, 20°C). At time zero, a sample was removed and the agitation switched on. Subsequent samples were then collected at regular intervals and analyzed following an optimized agar double layer method (Rossi, 1994).

2.2 Inactivation-adsorption kinetics
The desired volume of CCP, was added to the 4 ml of Tris-HCl buffer and the final volume was then adjusted to 200 ml with SF. A second flask without clay was also prepared as shown in 2.1. The pH of both media was adjusted as before. 2 to $4*10^5$ bacteriophages were added to both flasks, which were stirred manually. A sample was taken from the second flask, containing only SF (the number of phages in this sample is arbitrarily set as 100%). The flask containing the colloidal particles was put in the constantly agitated water bath. The timer was then started and samples were taken from this vessel at regular intervals and analysed as above. The experiments were performed in duplicate.

This technique allows us to detect the *virulent fraction* (V) of bacteriophages present in the agitated medium. But this fraction can be divided in further so to obtain:
A) The *free/virulent fraction* (FV) is detected by the analysis of the supernatant of centrifuged samples (1.5ml Eppendorf vials, 3 minutes, 10,000g, Sanyo Mini-Centaur tabletop centrifuge). Centrifugation causes the clay particles to sediment, along with the phages adsorbed onto them. The kinetics of this analysis are shown on figures 1 to 6 with open triangles.
B) The *adsorbed/virulent fraction* (AV) is obtained graphically by subtracting FV from V (AV=V-FV).
C) Finally, the *inactivated fraction* (I) is obtained graphically by subtracting V from the total number of phages initially present in the medium (100% - V) that does not contain added clay.

2.3 Bacteriophages
Phage T7 (host *Escherichia coli B*) is a small phage (head/tail length: 60/17nm) of the family *Podoviridae* (Ackerman and DuBow, 1987), with an icosahedral head and a short noncontractile tail. Phage f1 (host *Escherichia coli K12*) is a filamentous phage of the *Inoviridae* family (length/diameter: 100/6 nm). Phage Psf2 (host *Pseudomonas fluorescens*) is a phage isolated from a sewage treatment plant (Rossi, 1994). It has a flexible and noncontractile tail (dimension not determined) and belongs to the *Siphoviridae* family (B1 type). Phages H40/1 (host *H40*) and H6/1 (host *H6*) have been isolated by Moebus from the Atlantic Ocean (Moebus, K., Nattkemper, H., 1989; Moebus, K., 1992). H40/1 belongs to the *Siphoviridae* family (B1 type). It has an icosahedral head and a flexible noncontractile tail (head/tail length 84/41 nm). Phage H6/1 is a tailless phage (*Leviviridae* family) and has an icosahedral head of 30 nm in diameter. MS2 (host *E. coli K12*) belongs to the *Leviviridae* family. It has a quasi-isometric elongated tailless head (24-26nm in diameter).

2.4 Colloidal clay particles (CCP)
Attapulgite clay (Florida), also called Paligorskite, was provided by the "Source Clay Minerals Repository" of the University of Missouri. The CCP suspensions were prepared on the basis of the method proposed by Babich and Stotzky (1980).

2.5 Karstic aquifers
A. Areuse spring: Bacteriophages were injected into the water sink of the Moulin du Lac, also called water sink of the Lac des Taillères, which is located in the hydrogeological basin of the Areuse spring. It is the main outlet of the karstic aquifer that contains the synclinal of the Verrières. The catchment basin of the spring covers 127 km^2. From a geological point of view, the hydrogeological basin of the Areuse spring is situated in the folded mountains of the Jura. The aquifer is composed of karstified limestone of the superior Malm that have an average thickness of 300m. The aquiclude is formed by Argovian marls. The distance between the injection point and the Areuse spring is about 6.25km.

B. Bure aquifer: this aquifer is in Ajoie, in the Swiss part of the tabular Jura (NW Switzerland). From a tectonic point of view, this site consists of sub-horizontal Malm limestone layers (Upper Jurassic) which are fractured by systems of faults and fracture joints related to the Rhine basin to the NE and the formation of the folded Jura to the South. From a hydrogeological point of view, the reef limestone from the Rauracian is karstified, and is confined at its base by impermeable Oxfordian marl. The plateau of Bure is drained by an underground karstic network, called the Milandre network and the underground river is called the Milandrine, into which the bacteriophages were injected. The distance between the injection and the sampling points is about 1.2km (Doerfliger and Zwahlen, 1995)

3 RESULTS AND DISCUSSION

Several authors (Trouwborst et al., 1974; Bricelj and Sisko, 1992) have already shown that the inactivation of a viral suspension is increased by agitation or gassing of the medium. In our case a strong agitation of the SF medium is used to give faster inactivation (or adsorption) kinetic responses than classical techniques. The inactivation kinetics obtained without any addition of CCP (shown with empty circles in fig 1 to 6) perfectly follow a first order kinetic decrease:

$$C_t = C_0 * e^{-Kt} \qquad (1)$$

where C_0 = initial amount of phages in the suspension; C_t = amount of phages in the suspension at time t; and K = inactivation constant (min^{-1}).

Phage H6/1 shows a very low inactivation rate ($K = 6.92*10^{-4}$ min^{-1}). The other five bacteriophages uniformly stress a very high inactivation rate. After 180 minutes of experiment, between 2 and 15% only of the initial virulent population (V) are still virulent, giving respective K values ranging from $K = 1.59*10^{-2}$ min^{-1} for Psf2, to $K = 1.51*10^{-2}$ min^{-1} for T7. As any interaction with the coated walls of the glass recipient is unlikely to occur, agitation is believed to be the main reason of this sharp decrease.

Grant at al. (1993) have shown that the interactions between phages and adsorption sites on particles proceed following a dynamic equilibrium. According to Stotzky (1986), the equilibrium adsorption is reversible and exhibits low specificity between the phages and the substrate. Equilibrium is attained when the rate of adsorption equals the rates of desorption. As coverage of the clay surface increases, it becomes increasingly difficult for adsorption to proceed, and mutual repulsion of the phages particles may result in a reduction in the bond strength. The dynamic equilibrium, which is created between the phages and the non specific binding sites onto the CCP, is called *Quasi-Equilibrium Adsorption* (QEA) by Grant et al. The authors singled out three theoretical cases that will be reviewed later. The regressions on our experimental points of the adsorption kinetics were fitted with the empirical formula (2):

$$C_t = C_1 * e^{-K1t} + C_2 * e^{-K2t} \qquad (2)$$

where C_1 and C_2 = initial amount of phages in the suspension (C_1 and $C_2 > 0$); C_t = amount of virulent phages in the suspension at time t; and K_1 and K_2 = inactivation/adsorption constants (min^{-1}).

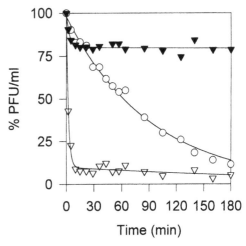

Fig 2: Inactivation (circles) and adsorption (triangles) kinetics of phage T7. QEARI-type reaction. See fig. 1 and text for details.

The use of this equation is much simpler than the three differential equations calculated by Grant, and also allows the calculation of the non-linear regression of biphasic adsorption kinetics. Just as with Grant's model, it makes it possible to graphically visualize and measure the evolution of the various fractions of a phage population.

A. *Quasi-Equilibrium Adsorption* (QEA):
In this first case, the model predicts that the stability of the viruses is not influenced at all by the presence of adsorbent material.

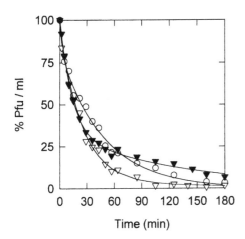

Fig. 1: Inactivation (circles) and adsorption (triangles) kinetics of phage Psf2. Filled triangles: V fraction; open triangles: centrifuged samples, FV fraction. QEA-type reaction, see text for details.

Adsorption, if it occurs, is reversible and is of no advantage to the phage for its survival. Therefore, the inactivation rate of the adsorbed viruses is equal to that of those still free.

Phage Psf2 shows a very low interaction with Attapulgite CCP, with an AV fraction of less than 10% at the end of the experiment (fig.1). Since the kinetic of the V fraction is similar to the inactivation curve, the presence of the CCP does not prevent phages from inactivation. This behavior is typically a QEA-type reaction. The weak interaction with the CCP shows that this phage is unlikely to adsorb onto the matrix of the aquifer or onto natural colloidal particles. Consequently the QEA-type reaction shown by phage Psf2 is certainly the best behavior required for an ideal biological tracer.

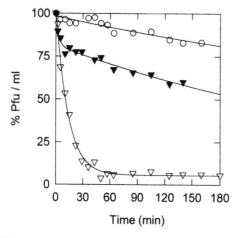

Fig 3: Inactivation (circles) and adsorption (triangles) kinetics of phage H6/1. QEARI-type reaction. See fig. 1 and text for details.

Psf2 has been used successfully in many tracing experiments, through karst and porous environments (Rossi, 1994, data not shown). It has always shown very high breakthrough rates, even after migrations through environments of very low permeability.

B. *Quasi-Equilibrium Adsorption and Reduced Inactivation* (QEARI):

The second theoretical case describes a lower inactivation rate for the adsorbed viruses than for the free floating ones. Adsorption onto suspended colloidal particles or solids protects the viruses from inactivation caused by the physico-chemical and biological conditions of the environment. This pattern of behavior is shown in our experiments with phages T7, H40/1 and H6/1 (fig. 2,3 and 4). In the presence of a small amount of CCP, the behavior of these

three phages changes radically from the inactivation kinetics (empty circles). The kinetics of non-centrifuged samples (filled triangles) and centrifuged samples (open triangles) show two distinct phases. The first is very rapid and occurs within the first few minutes, with a sharp decrease of the V fraction, and corresponds certainly to the installation of the dynamic equilibrium. The second phase is slower and occur during the last minutes of the experiments. It seems to correspond to the drift away from this equilibrium.

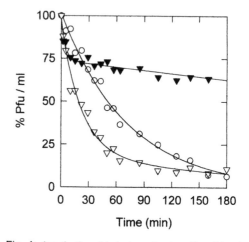

Fig 4: Inactivation (circles) and adsorption (triangles) kinetics of phage H40/1. QEARI-type reaction. See fig. 1 and text for details

Fig. 2 shows the results for phage T7. The kinetics belong to a strong QEARI-type reaction: more than 80% of the phages are adsorbed in less than 10 minutes. Despite this important adsorption, an important fraction of the population remains virulent (AV=80%). The second phase of the reaction shows that there is practically no further inactivation, despite the strong agitation.

This kind of behavior predicts a very high adsorption rate onto the particles within an aquifer, and therefore a poor breakthrough.

The results of phage H6/1, shown in fig.3, shows a rate of inactivation of the V fraction slightly higher than the inactivation rate. The interaction with the CCP is weaker than the T7 one, as the FV fraction stabilises within 30 minutes only.

Fig. 4 displays a lesser QEARI-type behavior for phage H40/1. The FV fraction (open triangles) stabilises slowly, showing a lesser interaction with the CCP, and consequently a weak protection from rapid inactivation. This weak kind of reaction is also interesting in our case, as the phages are less likely to react, and therefore be trapped, within an aquifer.

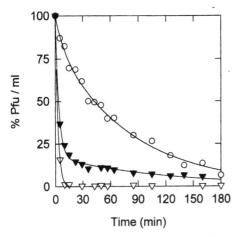

Fig 5: Inactivation (circles) and adsorption (triangles) kinetics of phage H40/1. QEASS-type reaction. See fig. 1 and text for details

Fig 6: Adsorption (triangles) kinetics of phage H40/1. Type of reaction non determined. See fig. 1 and text for details

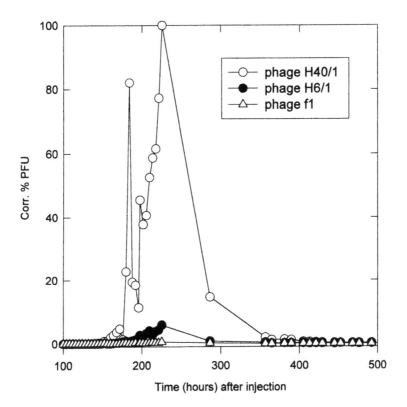

FIG 7: Breakthrough curves of phages H40/1, H6/1 and f1. These three phages were injected simultaneously in the Areuse aquifer. The three curves are corrected (Y axis) in order to take into account the different amounts of injected phages, and to display the maximal breakthrough concentration of phage H40/1 as 100%. The breakthrough rates of the three phages are respectively 21, 1 and 0.15%. the rate for sulphorhodamine (curve not shown) injected in the same time is 80%.

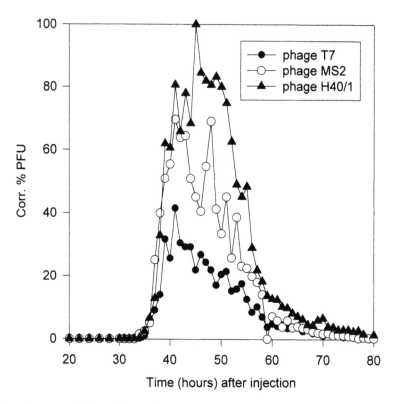

FIG 8: Breakthrough curves of phages H40/1, MS2 and T7. These three phages were injected in the same time in the Bure (Milandrine) aquifer. The three curves are corrected (Y axis) in order to take into account the different amounts of injected phages, and to display the maximal breakthrough concentration of phage H40/1 as 100%.

These three different behaviors can be compared to the results of the two tracing experiments shown in fig. 7 and 8. There is a great difference between the breakthrough curves and rates of H40/1 and H6/1 (fig. 7). Aftfer more than 100 hours of migration in the karstic aquifer, the breakthrough rate of H40/1 is considerably higher (21%) than the one of H6/1 with of 1%. See fig. 7 for details.

Fig. 8 shows that a few hours of migration within an aquifer can influence considerably the breakthrough rate of different phages. In this case, H40/1 has a higher breakthrough rate than T7, even if they both share the same QEARI-type of reaction. H40/1 shows a low QEARI-type reaction suggesting a low interaction with the aquifer particles, whereas T7 greater reaction certainly induces a low breakthrough rate.

C. *Quasi-Equilibrium Adsorption and Surface Sink (QEASS):*

In this case, the solid surface acts as a trap for the viruses. The inactivation rate of the adsorbed phage is higher than that of the free-floating phage. Phage f1 shows this reaction model as it adsorbs quickly and massively onto the CCP (fig. 5). After less than

5 minutes of experiment, there is no more free and virulent phage within the suspension, and most of the adsorbed phages lose their virulence.

As this phage is very likely to be adsorbed onto mineral particles and to loose its virulence, phage f1 is to be considered as the worst candidate as a biological water tracer. Fig.7 shows that this kind of behavior induces a very low breakthrough rate (breakthrough rate is less than 0.2% of H40/1 one).

A few tracing experiments have shown that this QEASS-type of behavior is suitable only for tracing on very short distances, and in water containing low concentrations of colloidal particles (Rossi, 1994, data not shown).

D. *MS2 behavior*

Fig. 6 shows the adsorption kinetics of phage MS2. The inactivation kinetic is similar to the kinetics already shown (curve not displayed fig. 6) with a K similar as Psf2.

The presence of CCP induces a completely different reaction. There is no decline of the number of virulent phages. To the contrary, the number of the V fraction increases and reaches more than 60% of the amount added to the SF medium (filled triangles). The AV fraction (open triangles) increases by about 10% only, giving a large FV fraction of about 110%.

One hypothesis for this phenomenon is that the presence of CCP within the water medium considerably helps the phage to reach the specific adsorption sites on the host-bacteria. Consequently, the number of MS2 phages obtained using a classical analytical method is underestimated. Therefore, the percentage of phages coming out of the aquifer may be overestimated by a factor of 1.6 or higher. A second consequence is that the behavior of this phage in an aquifer will vary certainly according to the amount and type of CCP present in the water. But as the AV fraction is fairly low in our experiments, this phage is expected to travel easily even through aquifers showing a very low permeability.

And as the fig. 8 shows, MS2 phage usually has high breakthrough rates when used in karstic aquifers. Further studies should confirm this behavior within porous aquifers.

4 CONCLUSIONS

Though the agitation used for these experiments may be much more rigorous than that encountered in underground aquifers, the method proposed efficiently simulates a massive loss of virulence. The inactivation kinetics show that five of the six phages studied react in an identical way: more than 85% of the phages are inactivated after 180 minutes, following a law of exponential decrease. Phage H6/1 resists much better.

The reactions of the phages in the presence of CCP are very rapid, taking place in a few minutes. Each virus exhibits a specific behavior, and no phage can be used for the description of a general pattern of adsorption, confirming the results of Moore et al. (1975).

But we can classify most of their reaction with Attapulgite CCP according to three types of behavior described by Grant et al. (1993) and more precisely define the migration potential of a phage, and therefore its value as a biological tracer. The phages that seem most interesting as biological tracers are those whose behavior is of the QEA type (Psf2) and of a low reacting QEARI type, such as H40/1. One should not forget, however, that laboratory experiments only give a very partial image of the behavior of bacteriophages in an aquifer.

In our experiments we use purified clays and selected inorganic compounds under standardized conditions. The experimental conditions are very far from those in a natural environment. Nothing allows us to assert that the influences of the CCP tested in the laboratory allow the reconstitution of the complete behavior of a phage in its natural environment.

As we have just mentioned, each of the phages we tested reacted individually to the presence of mineral particles.

This behavior does not seem to be linked to the physical characteristics measured: Zeta potential, hydrophobicity, shape and size (Rossi 1994). Other specific parameters must determine such a diversity of reactions (the protein composition of the capsid and the tail for example).

Because of this it appears to us more and more that it is risky to simulate the propagation of an eucaryotic virus with a bacteriophage, on the basis of what are still random suppositions. We must at best satisfy ourselves with taking only as example the propagation and migration of the best bacteriophage available.

The kinetics reported here show that the behavior of phages in freshwater environments is strongly influenced by the presence of colloidal particles, even in very low concentrations, showing very fast reactions. Furthermore, the presence of CCP within the liquid medium can modify radically their virulence or resistance to a strong inactivation process. This stabilization is certainly also occurring to different kinds of viruses in natural environments. This may have a dramatic importance on the transport and survival of viruses in waters. As a consequence stabilized viruses could migrate massively or stay in a natural environment for a longer amount of time than it has been previously predicted.

ACKNOWLEDGMENTS

This work was supported by a grant of the Swiss Council of the Research, Bern, Switzerland.
We would like to thank Markus Karner (USC) for the review of the manuscript.

REFERENCES

Ackermann, H.-W., DuBow, M.S. 1987. *Viruses of prokaryotes, vol.: General properties of bacteriophages.* CR. Press, Boca Raton, Florida, 202 pp.

Adams, M.H. 1959. *Bacteriophages.* Interscience publishers, N.Y. pp. 592

Babich, H., Stotzky, G. 1980. Reductions in inactivation rates of bacteriophages by clay minerals in lake water. *Water Res.* 14: 185-187

Bales, R.C., Gerba, C.P., Grondin, G.H., Jensen, S.L. 1989. Bacteriophages transport in sandy soil and fractured tuff. *Appl. Environ. Microbiol.* 55: 2061-2067

Bales, R.C., Shimin, L., Maguire, K.M., Yahya, M. T., Gerba, C.P. 1993. MS-2 and poliovirus transport in porous media: hydrophobic effects and chemical perturbations. *Water Resources Research*, 29: 957-963

Bricelj, M., Sisko, M. 1992. Inactivation of phage tracers by exposure to liquid-air interfaces. In: *Tracer Hydrology*, p. 71-75, Hötzl&Werner (eds), Balkema, Rotterdam.

Doerfliger,N., Zwahlen, F. 1995: *Action COST 65 - Swiss national report in the final report of the COST action 65*, Karst groundwater protection, European Commission, Rept EUR 16547 EN, 446p. pp279-304.

Grant, S.B., List, E.J., Lindstrom, M.E. 1993. Kinetic analysis of virus adsorption and inactivation in batch experiments. *Water Res*. Research 29: 2067-2085

Payment, P., Morin, E., Trudel, M. 1988. Coliphages and enteric viruses in the particulate phase of river water *Can. J. Microbiol*. 34: 907-910

Lance, J.P., Gerba, C.P. 1984. Virus movement in soil during saturated and unsaturated flow. *Appl. Environ. Microbiol*. 47: 335-337.

Moebus, K., Nattkemper, H. 1989. Bacteriophages sensitivity patterns among bacteria isolated from marine waters. *Helgoländer Meersunters*. 34: 375-385

Moebus, K. 1992. Laboratory investigations on the survival of marine bacteriophages in raw and treated seawater. *Helgoländer. Meersunters*. 46: 251-273

Moore, B.E., Sagik, B.P., Malina, J.F. 1975. Viral association with suspended solids *Water Research* 9: 197-203

Moore, R.S., Taylor, D.H., Reddy, M.M.M., Sturman, L.S. 1982. Adsorption of reovirus by minerals and soils. *Appl. Environ. Microbiol*. 44: 852-859

Moore, R.S., Taylor, D.H., Sturman, L.S., Reddy, M.M., Fuhs, G.W. 1981. Poliovirus adsorption by 34 minerals and soils. *Appl. Environ. Microbiol*. 42: 963-975

Rossi, P., De Carvalho-Dill, A., Müller, I., Aragno, M. 1994. Comparative tracing experiments in a porous aquifer using bacteriophages and fluorescent dye on a test field located at Wilerwald (Switzerland) and simultaneously surveyed in detail on a local scale by radio-magneto-tellury (12-240 kHz). *Environ. Geology* 23: 192-200

Rossi, P. 1994. *Advances in biological tracer techniques for hydrology and hydrogeology using bacteriophages.* Ph. D. thesis, University of Neuchatel, 200pp.

Stotzky, G. 1986 Influence of soil mineral colloids on metabolic processes, growth, adhesion, and ecology of microbes and viruses. In: Huang, P.M., Schnitzer, M. (eds) *Interactions of soil minerals with natural organic and microbes*, p.305-428, SSSA Special publication n. 17, SSSA, Madison, WI, USA

Trouwborst, T., Kuyper, S., de Jong, J.C. 1974. Inactivation of some bacterial and animal viruses by exposure to liquid-air interfaces. *J. gen. Virol*. 24: 155-156

Tracer Hydrology 97, Kranjc (ed.) © 1997 Balkema, Rotterdam, ISBN 90 5410 875 4

Performance and persistence of Indium-EDTA as activable tracer in field experiments

E. Gaspar, R. D. Gaspar & I. Paunica
Institute of Physics and Nuclear Engineering Bucharest, Romania

ABSTRACT: Indium-EDTA, owing its extremely low detection limit (10^{-12} g/ml in routine determinations) and a distribution coefficient, K_d = 0.003 ... 0.025 ml/g was successfully used in: karst hydrology investigations, determination of leakage and water loss from reservoirs, labeling of drinkable waters and geothermal waters in flow-rate distribution in the Danube Delta. It can be considered as a reference tracer for the evaluation of the other one tracers behaviour. Having a great stability Indium-EDTA complex used as hydrological tracer, the water samples containing In may be conserved for long time in laboratory, for evidence, in important field works.

1. INTRODUCTION

In-EDTA as a tracer is used in hydrological studies to label and investigate waters that travel through various media: running or still surface waters and underground waters traversing porous, fissured or karst media or passing through semipermeable barriers owing to diffusion, drainage, and other phenomena.

The physico-chemical stability, the low retention in the matrix, the persistence of the tracer long time after labeling, the great detection sensitivity the possibility to use the In data in flow models permit to consider In-EDTA as a very good tracer for hydrological and pollution studies.

Some example of field works performed with In-EDTA in hydrological and pollution studies demonstrates the performances of this tracer.

2. TRACER STABILITY

As ideal tracers do not exists, hydrological tracers should be at least conservative. The basic properties of tracer depend on their chemical and physicochemical structure. If this structure remains stable in experimental conditions, the tracer may be used in quantitative assessments. The performance of anionic tracers (like In-EDTA) is dependent on the stability of the complex. The dissociation of the complexes is characterized by the stability constant K_s which represent the ratio between the molar concentration of the complex and the molar concentrations of the metal ion and of the ligand, respectively.

A relative criterion for assessing the stability of complexes is the $K_s = 10^{20}$ limit. Therefore, In-EDTA, with $K_s = 10^{24.9}$ may be considered a very good tracer.

An example in this respect is the Toplita de Vida experiment. (GASPAR & all., 1985) Used to label karst waters in a period of low flow, the tracer can be measured 180 days after labeling, as Figure 1 shows.

Although the medium was polluted owing to the penetration of waters resulting from bauxite processing into the karst and the contact between these waters and the tracer was long enough to have favored retention, the tracer, however, was recovered in a proportion of 85 % (8.5 g from the 10 g used for labeling). The retention of In-EDTA may rather be accounted for by hydrological causes, frequent in the case of labelings in low-flow periods, when the tracer is trapped by the auxiliary subsystems of the karst.

Not all hydrological tracers are stable at temperatures higher than that of the environment. The most relevant case is that of the fluorescent tracers. So, fluorescence intensity varies inversely with temperature, though this rate depends on the dye. (SMART & LAIDLAW, 1997)

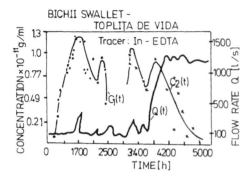

Figure 1. Behaviour of In-EDTA in a long term test. The recovery of 85 % of the amount of injected tracer 180 days after labeling proves a fine chemical stability of the tracer. As the hydrokarst medium was chemically polluted, the result is that In-EDTA behaved like a conservative tracer, the 15 % that was not recovered being due to tracer retention in the auxiliary subsystems of the karst, a phenomenon frequently encountered during labelings performed in conditions of very low discharges.

In-EDTA is indicated for geothermal water tracing also because in field experiments very little loss was observed. Thus, to determine the origin of the Hercules thermomineral spring from Baile Herculane spa, many ponors were labeled using In-EDTA as a tracer. Although the temperature of the spring is about 55^0 C, the tracer was recovered in good conditions (fig. 2) proving its stability at temperature and low retention in the matrix.

3. INTERACTION BETWEEN In-EDTA AND ROCKS IN THE UNDERGROUND MEDIUM

Interactions between tracers and the solid phase of the medium are mainly due to the following processes: filtration, physical adsorption, colloidal precipitation, ion exchange and chemical reactions. The water chemistry plays an important role both in migration and tracer analysis also. The interactions between the rocks in the underground and artificial tracers are the later's general reactions with the solid phase. Other chemical reactions between tracers and the elements of the solid phase are rare. An altogether different situation develops when a tracer comes into contact with the solid phase. Whereas, naturally, waters can have a pH value, a temperature, and a salt content that are greatly varied, owing to pollution the chemical composition of water may modify to such an extent that phenomena occur conductive to either the delay or the total reduction of the tracer.

The K_d distribution coefficient also proves of help

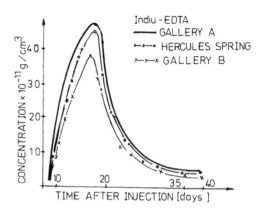

Figure 2. Experimental curves obtained in captation galleries and geothermal spring Hercules after injection of In-EDTA as a tracer in a sinkhole from waterfall valley.

in determination of tracer retardation compared to the labeled water. Thus, in the case of a flow through a porous medium where the solid particle of the porous bed are wholly penetrated by the tracer, the retardation factor depends on the K_d coefficient.

So, for instance, KLOTZ (1982) filtrated a number of fluorescent dye tracers on sands and gravels and established the following range of variation: uranine, K_d = 0.065 to 0.225; eosine, K_d = 0.585 to 0.93; and pyranine, K_d = 0.376 to 1.53. As for rhodamine B, K_d coefficients ranging from 1.42 to 9.8 have been registered; at times, rhodamine B is absorbed completely.

Indium, used under the form of In-EDTA as a tracer was successfully used because its distribution coefficient is very low. Thus, using natural waters and various rocks, the K_d for In-EDTA varied between 0.003 to 0.025 ml/g. In these determinations we have used the batch method.

4. OTHER FACTORS CONTRIBUTING TO LOWER TRACER CONCENTRATION IN TIME AND SPACE

Tracer migration in aquifer systems is a result of the combined action of the major mechanisms connection, molecular diffusion and dispersion. The main parameter resulted is the longitudinal dispersion. But in the case of hydrokarst structures due to auxiliary systems, solution channels and functioning of karst system, the transversal dispersion may atteint a great value, comparable with the longitudinal dispersion. As a result, a tracer dilution is produced.

Given these circumstances, what will the fate of a

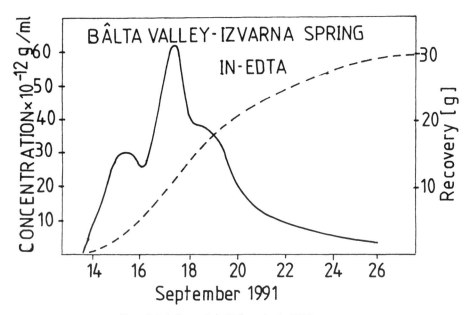

Figure 3. Labeling on Balta Valley using In-EDTA as a tracer

tracer pulse be according to the hydrodynamic behaviour of the karst? If during a labeling operation the tracer penetrates the auxiliary systems, the whole amount of tracer will be recovered after a long period: either continuously, a large dispersion of the tracer could being thus achieved, or intermittently, according to the length of the interval of high-flood succession, like in fig. 1.

In the Danube Delta experiment, a flow rate of 3 200 m³/s was labeled with In-EDTA, to determine the water distribution between channels and lakes through a great dilution has produced.

The karst Izvarna spring with a flow roughly 2.2 m³/s are a major source of potable water supply for Craiova town. After labeling the ponor from Balta valley (fig. 3) with In-EDTA, the tracer was recovered in Izvarna spring, 21 days after an underground a flow path of 20 km.

The determination of concentrated leakage, diffuse water loss and dam stability in the case of Dragomirna man made reservoir (10^6 m³ water) was performed using 100 g of In-EDTA as a tracer. The labeling of the whole body of water allowed the identification of the springs connected with leakage. The mean transit time and the permeability along the infiltration path were also determined (fig. 4).

5. THE OPTIMAL AMOUNT THE TRACER NEEDED IN ONE LABELING

The success of an experiment conducted in surface or underground water with artificial tracers primarily depends on the amount of tracer that is employed.

There is an optimal amount of tracer which meets both metrological and ecological criteria but it depends on a number of factors, most of them unknown.

The minimum detectable concentration of In complexed with EDTA, in routine conditions is 10^{-12} g/ml.

In the case of tracers detected with the help of chemical analysis or activation analysis, the relation of LEIBUNDGUT & WERNLI (1982) yields satisfactory results.

As for In-EDTA used as a tracer to investigate hydrokarst structure, the analysis of numerous trials (GASPAR & all., 1985) had led to an empirical relation:

$$M = QTPK \qquad (1)$$

where M is the In amount required for one labeling [g], Q is the sum of the flow rates of the emergency where the tracer might occur [m³/s], T is the time interval [d], as estimated by the investigator, needed for the tracer to pass through the monitoring point and for a value of at least 10 to be attained for the maximum - minimum concentration ratio, P is the loss coefficient and K is a safety coefficient.

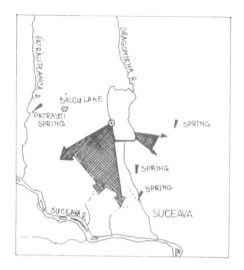

Figure 4. Hydrodynamical spectrum of leakage at Dragomirna reservoir determined with In-EDTA as a tracer.

In the case of polluted or strongly mineralized waters, neutron activation of contained elements creates great difficulties in tracer concentration measurement.

6. CONCLUSIONS

The use of Indium as a tracer which can be activated, proved to be a method characterized by high sensitivity and accuracy. The major advantage of this tracer is that, owing to its extremely low detection limit, it can label emergencies of very low flow rates an artificial reservoir or natural lake of 10^6 m^3 may be labeled using 100 g indium (complexed as In-EDTA). The detection limit can be lowered below 1.10^{-12} g/ml using a volume of water larger than 100 cm^3 for analysis and a larger neutron flux for irradiation. In-EDTA is an ecological tracer and may be used for labeling drinking water, as well as mineral and thermomineral waters in health resorts.

REFERENCES

Gaspar, E., Stanescu, S. P., Oraseanu, I., Farcasiu, O. M. & Spiridon, S. 1985. The behaviour of Indium as a tracer for karst water research. *Theoretical and Applied Karstology*. 2:149 - 156.

Klotz, D. 1982. Verhalten hydrologischer Tracer in ansgevahlten fluvioglazialen Keisen, Hangschutt - Kiesen und tertiaren Kies - Sanden aus Bayern. In: Leibubdgut, C. & Weingartner, R., (eds). *Proc. Int. Symp. Tracermethoden in der Hydrologie.* Geographischer Verlag, Bern, 245.

Smart, P. L. & Laidlaw, I. M. S. 1977. An evaluation of some fluorescent dyes from water tracing. *Water Res. Research.* 13, 1, 15.

Leibundgut, C. & Wernli, H. R. 1982. 2 ur Frage der Einspiesemenegenbuerechnung für Fluoreszentracer. In: *Tracermethoden in der Hydrologie*, Beith. 2. Geol. d. Schweiz - Hydrologie, 28/I, Bern, 119.

Tracer Hydrology 97, Kranjc (ed.) © 1997 Balkema, Rotterdam, ISBN 90 5410 875 4

Activation analysis in investigation of hydrokarst systems

E. Gaspar, S. P. Stanescu & R. D. Gaspar
Institute of Physics and Nuclear Engineering Bucharest, Romania

ABSTRACT: Activable tracers are used on large scale to investigate hydrokarst systems from Romania. The methodology of the activable tracer Indium-EDTA is presented. For other activable tracers: Dy-EDTA, La-EDTA, I, Br the methodology is similarly. The estimation of the tracer mass injected, the sample processing, sampling schedule, sample measurement technique and experimental transfer curves are presented.

1. INTRODUCTION

The intensive exploitation of aquiferous resources stored in karst zones, deforestation, climatic changes, water development works, the urbanization and location of various mining, industrial and farming units, the degradation of soils and the ecological aspects call for a massive research work into karst structures.

Pollution, salinization and over-exploitation of water resources are the most common symptoms of growing anthropogenic pressure on the hydrological cycle. Degradation of water resources from karst zones is not restricted to industrialized countries; it is of growing concern also throughout the developing world.

After 1960 y., constant attention has been paid to the supplementation of the classical methods with tracer techniques in order to determine the hydrological parameters and assess the vulnerability to pollution of aquiferous resources.

The best results were obtained with activable tracers. They have a great stability in time, slowly reacts with the environment, can be detected in very low concentrations and most of them (Indium especially) can be considered as ecological tracers.

2. METHODOLOGY OF ACTIVABLE TRACERS

2.1. Tracers

The principal nuclear and analytical parameters of the activable tracers (In-EDTA, Dy-EDTA, La-EDTA, I^- and Br^-) are included in Table 1.

A summary characterization of these tracers mark out the following:

- In-EDTA, the most used tracer by authors, is considered as a reference tracer. Indium is practically absent from the investigated waters. Under the In-EDTA complex form has a great chemical stability (K = 10^{24}.9), a small retardation;

Table 1. Analytical parameters of the tracer

Isotope used and relative abundance (percent)	Thermal neutrons activation cross section (b)	Measured gamma-ray (keV)
^{115}In (95.67)	155.0	417,0
^{127}I (100)	6.2	442,9
^{161}Dy (28.10)	800.0	94,6
^{139}La (99.91)	8.8	1596,2
^{81}Br (49.48)	3.3	554,3

^{115}In has a great neutron activation cross-section and ^{116m}In radioactive isotope emits intensive gamma-rays and has a small decay time.

- Dy-EDTA is a tracer less used because its presence in environment does not permit a complete valorification of the analytical facilities offered by its great neutron activation cross-section. It is more expensive than In-EDTA and has a stability constant of 17.6.

- La-EDTA has a small chemical stability, it is present in background and is more cheap than In.

- I⁻ under NaI or KI chemical compounds can be analyzed both by chemical and neutron activation. It is naturally present in water ($\sim 10^{-8}$ g/cm^3) and is kept in environment by enzymatic oxidation reactions.

- Br⁻ usually as NH$_4$Br, has similar properties with I⁻, it is naturally present in water and has a small K$_d$ coefficient in karst environment.

2.2. Labeling

Activable tracers satisfies partially the condition of mathematical idealization of the injection. Only tracers, with a low detection limit, as In-EDTA ($\sim 10^{-12}$ g/ml), permit an assimilate to "δ" function labeling. In this case, the investigated hydrologic phenomenon is not disturbed by labeling, even in the case of a karst systems having a small flow-rate of the sinkhole.

The success changes of an experiment with artificial tracers in karst depends on the quantity of tracer used. In the case of the activable tracers, we are using the relation:

$$M = 10.Q_{max}.t.C_{min} \qquad (1)$$

where: M[g] is the tracer mass; Q$_{max}$ [m^3/s] represents the sum of discharges of all possible emergencies connected to the labeled point; t[s] is the estimated transit time from hydraulic studies and C$_{min}$ [g/m^3] is the minimum detectable concentration in routine operations.

2.3. Sampling

To surprise the time of the first appearance, one can use a simple hydraulic calculation considering a plug flow model.

However, the prelevation of a great number of samples is noneconomically. Generally, for the goals of hydrogeologists in karst structures tracing, a number of 10 - 15 points on transfer curve are considered satisfactory for an experiment. But in this case, it should be known the total travel time of the tracer, T, through the monitoring point. This can be estimated using the relationship:

$$T = n.\tau \qquad (2)$$

where τ is the mean transit time. The coefficient n, for recent karst systems and plug flow model, has a value of 0.7 ... 6.0. But in general, the longitudinal dispersion parameter D/v$_x$ make that n takes values between 2 and 4. Taking into account these considerations, the sampling frequency in hydrokarst

Table 2. Sampling schedule in karst tracing

Distance (km)	Time after labeling			
	1st day	2nd day	3rd day	4 - 6 days
<1	1 - 2	3	4 - 6	8 - 12
1 - 10	2 - 4	6	8	12
>1	12	12	12	12 - 18

systems tracing can be as described in Table 2, where the time interval between two consecutive samples is given in hours.

To avoid the tracer loss by unexpected interactions and contaminations, the prelevation of samples must be made in polyethylene vessels of 250 ml, provided with double lid. One obtains satisfactory results even in the case of sample processing after some months from sampling without a previous chemical treatment of the vessels.

2.4. Sample processing

We have used two methods for interest samples identification: the selection by soundings and cocktail samples method. To realize a compromise between the number of processed samples and the possibility of an erroneous selection of interesting samples a simplified variant of Table 2 was used.

In the case of polluted or strongly mineralized waters, neutron activation of contained elements creates great difficulties in tracer concentration measurement.

The method used by authors are based on tracer preconcentration before irradiation. The general scheme of sample processing is: chemical complex separation, precipitate formation, coprecipitation, precipitate filtration, drying, precipitate separation and encapsulation.

At 100 ml water sample, 100 ml H$_2$SO$_4$ conc. and 2 ml Bi solution (10.7 g basic bismuth carbonate dissolved in a mixture of 150 ml NHO$_3$ with 350 ml H$_2$O and then diluted to 1000 ml) are added. After 1 h, 7 ml NH$_4$OH (25 % conc.) are added and the precipitate thus formed is filtered through a nuclear membrane filter (holes of 0.8 μm diameter) with a roughing pump and then dried. The precipitate is introduced in polyethylene vials of 10 x 10 mm that are irradiated in the nuclear reactor.

With a few modifications, this method is used in sample preparation for I, Br, Dy and La.

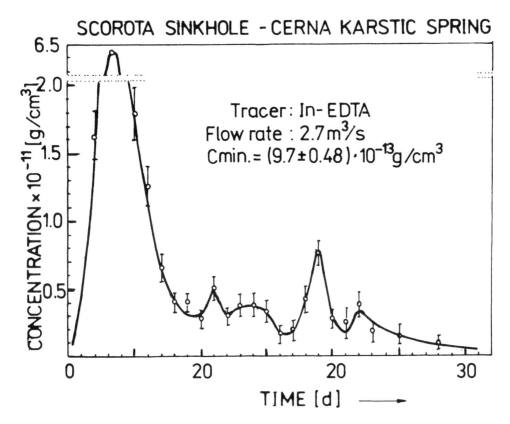

Figure 1. The concentration-time curve obtained during the investigation of a hydrokarst structure with In-EDTA as a tracer. One may note that a minimum detectable concentration of 9.10^{-13} g/ml was attained (with the limits of experimental errors) even in the case of field experiments.

2.5. Sample irradiation

The polyethylene capsule containing samples is introduced in the nuclear reactor at the neutron flux of 10^{12} n/cm^2.s with a pneumatic rabbit-tube. A number of 6 - 12 samples are irradiated simultaneously with a reference sample, irradiation time being between 20 and 60 minutes. After irradiation, follows a decay time and gamma-ray measurement. In table 3, t_{ir} = irradiation time, t_d = decay time and t_m = measurement time.

2.6. Sample measurement

The sample measurement consists in the relative to reference sample determination of the photopeak area for the representative gamma-ray emission of the radionuclides formed by neutron activation of the element used as tracer. The most used gamma-rays for ^{116m}In is 417,0 keV.

Considering the following notations: C_e, A_e, t_e - element concentration, photopeak net area, measuring time for reference sample; C, A, t_m - element concentration, photopeak net area and measuring time for unknown sample; t_d - decay time between reference and unknown sample measurement and λ - radioactive constant of measured radionuclid, then the tracer concentration can be calculated with the relation:

$$C = \frac{C_e . A . \left(1 - e^{-\lambda t_e}\right) . e^{\lambda t_d}}{A_e . \left(1 - e^{-\lambda t_m}\right)} \qquad (3)$$

As a rule, one use the same time for reference and unknowns samples and this value can be evaluated with the empirical relation; where N is the number of simultaneously irradiated samples.

$$t_m = \frac{\ln N / N - 1}{\lambda} \qquad (4)$$

53

2.7. Interpretation of the results

The minimum detectable concentration of the activable tracers acquired in the above mentioned conditions are presented in Table no. 3, where t_{ir} = irradiation time, t_d = decay time and t_m = measurement time. Thus, the activable tracers can be relatively measured with a concentration under 10^{-9} g/ml, supplies the conditions imposed to the ecological tracers, permitting the direct labeling of the drinkable waters and because of its variety they can be used in multitracing experiments. The use of a conservative tracer as In-EDTA which can be measured in concentrations of 10^{-12} g/ml in routine conditions, gives the possibility to obtain complete concentration-time curves. The knowledge of flow-rate variation of measured springs permits the calculation of the recovered tracer quantity for all points and tracer balance sheet. (fig. 1)

Defining on this basis one hydrokarst system, guides to hydrodynamic parameters knowledge (velocity, mean transit time, turnover time) and by the use of adequate flow models, one can determine the dynamic volume, static volume and the karst vulnerability to pollution.

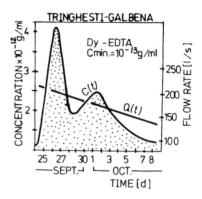

Figure 2. Dy-EDTA is a fine tracer for labelings of karstic waters. Thus distribution of concentrations in the Galbena spring, according to time, was obtained by labeling the Tringhesti ponor using 2.5 g Dy in the Dy-EDTA form.

tracer in routine determinations is presented in fig. 2. The karst subsystem Tringhesti - Galbena is a part of the great karst system Padis.

4. CONCLUSIONS

The activable tracers can be successfully used in investigation of hydrokarst systems, both by simultaneous and staggered in time tracings. In-EDTA can be considered a reference tracer, for the evaluation of the other tracers behavior. The major advantage of this tracer is that, owing to its extremely low detection limit, it can label emergences of very low flow rates which cannot be labelled by any other tracer. The use of La and Dy gives good results in the labelings of karst with a rapid circulation.

Table no. 3. Methodological parameters

Nuclide	t_{ir} [min]	t_r [min]	t_m [min]	C_{min} [g/ml]
^{116m}I	20	10	5	9.10^{-13}
^{165}Dy	60	60	5	2.10^{-12}
^{140}La	60	24 h	15	1.10^{-10}
^{128}I	20	5	3	1.10^{-10}
^{32}Br	60	24 h	15	1.10^{-9}

3. FIELD STUDIES

Activable tracers were used in multitracing and multisampling experiments performed in Romanian karst for: delimitation of aquifer systems, highlighting of partial captures and diffluence surfaces, investigation of geothermal systems stored in karstic structures, assessment of karst vulnerability to pollution, determination of regional geological structures, identification of sub-aquatic springs, influence of urbanization and water captures in the karst, determination of leaks and water loss from reservoirs, mine drainage investigation, labeling of potable waters and cave stream investigations.

The performance of In-EDTA as activable tracer is presented in fig. 1. We have used 100 g In obtaining 50 % recovery.

Also, the performance of Dy-EDTA as activable

REFERENCES

Gaspar, E., Farcasiu, O., Stanescu, S. P. & Spiridon, S. 1984. Nuclear methods for karst hydrology investigation. *Theoretical and Applied Karstology.* 1: 207 - 214.

Gaspar, E., Stanescu, S. P., Oraseanu, I., Farcasiu, O. & Spiridon, S. 1985. The behaviour of In-EDTA as a tracer for karst water research. *Theoretical and Applied Karstology.* 2: 149 - 156.

Behrens, H., Moser, H. & Wildner, E. 1977. Investigation of groundwater flow with the aid of Indium-EDTA complex using neutron activation for the determination of the tracer. *J. Analyt. Chemistry.* 38: 491 - 498.

Tracer Hydrology 97, Kranjc (ed.) © 1997 Balkema, Rotterdam, ISBN 90 5410 875 4

New fluorescent tracers

S. Hadi & Ch. Leibundgut
Institute of Hydrology, University of Freiburg im Breisgau, Germany

K. Friedrich
Institute of Organic Chemistry and Biochemistry, University of Freiburg im Bresgau, Germany

P. Maloszewski
GSF-Institute of Hydrology, Neuherberg, Germany

ABSTRACT: Succinylfluorescein disodium salt (SF) and 5(6)-carboxyfluorescein trisodium salt (CF) were developed and investigated as hydrological tracers. The physical-chemical properties were examined in comparison to those of uranine (UR). All of these xanthene dyes show similar behaviour, apparently due to their common structure. The various functional groups in the xanthene basis caused only limited changes in their behaviour. However, it would be advantageous to use these tracers for certain applications in which UR does not yield satisfying results. Considering the possible applications and the fact they are environmentally friendly, the new tracers extend the spectrum of artificial tracers.

1 INTRODUCTION

Fluorescent dyes have been widely used both for surface water and ground water studies. The main importance of the use of fluorescent dyes is in their relatively simple analysis, their low detection limit and, consequently, the small quantity of tracer needed to be used in field experiments. Furthermore, the background in their wavelength range is usually very low. Since the first tracer experiment using the fluorescent dye uranine in 1877 until approximately 1960 only a few fluorescent tracers were known. During the following two decades much effort was made in order to find additional tracers. A wide spectrum of artificial tracers was established. However, most of the tracers did not exhibit the right combination of properties and, therefore, are not well suited to be water tracers. Part of the fluorescent dyes, e.g. rhodamines, are being used in the textile industry due to their ability to be adsorbed on a substrate. Other fluorescent dyes can exhibit low solubility in water and/or degradation in the presence of light or oxygen or by microorganisms. After this phase (1960-1980), it seemed that the possibilities were exhausted. But in 1986 Leibundgut & Wernli suggested the tracer „naphthionate" (NA), which showed nearly conservative behaviour in short distances. Since then further research was executed, like that of Viriot & André (1989) and Netter & Behrens (1992), a few new tracers were suggested, but none of them have been further applied in the field. It was, therefore, of interest to search further for new ones.

The purpose of the present work was to carry out a systematic research, laying down the theoretical foundation for the development of a new, conservative fluorescent tracer to be used in the field of hydrology. To synthesize a new tracer and to investigate its various physical-chemical characteristics. Extensive attention was given to the investigation of the sorption property in the laboratory through batch and column studies, and in the field.

2 DEVELOPMENT AND SYNTHESIS

The relationship between tracer's structure and properties appears as an indispensable basis of future development of hydrological tracers. In order to initiate the development of a new fluorescent hydrological tracer and to increase the basis of existing research, an extensive search through the relevant literature for known fluorescent molecules was conducted concerning many characteristics such as sorption, toxicity, detection, light sensitivity, pH and temperature dependency and microbial degradation. It was inferred, that these properties are in mutual relation to each other. This search revealed that a deficiency in the literature has complicated the efforts in understanding the connection between the structure and behaviour of known tracers. Most authors have offered empirical observations regarding the behaviour of a fluorescent dye without offering theoretical explanations for the observed behaviour. Furthermore, in certain cases, it would be helpful to focus on a specific functional group, when accounting for certain molecular behaviours. At times, however, such an approach is less useful, proper inquiry requires examination of the molecule complex as a whole. A further complication arises from the fact that even a single functional group may promote various behaviours, the effects of which may or may not be helpful for the hydrologist. The conclusion from this search consists of a theoretical development of a new family of „ideal" molecules to the field of hydrology.

The family of molecules that seems to be suitable for water tracers includes the basis structure of xanthene dye (Figure 1 with $R_1=R_2=H$), which was chosen as the basis of the molecule to be synthesized. Xanthene is the basis of fluorescein which is considered to be nearly conservative. Therefore, it was expected that if this structural basis is chosen to be synthesized, the new tracers would show even better behaviour. In this context, in order to construct a fluorescent molecule, the entire basis of xanthene was required.

Hence, the prime molecules suggested as hydrological tracers were the sulfonated xanthene derivates, where R is CH_3, COOH or $CH_2CH_2CO_2H$. The idea was that sulfonated xanthenes would be suitable in either form which could be achieved in the synthesis. Possible combinations include mono- or multi-sulfonated compounds along with an alkyl functional group which could be oxidized (to a carboxylic acid group (COOH), for instance) and after isolation would be either in the acid form or the mono- or multi-sodium salt form. These structures were chosen mainly for two reasons, which enable an accurate simulation of the water's dynamic: the increase of the molecule's water solubility and the decrease in its dependence upon pH, in addition to how these factors influence sorption. The sulfonic acid functional groups as well as the conversion into their sodium salts increase the molecule's water solubility. Moreover, it is known that the substance's solubility is inversely correlated with adsorption (Baily & White 1970, Leibundgut & Wernli 1986, Shiau et al. 1992). The ability to describe the system accurately is, therefore, reduced when the tracer molecule has low solubility in water. The sulfonic acid functional groups are ionized below pH of 4 and, therefore, could reduce the tracer's dependency upon the pH of the water. It is known that certain tracers tend to be adsorbed in acidic water like uranine (Behrens 1982) and rhodamine WT (RWT) (Shiau et al. 1992) and, therefore, could not correctly simulate the water's dynamic. A computer search for known molecules with a xanthene basis was conducted, and 700 different molecules were found. Based on the conclusions of the report mentioned above, eight molecules, of these 700, were deemed

suitable for further research and were the basis for the „design" of the new molecules.

The preparation of the sulfonated xanthene derivates involved synthesizing and sulfonating xanthene dyes. Executing this stage in the research achieved two aims. The first was the synthesis of two new molecules, the 9-carboxy-6-hydroxy-5-sulfo-xanthen-3-one sodium salt and 6-hydroxy-9-methyl-xanthen-3-one-4,5-disulfonic acid disodium salt (Figure 1). The second was the improvement of the yield and the purity of these xanthene dyes before the sulfonation stage (like in the case of SF) by using ortho-phosphoric acid instead of concentrated sulfuric acid in the reactions. These sulfonated xanthene derivates could not be purified, in the time available, although much effort was given. Due to the ability of the different components in the unpurified compounds to fluoresce, no practical use was, therefore, possible. Moreover, the unpurified compounds constitute potential competitors to the water and to other components in the water (such as the tracer itself) for the active sites in the aquifer's matrix and could be adsorbed. That can lead to inaccuracies in the interpretation of the results achieved by tracer experiments.

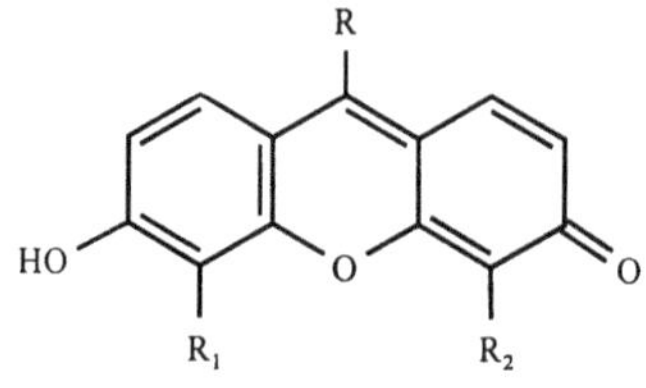

$$R = CH_3 , \quad COOH \quad or \quad CH_2CH_2COOH$$
$$R_1 = R_2 = SO_3Na \quad or \quad H$$

Figure 1. The crude sulfonated xanthene derivates.

Other alternative compounds of the xanthene family, which were expected to be nearly conservative as well, were tested further. These were 6-hydroxy-xanthen-3-one-9-propionic acid disodium salt, also known as succinylfluorescein disodium salt (SF), and 5(6)-carboxyfluorescein trisodium salt (CF) (Table 1). The former compound is the product of the first stage of the sulfonated molecule, i.e. 6-hydroxy-9-methyl-xanthen-3-one-4,5-disulfonic acid disodium salt, synthesized. The latter compound is, in its acid form, an industrial product which was converted to its trisodium salt. The substance in its acid form is known as a marker for biological assays (e.g. Weinstein et al. 1977).

3 PHYSICAL-CHEMICAL CHARACTERISTICS

The physical-chemical characteristics of the new hydrological tracers SF and CF were investigated in comparison to those of uranine (Table 1). All tracers fluoresce in the same spectral range, and have very high quantum yield of approx. 1. The extinction coefficient of CF is a little higher than that of UR but approximately twice the extinction coefficient of SF. Consequently, the detection limit of all tracers is similar, that of SF is a little higher than that of CF, which are 0.05 and 0.01 mg/m^3, respectively. The detection limit of UR is 0.02 mg/m^3. Similar to UR, the background fluoresence of other materials within the same spectral range of the new tracers in ground water and surface water is low. The solubility of CF is the highest and is 623 g/l at 20°C. That of UR is 500 g/l when heated (to approx. 80°C) and that of SF is 335 g/l at 20°C. The fluorescence intensity of all dyes depends similarly on the pH level of the system. Their fluorescence intensity is the highest above a pH value of 9, whereas SF reached its highest fluorescence intensity already above a pH of 8.5. In alkaline solution the tracers are in anionic state and in acidic solution in cationic state. Therefore, the tracers are expected to be adsorbed in acidic water. The highest resistance to temperature changes is shown by SF. However, all show high fluorescence intensity at low temperatures. The photochemical decay of these tracers was measured simultaneously on a sunny day out-of doors. The half life times are very short; 23.4 min for SF, 23.5 min for UR and 38.1 min for CF, at 100 mg/m^3. Hence, CF is approx. 1.6 times more stable than SF and UR. The direct comparison of the values of decay rate and half life time given in the literature with the values obtained in this study is difficult due to different test conditions (which were usually unspecified in experiments out-of doors). However, it should be noted that the half life times for UR obtained in this study are between 6.8 and 13.6 times shorter than those known in the literature which were achieved out-of doors on a sunny day. The new tracers and UR were not toxic to goldfish (*carassius auratus*) and crucians (*carassius valagaris*) in a time-to-death test over 5 days.

4 SORPTION STUDIES

The most important property for the use of fluorescent tracers in hydrology is sorption. The sorption behaviour of the new tracers was extensively investigated through batch, column and field studies. Since the new tracers and UR fluoresce at the same spectral range, they could not be injected simultaneously. The tracers bromide and NA (Ex$_{max}$=325 nm, Em$_{max}$=420 nm) were, therefore, used as references in the column and the field studies, respectively.

Batch studies were conducted using a horizontal batch table similar to the method of Klotz (1982) and Lang (1982), however, with a frequency of 25 strokes per minute. Four initial tracer concentrations (1, 10, 50 and 100 mg/m^3) and six different artificial substrates (Figure 2) were examined. The evaluation of the results from Batch studies was performed using the distribution coefficient, K$_d$, which is a measure of the maximum adsorption of a solute by a particular substrate.

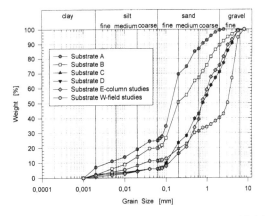

Figure 2. Composition of the six different artificial substrates used in the batch studies (at grain size <0.002 mm very fine silt grains (theoretically calculated) also exist).

The distribution coefficients obtained for the new tracers and UR were low and varied within a narrow range between 0.00 and 1.35 cm^3/g (Table 2). That is in comparison with other dyes such as the sulforhodamine B of which its K$_d$ value is about 30-40 cm^3/g. The general tendency of the K$_d$ values and, consequently, of the adsorption level, found in this study, indicates higher values for succinylfluorescein disodium salt than those for CF which were a little higher than those for UR (Table 2). Furthermore, although in some identical conditions the K$_d$ values of the different tracers varied greatly, they were low and were often in the same order of magnitude, considering the accuracy of batch tests.

The K$_d$ values were, in general, inversely dependent on the tracer concentration. That is similar to the observations given in the literature (Smart 1976, Sabatini & Austin 1991, Behrens & Zahn 1992). However, it is difficult to draw this tendency at very low distribution coefficients, because the K$_d$ values were so low (almost zero) they were within the error range. Moreover, the standard deviations calculated on the K_d values at low concentrations were usually higher than those at high concentrations. This represents a greater imprecision in the K_d values obtained and may cause a deviation from the general tendency, at low concentrations. A linear sorption isotherm (adsorbed mass versus tracer concentration in equilibrium) in the range of 0-100 mg/m^3 could be observed.

Table 1. Physical-chemical characteristics of SF, CF and UR.

Properties		CF	SF	UR
Structure				
Formula		$C_{21}H_9O_7Na_3$	$C_{16}H_{10}O_5Na_2$	$C_{20}H_{10}O_5Na_2$
Molecular weight [g/mol]		442.27	328.23	376.28
Solubility (temperature) [g/l]		623 (20°C)	335 (20°C)	500 (80°C)
Detection	Ex/Em max. [nm]	494 / 515	490 / 506	493 / 515
	interval, $\Delta\lambda$ [nm]	21	16	22
	quantum yield [-]	0.98	0.98	0.97
	extinction coefficient [mol^{-1}lcm^{-1}]	66,700	28,400	60,900
	detection limit [mg/m^3=ppb]	0.01	0.05	0.02
pH-dependency	pK$_a$ [-]	6.8	6.5	6.4*
	maximum fluorescence intensity at pH:	>9	>8.5	>9
Photodegradation	-k [min^{-1}] at 100 mg/m^3	$1.82 \cdot 10^{-2}$	$2.96 \cdot 10^{-2}$	$2.95 \cdot 10^{-2}$
	-k [min^{-1}] at 10 mg/m^3	$2.1 \cdot 10^{-2}$	$3.23 \cdot 10^{-2}$	$3.17 \cdot 10^{-2}$
	$t_{1/2}$ [min] at 100 mg/m^3	38.1	23.4	23.5
	$t_{1/2}$ [min] at 10 mg/m^3	33.3	21.5	21.9
Temperature dependency	maximum fluorescence intensity at:	low temperature	low temperature	low temperature
	temperature coefficient, n [1/°C]	0.00485	0.00329	0.00406
Toxicity	goldfish	not toxic	not toxic	not toxic
	crucians	not toxic	not toxic	not toxic
Background		low	low	low

* Calculated on the basis of the curve (relative fluorescence intensity versus pH) given by Käss (1992).

Table 2. The distribution coefficient (K_d) obtained in batch studies of SF, CF and UR at four initial concentrations (C_0) and six different artificial substrates (see Figure 2).

Tracer	C_0 [mg/m^3]	K_d [cm^3/g]					
		Substrate A	Substrate B	Substrate C	Substrate D	Substrate E	Substrate W
SF	1	1.07	0.00	1.18	0.92	0.33	0.04
	10	1.35	0.08	0.92	0.67	0.02	0.14
	50	1.30	0.02	0.43	0.37	0.18	0.07
	100	1.32	0.01	0.38	0.52	0.14	0.06
CF	1	0.49	0.25	0.37	0.26	0.53	0.15
	10	0.43	0.22	0.37	0.23	0.31	0.07
	50	0.33	0.12	0.16	0.22	0.25	0.03
	100	0.28	0.09	0.16	0.19	0.33	0.04
UR	1	0.21	0.00	0.31	0.14	0.00	0.16
	10	0.23	0.00	0.13	0.11	0.11	0.00
	50	0.21	0.00	0.08	0.08	0.02	0.01
	100	0.20	0.00	0.13	0.08	0.07	0.01

The adsorbent concentration, i.e. the matrix's adsorption sites, is also of importance in influencing adsorption of organic solutes (Smart & Smith 1976, Karickhoff et al. 1979). It can be assumed that within each of the substrates examined in this study, the quantity of adsorption sites is more or less constant. Therefore, a substantial influence of the adsorbent concentration, within each substrate, on the adsorption level of the different tracers could be excluded. Contrary, when comparing different substrates, the adsorbent concentration had a very strong impact on the adsorption level of the tracers. The clay content was mainly responsible for the adsorption of the different tracers. Though, this claim is probably true only when the substrate contains mainly fine material, from clay to coarse sand. At high content of coarse material, with insufficient adsorption sites, like gravel, other factors such as the water pH value and the ion content in the water may be of importance.

In the *column studies*, the column apparatus used in this study is similar to that of Klotz (1991) and was constructed from a 0.5 m Plexiglas cylinder with 0.09 m inner diameter. Two and three Plexiglas cylinder elements were combined together in order to obtain column lengths of 1.0 and 1.5 m, respectively (Hadi 1997). In the column studies the influence of the tracer injection mass (0.005, 0.01, 0.02, 0.05 and 1 mg), the flow rate, and consequently the Darcy flow velocity (between $2.39 \cdot 10^{-4}$ and $8.25 \cdot 10^{-4}$ m/min) and the flow distance (0.5, 1.0 and 1.5 m) on the sorption behaviour of the different tracers was investigated. The examination of NA in column studies was applied in order to enable

complementary evaluation of the results obtained from field studies.

In the column and field studies the evaluation of the results was performed using 1-D and 2-D dispersion models respectively, coupled with a reaction model of linear sorption with instantaneous equilibrium. The fitting of the measured breakthrough curves (BTCs) was conducted using the software programs „Fieldn" and „Field2", whereby the fitting parameters were found using the method of least squares (Maloszewski 1991). The evaluation of the results from the column and field tests was conducted using the retardation factor, R_D, dispersion factor (the ratio of the apparent dispersion parameters of the tracer and the reference tracer), P, and the relative recovery, RR.

Column studies indicate the variables characterizing the sorption behaviour (i.e. retardation factor, dispersion factor and relative recovery) of the new tracers, UR and NA to be low and the range in which these variables varied to be narrow. The retardation factors varied between 0.80 and 1.22. The retardation factors of the tracers SF and NA were mostly lower than 1.0 and those of UR were a little higher and approximately 1.0. The highest retardation factors, always above 1.0 was shown by 5(6)-carboxyfluorescein trisodium salt. The dispersion factors of SF, UR and NA were low and varied between 0.89-1.52. Those of CF were higher and varied mostly between 1.67-2.32. In general, the factors of retardation and dispersion of NA were greater than those of SF, lower than those of UR, but those of CF were the greatest. The steady increase of the relative recovery curves as they approach their maximum implies reversible adsorption reactions of all tracers, with the slowest desorption for CF. In some cases, CF's relative recovery was higher than that of SF. The tracer SF approached its maximum more quickly but usually with lower values than those of UR, which, mostly, were the highest. The relative recoveries of all tracers were usually more than 90%. By considering an error of the relative recovery between 4.0-5.4% the tracers could be considered as almost fully recovered.

The migration of CF in the column was slow. The form of CF's BTCs was usually flat, broad and contained a relatively long tailing, compared with an almost „ideal" form of the BTCs of SF, UR and NA. In few cases of CF a second peak appeared. Since it was observed only with CF and not by the simultaneously injected NA and bromide, it was clear that this was not due to a second flow path, but due to a specific characteristic of CF. The original CF is a mixture of the isomers 5-carboxyfluoerscein trisodium salt and 6-carboxyfluoerscein trisodium salt, i.e. the para- and meta- (positioned relative to the xanthene chromophor ring) isomers (Table 1). It was suspected that due to the quantity of adsorption sites, injection mass, contact time and/or other possible influencing factors the isomers are „chromatographic" separated. That is supported by Shiau et al. (1992) who investigated the two-step BTC phenomenon of RWT in column studies. Rhodamine WT has a similar structure to CF and it also comprises of a mixture of para- and meta-isomers. Similarly to RWT, the appearance of a two-step in the BTC is due to the separation of the isomers of the original tracer solution, resulting in faster movement of the para-CF isomer compared with the meta-CF isomer. The superposition of the two theoretical curves show very good compatibility with the BTCs of the original CF and, therefore, supports this explanation. In the field studies conducted in the current research no separation of the isomers was observed, similarly, as much it is known, to field tests with RWT (Jones 1976, Smart & Smith 1976, Smart & Laidlaw 1977, Trudgill 1987). This fact strengthens the expectation that CF's two isomers would not separate in other field experiments. Nevertheless, more experiments with various aquifers should be done in order to investigate further this phenomenon which was obtained in the laboratory. It is recommended that at least until more information is gained about CF, it should be used only in multi-tracer tests.

No relation or at least no simple relation between the tracers' relative recovery, retardation factors and dispersion factors and the experimental variables of injection mass, Darcy flow velocity and flow distance could be observed. Hence, the entire results lead to the conclusion that the sorption behaviour of

SF, NA and UR was similar and better than that of CF.

In *field studies* the sorption behaviour of the new tracers was tested in a porous aquifer in the well known test site Wilerwald in Switzerland, in comparison to UR and the reference tracer NA (Leibundgut et al. 1992). The tests were conducted at a short distance of 12.62 m, in two phases. The first test in each phase was always performed with UR and NA and the second one with NA and either CF or SF. Each experiment took about two weeks.

Naphthionate was used as a reference and was injected simultaneously with each of the tracers, CF, SF and UR, separately. The BTCs obtained were the most suitable for the comparison of NA's BTCs and for the comparison of the sorption behaviour of the xanthene dyes (Figure 3). Naphthionate behaved similarly within each of the phases and, therefore, enables the direct comparison between each of the two new tracers and UR within each phase, as well as an indirect comparison between CF and SF by means of the results for UR.

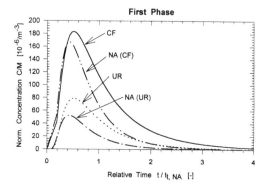

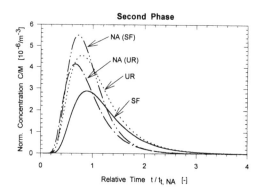

Figure 3. Fitted breakthrough curves of CF, SF, UR and NA in both phases.

Table 3. Tracer's injection mass (M), maximum observed concentration (C_{max}), dispersion factor (P), retardation factor (R_D) and relative recovery (RR*) of CF, SF, UR and NA in both phases of experiments.

	Tracer	M [kg]	C_{max} [mg/m^3]	P [-]	R_D [-]	RR* [%]
All Phases	NA	1.5 (3.0)	varied	1.00	1.00	50
1st Phase	CF	0.07	12.72	1.16	1.36	69
	UR	0.07	5.40	1.08	1.28	96
2nd Phase	UR	0.10	0.45	1.10	1.23	72
	SF	0.60	1.73	1.25	1.34	39

* theoretically calculated (see text)

Uranine, CF and SF were retarded in comparison with NA (Figure 3). The behaviour of the xanthene dye UR compared with NA was supported by the results of Sansoni et al. (1988). Succinylfluorescein disodium salt was retarded in comparison to the other tracers. The dispersion parameters of the xanthene dyes and NA varied in the range of 0.24 to 0.29 for the first phase, and 0.08 to 0.11 for the second phase. The dispersion factor (the ratio of the apparent dispersion parameters of the tracer and the reference tracer) and the retardation factor of NA are equal, per definition, to one. The dispersion factors in both phases for the xanthene dyes varied between 1.08 and 1.25, and their retardation factors varied between 1.23 and 1.36 (Table 3). The tracers' different parameters and factors should be considered similar and their variation range as very narrow, especially because it was obtained in a natural system. Furthermore, within this narrow range, these factors and parameter of the xanthene dyes were higher with SF than with CF and those of UR were the lowest. The relative recoveries of the tracers in the field studies were determined theoretically by assuming the recovery of NA to be 50%. That way a theoretical flow rate was calculated for each experiment and the theoretical relative recovery of CF, SF and UR could be calculated. Two different relative recoveries of UR were obtained in the two tests in which it was investigated, these were 96% and 73% in the first phase and in the second phase, respectively, both compared with 50% recovery of NA. It is, therefore, difficult to determine the relationship between the recoveries of CF and UR. However, it seems that the relative recovery of UR was similar or slightly higher than that of CF. The relative recovery of succinylfluorescein disodium salt was lower even than that of NA. The low recovery of NA compared with CF and UR could have resulted from various processes, such as irreversible adsorption and/or

microbial degradation (Behrens 1986, Leibundgut & Wernli 1986). The retardation of SF was probably due to colloidal precipitation and complexation of SF with divalent metal cations which exist in Wilerwald ground water. Finally, despite the slower migration of the xanthene tracers, i.e. CF, SF and UR, in comparison with NA, the results presented imply that the new tracers, CF and SF, could, as UR and NA, yield the aquifer parameters within an acceptable deviation range. Therefore, they can enable a satisfactory description of the aquifer.

5 CONCLUSIONS

In general, all of the investigated xanthene tracers, succinylfluorescein disodium salt, 5(6)-carboxyfluorescein trisodium salt and uranine show similar behaviour and the new tracers can, therefore, be used as hydrological tracers. The general similarity of their sorption behaviour is apparently based on their common structure. The various functional groups in the xanthene basis caused only limited changes in their behaviour. However, closer considerations of the results show small variations regarding the various influencing factors, among them the grain size distribution, the content of organic material and clay in the matrix, and the pH level and the ion content in the water.

A shortcoming of the new tracers is the fact they fluoresce at the same spectral range as UR and, therefore, they could not be injected simultaneously. However, consideration of the results obtained for CF and SF lead to the conclusion that it would be advantageous to use these tracers for certain applications in which UR does not yield satisfying results. For example, because of the greater stability of CF to light it is more suitable than UR to be applied for tracing surface water. Additionally, it was interesting to observe that in the column SF presented better sorption behaviour (the recovery was lower than UR but still above 90%) than UR, although it appeared with a delay, compared with CF and UR, in the field test. Also, the variable range of the factors of retardation and dispersion in the field imply that they all could yield the aquifer parameters with acceptable deviation range. Hence, SF might behave more conservatively than UR in certain aquifers. As much as a contaminant is similar in certain aspects to the new tracers, its transport in the water can be simulated by the latter. Further experiments are needed in order to increase the knowledge of the new tracers and to fully realize their potential applications. Considering the possible applications and the fact they are environmentally friendly, the new tracers extend the spectrum of artificial tracers.

As noted above, a further important accomplishment of this research was to establish a theoretical foundation for the development of a new conservative tracer. That was done by consolidation of the information regarding the relationship between structure and properties of the investigated tracers. Considering the nearly conservative behaviour of UR, it was expected that the new tracers would show even better behaviour. Hence, it was important to prove and to quantify these expectations through systematic research. The similarity of the tracers behaviour, obtained in this research, may lead to the conclusion that further search for other xanthene dyes is exhausted and that other fields of inquiry should be found. Any new tracers based on the structure of xanthene would probably yield similar behaviour to that of UR. The functional group will cause variations, however, these variations are expected to be limited. Nevertheless, the sulfonated xanthene derivates are theoretically „ideal" tracers and, therefore, could provide an ideal limit for the behaviour of any tracer based on the structure of xanthene. Therefore, it might be interesting to purify the sulfonated xanthene derivates, which were synthesized in this study, and to test their behaviour in the field. Another direction which should be followed involves the use of fluorescent dyes which fluoresce in the spectral range about 580-620 nm. These dyes have a different structure than xanthene and are recommended to be investigated as hydrological tracers. The main advantage of these would be that they could be used in multi-tracer experiments together with xanthene based tracers.

REFERENCES

Bailey, G.W. & J.L. White 1970. Factors influencing the adsorption, desorption and movement of pesticides in soil. *Residue Reviews*, 32: 29-92.

Behrens, H. 1982. Tracerverfahren zur Strömungs- und Abflussmessung. *14. DVWK*, 18-1 - 18-22.

Behrens, H. 1986. Water tracer chemistry-a factor determining performance and analytics of tracers. *Proc. 5th SUWT*, Athens, 121-133.

Behrens, H. & M.T. Zahn 1992. Quantitative investigations on the transport of fluorescein in fluvio-glacial gravels. *Proc. 6th SUWT*, Karlsruhe, 135-138.

Hadi, S. 1997. *New hydrological tracers-synthesis and investigation*. Diss. Thesis, Freiburg Uni., Germany.

Jones, W.K. 1976. Dye tracing in north America; A summary of techniques and results. *Papers 3. SUWT*, Ljubljana, 101-112.

Käss, W 1992. Lehrbuch der Hydrogeologie, Band 9, *Geohydrologische Markierungstechnik*. Gebr. Bornträger, Berlin, Stuttgart.

Karickhoff, S.W., D.S. Brown & T.A. Scott 1979. Sorption of hydrophobic pollutants on natural sediments. *Water Research*, 13: 241-248.

Klotz, D. 1982. Verhalten hydrologischer Tracer in ausgewählten fluvioglazialen Kiesen, Hangschutt-Kiesen und tertiären Kies-Sanden aus Bayern. *Beitr. Geol. Schweiz-Hydrol.*, 28(II): 245-256.

Klotz, D. 1991. Erfahrungen mit Säulenversuchen zur Bestimmung der Schadstoffmigration. GSF-Institut für Hydrologie, Neuherberg, *Bericht 7/91*.

Lang, H. 1982. Zur Bestimmung von Verteilungsgleichgewichten zwischen Lockergesteinen und wässrigen Radionuklid-Lösungen mittels verschiedener Batchverfahren. *DGM*, 26(3): 69-73.

Leibundgut, Ch. & H.R. Wernli 1986. Naphthionate-another fluorescent dye. *Proc. 5th SUWT*, Athens, 167-177.

Leibundgut, Ch., A. de Carvalho Dill, P. Maloszewski & I. Müller 1992. Investigation of solute transport in the porous aquifer of the test site Wilerwald (Switzerland). *Steir. Beitr. Hydrogeol.*, 43: 229-250.

Maloszewski, P. 1991. Bemerkungen über die Interpretation von Markierungsversuchen im Grundwasser. GSF-Institut für Hydrologie, Neuherberg, *Jahresbericht*, 1-18.

Netter, R. & H. Behrens 1992. Application of different water tracers for investigation of constructed wetlands hydraulics. *Proc. 6th SUWT*, Karlsruhe, 125-129.

Sabatini, D.A. & T. A. Austin 1991. Characteristics of rhodamine WT and fluorescein as adsorbing groundwater tracers. *Ground Water*, 29(3): 341-349.

Sansoni, M.; B. Schudel, T. Wagner & Ch. Leibundgut 1988. Aquiferparameterermittlung im Porengrundwasser mittels fluoreszierender Tracer. *Sonderdruck aus Gas, Wasser, Abwasser*, 1988/3 Zürich, 141-147.

Shiau, B. J., D.A. Sabatini & J.H. Harwell 1992. Sorption of rhodamine WT as affected by molecular properties. *Proc. 6th SUWT*, Karlsruhe, 57-64.

Smart, P.L. 1976. The use of optical brighteners for water tracing. *Trans. British Cave Research Assoc.*, 3(2): 62-76.

Smart, P.L. & D.I. Smith 1976. Water tracing in tropical regions, the use of fluorometric techniques in Jamaica. *J. Hydrol.*, 30: 179-195.

Smart, P.L. & I.M.S. Laidlaw 1977. An evaluation of some fluorescent dyes for water tracing. *Water Resources Res.*, 13(1): 15-33.

Trudgill, S.T. 1987. Soil water dye tracing, with special reference to the use of rhodamine WT, lissamine FF and amino G acid. *Hydrol. Processes*, 1: 149-170.

Viriot, M.L. & J.C. André 1989. Fluorescent dyes: A search for new tracers for hydrology. *Analysis*, 17(3): 97-111.

Weinstein, J.N., S. Yoshikami, P. Henkrat, R. Blumenthal & W.A. Hagins 1977. Liposome-cell interaction: Transfer and intracellular release of tapped fluorescent marker. *Science*, 195: 489-492.

Tracer Hydrology 97, Kranjc (ed.)© 1997 Balkema, Rotterdam, ISBN 90 5410 875 4

Can ^{222}Rn be used as a partitioning tracer to detect mineral oil contaminations?

Daniel Hunkeler, Patrick Höhener & Josef Zeyer
Institute of Terrestrial Ecology, Soil Biology, Swiss Federal Institute of Technology, Schlieren, Switzerland

Eduard Hoehn
EAWAG, Swiss Federal Institute for Environmental Science and Technology, Dübendorf, Switzerland

ABSTRACT

^{222}Rn has the potential to be used as a partitioning tracer since (i) it is present at a constant activity in many aquifers and (ii) it is known to partition between nonaqueous phase liquids (NAPLs) like mineral oils and chlorinated hydrocarbons and water. The aim of our study was to investigate the use of ^{222}Rn to detect and quantify light nonaquoeus phase liquid (LNAPL) contaminations in aquifers. ^{222}Rn activities were measured in a sand-and-gravel aquifer of the perialpine belt of central Switzerland which was contaminated with diesel fuel. Upgradient of the contaminated zone, the ^{222}Rn activity was about 9 kBq m^{-3} and decreased to about 5 kBq m^{-3} within the contaminated zone. In wells with a reduced ^{222}Rn activity, dissolved hydrocarbons were detected. This indicated that the decrease of the ^{222}Rn activity was due to the presence of diesel fuel. Laboratory experiments showed that the ratio of the ^{222}Rn activity at steady state in uncontaminated aquifer material to the ^{222}Rn activity in contaminated aquifer material is a linear function of the volumetric oil content. Based on these laboratory experiments, a mineral oil - water partition coefficient of ^{222}Rn was calculated and used to estimate the diesel-fuel content in the aquifer. The calculated diesel-fuel content was in the range of the measured content in aquifer samples. This suggests that ^{222}Rn activity measurements can be used to detect and quantify diesel fuel contaminations and potentially other NAPLs.

INTRODUCTION

Tracers which partition between the aqueous phase and non-aqueous phase liquids (NAPLs) have found increasing attention as a tool to detect subsurface con-taminations. So far, SF$_6$ (Wilson & Mackay, 1995; Nelson & Brusseau, 1996) and various alcohols (Jin et al., 1995) have been used as partitioning tracers to detect NAPL contami-nations. The aim of our study was to investigate whether the naturally occuring ^{222}Rn can be used to detect and quantify such contamination.

The radioactive isotope ^{222}Rn (half-life 3.8 days) is one of the products of the natural radio-active decay series of ^{238}U. The α-particle decay of ^{226}Ra produces ^{222}Rn which decays to a series of short-lived daughter products. As a noble gas, ^{222}Rn is chemically inert. However, ^{222}Rn is known to partition into organic liquids, e.g. toluene (Horiuchi & Murakami, 1981). From surfaces of minerals which contain ^{226}Ra, as found in aquifers, ^{222}Rn emanates to the surrounding gas or aqueous phase by recoil and diffusion (Andrews & Wood, 1972; Rama & Moore, 1984; Krishnaswami & Seidemann, 1988; Maraziotis, 1996). Surface waters usually contain little ^{222}Rn, since ^{222}Rn is released to the atmosphere. When water infiltrates from the surface to the saturated subsurface, the ^{222}Rn activity concentration (denoted as ^{222}Rn activity) in the ground water increases to reach a steady state between emanation and radioactive decay (Hoehn & von Gunten, 1989; Hoehn et al., 1992). The time required to establish a steady state acti-vity is about 20 days. ^{222}Rn is present at a con-stant activity in most aquifers, and is transported without retardation in aquifers which are not con-taminated with NAPLs (Hoehn & von Gunten, 1989). Therefore, it has the potential to be used as a tracer to detect subsurface contaminations of NAPLs such as mineral oil.

We studied the impact of diesel fuel on the ^{222}Rn activities in a contaminated aquifer. The field study included the measurement of ^{222}Rn activity and hydrocarbon concentrations in the ground water. Concomitant laboratory investigations in batch and column systems and a mathematical model describing the relation between ^{222}Rn activity concentrations and diesel fuel saturation in an aquifer are reported elsewhere (Hunkeler et al., 1997).

FIELD SITE

The field site (Figure 1) is located in central Switzerland, on 555 ma.s.l. and represents a typical hydrogeological situation of the perialpine belt of central Europe. The shallow valley-fill aquifer consists of glaciofluvial outwash deposits. Core material showed 3 - 9 m of interbedded layers of poorly sorted sands and gravels underlain by an aquitard which consists of a tightly packed till. The aquifer is covered by a 2 - 3 m thick layer of loamy sediments. The ground water is unconfined and exhibits a thickness of 2 - 8 m. The hydraulic conductivity of the aquifer is between 0.5 and 1 mm s^{-1}. The porosity was assumed to correspond to the that measured at a site of similar geology which was in the average 0.19 (Jussel et al., 1994). The water table is at 3 - 4 m below surface and fluctuates seasonnally by up to 0.4 m at natural gradient conditions. Flow lines point from south to north at a hydraulic gradient of 2 - 3 per cent (average flow velocity: 4.5 to 14 m d^{-1}). The average temperature of the ground water was 12°C. The ground water is carbonate rich (about 6 mM dissolved inorganic carbon).

The aquifer was contaminated in 1988 by 10 to 12 m^3 of diesel fuel from a leaking storage tank of a gas station. After discovery of the contamination, a total of 18 bore-holes were drilled and fitted with PVC piezometer tubes of a diameter of 11.4 cm (4.5 inch), screened in the saturated part of aquifer. Five of the wells (KB1, KB3, KB4, KB10 and KB14) reached the aquitard, all others penetrated the aquifer only partly. Furthermore, the contaminated zone in the vicinity of the tank (5-10 m) was excavated. Floating diesel fuel in free phase was found at a distance of up to 65 m downgradient from the storage tank (Figure 1). The area in which diesel fuel in free phase was detected is referred to as the initially contaminated zone. The vertical extension of the contamination found in cores was between 0.45 and 1.8 m. The hydrocarbon concentrations were between 0.9 and 2.5 mg kg^{-1} (S=1.5 to 4 %) in drilling cores in the initially contaminated zone, between 5.5 and 8.6 mg kg-1 (S=9 to 14 %) on

the northern edge of the excavated zone and up to 72 g kg^{-1} (S=100 %) below the storage tank. The diesel fuel saturation was calculated assuming a porosity of 0.19 and a density of the aquifer material of 2.5 kg L^{-1}. After removal of the diesel fuel in free phase by pumping and excavation of the contaminated unsaturated zone around the tank, an *in situ* bioremediation scheme was implemented between 1989 and 1995 (Hunkeler et al., 1995).

SAMPLE ANALYSIS

^{222}Rn activities were measured with liquid scintillation counting (LSC), except for October 29, 1996. Filling of the sampling bottles, extraction from the water into a scintillation liquid (toluene basis), measurement of the decay of ^{222}Rn and four daughter products with LSC, and the calculation of the ^{222}Rn activity at the time of

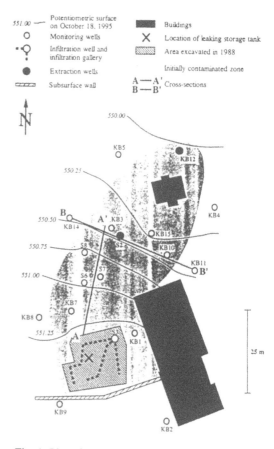

Fig. 1: Plan view of field site with potentiometric surface (ground-water level)

sampling followed the procedure given in Hoehn & von Gunten (1989). LSC instrument signals were converted into ^{222}Rn activities and expressed in kBq m^{-3}. The detection limit of the ^{222}Rn activity measurement was 0.02 kBq m^{-3}. The analytical uncertainty was in the range of ±20% (Hoehn & von Gunten, 1989).

On October 29, 1996, ^{222}Rn activities were measured with a Niton RAD7 instrument with a RAD-H$_2$O attachment (Niton Corporation, Bedford, MA, USA). The RAD-H$_2$O attachment transfers ^{222}Rn from the water sample to the measurement cell of the RAD7 by purging ^{222}Rn out of the sample during 5 min in a closed loop filled with air. According to the manufacturer, the stripping efficiency of the ^{222}Rn is 94% for 250 mL vials. Tests showed that dissolved hydrocarbons at concentrations measured in the field does not influence the efficiency of ^{222}Rn stripping. In the measurement cell of the RAD7, the ^{222}Rn activity was determined after electrostatic collection of alpha emitters on a solid state detector based on the ^{218}Po α-decay. After a secular equilibrium between ^{222}Rn and ^{218}Po was established, the decay of ^{218}Po ($T_{1/2}$=3.05 min) was counted 4 times during 45 min. The detection limit was 0.5 kBq m^{-3} for 250 mL vials. The analytical uncertainty was ±10%. Dissolved hydrocarbons were analysed with capillary GC as described in (Bregnard et al., 1996). The detection limit was at 0.01 mg L^{-1}.

RESULTS AND DISCUSSION

Distribution of ^{222}Rn activities and dissolved hydrocarbons

^{222}Rn activities were measured three times at natural-gradient flow conditions (Table 1). Wells KB2, KB8 and KB9 are located upgradient of the contaminated zone (see Figure 1). Furthermore, a regional flow model indicated that the ground water was more than 20 days in the aquifer before these wells were reached. Therefore, we assumed that the ^{222}Rn activities measured in these wells represent the emanation - decay steady state ^{222}Rn activity of the uncontaminated zone of the aquifer.

Wells S6, S7, S8 and KB7 are located within the initially contaminated zone. Here the ^{222}Rn activity was always at least 20 % smaller than the ^{222}Rn activities in the wells upgradient of the contamination (Table 1). In other wells within the initially contaminated zone (KB3 and KB15), the ^{222}Rn activities were occasionally smaller than in

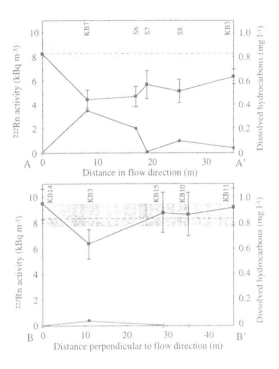

Fig. 2: ^{222}Rn activities (quadrangles; left vertical scale) and dissolved hydrocarbon concentrations (diamonds; right vertical scale; open diamonds: below detection) measured in monitoring wells on 10/18/95 along cross-sections A-A' (a) and B-B' (b) of Figure 1; Broken line: Average ^{222}Rn activity in wells upgradient of contaminated zone; hatched area: range between highest and lowest value

the wells upgradient of the contamination. Note that the extension of the initially contaminated zone does not necessarily correspond to the extension of the contaminated zone at the time of the ^{222}Rn activity measurements. ^{222}Rn activities that were lower than those from wells upgradient of the contaminated zone were only measured in wells within the initially contaminated zone. This indicates that the decrease of the ^{222}Rn activity was due to the diesel fuel contamination. The connection between the ^{222}Rn deficit and diesel fuel contamination was verified by measuring dissolved hydrocarbons on October 18, 1995. (Table 1). Dissolved hydrocarbons were detected only in wells with reduced ^{222}Rn activities. Furthermore, the ^{222}Rn activity was lower in wells with higher dissolved hydrocarbon concentrations. This confirms that the ^{222}Rn deficit is due to the presence of the diesel fuel.

Table 1: Measured ^{222}Rn activities and hydrocarbon concentrations

	^{222}Rn activity 12/9/1994 Bq. l^{-1}	^{222}Rn activity 10/18/95 Bq. l^{-1}	^{222}Rn activity 10/29/96 Bq. l^{-1}	^{222}Rn ratio [a] 12/9/1994 -	^{222}Rn ratio [a] 10/18/95 -	^{222}Rn ratio [a] 10/29/96 -	Dissolved hydrocarbons 10/18/95 mg l^{-1}
upgradient of initially contaminated zone							
KB2	6.0	7.5	9.0	1.00	0.89	0.95	n.d.
KB8	-	9.4	9.8	-	1.11	1.05	n.d.
KB9	6.9	7.8	10.1	1.15	0.93	1.07	-
in initially contaminated zone							
KB7	4.7	4.4	6.1	**0.79**	**0.52**	**0.65**	0.35
S6	3.7	4.7	5.2	**0.61**	**0.56**	**0.55**	0.20
S7	2.2	5.7	6.0	**0.37**	**0.67**	**0.64**	0.01
S8	4.2	5.2	-	**0.71**	**0.61**	-	0.10
KB1	5.7	6.9	7.7	0.95	0.82	0.82	0.21
KB3	5.2	6.4	8.0	0.87	**0.76**	0.85	0.04
KB10	-	8.6	10.9		1.02	1.16	n.d.
KB12	-	9.2	10.1		1.10	1.08	n.d.
KB15	4.5	8.8	9.5	0.75	1.04	1.01	n.d.
outside but not upgradient of initially contaminated zone							
KB5	7.7	10.5	12.2	1.29	1.25	1.30	-
KB11	-	9.2	8.7	-	1.09	0.93	-
KB14	-	9.5	11.4	-	1.13	1.22	n.d.

a)   Ratio of the ^{222}Rn activity to the average of the ^{222}Rn activity in monitoring wells KB2, KB8 and KB9.

bold ^{222}Rn deficit (^{222}Rn activity more than 20 % below the average of KB2, KB8 and KB9)

n.d.   not detected (detection limit dissolved hydrocarbons 0.01 mg l^{-1})

-   not measured

Figure 2 shows the correlation between the ^{222}Rn deficit and hydrocarbon concentrations along a ground-water flow line and in a cross section perpendicular to the flow direction of the ground water.

Estimation of the diesel fuel saturation in the aquifer

The ratio of the emanation-decay steady state ^{222}Rn activity in the uncontaminated zone of the aquifer (S=0) to that in the contaminated zone corresponds (S>0) to the retardation factor, R (Hunkeler et al., 1997):

$$R = \frac{A_e^{S=0}}{A_e^{S>0}} = 1 + S * K \qquad (1)$$

S is the diesel fuel saturation (ratio of volumetric diesel fuel content to total pore space). K is

the dimensionless diesel fuel - water partitioning coefficient for ^{222}Rn which was found in laboratory experiments of Hunkeler et al. (1997) to be 40±2 at 12 °C. The equation relies on the assumption that the partitioning of ^{222}Rn between the diesel fuel and the water phase is in equilibrium. This assumption was tested in laboratory experiments (Hunkeler et al., 1997). Note that Equation 1 is valid only for small diesel fuel saturations (< 5%). Based on Equation 1, the remaining diesel fuel saturation down-gradient of the location of the former tank after the end of the remediation was estimated. An average value of R of 0.61 was taken from Table 1 using the data from wells S6, S7, and S8, on October 10, 1995, and October 29, 1996. Using Equation 1, an average diesel fuel saturation of 1.6 % is obtained. The calculated value is in the same range as what was measured in an aquifer sample, excavated from the contaminated zone downgradient of the former tank in November 1992. In the aquifer material sample, a diesel fuel saturation of 1.9 % was found.

Prerequisites for the assessment of the diesel fuel saturations in aquifers using ^{222}Rn

Our results suggest that ^{222}Rn can be used as a partitioning tracer to detect and to quantify diesel fuel and other NAPL contamination, if the following prerequisites are met (besides the ones given by Hoehn & von Gunten, 1989):

1) The NAPL - water partition coefficient, K, which is used to estimate the NAPL saturation, S, is known.

2) The emanation - decay steady state ^{222}Rn activity, A_e, in uncontaminated ground water is also known.

3) The sampled ground water has a residence time in the contaminated zone of the aquifer of at least 7 days. After this time, the emanation-decay steady state ^{222}Rn activity in the contaminated zone is reached (Hunkeler et al., 1997).

4) The monitoring well used to sample water that has passed through the contaminated zone is within or close to the contaminated zone. After the ground water has left the contaminated zone, the ^{222}Rn activity increases which leads to an underestimation of the actually present diesel fuel saturation.

Compared to an artificially injected partitioning tracer, ^{222}Rn has the advantage to be naturally present in many aquifers. Tracing large groundwater volumes with an artificial tracer is expensive and may require a ground water injection scheme. Additionally, ^{222}Rn is not degradable and can easily be measured on site.

ACKNOWLEDGMENT

The authors thank G. Wyssling (Geologisches Büro Dr. Wyssling, Zürich), A. Stöckli (Baudepartement des Kantons Aargau, Aarau, Switzerland), W. Jucker and Ch. Herzog for cooperation at the field site, A. Häner and H.R. Zweifel for analysing samples and K. Häberli for stimulating discussions. The work was supported by the Swiss National Science Foundation (Priority Programme Environment).

LITERATURE

Andrews, J.N. & Wood, D.F. 1972. Mechanism of radon release in rock matrices and entry into groundwaters, *Transactions of the Institution of Mining and Metallurgy*, 81: 198-209.

Bregnard, T.P.-A., Höhener, P., Häner, A. & Zeyer, J., 1996. Degradation of weathered diesel fuel by microorganisms from a contaminated aquifer in aerobic and anaerobic microcosms, *Environ. Toxicol. and Chem.*, 15,: 299-307.

Hoehn, E. & von Gunten, H.R., 1989. Radon in groundwater: A tool to assess infiltration from surface waters to aquifers, *Water Resour. Res.*, 25: 1795-1803.

Hoehn, E., von Gunten, H.R., Stauffer, F. & Dracos, T., 1992. Radon-222 as a groundwater tracer. A laboratory study, *Environ. Sci. Technol.*, 26: 734-738.

Horiuchi, K. & Murakami, Y., 1981. A new procedure for the determination of radium in water by extraction of radon and application of integral counting with a scintillation counter, *International J. of Appl. Radiation and Isotopes*, 32: 291 - 294.

Hunkeler, D., Höhener, P., Häner, A., Bregnard, T. & Zeyer, J., 1995. Quantification of hydrocarbon mineralization in a diesel fuel contaminated aquifer treated by *in situ*–biorestoration, in *Groundwater Quality: Remediation and Protection, IAHS Publication No. 225*, edited by K. Kovar and J. Krasny: 421-430, IAHS Press, Wallingford, Oxfordshire.

Hunkeler, D., E. Hoehn, P. Höhener, & J. Zeyer, 1997. ^{222}Rn as a partitioning tracer to detect mineral oil contaminations: laboratory experiments and field measurements, submitted to *Environ. Sci. Technol.*.

Jin, M., Delshad, M., Dwarakanath, V., McKinney, D.C., Pope, G.A., Sepehrnoori, K., Tilburg, C.E. & Jackson, R.E., 1995. Partitioning tracer test for detection, estimation and remediation performance assessment of subsurface nonaqueous phase liquids, *Water Resour. Res.*, 31: 1201-1211.

Jussel, P., Stauffer, F. & Dracos, T., 1994. Transport modeling in heterogeneous aquifers .1. Statistical description and numerical generation of gravel deposits, *Water Resour Res*, 30: 1803-1817.

Krishnaswami, S. & Seidemann, D.E., 1988. Comparative study of ^{222}Rn, ^{40}Ar, ^{39}Ar and ^{37}Ar leakage from rocks and minerals: Implications for the role of nanopores in gas transport through natural silicates, *Geochim. Cosmochim. Acta*, 52: 655-658.

Maraziotis, E.A., 1996. Effects of intraparticle porosity on the radon emanation coefficient, *Environ. Sci. Technol.*, 30: 2441-2448.

Mayer, A.S. & Miller, C.T., 1992. The influence of porous medium characteristics and measurement scale on pore-scale distributions of residual nonaqueous-phase liquids, *J. Contam. Hydrol.*, 11: 189-213.

Nelson, N.T. & Brusseau, M.L., 1996. Field study of the partitioning tracer method for detection of dense nonaqueous phase liquid in a tetrachlorethene-contaminated aquifer, *Environ. Sci. Technol.*, 30: 2859-2863.

Rama & Moore, W.S., 1984. Mechanism of transport of U-Th series radioisotopes from solids into ground water, *Geochim. Cosmochim. Acta*, 48: 395-399.

Wilson, R.D. & D.M. Mackay, 1995. Direct detection of residual nonaqueous phase liquid in the saturated zone using SF_6 as a partitioning tracer. *Environ. Sci. Technol.* 29: 1255-1258.

Tracer Hydrology 97, Kranjc (ed.)© 1997 Balkema, Rotterdam, ISBN 90 5410 875 4

A contribution to toxicity of fluorescent tracers

Ch. Leibundgut & S. Hadi
Institute of Hydrology, University of Freiburg im Breisgau, Germany

ABSTRACT: The study summarizes and discusses the methodology and mainly the new human and ecotoxicological information of fluorescent hydrological tracers. Most of the dyes can be considered harmless under normal field applications. It is recommended that, although the usual maximum tracer concentrations of 10 mg/m^3 for ground water and 100 mg/m^3 for surface water are below the levels of toxicological relevance, dye concentrations should be maintained as far below those values as possible.

1 INTRODUCTION

In modern hydrology the tracer technique is one of the most powerful tools employed for solving problems associated with resources and management of natural water as well as with pollutants in water. Among others, tracer studies are widely used to calibrate flow and transport models in both ground water and surface water. The importance of practical work, i.e. field experiments, in hydrology is steadily increasing. It is obvious that in such studies artificial tracers, i.e. chemical substances, have to be injected into the natural water which is being researched.

Growing consciousness for the environment has increased the need for toxicological information of the tracers introduced into natural water as well as their possible metabolic effects. This consciousness has increased the suspicion of the general public and, therefore, of the authorities regarding the use of fluorescent tracers. That could lead to the limitation and even to the prevention of the utilization of artificial tracers, even in cases in which such experiments are necessary and are the best technique available. Furthermore, the research in the field of tracer hydrology could be stopped. Although the toxicity of the dyes used in natural water must be taken into account, it should be considered in the right proportions. It is important to weigh the relation between the potential danger caused by the tracers use and the benefit of the information achieved. It should be realized that, compared with the occurrence of extensive daily contaminations of water, the masses of the tracers used are usually very low (in grams for short distances), due to very low detection limits.

Two aspects of dye toxicity are important in the field of tracer hydrology. First, the possible deleterious effects on aquatic and marine life and second, the limitations which should be considered where human consumption of the labeled water is a possibility. The *human toxicological* characteristics of tracers are of interest when they are applied in water which is used as drinking water, thus mainly in ground water. The tracers' *ecotoxicological* characteristics are of importance when they are applied in surface water, in which they influence aquatic life. However, it should not be forgotten that in some cases ground water interconnects with surface water, for example in springs or along infiltrating streams.

Detailed reviews regarding toxicology of hydrological tracers are given by Smart & Laidlaw (1977), Smart (1982) and Käss (1992). Further investigations concerning the toxicity of tracers have been conducted by the German Environmental Ministry („Umweltbundesamt"). The toxicity of uranine and the two new hydrological tracers, succinylfluorescein disodium salt (SF) and 5(6)-carboxyfluorescein trisodium salt (CF), was investigated with fish (Hadi 1997).

The purpose of this study is to summarize and discuss the methodology and mainly the new human and ecotoxicological information of fluorescent dyes, and to provide tools for evaluation of potential dangers in tracer experiments.

2 METHODOLOGY

The methodology of toxicity tests is very important as it permits comparison and reliable judgment of results from different sources. The standardization of toxicity tests is, therefore, of great importance. However, difficulties arise when, for instance, comparisons of a dye from different suppliers and even different batches from the same supplier are required. This is because the dye may vary in degrees of purity and include different impurities. Moreover, when a tracer is used in the environment, chemical reactions with materials within the water, photochemical and/or biological degradation of the tracer may occur, resulting in decomposition products which may be more toxic than the parent compound. The methods used to measure biochemical changes are often very sensitive but may require an understanding of the mode of toxic action, which is in many cases not available. Furthermore, it is difficult to measure behavioural changes in biological species, e.g. fish, while integrating all biochemical and physiological changes.

In toxicology studies two exposure patterns are acute, referring to a single, often large, dose, and chronic (sub acute), referring to prolonged ingestion through time. In acute studies lethality or pathological change are the main criteria of toxicity in the tested subject. The former is a rather crude but readily identified measure, while the latter is often more sensitive but requires the animals to be sacrificed and examined for pathological abnormalities and weight loss (Smart 1982).

Comparison of the results from toxicity tests can be performed using the LD_{50} (lethal dosis, in mg/kg) and the LC_{50} (lethal concentration, in mg/l). The former is defined as the amount of compound in grams per kg of body weight, which causes the death of half the controlled animals. Whereas the latter is the concentration in solution which causes 50% mortality in the tested species after a specified period of exposure.

In the following the methodologies of human toxicological tests and ecotoxicological tests, are briefly presented.

2.1 Methodology of human toxicological tests

Fluorescent dyes can be tested either in-vitro or in-vivo (applied with living animals). Toxicologists generally agree that a hierarchical approach to the identification of potential carcinogens should be used. Mutagenicity and carcinogenicity of chemicals are generally correlated. Mutagenicity tests are usually used as a simple screening procedure for the carcinogenicity tests, because they are much cheaper

and more easily performed. However, there is not a complete correspondence between the two, nor is the mutagenic response of different organisms consistent (Smart 1982). The use of microorganism tests such as the Ames test can be applied initially for the rapid screening of many chemicals. However, it should be supported by the results of a further test. Two important in-vitro tests to determine the mutagenic activity of substances, indicating the genotoxicological potential, are the Salmonella/microsomes-test and the Cytogenetic analysis in mammalian cells (Chinese hamster cells) (UBA 1996). In-vivo mutagenity tests are carried out with rats, mice or rabbits. Carcinogenicity can be proven using these in-vitro and in-vivo tests.

Direct experiments on human were performed with both fluorescein and optical brighteners. Fluorescein is used in medicine, for example, as a marker for cells, similarly to 5(6)-carboxyfluorescein (Weinstein et al. 1977, Graber et al. 1986). Optical brighteners are used in the detergent industry. Among many other tests, the toxicity of optical brighteners was examined through skin photosensitization, oral exposure (e.g. through introduction of sewage sludge into croplands which may bring optical brighteners into contact with vegetables), exposure through inhalation (e.g. by workers in plants where these are manufactured) and exposure on human skin (e.g. with women who used products containing optical brighteners for washing dishes) (Burg et al. 1977).

2.2 Methodology of ecotoxicological tests

Standardization of ecotoxicological tests is available using the OECD-test guideline. It comprises internationally standardized tests on algae, daphnia and fish. The species are exposed to different substance concentrations in water during various periods of time. The substances are considered ecologically harmless when their PE-value is below one. The PE-value is a quotient of the predicted environmental concentration and the predicted no-effect concentration. Furthermore, the compounds' toxicity are judged on the basis of a 'worst case scenario', i.e. at 100 mg/m^3 and exposure duration of 20 days (UBA 1996). The behaviour of the fish is also considered. Time-to-death toxicity tests are conducted similarly to those described above. Fish are exposed to a particular dye concentration in the water and their behaviour as well as the number of dead fish are noted during the time. Five days of exposure are sufficient when a factor of approximately one thousand times the usually used dye concentration is tested. Tests with eggs and larvae of shellfish are significant in indicating the toxicity to the most sensitive portion of the life cycle. The ecotoxicological test procedure with daphnia magna is conducted in the following manner. Young daphnia magna are exposed to different dye concentrations prepared with artificial sweet water. After 24 hours and 48 hours the number of flea that are unable to swim is evaluated. The pH value and the content of the oxygen is controlled during the experiment.

3 RESULTS

The presented toxicological results for different tracers are subdivided into human toxicology and ecotoxicology.

3.1 Human toxicology

Fluorescein, as noted above, is widely used in medicine, where it is used in very high concentrations, for colouring human cells. Käss (1992) indicates that in ophthalmology fluorescein is applied for vessel tracing, when illness of the chorioid of the optic nerve and the retina appears. In those cases, 5 ml of a 10% (wt/v) solution of fluorescein is intravenously injected. That is despite the appearance, in some cases, of reactions (allergy, difficulties in breathing for a certain time).

The toxicity effect of optical brighteners on humans has been extensively examined, mainly due to their continuos exposure in daily life. A literature review concerning optical brighteners was conducted by Hadi (1993). Because of the lack of application of optical brighteners as hydrological tracers in the last years, the various tests and their detailed results are not included. However, the reviewed studies indicate no detrimental effects even under high concentrations and dosis levels. Therefore, it is concluded, that these compounds are substantially safe in the concentrations used in tracer experiments (Hadi 1993).

The German Environment Ministry carried out two in-vitro tests (i.e. the Salmonella/microsomes-test and the Cytogenetic analysis) for several commonly used hydrological tracers, such as uranine, amidorhodamine G (ARG), sulforhodamine B (SRB), rhodamine B (RB), rhodamine WT (RWT), pyranine, naphthionate, tinopal CBS-X and tinopal ABP. The results were evaluated with respect to their applicability in the environment, where human consumption is a possibility, thus mainly ground water (Table 1). Only RWT, RB and Tinopal ABP showed positive results in those tests, and their applicability for ground water is questionable. However, RB is usually strongly adsorbed and, therefore, it is not used for ground water. The positive result of tinopal ABP might be an exception

from the above mentioned results. Eosine seems to be toxicologically harmless to humans, though, it was not tested in in-vitro tests (UBA 1996).

Different fluorescent tracers have been investigated with in-vivo tests with mammals by several researchers. The LD_{50} of the most applied tracers are presented in Table 2 for different animals. It seems that these dyes are only toxic in very high doses, which far exceed those used in hydrological tracing. Consideration of further studies concerning human toxicological tests supports, in principle, the latest results obtained. A brief summary is given in the following. Uranine is considered harmless (Käss 1967). Many researchers have investigated the toxicity of the rhodamine family, e.g Nestmann & Kowbel (1979), Smart (1982), Behrens (1991) and Hofstraat et al. (1991). It appears that the rhodamine group as a whole is suspected to be toxic. Rhodamine WT is usually considered less toxic than RB. Rhodamine B is generally the most toxic of the tracers investigated, and proved to be a mutagen/carcinogen, its commercial sources contain mutagenic impurities. Rhodamine WT might be mutagenic. It was assumed that SRB and ARG, are also mutagens, however, Table 1 indicates that these dyes showed negative results in the two in-vitro mutagenicity tests.

3.2 Ecotoxicology

Rhodamine dyes were tested with daphnia magna with exposure times of 24 hours and 48 hours and at concentrations up to 100 mg/l (UBA 1996). Table 3 presents the EC_0 and EC_{50} values after 48 hours. The EC value is the concentration in which none, 50% or 100% of the daphnia show inability in swimming when they are exposed to different dye concentrations. The results indicate that daphnia were affected more in the presence of RB and SRB than in the presence of RWT, rhodamine 6G and ARG extra, which have no effect up to a measured concentration of 100 mg/l. It was observed that SRB was more toxic than RB to daphnia. Furthermore, a clear relationship between the tracer concentration and the toxical influence on the daphnia was observed with RB and SRB. The ability to swim decreased with increasing dye concentrations as well with the exposure time. It is, therefore, recommended not to use the latter compounds for tracing surface water.

Toxicity tests on aquatic life were widely performed with fluorescent dyes between the years 1885 and 1979. Smart (1982) claims that even prolonged exposure of aquatic organisms to most tested tracers (e.g. eosine, UR, RWT, RB, SRB, ARG, pyranine and lissamine) did not cause lethality until concentrations well in excess of those commonly used in tracer experiments. Nevertheless, much lower concentrations were found to affect the development of eggs and larvae of shellfish. Quednau (1930) tested the toxicity of UR on fish food such as water-isopods, Amphipoda, pond-snails and dragonfly larvae up to concentrations of 1.2 mg/l. Only the amphipoda felt inconvenient during the first three minutes but then tolerated the medium as the rest of them.

Table 1. Results from in-vitro tests for the genotoxic potential of hydrological tracers and the assessment of their applicability in ground water (UBA 1996).

Tracer	In-vitro Tests		Applicability in ground water
	Salmonella/ microsomes-test	Cytogenetic analysis	
Uranine	negative	negative	ok
Sulforhodamine B	negative	negative	ok
Amidorhodamine G	negative	negative	ok
Rhodamine WT	positive	positive	questionable
Rhodamine B	-----	positive	questionable
Pyranine	negative	negative	ok
Na-Naphtionate	negative	negative	ok
Tinopal CBS-X	negative	negative	ok
Tinopal ABP	-----	positive	questionable

Table 2. LD_{50} values of several hydrological fluorescent dyes (cf. Käss 1992).

Tracer	LD_{50} [mg/kg]	Tested Animal	Reference
Uranine	4,740	mouse, oral	Smart (1982)
	300	mouse, i.v.	Smart (1982)
Eosine	>1,000	rat, oral	Smart (1982)
	550	mouse, i.v.	Lutty (1978)
Rhodamine B	89.5	rat, i.v.	Webb et al. (1961)
	120	rat, i.p.	Rochat et al. (1978)
	95	mouse, i.p.	Rochat et al. (1978)
	890	mouse, oral	Rochat et al. (1978)
Sulforhodamine B	>10,000	rat, oral	Smart (1982)
Amidorhodamine G	>10,000	rat, oral	Smart (1982)
Rhodamine WT	>25,000	rat, oral	Smart (1982)
	430	mouse, i.v.	Lutty (1978)
Pyranine	>5,000	rat, oral	Bayer-Sicherheitsblatt (1983)
	1,050	mouse, i.v.	Lutty (1978)
Amidoflavin FF-PW	8,524	rat, oral	Hoechst- data sheet
Lissamine	8,560	rat, oral	Smart (1982)
	110	mouse, i.v.	Lutty (1978)
Tinopal CBS-X	7,800	mouse, oral	Ciba-Geigy - data sheet
Triazine-stilbene	7,000	mouse, oral	Akamatsu & Matuso (1973)
	>10,000	rat, oral	Gloxhuber et al. (1962)

*i.v. is intravenous injection; i.p.is intraperitoneal injection

Table 3. Tracer concentrations, EC_0 and EC_{50}, in which none or 50 % of the daphnia were unable to swim after exposure time of 48 hours (UBA 1996).

Tracer	EC_0 (48 h) [mg/l]	EC_{50} (48 h) [mg/l])
RB	16.2	33
SRB	0.16	0.7
RWT	≥100	>100
Rhodamine 6G	≥100	>100
ARG extra	≥100	>100

The first research concerning the toxicity of UR to fish was published already in 1885 by Rupp. Rupp tested the toxicity of UR to tench at concentrations between 0.05 and 4 mg/l for eight days. The fish were fed during the test with termite eggs. Under these conditions, the fish moved normally in the green coloured solution and no toxic influence could be seen. In 1929 Czensny (in Egger 1929) claimed that trout tolerated 50 mg/l of UR despite a partial appearance of dye on their back. Quednau (1930) performed toxicity tests with crucians, pungitius pungitius, tench and anguilla anguilla with UR up to concentrations of 1.2 mg/l. Difficulty in breathing was seen only in the beginning but no reactions or changes in tissue could be observed. In 1957 it was found that UR and eosine were not toxic to trout and roaches up to concentrations of 100 mg/l (Bandt 1957). Käss (1976) examined the reaction of trout to UR up to a maximum concentration of 5 mg/l in a stream for 10 hours. The author claimed that despite the sensitivity of trout to impurities in water they tolerated UR. Leibundgut (1973) observed the behaviour of trout during an infiltration experiment into a stream with ARG extra. The average concentration was 5 mg/m^3 and the highest concentrations were more than 10 g/m^3, however, only for a very short time. With the appearance of the tracer cloud, the fish felt inconvenient apparently due to the sudden change of environment. Other abnormalities could not be observed and ARG extra was not found in the bodies of a few sampled fish.

Table 4 summarizes the lethal concentration, LC_{50}, for fish at different exposure times given in the literature. It can be concluded that uranine can be considered to be non toxic to fish. Similar to the observation of human toxicology, the rhodaimes (excluding RWT) are suspected to be more toxic than the other tracers presented. It is apparent that RB is more toxic to aquatic organisms than both RWT and fluorescein, probably because it is readily adsorbed on living tissue owing to its cationic nature (Little & Lamb 1973).

The toxicity of UR to fish in the presence of a fish toxicant was examined by Sowards (1958) and Marking (1969). Sowards (1958) observed that visible concentrations of this dye (no specific concentration was given) did not affect the toxicity of the fish toxicant, Pronoxfish, at 0.053 μg/l to longnose dace (Rhinicthys cataractae). Furthermore, Marking (1969) observed that the toxicity of the fish toxicant, antimycin, at a concentration of 0.05μg/l to rainbow trout for 96 hours was slightly decreased in the presence of 100 mg/l UR.

One of the latest toxicity tests, time to death, was conducted by Hadi (1997). In this study, the toxicity of new hydrological xanthene tracers, succinylfluorescein disodium salt and 5(6)-carboxyfluorescein trisodium salt, was tested in comparison with that of UR using goldfish (carassius auratus) and crucians (carassius valagaris) for 120 hours (five days), at 18°C and a pH of approx. 7.4. Twelve fish (six of each sort) were put into each of three aquariums. Two aquariums were filled with an identical tracer solution in order to enable reproducibility. The third aquarium was filled with water and used for control. The volume of each aquarium was 25 liters. The concentration of each tracer solution was 100 mg/l (100,000 mg/m^3). The water used for this experiment was ground water. The concentration of the dye in the aquariums as well as the temperature, the pH and the conductivity was measured during the test. Oxygen was supplied to the fish during the experiment but no food was given.

The constant conditions (concentration, temperature, pH and conductivity) held in the aquariums during these tests enable the comparison between these tracers. At the concentration tested, the solution's colour of CF and UR was green and was translucent. In these cases the fish behaved normally, like the fish in the control aquarium (without a dye) and showed neither disturbance nor discomfort. The solution colour in the test with SF was dark green and it was barely translucent. The fish showed, therefore, a little discomfort during the first day. They moved more slowly in comparison to the fish in the control aquarium. Nevertheless, they tolerated the dye. None of the fish died during these tests. After the experiments, the fish were put into fresh water aquariums and fed. Moreover, more than three months after the experiment, no death or disturbance, from all tests done could be seen in the fish. It can be, therefore, concluded that CF, SF and UR are not toxic to goldfish and crucians even at such high concentrations. Furthermore, it is expected that the new tracers will react similarly to UR in other toxicity tests.

Table 4. Summary of the lethal concentration, LC$_{50}$, for fish at different exposure times (cf. Käss 1992)

Tracer	LC$_{50}$ (t) [mg/l]	t [h]	Test-Species	Reference
Uranine	1372	96	rainbow trout	Marking (1969)
	2267	96	channel catfish	Marking (1969)
	3433	96	bluegill	Marking (1969)
	>752	24	bluegill	Benoit-Guyod (1979)
	>100	120	goldfish	Hadi (1997)
	>100	120	crucians	Hadi (1997)
CF	>100	120	goldfish	Hadi (1997)
	>100	120	crucians	Hadi (1997)
SF	>100	120	goldfish	Hadi (1997)
	>100	120	crucians	Hadi (1997)
Eosine	>138	24	bluegill	Benoit-Guyod (1979)
Rhodamine B	217	96	rainbow trout	Marking (1969)
	526	96	channel catfish	Marking (1969)
	86	24	bluegill	Benoit-Guyod (1979)
Sulforhodamine B	100-500	48	trout	Hoechst-data sheet
	60-120	24	bluegill	Benoit-Guyod (1979)
Amidorhodamine G	>88	24	bluegill	Benoit-Guyod (1979)
Rhodamine WT	>1360	24	bluegill	Benoit-Guyod (1979)
Rhodamine 6G	0.5	24	bluegill	Benoit-Guyod (1979)
Pyranine	>500	96	fish (no data)	Bayer- data sheet
Amidoflavin FF-PW	>500	96	guppy	Hoechst-data sheet
3-Triazine-stilbene	108-1780	96	rainbow trout	Keplinger et al. (1974)
	86-1060	96	channel catfish	Keplinger et al. (1974)
Tinopal CBS-X	130	96	rainbow trout	Keplinger et al. (1974)
	126	96	channel catfish	Keplinger et al. (1974)
Triazine-stilbene	2000	96	goldfish	Akamatsu & Matuso (1973)

4 CONCLUSIONS

This study increases the information gained in the last few years. Consideration of the results obtained for the different fluorescent hydrological tracers leads to the conclusion that most of the tracers can be considered harmless in the concentrations usually found in tracer experiments. Under normal field applications these concentrations are transient because of rapid dilution following injection. Nevertheless, the use of the rhodamine group should be limited and restricted to exceptional cases. Although the usual maximum tracer concentrations of 10 mg/m^3 for ground water and of 100 mg/m^3 for surface water are below levels of toxicological relevance, it is recommended that for hydrological aims dye concentrations should be maintained as far below those values as possible. Additionally, the hydrologists should minimize the mass of tracer used in a tracer experiment as much as possible. The needed mass can be calculated with known equations (e.g. Leibundgut & Wernli 1982), under the consideration of the very low detection limit of the dyes. It is further important that artificial tracer experiments be performed only by tracer hydrology specialists and institutions, where modern measurement techniques are available.

Finally, the toxicity of dyes used in natural water must be taken into account and should be considered in the right proportions. It is important to weigh the relation between the potential danger caused by the tracer used, which are seen to be low, and the gain of the information achieved from the application of tracers in hydrological research.

REFERENCES

Akamatsu, K. & M. Matsuo 1973. Safety of optical whitening agents. *Senryo To Yakuhin*, 18(2): 2-11 (English translation. *Translation Programme* RTS 9415, Brit. Libr., Boston SPA, Yorkshire, England, June 1975).

Bandt, J.H. 1957. Giftig oder Ungiftig für Fische. *Deut. Fischerei Zeitung*, 4(6): 171-179.

Behrens, H. 1991. Fluoreszenzfarbstoffe-Eigenschaften und Anwendbarkeit. *9. DVWK*. 3: 1-16.

Benoit-Guyod, J.L., J. Rochat, J. Alary, C. Andre & G. Taillandier 1979. Correlations between physiochemical properties and eco-toxicity of fluorescent xanthenic water tracers. *Toxicol. Eur. Res.*, 2(5): 241-246.

Burg, A.W., M.W. Rohovsky & C.J. Kensler 1977. Current status of human safety and environmental aspects of fluorescent whitening agents used in detergents in the US. *Crit. Env. Control*, 7: 91-120.

Egger, F. 1929. Arbeitserfahrungen beim Nachweis des Zusammenhanges von Wasservorkommen durch Fluoreszeineinfärbung. *Vom Wasser*, 3: 22-33.

Gloxhuber, Chr., G. Hecht & G. Kimmerle 1962. Toxikologische Untersuchungen mit Aufhellern (Blankophor[R] -Typen). *Arch. Toxikol.*, 19: 302-312.

Graber, M.L., D.C. Dilillo, B.L. Friedman and E. Pastoriza-Muñoz 1986. Charactistics of fluoroprobes for measuring intercellular pH. *Anal. Biochem.*, 156: 202-212.

Hadi, S. 1993. *Summary of physical-chemical characteristics of fluorescent tracers and other organic solutes*. Unpublished.

Hadi, S. 1997. *New hydrological tracers-synthesis and investigation*. Diss. Thesis, Freiburg Uni., Germany.

Hofstraat, J.W., M. Steendijk, G. Vriezekolk, W. Schreurs, G.J.A.A. Broer & N. Wijnstok 1991. Determination of rhodamine WT in surface water by solid-phase extraction and HPLC with fluorescence detection. *Wat. Res.*, 25(7): 883-890.

Käss, W. 1967. Erfahrungen mit Uranin bei Färbversuchen. *Steir. Beitr. Hydrogeol.*, Jahrgang 1966-1967, 123-134, Graz.

Käss, W. 1976. 100 Jahre Uranin. *Papers 3. SUWT*, 113-122, Ljubljana.

Käss, W 1992. Lehrbuch der Hydrogeologie, Band 9, *Geohydrologische Markierungstechnik*. Gebr. Bornträger, Berlin, Stuttgart.

Keplinger, M.L., O.E. Fancher, F.L. Lyman & J.C. Calandra 1974. Toxicological studies of four fluorescent whitening agents. *Toxicol. Appl. Pharmacol.*, 27: 494-506.

Leibundgut, Ch. 1973. *Anwendung von Markierstoffen in der Hydrologie*. Phil. Nat. Fak., Uni. Bern, Switzerland.

Leibundgut, Ch. & H.R. Wernli 1982. Zur Frage der Einspeisemengenberechnung für Fluoreszenztracer. *Beitr. Geol. Schweiz-Hydrol.*, 28(I): 119-130.

Little, L.W. & J.C. Lamb 1973. Acute toxicity of 46 selected dyes to Fathead minnow (Pimephales promelas). *Dyes and the Environmen*, Vol. 1, chap. 5, American dye manufacturers institute, New York.

Lutty, G.A. 1978. The acute intravenous toxicity of biological stains, dyes and other fluorescent substances. *Toxicol. Appl. Pharmacol.*, 44: 225-249.

Marking, L.L. 1969. Toxicity of rhodamine B and fluorescein sodium to fish and their compatibility with antimycin A. *Prog. Fish Cult.*, 31: 139-142.

Nestmann, E.R., D.J. Kowbel & J. Ellenton 1980. Mutagenicity in Salmonella of fluorescent dye tablets used in water tracing. *Water Res.*,14 (7): 901-902.

Quednau, W. 1930. Die Unschädlichkeit des Farbstoffes Uranin-Kali für Fische und Fischnahrung und seine Bedeutung im landwirtschaftlichen Betrieb. *Mitteilungen der Deutschen Fischereivereine für die Provinzen Brandenburg Ostpreußen, Pommern, Oberschlesien und für die Grenzmark Posen-Westpreußen*, 34: 72-74.

Rochat, J., P. Demenge & J.-C. Rerat 1978. Contribution à l'étude toxicologique d'um trasceur fluorescent: la Rhodamine B. *J. Hydrol.*, 26: 277-293.

Rupp 1885. Die Verwendung von Uranin und Fluorescein zum Färben von Wässern betreffend. Ber. des Chem. Laboratoriums des Techn. Hochschule Karlsruhe vom 29. August 1885. Badisches Generallandesarchiv Karlsruhe, 237/39752.

Smart, P.L. & I.M.S. Laidlaw 1977. An evaluation of some fluorescent dyes for water tracing. *Water Resources Res.*, 13(1): 15-33.

Smart, P.L. 1982. A review of the toxicity of 12 fluorescent dyes used for water tracing. *Beitr. Geol. Schweiz-Hydrol.*, 28(I): 101-112.

Sowards, C.L. 1958. Sodium Fluorescein and the toxicity of Pronoxfish. *Prog. Fish Cult.*, 20(1): 20.

UBA (Umweltbundesamt-Berlin) 1996. *Toxicological tests for fluorescent tracers*. Unpublished.

Webb, J.M., W.H. Hansen, A. Desmond & O.G. Fitzhugh 1961. Biochemical and toxicological studies of rhodamine B and diaminofluoran. *Toxicol. Appl. Pharmacol.*, 3(6): 696-706.

Weinstein, J.N., S. Yoshikami, P. Henkrat, R. Blumenthal & W.A. Hagins 1977. Liposome-cell interaction: Transfer and intracellular release of tapped fluorescent marker. *Science*, 195: 489-492.

Tracer Hydrology 97, Kranjc (ed.) © 1997 Balkema, Rotterdam, ISBN 90 5410 875 4

Sorption of Uranine and Eosine on an aquifer material containing high organic carbon

C. Mikulla, F. Einsiedl, Ch. Schlumprecht & S. Wohnlich
Department of Hydrogeology, Institute of General and Applied Geology, University of Munich, Germany

ABSTRACT: Eosine and especially Uranine usually are used in tracer experiments as conservative tracers. In many tracer tests they however have shown significant retardation. Starting from a negative tracer test in a bank infiltration test, batch partition experiments and column tests were performed for a aquifer material containing a high content of organic carbon (31.2 %). The *FREUNDLICH*-isotherms for Uranine and Eosine, done by batch experiments, show sorption coefficients from 10.2 (Uranine) and 102.3 (Eosine). The column experiments were carried out in a stainless steel-teflon-brass permeameter system. Breakthrough curves were obtained and the retardation factors R_f were evaluated by curve fitting. R_f ranged from 10 - 39.3 for Uranine and 40 - 214 for Eosine.

1 INTRODUCTION

The most frequently used fluorescent dyes in groundwater tracer tests are Uranine and Eosine (BEHRENS & TEICHMANN 1988). Normally Uranine is considered to have low sorption rates and therefore is oftenly used as a reference dye. (BEHRENS & TEICHMANN 1988).

In several studies the sorption of various fluorescent dyes like Uranine, Pyranine, Rhodamine WT, Sulpho Rhodamine B, Lissamine FF etc. is discussed (SMART & LAIDLAW 1977, OMOTI & WILD 1979, TRUDGILL 1987, SABATINI & AUSTIN 1991).

Because of this Uranine and Eosine were used in a tracer experiment at the lake "Obinger See", 60 km East of Munich. The aim of the test was to prove the bank infiltration of lake water into the groundwater. Laboratory tests were executed subsequently in order to obtain characteristics for the adsorption of the fluorescent dyes under saturated flow conditions through the reed zone bank sediment with a high content of organic matter.

2 BANK INFILTRATION OF THE OBINGER SEE

2.1 Groundwater monitoring network

The "Obinger See" is a postglacial lake in the northern alpine foreland of Bavaria, Southern Germany. It originates from the smelting of a block of glazial ice, which was surrounded and partly covered by glacial gravel. It is surrounded by a reed zone, especially at the outflow area in the East. The regional groundwater level was known to be significantly lower than lake water level, but in addition a near to the surface aquifer was indicated by shallow groundwater levels in the vincinity of the in- and outflow.

Therefore investigations were started with the installation of two groundwater monitoring networks in the East and the West of the lake (fig. 1). The monitoring network at the outflow of the lake includes 7 shallow groundwater monitoring wells (RP4 - RP10) which were designed to monitor the storage within the bank zone and 3 surface water measuring points (LP1 - LP3). The casing diameter of the groundwater monitoring wells varies from 0.5 to 2 inches.

2.2 Hydrogeology

The structure of the shallow underground can be shown from the geological profiles of the wells in the Eastern area which is given in fig. 1. The structure of the area is geologically rather heterogenous, consisting out of various layers of peat, gravel and lacustrine clay. The thickness of the layers as well as their horizontal distribution varies significantly. The piecometric head fluctuates between few centimeters up to 2.5 m below ground surface. During the whole time of observation (31st May to 12th December

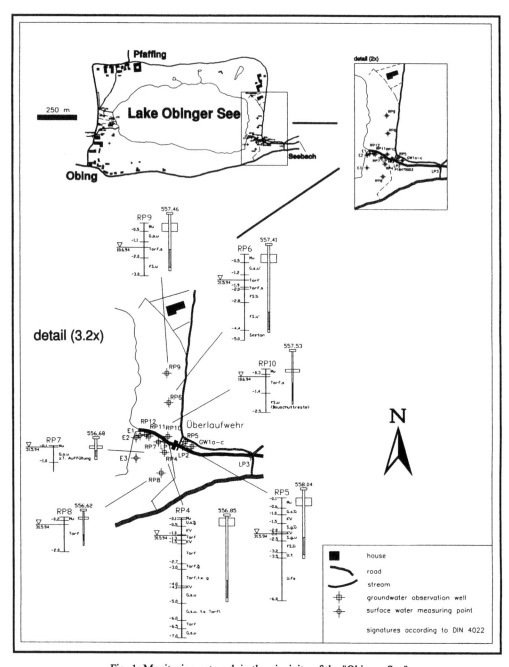

Fig. 1: Monitoring network in the vincinity of the "Obinger See"

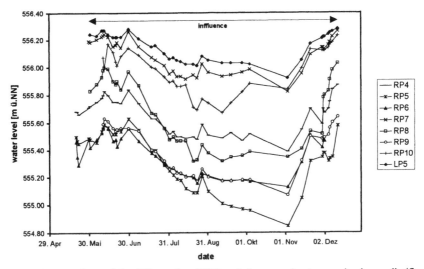

Fig. 2: Water level fluctuations of the Obinger See (LP5) and the groundwater monitoring wells (fig. 1)

1994) the hydraulic gradient indicates an infiltration of surface water into the groundwater. The water level fluctuation of all the observation points are more or less parallel (fig. 2).

2.3 Experimental setup

The area was selected for a tracer test, because it showed constant infiltration into the groundwater. The upstream inflow area on the contrary was characterized by a change of inffluent and effluent conditions.

The amount of tracer used was calculated according to LEIBUNDGUT & WERNLI (1982) with the aquifer cube method. The fluorescent dyes Uranine (0.4 kg) and Eosine (1.6 kg) were dispersed in water and injected into three specially designed injection probes (diameter 5/4 inches) in the bank area of the lake (E1 - E3). The injection into 3 different wells was done in order to distribute the tracer over a larger area (fig. 1). The injection wells also had a special outer casing which was designed to prevent boundary flow and direct infiltration of surface water. Within all the monitoring wells RP4 to RP10 sampling hoses were installed, in order to prevent the contamination of the each other. Sampling was done from downstream to upstream and in intervalls of 72 hours. On the occasion of tracer breakthrough, this intervall was to be shortened to 24 hours.

2.4 Result of the field tracer test

No fluorescent dye could be detected in eather of the monitoring wells and surface water sample locations (RP4 - RP10, LP 1 - LP3). The observation time was 201 days (18.8.94 - 6.3.95). The tracer test was abandoned, when a rising lake water level flooded large areas of the monitoring area in March 1995.

On 11th Oct. and 1st Nov. 1994 two additional monitoring wells were installed 10 m (RP 12) and 15 m (RP11) downstream of E1. In both wells the tracers could be detected, but the maximum concentration already had passed (fig. 3 and fig. 4) only show the decreasing part of the breakthrough curves.

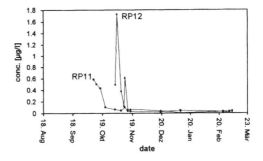

Fig. 3: Uranine distribution in the monitoring wells RP11 and RP 12, which were installed subsequently

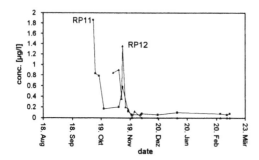

Fig. 4: Eosine distribution in the monitoring wells RP11 and RP 12, which were installed subsequently

The results of the tracer test show a very slow downstream distribution of the tracer, very low concentration of the tracer dyes in comparison to the input and a significant tailing of the breakthrough curves. These factors indicate adsorption processes during the transport of the tracers. It had to be suspected, that the possible adsorption was to be seen in connection with the relative high organic carbon contents ($C_{org.}$) of the sediment in the bank zone.

3 LABORATORY TESTS

3.1 *Soil characteristics*

Undisturbed soil samples were taken from the reed zone of the lake. The soil analyses done and their results are given in table 1. The mineralogical determination of the matrix was done with the particle fraction < 2 μm, which was separated by using the Atterberg sedimentation process (DIN 18123). The particle fraction < 2mm was used for the investigations concerning the organic carbon contents. The permeability tests were using the complete particle fraction.

Tab. 1: Results of the soil investigations

characteristic	result	n
mineralogy	Quarz, Illit, Clinochlor	2
total carbon (mass %)	35.3 ± 0.57	3
organic carbon (mass %)	31.23 ± 2.26	2
saturated hydraulic permeability K (m/s)	$2.65 \cdot 10^{-6} \pm 1.73 \cdot 10^{-6}$	5

3.2 *Sorption isotherms*

The principles on the execution of batch tests is found in a large number of publications (KLOTZ & Lang 1992, BAUMANN et al. 1994)

For statistic assurance 6 adsorption tests were done for each concentration and included also 3 blind tests (without sediment). The ratio between solid/liquid phase was 1:2 for Uranin and 1:50 for Eosine. The samples were turned around for 24 hours at a roomtemperature of 18 °C (KUKOWSKI & BRÜMMER 1987), centrifugated and subsequently Uranine and Eosine were analyzed. All the tests were done in a dark room in order to avoid the photochemical decay of the fluorescent dyes.

For the sediment containing a high $C_{org.}$ 6 different concentrations were applied. A significant sorption was determined for Uranine as well as for Eosine. The concentrations used were not high enough to load all the available particle surfaces. It can be assumed that the surface of the sediment is equal to the sorbing surface, having different adsorption energy at different adsorption locations (SONTHEIMER et al. 1985), the results were analyzed using the isotherm model by *FREUNDLICH*. The different parameters of the Freundlich isotherms as well as the coefficients of regression are listed in table 2 and figure 5. From the *FREUNDLICH* constants a clear adsorption of the two dyes under investigation can be shown. Due to the different solid/liquid ratio of the dyes, a comparison of the *FREUNDLICH* constants can be done only be estimated. As a result a significantly higher sorption (about 10 times) of Eosine to the sediment can be started.

3.3 *Transport parameters under laboratory conditions*

Figure 6 demonstrates the experimental setup of the column apparatus used. In order to avoid to adsorption to the walls of the test cells as well as photochemical degradation of the fluorescent dyes, all technical parts were constructed from stainless steal/brass/teflon/plexi-glass and the experiments were executed in a darkroom.

The column tests were run with a constant hydraulic head. The tracer input was achieved by pulse injection. The injection is situated directly in front of the column, where the tracer is distributed as a thin film across the whole column section.

In addition to Uranin and Eosin Bromide was used as a reference tracer. The desturbed soil samples were placed into tubes of plexi glass (diameter 0.1 m;

Tab. 2: *FREUNDLICH* constants K, *FREUNDLICH* exponents n and coefficients of regression for the sorption of Uranine and Eosine at a sediment with high $C_{org.}$

	Equilibrium constant according to *FREUNDLICH* $\log K \pm T$ (p = 0.95)	Equilibrium constant according to *FREUNDLICH* K	Slope $n \pm T$ (p = 0.95)	correlation coefficient r (n = 6)
Uranine	1.01 ± 0.04	10.2	0.93 ± 0.03	0.995
Eosine	2.01 ± 0.03	102.3	0.88 ± 0.05	0.986

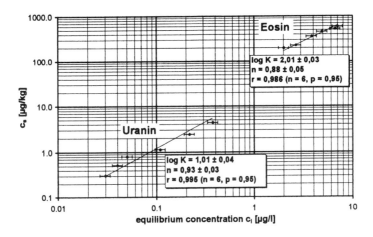

Fig. 5: *FREUNDLICH* isotherms for the sorption of Uranine and Eosine to sediments containing high organic carbon

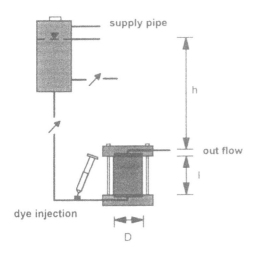

Fig. 6: Experimental setup of the the column tests

length 0.15 m). The wet soil samples were compacted with the help of a round stem in several thin lifts. This method can result into a horizontal layering. Subsequent to the installation of the soil, the columns were exposed to one week of flow with a hydraulic gradient of 7.5 (test 1 and 2) respectively 5 (test 3 and 4). The temperature of the deareated water was 18 °C. The water used had a similar mineralization as the groundwater in the field (calcium-magnesium-hydrogen-carbonate type). In order to estimate the influence of desolved organic carbon (DOC) on the detection of fluorescent dyes, the DOC contents of the effluent was analyzed prior to the application of the tracer (tab. 3). In tracer test 4 humic acid (Supplier: ROTH) was added in a concentration of 58.7 mg/l. No disturbances of the detection of the dyes were observed. The mean flow velocities were calculated from the focus of the breakthrough curves by the help of the computer program FIELD (MALOSZEWSKI 1990). Because of the unstable breakthrough curve for Eosine it was not possible to adapt a best-fit. All the results of the

Tab. 3: Experimental parameters for the column tests

Experiment		1	2	3	4
experimental time	[d]	3,4	3,1	47,9	2,3
hydraulic gradient i	[-]	7,5	7,5	5,0	5,0
saturated hydraulic permeability K	[m/s]	$2,72 \cdot 10^{-6}$	$1,1 \cdot 10^{-6}$	$1,68 \cdot 10^{-6}$	$2,2 \cdot 10^{-6}$
effective porosity n_{eff}	[-]	0,74	0,75	0,38 (1) 0,56 (2)	0,83
quanitity Uranine	[µg]	7,7	7,7	7,7	7,7
quanitity Eosine	[µg]	69	69	69	69
quanitity Bromide	[mg]	5,05	5,05	5,05	5,05
DOC	[mg/l]	1,21	2,5	0,65	58,7

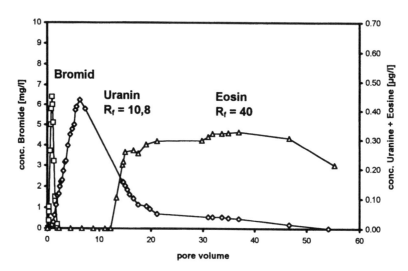

Fig. 7:   Breakthrough curves of Uranine, Eosine and Bromide for an aquifer material with high contents of C_{org}.

Tab. 4: Results of column tests

Experiment		1	2	3	4
v_a Bromide	[m/s]	$2.8 \cdot 10^{-5}$	$1.1 \cdot 10^{-5}$	$1.3 \cdot 10^{-5}$ (1) $1.1 \cdot 10^{-5}$ (2)	$1.3 \cdot 10^{-5}$
v_a Uranine	[m/s]	$2.6 \cdot 10^{-6}$	$1.1 \cdot 10^{-6}$	$5 \cdot 10^{-7}$ (1) $2.8 \cdot 10^{-7}$ (2)	$1.3 \cdot 10^{-6}$
v_a Eosine	[m/s]	$6.8 \cdot 10^{-7}$	$\ll 5.7 \cdot 10^{-7}$ (Eosin n.d.)	$1.9 \cdot 10^{-7}$ (1) $7 \cdot 10^{-8}$ (2)	$\ll 7.5 \cdot 10^{-7}$ (Eosin n.d.)
R_f Uranine	[-]	10.8	10.0	26.0 (1) 39.3 (2)	
R_f Eosine	[-]	40	$\gg 19.3$ (Eosin n.d.)	116 (1) 214 (2)	$\gg 17.3$ (Eosin n.d.)
α_l Bromide	[m]	0.0094	n.c.	n.c.	0.00944
α_l Uranine	[m]	0.030	0.021	n.c.	0.0086
recovery Bromide	[%]	87	86	72.6	81.9
recovery Uranine	[%]	47.5	37.6 (*)	106.5	26.6
recovery Eosine	[%]	15 (*)	n.c.	34 (*)	n.c.

n.d.: not detected, n.c.: not calculated, (*): recovery until end of monitoring, (1): parameter 1st maxima; (2) parameter 2nd maxima column tests are given in table 4. The retardation factor were calculated in reference to the mean flow velocity v_a of Bromide. Since also Bromide had no total recovery of the tracer, it can not be calculated that Bromid also shows a low sorption of the soil material used in the tests. In this case the retardotion facctors of Uranine and Eosine would be even higher. In contrary of the tests No. 1,2 and 4 shows a double peak tracer breakthrough curve for all tracers applied. In this case it has to be assumed, that preferential flow paths developed within the column, causing a double porosity system.

4 RESULTS

Both fluorescent dyes Uranine and Eosine indicate a clear retardation compared with Bromide. In all tests the retardation of Eosine was higher compared with Uranine. The retardation coefficiant R_f for Eosine was determined between 40 and 214, whereas the R_f values for Uranine varied between 10 and 39.9. Due to the disturbed placement of the materials it was not possible to correlate the retardation with other parameters like mean travel time. The *FREUNDLICH* constants from the batch tests were not directly comparable to the retardation values. These show however the same tendency as the R_f values.

REFERENCES

BAUMANN, T., MIKULLA, C., ZEYN, A., NIESSNER, R. (1994): *Hydrogeologische und hydrochemische Untersuchungen im Einflußbereich einer Hausmülldeponie.* Teil 2: Transport von LHKW und PAK in mineralischen Abdichtungen. - Vom Wasser, Bd. 82, 145-162, Weinheim (VCH)

BEHRENS, H., TEICHMANN, G. (1988): *Lichtempfindlichkeit von Pyranin, Uranin und Eosin in Abhängigkeit vom pH-Wert.* - GSF-Jahresbericht 1988: 235 - 242; Neuherberg bei München

DIN 18123: *Bestimmung der Korngrößenverteilung.* - Berlin/Köln (Beuth)

DIN 4022: *Benennen und Beschreiben von Boden und Fels.* - Berlin/Köln (Beuth)

KLOTZ, D., LANG, H. (1992): *Experimentelle Untersuchungen zur Migration ausgewählter Radionuklide im Deckgebirge des Endlagerortes Gorleben, Untersuchungsprogramm IV.* - GSF-Bericht, 20/92: 164 S., Neuherberg

KUKOWSKI, H., BRÜMMER, G. (1987): *Untersuchungen zur Ad- und Desorption von ausgewählten Chemikalien in Böden.* - Umweltforschungsplan des Bundesministers für Umwelt, Naturschutz und Reaktorsicherheit; Forschungsbericht 106 02 045/II Teil 2, Kiel

LEIBUNDGUT, CH., WERNLI, H.R. (1982): *Zur Frage der Einspeisemengenberechnung für Floureszenztracer.* - Beiträge zur Geologie der Schweiz - Hydrogeologie, 28 (1): 119-130, Bern

MALOSZEWSKI, P. (1990): Computerprogramm „Field"

OMOTI, U., WILD, A. (1979): *Use of Fluorescent Dyes to Mark the Pathways of Solute Movement through Soils under Leaching Conditions: 1. Laboratory Experiments.* - Soil Science, 128 (1), 28-33

SABATINI, D.A., AUSTIN, T. A. (1991): *Characteristics of Rhodamine WT and Fluorescein as Adsorbing Ground-Water Tracers.* - Ground Water 29 (3), 341-349

SMART, P.L., LAIDLAW, I.M.S. (1977): *An Evaluation of Some Fluorescent Dyes for Water Tracing.* - Water Resources Research, 13 (1), 15-33

SONTHEIMER, H., FRICK, B.R., FETTIG, J., HÖRNER, G., HUBELE, C., ZIMMER, G. (1985): *Adsorptionsverfahren zur Wasserreinigung.* - DVGW-Forschungsstelle am Engler-Bunte Institut der Uni. Karlsruhe: 124-312

TRUDGILL, S.T. (1987): *Soil Water Dye Tracing, with Special Reference to the Use of Rhodamine WT, Lissamine FF and Amino G Acid.* - Hydrogeological Processes, 1: 149-170

Tracer Hydrology 97, Kranjc (ed.) © 1997 Balkema, Rotterdam, ISBN 90 5410 875 4

^{222}Rn concentration variations owing to dynamic condition changes induced in the waters of the Apulian karst aquifer (Southern Italy)

T. Tadolini, M. Spizzico & D. Sciannamblo
Politecnico di Bari, Italy

ABSTRACT

The Apulian carbonate rocks, that constitute the largest coastal aquifer in the region, often include varying amounts of a residual product of weathering known as "terra rossa".

"Terra rossa" is widely and variously distributed and, being rich in ^{226}Ra, labels the water of this karst aquifer with ^{222}Rn concentrations ranging from a few dozen pCi/l to over 10,000 pCi/l.

Such natural labelling provides a good tracer for studies in hydrogeology as it does not only help detecting for conditions of permeability of the water-bearing system but also outlines the main water circuits.

Ground water flow is naturally affected by such factors as periodic and non-periodic sea level variations, in addition to the processes of groundwater supply and depletion. These phenomena are enhanced, often quite heavly, by pumping water for the various branches of human activity as a result of which groundwater flows with velocity gradients and along directions that may be quite different from those of naturally flowing water.

Many water points were surveyed: it was found that, although water is withdrawn at regular intrvals, ^{222}Rn concentrations often show a pulsatile behaviour due to the fact that water is drawn from different pervious levels that contribute to supply the required discharge in various proportions.

1. INTRODUCTION

Methodological investigations into the use of ^{222}Rn as an environmental marker have provided a further valid basis for the hydrogeological characterization of the Apulian Cretaceous carbonate aquifer [Tadolini and Spizzico, 1996].

Radon, as well as other parametres, has yielded useful indications of the phenomena governing groundwater changes within a fissured and karst carbonate aquifer with variously occurring "terra rossa". Radon-222, in particular, has produced evidence of the sensitivity of the Apulian coastal karst aquifer to any stress, liable to modify the hydrodynamic, chemical and physical conditions thereof, caused by direct actions, namely recharge, runoff, extractions and boundary events, such as periodic and aperiodic sea-level variations.

^{222}Rn concentration in groundwater varies greatly with the area and, within a given water-point, with the depth. This is due to the different concentration of Radium-226 salts in the aquifer bedrock and "terra rossa" and to the various degree of fissuring and karst features of the aquifer.

This paper is an additional contribution to the characterization of the Apulian carbonate aquifer. It presents the results of surveys executed at many water-points in the Murgian and Salentine areas (Fig. 1), encompassing test-withdrawals from the aquifer at different times, aimed at determining ^{222}Rn concentration variations in pumped water.

2. SURVEYS AND RESULTS

During a number of trials, with constant withdrawals over time corresponding to equally constant dynamic depressions of the water table, ^{222}Rn concentrations were not constant. They varied greatly reaching a max 100% ratio over limited time lags, some ten minutes (Figg. 2, 3).

Parallel observations were made on the TDS content of pumped water 50% variations were reported.

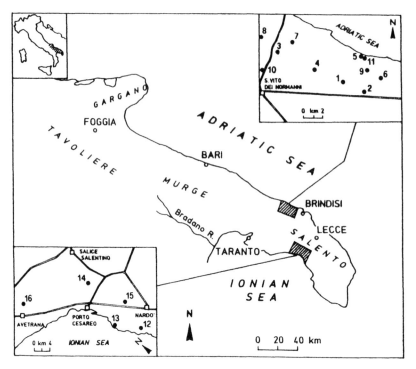

Figure 1: Apulian areas covered by surveys and location of tested wells.

Variations in the concentration of ^{222}Rn measured during the trials cannot be ascribed to a different solution in water of the radon produced by the aquifer bedrock and/or "terra rossa". Given a confined aquifer, ^{222}Rn requires a given time to reach the normal radioactive equilibrium and thus be found significantly in water. This is related to boundary physical conditions, such as temperature, environmental pressure, etc...

As to filtering velocity, under environmental flow conditions, it enables groundwater to be transported downhill below naturally found water gradients. As a result of pumping, filtering velocity is increased, mostly near the catchment area. The flow direction is often reversed and water contributing slightly to the overall flow of the watertable is withdrawn.

In the investigated aquifer, water circulation follows preferential water levels, as shown by field tests. The distribution of salt content along the water column of the drilled wells was correlated with the salt concentration of groundwater pumped out [Tadolini and Tulipano, 1977; Tadolini and Tulipano 1981]. Results also showed that the location of the deep-well motor pump did not affect the prevailing drainage area. Furthermore, following an increase in

the flow rate, once the capacity of a given preferential water level was exceeded, draning from other water levels was started.

Nevertheless, observations derived from measurements taken during test-withdrawals do not account for the "pulse" effect of ^{222}Rn concentrations. Diagrams in figures 2 and 3 indicate that, during pumping, the initial concentration increase is followed by a decrease which is subsequently retrieved. Hence, a "quasi-pulse" pattern is reported in all trials run.

Salt contents were measured as well. Figures 2 and 3 shows that also salt concentration variations comply with ^{222}Rn concentrations. However, the percentage is lower, as salt pollution exhibits a more homogeneous distribution in the portion of the aquifer where various water circuits develop. This confirms the view that, during pumping, different water pathways are opened. They exhibit a different permeability depending on the extent to which they are affected by karst phenomena and the occurrence of "terra rossa".

Salt concentration in these water levels varies. For under normal groundwater flow conditions, they are governed by pseudostatic equilibria, whereas during draining, by means of the

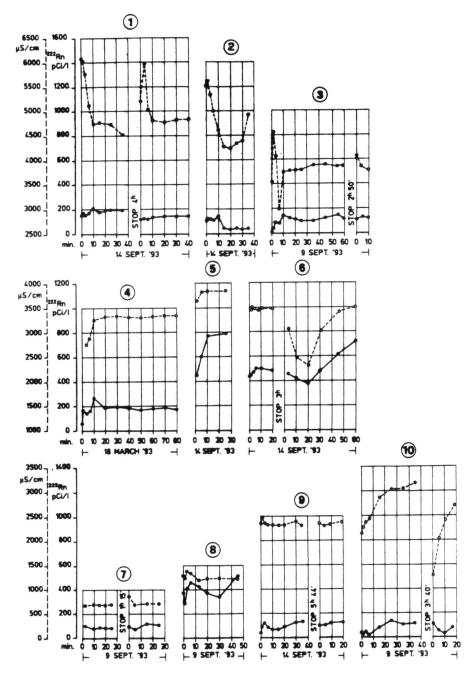

Figure 2: Murgia area) showing the trend against time of ^{222}Rn concentrations (pCi/l) and conductivity (μS/cm) derived from water pumped under trials. Test stops are indicated.

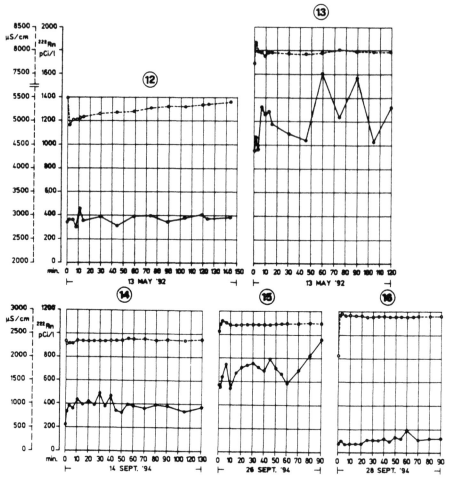

Figure3: Diagram (Salentine area) showing the trend against time of ^{222}Rn concentrations (pCi/l) and conductivity (µS/cm) derived from water pumped under trials. Test stops are indicated.

deep-weel motor pumps, they are affected by dynamic equilibria. Pumping can cover large portions of an aquifer, over time, having different concentrations of ^{226}Ra salts and thus water levels with a broadly differentiated salinity.

Despite the apparent uniformity of the investigated hydrogeological system, a marked local anisotropy may be recorded which characterizes even a single water-point with the depth.

The above is a behaviour generally found during T. Guaceto surveys in the Southern Murgian area [Sciannamblo et al., 1994] as well as in the middle and lower Salentine area (Fig. 1) [Cotecchia, 1977], despite the obviously different characterizations of the two hydrogeological environments [Grassi and Tadolini, 1985].

Moreover, per cent variations in ^{222}Rn concentrations are higher in the Salentine region. This is due to highly fissured and karst rocks significantly found in the area, occasionally and abundantly interbedded with "terra rossa".

Despite the apparent uniformity of the investigated hydrogeological system, a marked local anisotropy may be recorded which characterizes even a single water-point with the depth.

The above is a behaviour generally found during T. Guaceto surveys in the Southern Murgian area [Sciannamblo et al., 1994] as well as in the middle and lower Salentine area (Fig. 1) [Cotecchia, 1977], despite the obviously different characterizations of the two hydrogeological

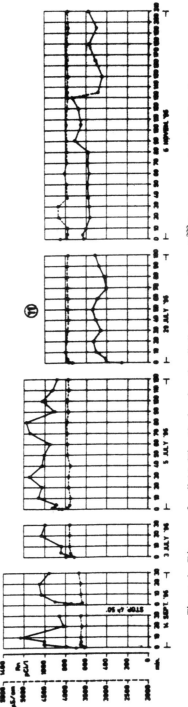

Figure 4: Diagram of well 11 (Fig. 1) showing the ditribution over time of ^{222}Rn concentration (pCi/l) and conductivity (µS/cm) of water pumped in different years and at different times of the year.

environments [Grassi and Tadolini, 1985].

Moreover, per cent variations in ^{222}Rn concentrations are higher in the Salentine region. This is due to highly fissured and karst rocks significantly found in the area, occasionally and abundantly interbedded with "terra rossa".

Diagrams of well 11 in Torre Guaceto (Fig. 4) show similar behaviours over different time periods with intervals ranging from some months to a few weeks. Observations were made while the aquifer was intensely drained to supply farmers with water. On July 5th, the average ^{222}Rn concentrations equalled 1000 pCi/l. By late July, they had lowered to 400 pCi/l. Finally, the effect of drainage was reported on water characterized by a radioisotope enrichment time lower than that of radioactive equilibrium.

3. FINAL ASSESSMENT

It may be concluded that in the Apulian carbonate aquifer, along the vertical axis of the tested water-point, preferential circulation levels can be found.

Within each preferential level, specific water circuits may develop exhibiting distinct radioisotope features, due to variously occurring karst phenomena and "terra rossa". This implies different degrees of permeability, which, however, met the pumped discharge, during test-withdrawals, by a differentiated entering into play of water circuits.

As a result of withdrawals, the catchment basin water is recharged and, in the specific portion of the aquifer, water exhibits a different residence time from that formerly extracted.

Considerations applying to a single catchment area are broadened to the whole portion of the aquifer covered by the survey, as this behaviour can be detected in all tested wells.

The pulse-like pattern of ^{222}Rn concentrations is found in the most internal sections of the aquifer, where they are low, as well as along the coastal areas where karst phenomena are widespread, "terra rossa" largely occurring, and radioactivity levels much higher.

Based upon data gathered to date, differences in ^{222}Rn concentrations found under static and dynamic conditions, within the test-well and the surrounding ones, will help to identify the portions of the aquifer which contribute to the catchment basins.

This work was financialy supported by MURST (40% & 60%) and by CNR-CE.RI.S.T.

REFERENCES

COTECCHIA V., 1977. Studi e ricerche sulle acque sotterranee e sulla intrusione marina in Puglia (Penisola Salentina). *Quad. Ist. di Ricerca sulle Acque* 20: 296-302.

GRASSI D., TADOLINI T., 1985. Hydrogeology of the mesozoic carbonate platform of Apulia (South Italy) and the reason for its different aspects. *Int. Symp. on karst water resources*, Ankara.

SCIANNAMBLO D., SPIZZICO M., TADOLINI T., TINELLI R., 1994. Lineamenti idrogeologici della zona umida di Torre Guaceto (BR). *Geologica Romana*, 30: 775-760, Roma.

TADOLINI T., TULIPANO L., 1977. Hydrodinamic by means of discharge tests of water-bearing layers in fractured and karstic aquifers through the analysis of the chemico-physical properties of pumped waters. *Symposium on: Hydrodinamic diffusion and dispersion in porous media. International Association for Hidraulic Research*, Pavia.

TADOLINI T., TULIPANO L., 1981. The evolution of fresh water/salt-water equilibrium in connection withdrawals from the coastal carbonate and carstic aquifer of the Salentine Peninsula (Southern Italy). 69-85 Hannover.

TADOLINI T., SPIZZICO M., 1996. ^{222}Rn as a powerful tool for solving hydrogeological problems in karst areas. International Symposium Groundwater discharge in the coastal zone. Russian Academy of Science, Moscow, Russia.

Tracer Hydrology 97, Kranjc (ed.)© 1997 Balkema, Rotterdam, ISBN 90 5410 875 4

Use of optical brighteners in applied hydrogeology

A.Uggeri
Idrogea, Varese, Italy

B.Vigna
Dipartimento Georisorse e Territorio, Politecnico di Torino, Italy

ABSTRACT: The artificial tracers tests can provide very useful information in applied hydrogeological researches, but potential pollutant effects must be taken in into consideration. To avoid such problems, two fluorescent dyes (Leucophor BCF and Tinopal CBS-X) were tested in laboratory and experimental fields and then used in different aquifers (karst, fractured, porous) with focus on various issues: location of protection areas, detailed flow identification for pollution problems, testing well cementation, average flow rate for engineering projects, calculation of kinematic porosity, estabilishing the relationship between surface and ground waters. In spite of some problems due to low detectability, high photosensivity and lack of data on tracers'transfer speed related to aquifer flow rate, these substances are useful and well suited for hydrogeological research.

1 INTRODUCTION

In applied research on environmental and hydrogeological problems, the use of optical tracers has prooved to be a very valuable tool and has frequently provided reliable evidence of the relationship between two points, times of arrival and modalities of retrieval of the substances used. In Italy tests with tracers have not yet gained general recognition and are little used because of scepticism and lack of familiarity with these tecniques.

In many studies on aquifers providing drinking water, the use of common tracers, even in low dosages, is forbidden. Tracers are thought to be too strong colourants or potentially toxic, even in waters that sometimes are already compromised by more serious polluttants.

In order to overcome such unwarrented reluctance, two fluorescent tracers were proposed: Tinopal CBS-X and Leucophor BCF, both belonging to the generic category of "optical brighteners". Thanks to their particular characteristics (low toxicity and colouring) these substances have been accepted by the authorities concerned, their use being allowed also in pubblic drinking water systems.

The bibliography on the chemical and physical characteristics and the use of these tracers is quite limited (Crabtree 1970, Glover 1972, Behrens 1976, Smart 1984, Parriaux et al., 1988, Fay et al., 1995) with few operative examples and determinations of the delay factors (Merdingen test site, De Carvalho Dill et al., 1992). Therefore, a series of tests was made both in the laboratory and in the field, in order to establish correct methods for their use. For measuring the fluorescence value (as transmittance) an appropriately modified spettrophotometer (Beckmann UV 5270 - property of Water Geochemical Laboratory of Dipartimento di Georisorse e Territorio of Turin Polytechnic) was used.

2 LABORATORY EXPERIMENTS

For Leucophor BCF, the calibration curve showed quite a low limit of detectability (Tab.1); below a concentration of 0.1 ppm the measures can be considered merely semiquantitative. For the Tinopal CBS-X, the limit was found to be less than 0.01 ppm. The analysis of photosensitivity (Uggeri & Vigna 1993) showed a sudden photochemical degradation of the Leucophor BCF in water solutions, with a sharp fall of the fluorescence in the first 60 minutes of light exposure, whereas several water solutions of Tinopal CBS-X showed manifestly lower values (Tab. 2 and 3). Consequently, it seems advisable to make injections into surface watercourses during the dark hours and to keep the samples in opaque containers.

The solubility limits of these substances are low:

the Leucophor BCF starts to precipitate and flocculate at the concentration (in distilled water) of 1 g/l, while for the Tinopal CBS-X the limit is 10 times lower (0.1 g/l). The filtration test in fine sediments indicates a notable retention of the Tinopal solution while the values of the Leucophor are similar to those of Fluoresceine.

Table 1. Leucophor and Tinopal calibration data (fluorescence).

	1 ppm	0.1 ppm	0.01 ppm
Tinopal CBS-X	282	32.5	4.2
Leucophor BCF	9.3	1.6	0.9
distilled water		0.7-0.8	

Table 2. Tinopal CBS-X in different concentrations: fluorescence (transmittance) decay after exposure to light.

	0 h	1 h	4 h
1 ppm	282	248	232
0.1 ppm	32.5	28	26
0.01 ppm	4.2	4.0	3.2

Table 3. Leucophor BCF in different concentrations: fluorescence (transmittance) decay after exposure to light.

	0 h	1 h	4 h
1 ppm	9.3	2.2	1.5
0.1 ppm	1.6	1.1	1.0

3 FIELD EXPERIMENTS

A series of field experiments was made, comparing the brighteners with other tracers in karst or alluvial aquifers, characterized by relatively high permeability. In the carbonatic aquifer, with high development of karst, the times of arrival and the modalities of retrieval were found to be similar to those of fluoresceine. For the Tinopal, the limits are represented by the low solubility which causes some difficulties of dilution at the point of injection with very low flow rate. For the Leucophor BCF, the limits are represented by the low detectability which implies the need for substantially higher quantities than those used with traditional tracers.

The behaviour of the brighteners in porous media was studied in the C.N.R. of Venice (Istituto per lo studio delle grandi masse) experimental field, where a series of tests was made (Bernardi et al. 1995) to verify the spreading modalities of different tracers.

The field, carried out in alluvial deposits (coarse gravel with sandy matrix, K: 0.2-0.6 $*10^{-2}$ m/s), consists of one well of injection and a series of piezometers downward, 5-25 m distant, with an average depth of the water level of 2.5 m. The injection was made istantaneously using, in three

different tests, respectively 25 mCi of Iodium-131 diluted in 2 l of water, 200 g of LiCl and 4 kg of Leucophor BCF: the measurements in the piezometers were made with 5 scintillation detectors located at different depths for the analysis of Iodium-131. For the other substances, the sampling was made through a peristaltic pump, placed in relation to the levels of preferential flow. The values of the modal speed (V_M) varies between 16 and 48 m/day for the Iodium-131 (NaI in water solution), 34 and 59 m/day for LiCl and 30 and 48 m/day for Leucophor BCF.

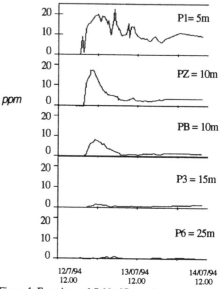

Figure 1: Experimental field of Paese: breakthrough curves

The breakthrough curves for the Leucophor BCF (Figure 1) show in the first piezometer a series of secondary peaks. This is probably linked to the considerable quantity of solution injected and to its high diffusion in the aquifer, whereas in the most distant wells (P3 and P6) the different times of arrival seem to be linked to the eterogeneity of the aquifer with preferential flow. On the other hand, a tracing made with Tinopal CBS-X provided ambiguous and inconsistent results, due to the low solubility of the substance and to a considerable matrix retention.

4 TEST RESULTS

In the following, we will briefly discuss the results of a series of tests made in different aquifers (alluvial, volcanic and carbonatic) used for drinking water. The results have proved to be extremely valuable in applied researches, where other current methodologies have only provided general indications. The injection of the tracer has always taken place istantaneously, preparing a water solution on site, while in the control points programmable automatic samplers have been used and the captors have been replaced at different intervals.

Case 1

Objective: To verify the relationship between surface and groundwaters in a porous aquifer used by a public water system.

Aquifer system: Porous aquifer in alluvial gravel and sand (Rio Ranza, Varese, Northern Italy).

Test conditions (Fig.2): 5 kg of Leucophor BCF were injected into a surface watercourse (R.Ranza) with an estimated flow of 1 m³/s. The injection was made at night, because of the photochemical degradation of the tracer. The river runs through the well field on a gravel bed, at a linear distance of 440 m from the point of injection. Pumping tests on the wells demonstrated delayed drainage, but not any recharge boundary within cone of depression.

The output consists of the mixing tank of the well field water, which was sampled with varied frequency (every hour from the time of injection during the first 24 hours, every 4 h between 24 and 110 h, every 18 h between 110 and 578 h).

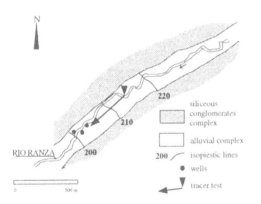

Figure 2: Rio Ranza test: aquifer system

Results: The analysis of the samples and the captors provided concordant indications. The T_C (first appearance time) occurs between 12 and 14 h from the time of injection, the T_M between 14 and 16 h, the breakthrough curve (Fig.3) shows the presence of some minor peaks. Considering a Q average of 30 l/s from the wells, the percentage of

retrieved tracer is 5%; a quantity compatible with the volume of the inflow since a large part of the river water does not infiltrate into the riverbed. Hence, this test has provided important indications to the Water Supply Authority with regard to protection of the water resources and the warning time in case of pollution of the Rio Ranza.

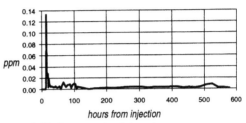

Figure 3: Rio Ranza test: breakthrough curve

Case 2

Objective: To verify the relationship between surface and ground water in a volcanic aquifer, in order to protect an important water resource (which was showing signs of microbiological pollution) piped as communal water supply.

Aquifer system: Volcanic and volcanoclastic aquifer, with a primary and secondary permeability, recharged by water courses and precipitation, the outflow being a linear front of springs (San Savino, volcanic district of Bolsena, central Italy).

Test conditions (Fig.4): The tracing test was made on F. Acquarella, which crosses the S.Savino spring front. The injection was made at night using a cavity (in the recent alluvial deposits) of 70 cm depth, into which a solution was poured: 100 l of water and 1500 g of tracer (Leucophor). At the time of injection, the water course, frequently torrential, had a flow of less than 1 l/s. The outflow consists of

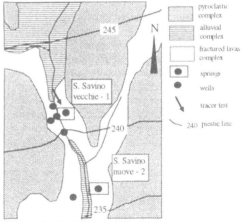

Figure 4: San Savino test: aquifer system

a 700 m long front of springs, subdivided into two groups ("Vecchie" and "Nuove"). The water is utilized through small wells dug into the lava. Several sectors of the springs were fitted with fluocaptors (replaced after one and 10 days), while the automatic sampling in the pipe system of the "Vecchie" was made hourly in the first 24 h and every 8 h in the following 15 days. The distance between the point of injection and retrieval points vary from 110 to 330 m.

Results: The tracing results were positive on both the cotton fabric detectors and the liquid samples. Among the detectors, the only ones tested positive were those taken in two sectors of the "Vecchie" with a maximum concentration in the first 24 h. In the liquid samples the first arrival occured after 5 h and it was followed by other peaks with an ever higher concentration, difficult to account for because of too low sampling frequency (Fig.5) . The test has provided important indications as to the protection of the water resource, showing an extremely high intrinsic vulnerability of the F. Acquarella bed and completing the information relative to the recharge of the aquifer system.

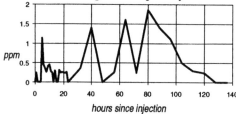

Figure 5: San Savino test: breakthrough curve

Case 3

Objective: To determine the configuration of the flow field of the groundwater in a case of chromium pollution.

Water system: Unconfined alluvial (gravel, sand, conglomerates) aquifer overlaying a hard rock substratum, partially permeable because of fracturing, in the mid Lombard plain (Ponte San Pietro, northern Italy).

Test conditions (Fig.6): Investigations were necessary because of the finding of chromium concentrations in a pubblic well used for drinking water. Upward there are several centres of danger. In the investigated area there are several industrial wells, which worked permanently, even if discontinuosly, since their activity depends on production demands. Pumping produces strong variations in the flow field. Also, there is a discharge boundary (Brembo River). Almost all the wells and the piezometres match both the porous and

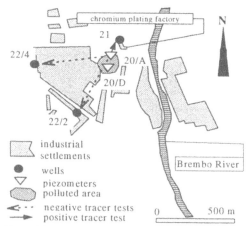

Figure 6: Ponte S. Pietro test: aquifer system

the fractured aquifer. The piezometric measurements do not allow a definition of the flow field accurate enough to ascertain the responsibility of pollution and remedial planning.

A multiple test was made with the following injections:

A) 1 kg of fluoresceine in the sewer of a factory using chromium, a potential centre of danger along the drainage system.

B) 5 kg of Tinopal CBS-X, diluted in 25 l of water, in a piezometer (20/A) located near the most seriously contaminated area close to the sector in which the reconstruction of the flow field was shown to be particularly uncertain.

The following outflows were set up:

	Distance from A (m)	Distance from B (m)	Flow rate (l/s)	Pumping	Sampling frequence
20/D	180	120	0.25	continuous	water: every 840 min; captors after 1, 2 and 4 weeks
21	10	70	3	working hours	captors after 1, 2 and 4 weeks
22/2	600	530	2	continuous	captors after 1, 2 and 4 weeks
22/4	570	530	4 - 6	working hours	captors after 1, 2 and 4 weeks

A special device was prepared for taking samples, in complete darkness, in the piezometer 20/D, which was fitted with a pump.

Results: The fluoresceine was absent in all samples and detectors. Heavy concentrations of Tinopal were found in all the detectors of point 21, appearing in less than 1 week, while it was absent in all the other check points. The tests showed the presence of a mobile ground water divide between the injection piezometre 20/A and the heavily contamined area (20/D), whose geometry is influenced by the presence and the shape of the cones of depression of the industrial wells. The responsibility of the first factory (21) is ruled out.

Case 4

Objective: To verify the efficacy of the cementation in a well in order to restore the isolation of a porous confined aquifer from the unconfined aquifer contaminated by pesticides.

Aquifer system: Alluvial multiaquifer system (gravel, sand, conglomerates) in the upper Lombard plain (Verdellino, Bergamo, northern Italy), divided by clay layers 10-30 m thick.

Test conditions (Fig.7): The traced system consist of a well 292.8 m deep with filters starting at 155 m below the surface, fitted with piezometers monitoring the uppermost water lavels. The continuity of low permeability levels was restored with compactonit (heaving clay). 200 g of tracer were injected into the piezometer of the uppermost level. A step-drawdown test was then made, with

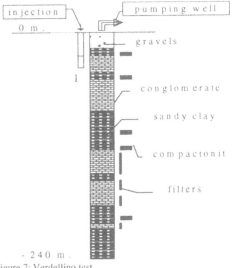

Figure 7: Verdellino test

measurement of the levels in the piezometers and periodical sampling of the pumped water for analysis of the fluorescence and the measurement, *in situ*, of the specific electric conductivity and of the temperature.

Results (Fig.8): A slight increase in the depth of water in the unconfined aquifer was verified; at the same time a decrease in the temperature of the pumped water and an increase in the specific electric conductivity, compatible with the inflow from the uppermost, most mineralized aquifer were registered. Analysis showed the presence of the

tracer in significant concentrations starting from the 4th (Q:80 l/s) of the 5 steps. Hence, the test demonstrated a certain lack of efficacy of the plugging, suggesting the necessity of corrective intervention by the building contractor.

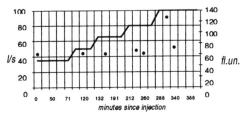

— Q. pumped   • fluorescence

Figure 8: Verdellino test: step drawdown and fluorescence data

Case 5

Objective: To measure flow speed in order to ascertain the modalities of waterproofing concrete injections before building a gallery below the water level.

Aquifer system: Unconfined aquifer consisting of coarse gravel and sand, with a lower confining bed in hard rock (Rio Varenna, Genova, northern Italy).

Test conditions: Injection of 3 kg of Tinopal CBS-X diluted in 100 l of water in a piezometer with static level at 10 m of depth. The aquifer has an avarage thickness of 7 m and the piezometric surface slope of 1%. Manual samples were taken every 120 minutes in a piezometer located at a distance of 35 m downstream, in undisturbed conditions. Another test was then made, injecting a solution of 50 kg of NaCl adequately diluted; the modalities of arrival were continuosly registered using a data logger that measured the specific electric conductivity.

Results: The Tinopal breakthrough curve is characterized by a very rapid first arrival, with the registration of the tracer in the first sampling two hours after the injection (Fig.9), indicating a Vc (first appearance velocity) higher than 17 m/h.

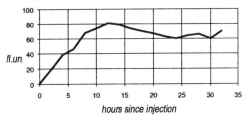

Figure 9: Rio Varenna test: breakthrough curve

The test showed extremely fast flow speeds in highly permeable horizons which, if not identified, could have compromised the impermeabilization of the gallery being built. The measurements of electrical specific conductivity (for the detection of NaCl) showed a much higher Tc (Vc: 7.9 m/h), pointing out the limited reliability of this method and tracer.

Case 6
Objective: To determine the hydraulic parameters for the remediation planning of polluted sites.

Aquifer system: Unconfined aquifer in gravel and sand in the mid Lombard plain (Agrate Brianza, Milan).

Test conditions (Fig.10): The tracing was made during a pumping test with constant flow of 6.2 l/s (Ghezzi et al. 1993). The injection piezometer and the well under test are aligned according to the flow and are 8 m apart, fully crossing the aquifer (about 35 m). The water level in static conditions measures 19 m. 200 g of Fluoresceine mixed with 30 l of water was injected into the piezometer. The sampling was made near the well at intervals of 10 and 60 minutes over a time of about 20 h.

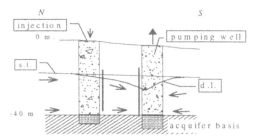

Figure 10: Ponte S. Pietro test: acquifer system

Results: The breakthrough curve (Fig.11) is characterized by a single impulse. the measured speeds are the following: Vc = 3.3 m/s $*10^{-4}$, VM = 2.72 m/s $*10^{-4}$, V50 (average velocity) = 2.59 m/s $*10^{-4}$. The data from the pumping test were interpreted using several methods (Neumann, Jacob, Jacob recovery, Jacob piezometer); the transmissivity value with seemed most accurate was 5.07 m²/s $*10^{-3}$.

By combining the data from the sampling the value of n_e, which otherwise would have been

difficult to measure, was determined. By applying the law of Darcy the result was 5.53 and 6.15 according to the formula $n_e = Q \, T_{50} / \pi r^2 b$ (b: aquifer thickness). The longitudinal dispersivity α_L was obtained by using the formula of Sauty (1978) with a value of 0.42 m.

Case 7
Objective: To verify the characteristics and the modalities of the recharge and protection of an aquifer system feeding a commercially used oligomineral water source.

Aquifer system: intravallive alpine (Julian Alps, Italy) unconfined aquifer in continental (slope, glacial, alluvial) deposits covering a metamorphic complex of low pemeability. The maximum thickness of the aquifer was estimated to be around 40 m (geophysical data).

Test conditions (Fig.12): The injection of 3 kg of Tinopal CBS-X was made into a small surface watercourse (Q: 2-3 l/s) near a leakage in the slope, coarse, deposits. The test was made under conditions of intense cold with a partial covering of snow (max 10 cm) and no precipitation. The springs are 490-520 m from the injection point. The sampling was made at variable frequency: every 255 minutes in the first 5 days after the injection and every 780 minutes in the following 12 days.

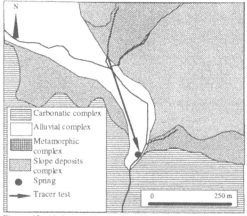

Figure 12: Alpine aquifer test: aquifer system

Results: The results of the test were positive (both samples and detectors) in 2 of the 3 springs, despite the low concentrations (field of the semiquantitative determinations). The Tc (Fig.13) was 125h30' (Vc:4.0 m/h), TM was 138 h25' (VM: 3.6 m/h) and T50 was 263 h (V50: 1.9 m/h). The test confirmed the recharge modalities of the system with secondary recharge by infiltration of water courses and high flow rates. Hence, the possibility

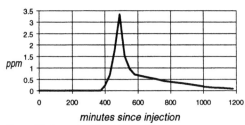

Figure 11: Agrate test: breakthrough curve

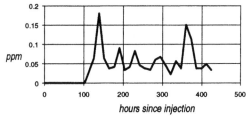

Figure 13: Alpine aquifer test: breakthrough curve

of microbiological contamination by surface watercourses was demonstrated.

Case 8

Objective: To determine the limits and the characteristics of a prealpine and karst hydrostrucure with aim of protecting the water resource piped for drinking water.

Aquifer system: Karst (M.Campo dei Fiori, Varese, northern Italy), partially covered by tills of low permeability in the altimetric lower sector. The main centres of danger are found on the tills. The thickness of the deposits decreases or goes towards zero corresponding to the valley incisions. The hydrostructure feeds two underground karstic springs (total Q at 200-250 l/s) exploited for drinking water by ASPEM, Varese.

Test conditions: Four tests were made (Civita et al., 1994), two or which are described in the following:

Case 8A (Fig.14)
Injection of 5 kg of Tinopal CBS-X into an underground river (Q: 3 l/s) in the Marelli cave. The outflows consist of the Luvinate, Barasso and Valle Luna springs (distance: 2850-3200 m), which were fitted for a period of 1 month with automatic samplers (daily sampling) and cotton fabric

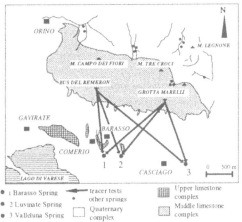

Figure 14: Campo dei Fiori M. test: aquifer system

detectors (weekly replaced). During the test there were some considerable precipitations which significantly modified the hydrodynamic conditions.

Case 8B
Injection of 200 g of Leucophor BCF, mixed with 30 l of water, into a piezometer located 46 m from the Fontanone di Barasso. The piezometer, 31.7 m deep, passes through the surface (low permeable) deposits for the first 28 m and then enters the limestone.

The outflow consists of the Fontanone di Barasso, a buried karstic spring with a Q varying between 60 and 100 l/s. The spring was fitted with an automatic sampler and sampling was made at varied frequence(every 3 h in the first 72 h, every 6 h in the following 120 h and every 12 h in the following 11 days) and with detectors.

Results:
Case 8A
All the springs were positive (captors and samples). The shape of the breakthrough curve (single impulse, Fig.15) indicates the presence of karstic drains of primary importance. The times and the modalities of the tracer seem to be significantly affected by the hydrodynamic regime, with a pressure effect induced by the precipitation. The V_C vary from 7.4 (V.Luna) to 13.3 (F.Barasso) m/h, and the V_M from 6.6 (V.Luna) and 10.3 (F.Barasso) m/h. The test showed an extremely high intrinsic vulnerability in the upper zone of the massif (Civita et al. 1994), with possibilities of pollution of the spring and, hence, a necessity for urbanistic and civil engineering intervention (connection of all the settlements to a sewer with outflow out of the karst).

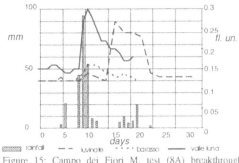

Figure 15: Campo dei Fiori M. test (8A) breakthrough curves

Case 8 B
The test was positive, with a breakthrough curve (Fig.16) with multiple peaks and low maximum concentration. The most significant implication is of an urbanistic kind: it seems clearly necessary to

avoid building activities which may reduce the thickness or destroy the surface deposits protecting the water systems. It is likewise clearly necessary to eliminate pollution in the surface watercourses in the area near the springs as they run in valleys directly enclosed by the carbonatic substratum and marginally feed the karst aquifer.

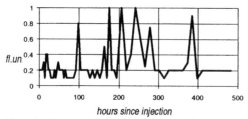

hours since injection

Figure 16: Campo dei Fiori M. test (8B): breakthrough curve

5 CONCLUSIONS

The laboratory and field tests have shown some drawbacks of the optical brighteners: for the Leucophor BCF low detectability and high photosensitivity, for the Tinopal CBS-X low solubility and high retention by the fine matrix in porous aquifers. Conversely these substances, frequently used in common detergents, have undergone rigorous tests which have shown their properties to be generally acceptable. Furthermore, they are inexpensive and easily analyzable and, above all, their use is allowed by the concerned authorities.

If correctly used, optical brighteners can provide very useful indications in applied research. Their use can solve a series of problems, particularly with regard to the protection of aquifers (delimitation of the protection zone, interaction between surface and ground waters and verification of the cementation of wells used for drinking water, etc.).

The authors wish to thank: A. Santoro (translation), Studio Ghezzi, Milan (cases 2, 3, 4, 6), HPC Envirotec, Milan (case 7), P.Mauri and P.Scarato (graphics), M.Civita (supervision).

Pubblication n.1632 of the C.N.R National Group against Hydrogeological Disasters - Task Unit 4.1. - Prof. Massimo Civita responsible.

6 REFERENCES

Behrens H., Martina Zupan & M.Zupan M. 1976. Tracing with fluorescent tracers. In: R.Gospodaric & P.Habic (ed.) *Underground water tracing. Investigations in Slovenia 1972-*

1975: 139-164.

Bernardi A., P.M.Cantori, G.F.Ciancetti, R.Dazzi, F.Fumagalli, G.Gatto, B.Matticchio, G.Mozzi, G.S.Tazioli, B.Vigna & G.Zambon 1995. Confronto di metodologie di misura dei flussi idrici sotterranei in terreni disciolti. *Quaderni di geologia applicata* n.1/95, suppl. 1: 23-32.

Civita M., M.Parmigiani, A.Uggeri & B.Vigna 1994. Protezione delle sorgenti sepolte di M.Campo dei Fiori (Varese): quali aree di salvaguardia? *IV Convegno internazionale di Geoingegneria "Difesa e valorizzazione del suolo e degli acquiferi"*, Torino, Marzo 1994: 427-433.

Crabtree H. 1970. Water tracing with optical brightening agents. *Univ. of Leeds Spel. Ass. Rew.* n.7: 26-28

De Carvalho Dill A., K.Gerlinger, T.Hann, H.Hötzl, W.Käss, Ch.Leibundgut, P.Maloszewsky, I.Müller, S.Oetzel, D.Rank, G.Teutsch & A.Werner 1992. Porous aquifer - Test Site Merdingen (Germany). *Steir. Beitr. z. Hydrogeologie* 43: 251-280.

Fay S., C. Spong, A.S.Alexander & C.Alexander Jr. 1995. Optical brighteners: Sorption Behavior, detection, Septic System Tracer Application. *Proceedings of the International Association of Hydrogeologists XXVI International Congress, Alberta, Canada* (in press).

Ghezzi E., M.Parmigiani M. & A.Uggeri 1993. Prove di pompaggio per il collaudo di pozzi: modalità operative, miglioramenti strumentali ed esperienze con traccianti artificiali. *Ingegneria e Geologia degli Acquiferi* 2: 109-125.

Glover R. 1972. Optical brighteners - a new water tracing reagent. *Cave Res. Group of Great Britain*, Trans.14: 84-88

Parriaux A., M.Liszkay, I.Müller & G.Della Valle 1988. Guide pratique por l'usage des traceurs artificiels en hydrogéologie. *Société Geologicque Suisse, Groupe des Hydrogéologues*: 33-34

Smart P.L 1984. A review of the toxicity of twelve fluorescent dyes used for water tracing. *The NSS Bulletin*, October 1984: 21-30

Sauty J.P. 1977. Contribution à l'identification des paramètres de dispersion dans les aquifères par l'interpretation des expériences de traçage. *Theses Univ. Sci. et Méd. Grenoble.*

Uggeri A. & B.Vigna 1993. Nuovi traccianti ed esperienze di valutazione della velocità di flusso in acquiferi carbonatici. Atti del Convegno "ricerca e protezione delle risorse idriche sotterranee delle aree montuose", *Quaderni di sintesi* 42: 29-51

Tracer Hydrology 97, Kranjc (ed.)© 1997 Balkema, Rotterdam, ISBN 90 5410 875 4

Fluorescent polystyrene microspheres as tracers of colloidal and particulate materials: Examples of their use and developments in analytical technique

R.S.Ward, I.Harrison & R.U.Leader
Fluid Processes Group, British Geological Survey, Nottingham, UK

A.T.Williams
Hydrogeology Group, British Geological Survey, Wallingford, UK

ABSTRACT: Fluorescent polystyrene latex microspheres are utilised widely in biological and medical applications but they also offer potential as tracers of particle movement in the sub-surface. Microspheres, manufactured in a variety of sizes, are available dyed with any one of a number of fluorescent dyes. This allows several differently sized particles to be investigated simultaneously. Comparatively few investigations of their transport through geologic media have been reported but where they have been used they have been enumerated by epifluorescence microscopy to permit the calculation of their concentration in the aqueous phase. This procedure is both laborious and time consuming and provides no information on the disposition of the microspheres filtered or sorbed by the geologic matrix. A method has therefore been developed to allow a much more rapid and sensitive determination of the microsphere content in waters and in solid geologic matrices. It has proved useful in both laboratory column experiments and field investigations concerning migration. Two example applications of microspheres in hydrogeological investigations are described in the paper.

1. INTRODUCTION

Tracer testing has been shown in countless cases to be an extremely useful practical technique for investigating hydrogeological problems and in general suitable tracers can be selected to satisfy operational requirements. Most readily available tracers are however solute tracers and these are not ideal for colloid or particulate migration studies except for comparative purposes. The range of colloids/particles which are available for groundwater tracing is somewhat limited. Those which have been used include colloidal silica particles, radio-labelled polymeric materials, bacteria and bacteriophage. In most cases, the tracers are either not easy to prepare or are very time consuming to analyse.

Another potential suite of particulate tracers is fluorescent polystyrene latex microspheres. Although these tracers are not new, their application in medical and microbiological research has been widely reported, application in the field of hydrogeology appears to be limited to tracing groundwaters to investigate the migration of bacteria (Harvey et al, 1989 and 1993), evaluating drilling fluid contamination (Chapelle and McMahon, 1991) and measuring fracture apertures in clay (Hinsby et al, 1996). In each of these cases, analysis and quantification of the microspheres was by epifluorescence microscopy - a relatively slow and time consuming technique which results in very little advantage over the other particulate tracers.

However, the recent developments in the analytical techniques which are reported in this paper for fluorescent polystyrene latex microspheres have resulted in a considerable improvement in detecting and measuring the tracer. It has resulted in a suite of tracers which are now more cost-effective, versatile, sensitive and rapidly analysed. To demonstrate the use of the tracer, examples of laboratory and field experiments for studying specific natural processes are described.

2. FLUORESCENT MICROSPHERES

The microspheres used in the work described in this paper are manufactured by Polysciences Inc. and are available in a variety of diameter sizes ranging typically from 0.05 - 90.0 μm. Prepared by surfactant-free emulsion polymerization, the polystyrene latex beads have hydrophobic fluorescent dyes incorporated into their structures during synthesis. The fluorescent dyes are similar to those used commonly in groundwater tracing. The microspheres can be supplied as a suspension in pure de-ionised water at a concentration of 2.5% w/v of beads in aqueous suspension. The density of the microspheres is 1.055 g/ml.

Two types of bead are available - plain microspheres and carboxylated microspheres. Both types of bead have a low level of negatively charged sulphate groups associated with them but in the case of the carboxylated microspheres, the surfaces are deliberately functionalised by negatively charged carboxyl groups as well. This offers a wider choice of tracer. Different surface charge particles can be selected to simulate different particle properties or in response to rock matrix properties depending on the nature of the experiment.

2.1 Previous methods of analysis

Analysis of samples for the detection and quantification of fluorescent microspheres has until now been performed by epifluorescence microscopy. This technique, which has been reported widely in the literature (Harvey et al, 1984 and Jones, 1979), involves filtration of the sample on to a black membrane filter which is then placed under a microscope. A light source of suitable wavelength (usually UV) is then selected and deflected by a dichroic mirror down through the objective on to the filter surface. The light from the fluorescing particles, which is of lower energy than that used for excitation, passes back up the objective, through the dichroic mirror to the eyepiece. Suitable filters can also be used to reduce interference and improve detection. Quantification is made by manually counting the microspheres for a known area of filter paper, determining a total number for the whole paper and a concentration by relating to the volume of sample filtered. Variability in the properties of the filter paper may however result in erroneous results because only small areas of the paper are viewed at any one time and the more areas which are viewed to improve confidence is very time consuming, tedious and impractical where large numbers of samples require analysis.

2.2 Development of new method for analysis

In order to improve efficiency and allow greater numbers of samples to be analysed rapidly, we have developed a new technique for identifying and quantifying fluorescent microspheres. The technique can be used for analysing both water samples and solid samples (to investigate sorption or straining) (Harrison, 1996).

The nature of the dyes used in the manufacture of the microspheres is similar to the fluorescent dyes commonly used in groundwater tracing. It therefore became clear that similar techniques, as those used for fluorescent dye tracers, should be applicable for the microspheres. However, whilst truly colloidal particles which are kept in suspension by Brownian motion should be analysable without any sample preparation, those particles which were larger (> 1 μm) and had a tendency to settle may present problems when instrument calibration and subsequent sample analyses is performed.

To overcome this problem, a known volume of sample was filtered to trap the microspheres (filter size was selected on the basis of microsphere size). The beads were then preferentially eluted by passing a known volume of Acetone AR through the filter. Due to the nature of the microspheres, the Acetone effectively dissolved the polystyrene matrix of the microspheres releasing the fluorescent dye into solution. Both nylon 66 and inorganic membranes were found to retain and elute the beads quantitatively and reproducibly.

This methods was found to have a number of clear advantages. Firstly, any extraneous insoluble particulate material which may have had an interfering effect could be filtered out. Secondly, the microspheres are manufactured with the dye incorporated within the whole bead. Dissolution resulted in a release of all of the dye which in solution was then able to fluoresce. Whilst the bead remained intact, only the dye at or close to the surface of the microsphere was available. Analysis of the dissolved microspheres resulted in a significant increase in sensitivity. The difference for two differently dyed beads is shown in Figure 1. Analysis of samples can be performed using any fluorescence spectrophotometer which is suitable for groundwater dye tracing, however care must be taken to avoid evaporation of the acetone. In the experiments reported here, an Hitachi F2000 instrument was used with 5 ml quartz glass cuvettes. The measured detection limits of the microspheres in water and acetone with this instrument (without pre-concentration) were better than 10 μg/l for blue and green microspheres in water and 1 μg/l in acetone.

The third advantage of the analytical method outlined above is the ability to concentrate samples in order to improve detection. Large volumes of water sample may be filtered and then eluted with acetone using a relatively small volume of solvent. The amount of sample concentration is effectively restricted only by the amount of 'raw' sample available.

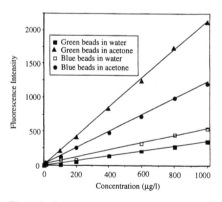

Figure 1. Calibration curves for two microsphere types in water and acetone.

2.3 *Analysis of solid samples*

Where water-rock interaction processes are of interest tracer tests can play an important part in developing an understanding. Breakthrough curves can yield important information but do not enable the distribution of tracer within the water-rock system to be physically investigated. In order to do this samples of the matrix are required to be examined. This can prove extremely difficult in many cases but where fluorescent microspheres are used detailed analysis is possible. Cores of rock which have been exposed to the tracer, fracture faces and adjacent matrix material can all be examined by mixing the solid material with acetone to dissolve and release the dye from the sorbed or trapped microspheres. Fluorimetric analysis can then yield valuable information on water-colloid/particle-rock interaction and quantitative results.

3   EXAMPLES OF MICROSPHERE USE

Two examples of the use of fluorescent microspheres as tracers to investigate particulate movement in aquifer material are presented below. In the first case, the test reported was part of an investigation to study the rôle of colloids in radio-nuclide transport (Harrison, 1996). In the second case, the microspheres were used to investigate the potential for pathogen migration from the surface, through the unsaturated zone, to the water table and a major water supply aquifer.

3.1 *Migration through glacial sand*

This laboratory scale experiment used 6.5 cm long glass columns packed with saturated glacial sand. To simulate groundwater flow, native filtered groundwater was pumped through the columns until steady-state conditions were obtained, i.e. a constant flow rate. Groundwater was pumped through the columns for at least 48 hours prior to any tracer testing.

Once steady-state conditions had been established, a conservative tracer, 0.2 ml of $Na^{36}Cl$ (ß-emitter) was injected on to the columns at the inlet end by means of a sample loop (Figure 2). A fraction collector was used to sample the effluent from the columns for subsequent analysis. Scintillation counting was then performed on the samples to determine the breakthrough curve of the conservative tracer for each of the sand columns used.

After completion of the tests using a conservative tracer, a tracer test using the fluorescent microspheres was performed. Bead suspensions containing 2.5% w/v (as supplied by the manufacturer) were diluted ten times using groundwater. Again 0.2 ml of the diluted tracer was injected via sample loop on to the column and the

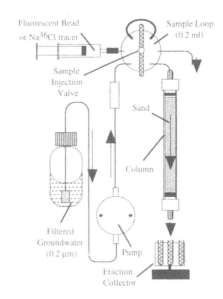

Figure 2. Column experiment apparatus.

effluent sampled by fraction collector.

The 5 ml of the collected fractions were then filtered and the filters eluted with 5 ml of acetone, as described above, prior to analysis. Where analysis revealed low concentrations of beads, preparation of samples was repeated using a concentration step to improve detection and quantification.

Once a complete breakthrough curve had been obtained, the flow through the column was stopped and the column material (core) extruded. As the core emerged from the glass column it was sectioned, with each section sample placed in a glass bottle of known weight. A measurement of the length of core extruded in each section/sample was also recorded.

Samples were then dried for 24 hours (at 100° C) before being weighed. The samples were then broken up by hand and 5 ml of acetone AR added to each bottle before sealing. The bottles were shaken for 2 hours to allow complete dissolution of the microspheres before a small volume was filtered and analysed using the fluorescence spectrophotometer.

Figure 3 shows the comparison between conservative and microsphere tracer breakthrough curves for a column experiment and the distribution of the 'trapped' colloids unaccounted for by the breakthrough curve within the column. Results of these column tests were subsequently used to design field scale tracer test and enabled a study of the potential rôle that colloids can play in contaminant (radionuclide) migration.

3.2 *Migration through unsaturated fissured chalk.*

This field experiment was aimed at investigating the

controlling mechanisms of pathogen migration. An area of unconfined chalk was selected with an unsaturated zone approximately 20 m deep. The top soil of this area measuring 7 m x 7 m was removed to expose the surface of the chalk. At the centre of this area, a borehole (cased to 20 m) was drilled to 24 m below ground level and a sampling pump installed with an intake at 23 m below ground level.

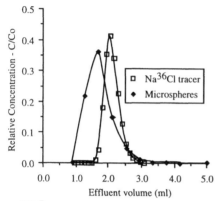

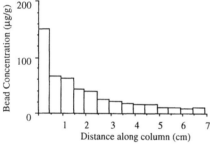

Figure 3. Column breakthrough curves and distribution of trapped beads.

The design of the tracer test was to inject a series of tracers evenly over the exposed area of chalk and monitor their arrival at the water table. Subsequent drilling within the area would then allow core to be recovered and pathways of tracer migration within the unsaturated zone to be identified. A range of tracers were chosen which included LiBr (conservative solute tracer), bacteriophage (0.1 μm) and three different sized and dyed microspheres (1, 6 and 10 μm).

Prior to injection, the site was irrigated with water to saturate and flush through the unsaturated zone. Tracer injection was then completed and further light irrigation of the site continued to simulate recharge. After injection the site was covered to reduce evaporative losses and to shade the site to reduce photochemical decay of the fluorescent dyes used in the manufacture of the microspheres.

Groundwater sampled from below the water table was analysed for each of the tracers. For the

microspheres, a volume (1 litre) of groundwater was filtered through a 0.45 μm nylon filter and the presence of beads determined by both epifluoresence microscopy and fluorescence spectrophotometry. Presence of naturally fluorescent solid particles in the groundwater which were trapped on the filter paper made analysis by microscopy difficult. However the acetone elution significantly improved analysis by separating insoluble material and allowing more controlled fluorescence analysis.

Interpretation of the microsphere results and comparison with the other tracers showed that particles of 6 μm diameter and below could migrate rapidly through the unsaturated zone to the water table. The calculated percentage recovery of each of the tracers at the water was 0.01% for LiBr, 0.00001% for the bacteriophage, 0.01% for the 1.0 μm microspheres and <0.001% for the 6.0 μm microspheres. The approximate travel times for the tracers from the surface to the water table are shown in Table 1.

Table 1. Travel times of tracers through unsaturated zone to water table.

Tracer	Travel time (h)
Li^+ (as LiBr)	72.0
Bacteriophage 0.1 μm	18.5
Microspheres 1.0 μm	15.0
Microspheres 6.0 μm	18.5

Vertical distance between injection point and water table = 20 m.

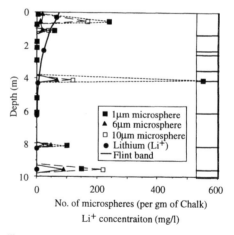

Figure 4. Distribution of microspheres, Li^+ and flint bands in the unsaturated zone.

After sufficient time had elapsed and no further tracer was measurable at the water table, two additional boreholes were drilled in the test area to a depth of 10 m. Core was recovered for each of these to investigate vertical distribution of tracers within the unsaturated zone. Selected samples were examined for LiBr and microspheres. For LiBr (Li^+), pore water samples were extracted by centrifugation. For the microspheres, samples of chalk were prepared and analysed using the method described earlier for the solid material analysis in the column experiments. For this experiment, larger samples ranging from 0.6 - 1.6 kg were used. Comparison of the results for the microspheres showed that the vertical distribution was not uniform. Specific horizons, which appeared to coincide with flint bands, revealed the presence of microspheres but in between no beads were detected. The distribution of microspheres in one of the cores is shown in Figure 4. In the case of the LiBr tracer an exponentially decreasing concentration with depth was observed.

Interpretation showed that particle migration can occur via discrete vertical pathways within the unsaturated zone which intersect horizontal bedding plane features associated with flint bands. Migration of the particles into the matrix of the chalk is precluded because of the small pore throat apertures (<1 µm) whereas the LiBr, as a solute, can move into the matrix. This explains the observed Li^+ concentration distribution.

4. CONCLUSIONS

Developments in analytical techniques for the quantitative determination of fluorescent microspheres in both aqueous and solid samples have improved efficiency, confidence and versatility in the use of these materials as groundwater tracers The microspheres are available in a wide range of sizes, with different surface charge properties and dyed with different coloured fluorescent dyes enabling their use in a wide-range of situations alone or in combination.

Two examples of microsphere use for investigating highly relevant, but different, environmental problems have been described. Their use in both cases was successful clearly demonstrating their relevance and suitability as an investigative tool for examining groundwater and contaminant transport and behaviour.

5. ACKNOWLEDGMENTS

We are grateful to staff of the Fluid Processes Group and Hydrogeology Groups for their assistance in various elements of this work. This paper is published by permission of the Director of the British Geological Survey (Natural Environment Research Council).

REFERENCES

Chapelle, F.H & P.B. McMahon 1991. Geochemistry of dissolved inorganic carbon in a Coastal Plain aquifer. 1. Sulfate from confining beds as an oxidant in microbial CO_2 production. *Journal of Hydrology*. 127: 85-108.

Harrison, I. 1996. A study of colloidal mobility in a glacial sand formation. *PhD Thesis*, Nottingham Trent University, June 1996.

Harvey, R.W., N.E. Kinner, D. MacDonald, D.W. Metge & A. Bunn. 1993. Role of physical heterogeneity in the interpretation of small-scale laboratory and field observations of bacteria, microbial-sized microsphere and bromide transport through aquifer sediments. *Water Resources Research*; 29(8): 2713-2721.

Harvey, R.W., R.L. Smith & L.W. George. 1984. Effect of organic contamination upon microbial distributions and heterotrphic uptake in Cape Cod, Mass., aquifer. *Appl. Environ. Microbiology*; 48: 1197-1202.

Harvey, R.W., L.W. George, R.L. Smith & D.R. LeBlanc. 1989. Transport of micropheres and indigenous bacteria through a sandy aquifer: results of natural and forced gradient tracer experiments. *Environ. Sci. Technol.*; 23: 51-56.

Hinsby, K. , L.D. McKay, P. Jørgensen, M. Lenczewski & C.P. Gerba 1996. Fracture Aperture Measurements and Migration of Solutes, Viruses and Immiscible Creosote in a Column of Clay-Rich Till. *Groundwater*, 34(6): 1065-1075.

Jones, J.G. 1979. A guide to methods for estimating microbial numbers and biomass in freshwater. *Pulication of Freshwater Biological Assoc*. Scientific Publication no. 39.

Tracer Hydrology 97, Kranjc (ed.) © 1997 Balkema, Rotterdam, ISBN 90 5410 875 4

The importance of tracer technology at combined borehole investigations

H. Zojer
Institute of Hydrogeology and Geothermics, Joanneum Research, Graz, Austria

ABSTRACT: In an experimental field for a waste disposal site investigations have been performed applying methods from geology, hydrogeology, hydrology, hydrochemistry, environmental isotope hydrology and tracer technology. All data obtained result to a dynamic drainage model of groundwater. The combined interpretation of borehole data guarantees a high-grade knowledge of groundwater exfiltrating to the surface drainage, which enables proper control measures of the disposal site and an effective groundwater protection.

1 HYDROGEOLOGICAL BACKGROUND

For a proper interpretation of borehole data there exists a big need of combined investigations including tracer application, since conventional data are limited and only available from the geological description of strata and the performance of pumping tests, in some selected cases from some geophysical logging.

In an experimental field, investigated for a waste disposal site in Upper Austria a reasonable number of boreholes has been drilled (fig. 1). The geological structure is defined by Neogen layers, mostly consisting of sand and clay sequences, the latter are wide-spread. In almost all geological units fissures and joints appear locally, they reach widths up to some centimeters. The main tectonic fractures are directed WNW-ESE, accompanied by secondary joints indicating tension processes along that fault zones. Therefore the main pre-conditions for the permeability are caused by the net of joints and fissures and only subordinated by the porous groundwater body.

There is no doubt that permeability as a main aquifer parameter gives evidence for the understanding of solute transport phenomena. Investigations of permeability show values in the magnitude of 10^{-11} and 10^{-3} m/s. The wide scattering of data is based on an interfingering of porous and fissured parts of the aquifer and on different evaluation methods like pumping, open end, flow meter, packer and tracer tests as well as on geophysical borehole logging and soil mechanic indications.

Regular measurements of groundwater level and spring discharge were performed and result to flow pattern of groundwater (fig. 1), but they must be considered not sufficient for the determination of the local drainage basis. Low seasonal fluctuations of groundwater level have been observed near spring outlets representing the discharge zone and at boreholes with a considerable thickness of the unsaturated zone, thus damping vertical flow processes. However, these effects have not been recognized overall but locally indicating the hydrogeological importance of preferential flow conditions in the infiltration part of the aquifer.

2 AQUIFER DYNAMICS

The water cycle implies the conception, that water, infiltrated to the underground, is reemerging at the surface after a certain turnover time and in an unknown distance from the recharge location.

The groundwater circulation is adjusted to the natural drainage (fig. 2). It was therefore essential to determine whether groundwater and the associated dissolved solids of the investigated aquifer are discharging to the local streams (fig. 1) or more downstream to the same surface drainage or maybe to another river of higher hierarchy within the drainage net. For such a conceptual model the unsaturated zone is of minor importance.

Between infiltration and exfiltration of groundwater there exists a dynamic equilibrium. It is determined by hydraulic as well as by solute parameters resulting to the fact, that in the groundwater body of

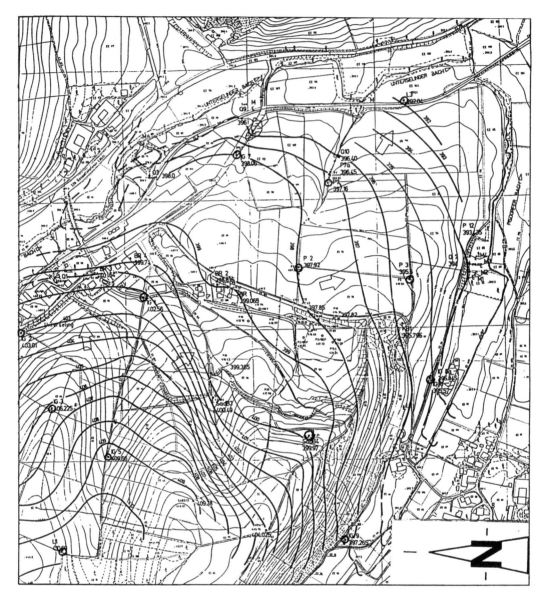

Figure 1. Location map of boreholes and groundwater contours.

the recharge zone a downward movement of water molecules and solute parameters occurs, while in the exfiltration zone the water movement is directed upwards (fig. 3). This is the most important consequence for the siting of waste disposals, since the knowledge of the hydrogeological conditions for groundwater exfiltration allows finally the control of contaminants and furthermore the setting of measures for the rehabilitation of possible disposal damages.

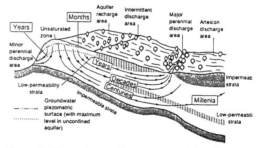

Figure 2. Groundwater flow systems (after Chapman 1992).

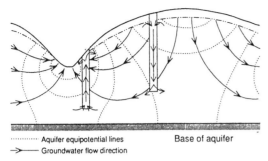

····· Aquifer equipotential lines   Base of aquifer
──▷── Groundwater flow direction

Figure 3. Vertical flow in boreholes (after Chapman 1992).

3 TURNOVER TIME OF GROUNDWATER

The age distribution of groundwater will be defined by the tritium content. The investigated area is located between Unterseliger and Pisdorfer Bach (fig. 1). A rough overview of the data show a large scattering between almost 0 and 70 TU. Fig. 4 offers an insight to the relation between tritium concentration and groundwater level, tritium profiles are not included in these considerations.

Originating from the fact, that the tritium content of precipitation and infiltration water varies between

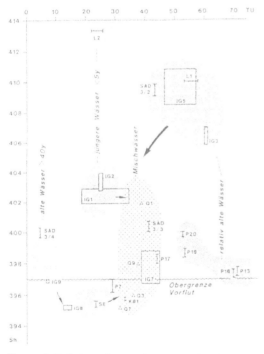

Figure 4. Relation of groundwater level and tritium content.

15 and 20 TU, the oldest groundwater, not exceeding 5 to 10 TU, can be easily distinguished having a turnover time of more than 40 years. Such groundwater is exfiltrating along a fractured zone parallel to the Pisdorf Valley to the surface drainage, represented by the boreholes SAD ¾ and IG 9. In the latter the sodium content is slightly increased, which points to a deep circulation of groundwater

The youngest groundwater is exposed by the boreholes L 2, IG 2, IG 1, P 3 and P 7. In the most cases the groundwater is locally recharged through a very thin zone overlaying the groundwater. Fig. 3 shows furthermore that the groundwater in the southern part (near Pisdorf Valley) is composed by groundwater of different origin and age, simplified by a young/shallow and a deep circulating component of higher age as mentioned above.

The classification of the remaining relatively old groundwater (mean residence time between 20 and 30 years) is caused by the sedimentology and by the thickness of the unsaturated zone, which locally can reach 40 m.

4 VERTICAL MOVEMENT OF GROUND-WATER

In order to gain optimal results for a dynamic infiltration-exfiltration model a programme of combined borehole investigations has been realized:
- geophysical borehole logging
- flowmeter measurements
- tracing the whole water column to obtain preferential flow paths
- measurement of filter velocity by one point dilution method
- point-injection of tracers to receive information on the vertival water movement (downwards or upwards)
- vertical profiles of selected parameters, especially of natural tritium (fig. 5-7)

In the upper part of Unterselig Valley the boreholes L 1, L 2, IG 5, IG 3 and IG 2 has been drilled for distinct measurements. Borehole profiles of tritium are shown in fig. 5. The tritium content of groundwater from the drillings L 2 and IG 2 is varying between 20 and 25 TU and does not considerably change by depth, which indicates obviously to young groundwater, infiltrated at the flanks and not circulating deeply. A tracer, injected in borehole IG 2 at a depth of 21 m was transported upwards, thus recording clear exfiltration conditions to the surface drainage (Unterseliger Bach). Tritium profiles at L 1 and IG 3 show likewise no changes of the concentration by depth but with a considerable high content of 50 to 60TU, which corresponds to a mean transit time of 20 to 30 years. Both boreholes are located in the recharge zone indicated by a slightly downwards movement of groundwater derived from

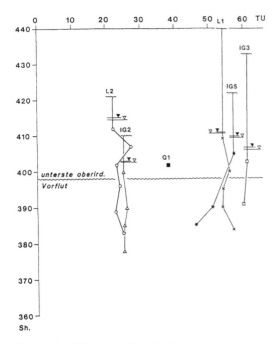

Figure 5. Tritium profiles in boreholes of upper Unterselig Valley.

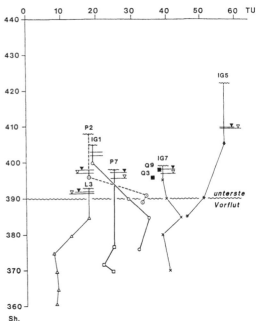

Figure 6. Tritium profiles in boreholes of lower Unterselig Valley.

point injections of tracer. The same is supposed for borehole IG 5. However, due to the tritium profile a structuring of groundwater turnover time is remarkable: the upper part of groundwater seems to be older than the lower one. It is evident, that the vertical movement of water in the unsaturated zone is slow down because of less permeability. On the other hand the borehole is touching in the deeper part of the aquifer a fractured net system causing a fast groundwater exchange. These processes are confirmed by flow meter measurements.

The boreholes IG 1, P 2, IG 7, P 7 and L 3 are located in the lower part of Unterselig Valley. With regard to the tritium profile (fig. 6) the groundwater of IG 7 is identified as well mixed in the whole aquifer (^{3}H about 40 TU). In contrary boreholes IG 1 and P 2 show the same mixed groundwater, but overlayed by groundwater of young age, caused by a quick infiltration through a thin unsaturated zone. The well mixed groundwater at borehole IG 7 shows a downwards movement derived from a tracer injection about 5 m below groundwater level. That seems to be a surprising result since the borehole is situated only some 100 m near to the local drainage. It is evident, that groundwater from this area is not exfiltrating along the shortest course to the local surface drainage as indicated by the groundwater contour map (fig. 1), it is discharging much more downstream into the same drainage stream after the junction of Unterseliger and Pistorfer Bach.

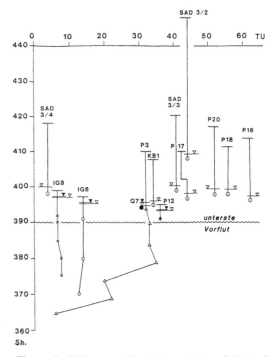

Figure 7. Tritium profiles in boreholes of Pisdorf Valley.

In borehole L 3 tracer experiments have not been performed, but the tritium profile in fig. 5 proves a young water portion in the upper section of the aquifer, which is very locally recharged, but deeper than 15 m below groundwater level a rather old water with a tritium content less than 10 TU has been determined. No upwards movement of groundwater in this borehole can therefore be assumed.

Fig. 7 is reflecting the horizontal and vertical distribution of tritium in the recharge area of Pisdorfer Bach. The immediate function of local drainage basis is clearly pronounced, expressed by an upwards water movement in boreholes IG 9, IG 8 and P 3 from tracing experiments. The tritium profile of P 3 indicates furthermore mixing processes of different water portions along the whole borehole.

5 UNDERGROUND DRAINAGE MODEL

The most important results of the investigation are the determination of recharge and discharge zones within the fissured and porous aquifer. They are essential for the hydrogeological assessment of a waste disposal:

1. The groundwater in the proposed disposal area is overlayed by a considerable infiltration zone with different permeability, which provides a rather high storage capacity.

2. In the North, groundwater is flowing fast towards Unterseliger Bach, more downstream of the creek the groundwater is mixed from more components of different recharge and age.

3. The steep slope in the South, towards Pisdorfer Bach, is tectonically determined. Old groundwater points to a deep reaching circulation, also mixed groundwater from ascending portions and from shallow origin have been found.

4. The most Eastern boreholes (IG 7, L 3), referring to their location and groundwater pattern, give the impression being in the groundwater exfiltration zone to the nearby local drainage. This is decidedly not the case, based on tritium profiles and tracer tests in boreholes.

5. The exfiltration of groundwater occurs in the more downstream valley field and effects a „natural barrier" for the neighbouring regions and confirms the recharge-discharge relationship based on the data obtained.

6 DISPOSAL SITE ASSESSMENT

Each scientific discipline is developing its own assessment criteria. This is true particularly for hydrogeology considering the term „barrier", which is not solely remaining to the geological view. This term should be extended to „natural barrier" in order to include all interdisciplinary aspects originating from geology, hydrogeology, chemistry and biology. For the field of hydrogeology the knowledge of drainage conditions are of prime importance, on which groundwater flow dynamics are adjusted.

With reference to the conventional definition of the term „geological barrier" it is admittedly difficult to come to an overall agreement, since in the investigation area it will not be possible to reach a „geological barrier" in a technically attainable depth. Furthermore it must be noticed that the disposal site is located above a fractured aquifer with a reasonable permeability. Nevertheless a positive hydrogeological assessment can be expressed because the drainage conditions are known and effective with respect to hydraulics and solute transport in groundwater.

For the case of a contaminant intrusion to groundwater the flow paths can be followed. It is therefore essential to establish proper measures for the control of the disposal site and the protection of groundwater quality. For that reason the development of multi control systems depending on a very well defined infiltration-exfiltration relationship is necessary.

ACKNOWLEDGEMENT

The autor would like to express his gratitude to the Austrian Research and Testing Center Arsenal, which carried out isotope analyses.

REFERENCES

Chapman, D. (Ed.) 1992. Water quality assessments. 585 pp. London: Chapman & Hall.
Foster, S.S.D. & R. Hirata 1988. Groundwater pollution risk assessment. *Pan American Centre of Sanitary Engineering and Environmental Sciences*. 73 pp. Lima.

Surface waters

Tracer Hydrology 97, Kranjc (ed.) © 1997 Balkema, Rotterdam, ISBN 90 5410 875 4

Natural tracers – Indicators of the origin of the water of the Vrana Lake on Cres Island, Croatia

Božidar Biondić, Sanja Kapelj & Saša Mesić
Institute of Geology, Zagreb, Croatia

ABSTRACT: A complex hydrogeological exploration has been made of Cres island in Croatia. It included macrostructural analyses, geophysical exploration, exploratory drilling, exploration of lake bottom lithology and morphology by divers, groundwater tracing, various analyses of recent lake bottom deposits and hydrological and biological exploration. This paper deals with only one part of that complex work, exploration by means of natural tracers. As natural tracers, the heavy mineral composition in the recent lake sediments and the isotopic and chemical composition of lake water were used. Some hypotheses on the lake water origin were analyzed by means of geochemical models. The analyses indicate a predominant rainfall contribution in the recharge of lake water, with a possible influence of a deep karst underground and fresh/sea water interface on the dynamics of lake water and its composition.

1. INTRODUCTION

The Vrana Lake (Vransko jezero) on Cres island is one of the most important and interesting water phenomena within the Dinaric karst and, most probably, even within the Mediterranean region. The lake is a cryptodepression and occurs within an elongated island composed mostly of karstified carbonate rocks. With its volume of 220 million m^3, the lake is the only source of potable water for the islands of Cres and Lošinj. A gradual decrease of the water level in the lake up to 1991 was the main impetus for the start of a complex interdisciplinary exploration. Particular attention was paid to a hydrogeological study which had to solve the problem of the genesis of the lake and account for the hydrodynamic changes that have occurred in order to determine how to protect the lake and estimate its potential for public water-supply use.

The exploration of Vrana Lake started as early as in the last century (Lorenz, 1859; Mayer, 1874) but this was a matter of mere geographical descriptions with some ideas on the origin of the lake water. It was followed by numerous limnological and hydrological analyses of the lake's water (Gavazzi, 1902; Cecconi, 1940; Nümann, 1949; Petrik, 1957, 1960). Poljak (1947) was the first geologist to explore the Lake of Vrana from a hydrogeological point of view.

In general, about Vrana Lake, two ideas of the origin (in fact, of the current recharge) of its water have been launched: recharge from the local lake drainage area or from a distant area, out of the island, from the continental Dinarides.

From the beginning of the research considered in this paper, (1989), the hydrogeologists tried to keep both ideas open, trying to find a proper answer by means of complex exploration. The local drainage area influence is entirely obvious and, especially, of the lake area proper, of its 5.5 km^2. However, the karst of the Dinarides comes into being as the result of tectonic and depositional facts from a large region, thus, the genesis of the Lake of Vrana has to be linked with the genesis of a larger area, of the development of the northern Adriatic space, meaning also of the genesis of the River Po delta. In order to support this opinion, it may be said that a depression in the southern part of Vrana Lake indicates a dynamic hydraulic connection with a deep karst underground.

Within the hydrogeological exploration considered in this paper, geological, hydrological and hydrogeological methods were used. An attempt to use the classical tracing from the close drainage area did not give any useful result due to the lack of water discharge at the lake edge and the great lake depth, 70 m, which makes visual observation of groundwater flowing into the lake from a deep karst underground impossible. Because of that, particular attention was paid to the analysis of recent deposits from the lake bottom and hydrogeochemical changes of lake water as possible natural tracers. This paper is dedicated entirely to that segment of the complex research work described. The authors would like to acknowledge the contribution of Dr. Ede Hertelendy, a colleague from the Institute for Nuclear Research in Debrecen, Hungary, who - within a European research project (COST Action 65) - explored the natural isotopes of this water.

2. GEOLOGY AND HYDROGEOLOGY

The Cres island is composed mostly of karstified Cretaceous carbonate rocks (Fig. 2.1). Limestones prevail over dolomites. Numerous occurrences of Palaeogene flysch sediments are a result of tangent tectonic movements, which was confirmed also by geophysical exploration around Vrana Lake. Of Quaternary deposits, only piedmont debris was found within the lake area. During the exploration of the lake bottom, Pliocene-Pleistocene lacustrine silty sediments, torrential slope detritus and Recent lacustrine sediments were discovered. From the tectonic aspect, the Cres island with its Vrana Lake belong to the Adriatic, i.e. to the Adriatic carbonate platform (Herak 1986, 1992).

The basic structural image is outlined by overthrusts that are characteristic of the platform edges. If the formation of present structural units is to be reconstructed, several tectonically active phases have to be defined. The basic tangent forms - overturned folds, overthrusts, reverse faults - happened during the Tertiary age. Neotectonic movements resulted in the rotation and mutually horizontal movement of structural blocks depending upon the regional stress direction. The general structural trend is that of the Dinarides, which is also the lake longitudinal axis line. The Vrana Lake depression was formed along a longitudinal relaxation

fault after the stress changed its direction. This depression now contains 220 million m^3 of fresh water. The depression reaches a depth of 60 m below sea level and comprises an area of about 5.5 km^2.

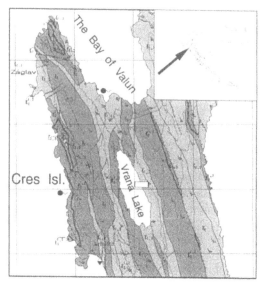

Geologic boundary	High permeable rocks
Erosional unconformity	Low permeable rocks
Reverse fault	
Normal fault	Pumping station
Comparatively subsided block	Brackish spring
Anticline and syncline axis	Submarine spring
Overturned fold axis	

Figure 2.1.   Schematic hydrogeological map of the Vrana Lake area

Taking into account sediment analyses, one may reconstruct the succession of events up to the formation of the lake as it currently exists. The depression, 7 km long, 3 km wide and almost 400 m deep, was formed by very heavy tectonic movements occurring from the Palaeogene till the beginning of the Quaternary and still active. The depression wah formed in the Pliocene, when numerous lakes within the Dinaric karst were formed. During the Pleistocene, the depression was still a karst polje, confirmed by the finds of fluvial sediments. In that time, sea-water level was some 100 m under its present elevation (Šegota, 1983) and the whole northern Adriatic region up to Dugi otok (Long Island), thus including the Cres island area, was land, i.e. the river Po mouth. During

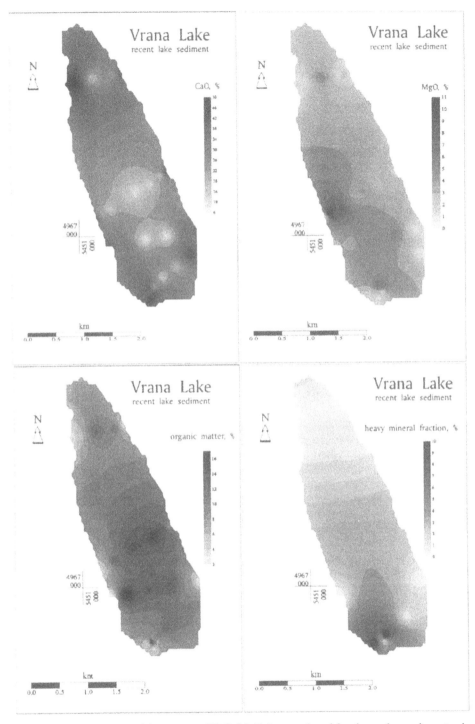

Figure 2.2.   Distribution maps of the content of CaO, MgO, heavy mineral fraction and organic matter in recent lacustine sediments

the glacial stage, almost desert conditions occurred in the area. There was no surface water and sand storms deposited sandy sediments on the edges of river mouth terraces. The remnants of those deposits can still be found on the slopes of the hills on surrounding islands. The rise of sea level of the Pleistocene resulted in the rise of fresh water in the karst underground and creates the present hydrological features of that space: a relatively shallow sea and numerous islands. The depression filled with fresh water. The present lake relief, with an open doline-type depression where surface water used to sink before the sea-water rise, makes possible a hydraulic connection between the lake water and a deep karst underground.

Analyses of the Recent lake bottom deposits offered interesting data for the understanding of the genesis of the lake water and its present dynamics. The largest concentration of heavy mineral fractions was found at the bottom of the submersed depression and its close surrounding (Fig.2.2). Its origin cannot be understood on the basis of the exploration performed but there is no doubt that such a mineral composition is not common in the close local lake catchment area. This one of elements indicating an origin for deposited material away from the close area. Future exploration has to show whether the determined heavy mineral fraction composition is that of distant flysch deposits, occurring away from the island, of flysch deposits of the island of Cres, or whether their origin has to be linked with the river Po deposits. The CaO and MgO contents are in agreement with the occurrence of limestones and dolomites on the slopes of the considered lake depression.

3. HYDROGEOCHEMISTRY

The analysis data and hydrogeochemical reflections in this paper are the result of long term monitoring and analyzing of the Vrana Lake water under different seasonal conditions. From a general point of view, this lake displays an obvious thermal stratification with one entire annual water overturn - which is a characteristics of monomictic lakes. A small amount of decomposed organic matter, a good aeration of lake water during a greater part of the year and an extremely low concentration of nutrients reflect the oligotrophic feature of the lake system (Lerman,

1979). The lake water alkalinity is to some degree lower than in other Dinaric surface streams. The lake water, rainfall and local groundwater are of a mixed $CaNaHCO_3Cl$ to $NaCaClHCO_3$ hydrochemical type and reflect a marine impact (Petrik 1960, Biondić et al. 1994) as can be seen in Fig.3.1.

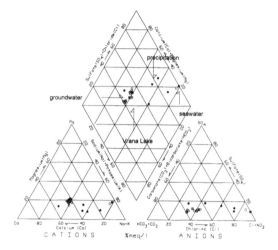

Fig.3.1. Piper diagram of water from the Vrana Lake catchment area

The thermal stratification regime followed by the distribution of parameters which depend on the seasonal biological activity and temperature (O2, pH, CO2, alkalinity, nutrient solids). The distribution of other components - such as chlorides, sulphates, macro, micro and trace elements at traces levels is quite equal during the annual seasonal changes in the lake water, except in the water column along the submerged depression, where low variations in the chloride composition were observed during different seasonal conditions. Although most of the lake is a well aerated environment, having a small quantity of decomposed organic matter during developed thermal stratification in late autumn, anoxic conditions were found 1 m above the bottom of the central part of the lake (Fig. 2.2). Simultaneously with a decrease of oxygen concentration at the lake bottom between the summer and late autumn, the concentration of metals increases due to the dissolution of bedrock sediments. This was recorded in November 1992, October 1993 and October 1994. At the same time, near the bottom of the submerged depression which is 20 m lower than the remaining part of the lake, the oxygen concentration never dropped under 4.6 mg/l, while the organic matter content in Recent lacustrine sediments was approximately equal in both sites, 10%.

During particularly wet years, in 1994 and 1995, the lake level rose and simultaneously the redox barrier at contact with Recent sediments decreased considerably. A mild fall of mean chloride concentration in the lake water was also recorded, from 65 mg/l to about 60 mg/l in 1996.

The distribution of lake water temperature along the depth follows the annual thermal stratification regime not exceeding 8.3 to 8.5 °C at the lake bottom during most of year. The isothermic state is usually established by the end of February, at about 7 °C. At same time, groundwater temperature is between 12.8 and 15 °C at the lake sides and this corresponds to the Cres island climate. All this leads to the conclusion that two different thermic media exist there. Electrolytic conductivity, alcalinity and calcium, magnesium, sodium and nutrient solids content are higher in the groundwater. Sulphate concentration is about equal while chloride content is about 10 mg/l lower than in the lake water.

During the thermal stratification period, the epilimnic space is usually saturated with carbonate minerals: calcite, aragonite, dolomite while the hypolimnic is either in balance with those mineral phases or slightly unsaturated. In the latter case the unsaturation rises with depth. After the vertical turn of lake water by the end of winter and the establishment of an isothermic state, the whole lake becomes unsaturated with carbonate mineral phases. The groundwater of the close catchment area is generally similar to the hypolimnic water with regard to saturation conditions, thus, it is slightly unsaturated but has a somewhat lower partial CO_2 pressure.

Temperature (°C)
DEPTH

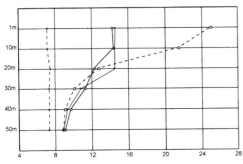

Oxygen (mg/l)
DEPTH

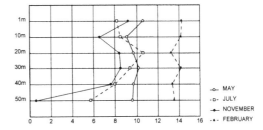

Figure 3.2 Vertical distribution of temperature and oxygen in the central part of Vrana Lake

One of the ways in which the proposed hypotheses about the lake water origin were researched into was the application of geochemical mass balance models - NETPATH (Plummer et al., 1994) and reaction path models - PHREEQE (Parkhurst et al., 1990). The input data for particular computations were selected and set during tests taking into account events and reactions in the lake catchment area that were observed or were possible. The input data consisted of the results of analyses of the lake water, precipitation and groundwater nearest to the mean values for the selected springs (Table 3.1).

Table 3.1. Hydrochemical data used in geochemical modelling

Sample	Vrana Lake	Groundwater- local	Groundwater- long MRT	Precipitation	Seawater 35‰
T - °C	8.3	12.8	8.3	8.3	14
pH	7.9	7.7	8.08	5.75	8.1
	mg/l	mg/l	mg/l	mg/l	mg/l
Ca	36	43	27	5.51	415.2
Mg	10.6	8.3	22	1.60	1305.3
Na	35	40	40	8.8	10864.2
Cl	63	53	53	16	19540
SO_4^{2-}	22	22	22	5.48	2735.6
HCO_3^-	119.6	146.4	193	11.00	141.1
^3H - TU	22	8	0	7	
Ionic Str.	0.055	0.006		0.001	0.670
log pCO_2	-2.925	-2.615	-2.991	-1.271	-3.282
SI calcite	-0.147	-0.119	0	-4.179	0.521
SI dolomite	-0.784	-0.780	0	-8.984	1.785
SI aragonite	-0.304	-0.271	-0.156	-4.343	0.368

Table 3.2.   Selected results of simulations compared to observed values of Vrana Lake and mass transfer obtained by NETPATH (*) in mmol/kgH$_2$O

	A		B			C		D		Vrana Lake
	*	SIMULATION	*	SIMULATIONS		*	SIMULATION	*	SIMULATION	Observed
% precipitation	0		0.99	0.99	0.99	1			0.93	
% groundwater	1		0.006	0.006	0.006	/			0.07	
% seawater	/		0.002	0.002	0.003	/		/		
Evaporation factor	1.189		1	1	1	3.938			3.383	
Mass Transfer										
mmol/kg H2O										
Calcite	-0.507	-0.315	0.422	0.497	0.559	0.055	0.393	0.084	1.267	
Dolomite	-0.013	-0.11	0.195	0.196	0.161	0.033	-0.081	-0.009	-0.813	
FixpCO2	-0.279	-0.614	0.177	-0.875	-0.881	-0.624	-2.977	-0.641	-0.278	
Ca		0.851		0.795	0.833		0.852		0.892	0.898
Mg		0.296		0.346	0.363		0.422		0.388	0.391
C		1.821		2.084	2.066		1.949		1.905	2.017
S		0.272		0.059	0.089		0.225		0.18	0.229
Na		2.069		0.976	1.464		1.507		1.214	1.523
Cl		1.778		1.116	1.674		1.777		1.562	1.777
pH		7.885		7.917	7.917		7.884		7.92	7.9

The tests and simulations were set in accord with events and reactions actually recorded in the lake catchment area (dissolution of carbonates, evaporation, sea-water influence) or conceivable (mixing with groundwater). The accuracy of results obtained by means of the mass balance models was tested by simulating the obtained participating values of individual contributions, evaporation factors, given partial CO$_2$ pressure and saturation conditions of carbonate minerals in the final Vrana Lake water. For the final lake water, water from the hypolimnic space was chosen because it is not influenced by seasonal changes. The evaporation effect was simulated by the precipitation concentration method, by means of so-called negative titration (Parkhurst et al.,1990; Appelo & Postma, 1995) while the evaporation factor was calculated by means of chlorides and/or tritium. A good correlation of evaporation factor values obtained by both ways was found. On the basis of results of mass balance models, several simulations were made but in the tables only those which offer the best correlation among the calculated mass transfer, mass balance models and analyzed composition of the Vrana Lake water are shawn.

As shown in Table 3.2, these models were tested as follows: A - mixing of precipitation and groundwater from a near catchment with the concentration of precipitation by evaporation; B - mixing of precipitation, groundwater from a nearby catchment area and sea water; C - evaporation of precipitation by negative titration; D - evaporation of precipitation in a medium composed of limestones and dolomites with the contribution of groundwater of a long average residence time. In all three cases, the mass transfer during chemical reactions occurring after the mixing of water and evaporation is small. In order to evaluate the contribution of groundwater of a long residence time, it was necessary to determine its composition. That composition was modelled by using the local groundwater under equilibrium conditions with calcite and dolomite as well as equal chloride content and tritium inactivity.

The results of computations performed show that a critical amount of dissolved solids, saturation conditions and mass transfer in the lake water can be achieved by means of several approaches. The first, model A, requires a prevailing entrance of local groundwater and about 16% of evaporation. However, the good correlation among the mass transfer data obtained both by mass balance models and simulation does not fit the saturation conditions of the final Vrana Lake water. According to model B, in the recharge of lake water the largest portion comes from precipitation. The mixing of precipitation, local groundwater and sea water occurs without any influence of evaporation and the mass transfer for carbonate mineral phases is in accord with the saturation conditions of lake water. In the remaining two models, C and D, the concentration of precipitation composition by evaporation is assigned and the computation results show that, with a prevailing share of precipitation in the composition of lake water, taking into account evaporation of about 75 or 70%, it is possible to obtain at least a small mass transfer under the given saturation state of lake water. The share of groundwater of long residence time from a deep karstified underground could reach some 7%.

4. ISOTOPIC EXPLORATION

Within the isotopic exploration (Hertelendi et al., 1994, 1995) lake water was sampled twice and sampling was later continued through weekly sampling of lake water from the surface and from the depth of 40 m. The precipitation samples were also collected at monthly intervals. The lake and precipitation water was analyzed to determine the stable isotopes ratio ($\delta^{18}O$, δ^2H and $\delta^{13}C$) and activity of tritium and ^{14}C activity.

The values obtained showed that there is a large increase of the content of stable isotopes ^{18}O and ^2H in the lake water - explained as the result of evaporation and lack of water losses by sinking from the lake. The obtained equalized lake water isotopic composition in the whole lake was explained by an effective mixing of the entire lake water. Taking into account the concentration of lake water composition by evaporation, the mean residence time of lake water is estimated to be 30 to 40 years. The results of $\delta^{13}C$ determinations indicate a prevailing atmospheric origin of CO_2. With regard to the obtained results of isotopic analyses, the answer is that the water of the Lake of Vrana derives mainly from precipitation.

and hydrogeochemical exploration proved to be really justified in the solving of the problem. This relates firstly to the discovery of the sites of possible hydraulic connection of the lake with the karstified underground, the best possibilities being at the site of the submerged depression. The indicator of its still active role is the occurrence of heavy mineral fraction in the Recent sediments, whose origin is not in the local lake catchment area. This is confirmed also by a good aeration of that deepest part of the lake during the thermal stratification period as well as by a small variation in chloride content along the lake water depth during different seasonal conditions. Hydrochemical data proceeded by geochemical models made possible the testing of hypotheses about the origin of the Vrana Lake water. In general, the results of isotopic analyses are confirmed, that is to say, the largest portion in lake water recharge comes from precipitation and percolated water of short residence time in the underground while a small portion could come from the karstified underground. The geochemical modelling showed that the portion of lake water recharge from long residence time could amount to 7%. The karstified underground term refers to a deep karstified aquifer because no hydraulic gradient directed toward the lake within its local catchment area has been recorded and the consequence is the complete separation of the lake and underground water systems of local catchment in thermic and geochemical sense.

Vrana Lake is in hydraulic balance with the other media, with its deep karstified underground and the Adriatic Sea. Their relationship is difficult to evaluate precisely, however.

5. CONCLUSION

The results of complex geological and, particularly, hydrogeological exploration show that Vrana Lake on Cres island, from a morphogenetic point of view, is a submerged karst polje. The sea-water level rise in the Mediterranean region, thus also in the Bay of Kvarner, after the melting of glacial ice, slowed down deep siphonal flows of fresh water. That is, the sea level rise either formed an open "interface" zone which reflects the equilibrium in the fresh-water lake system or created an underground barrier composed of deposited clayey sediments.

The application of results of depositional, isotopic

6. REFERENCES

Appelo, C.A.J. & D. Postma 1994. Geochemistry, groundwater and pollution. Balkema, Rotterdam.

Biondić, B., Ivičić, D. & E. Prelogović 1992. The Hydrogeology of the Vrana Lake - The Cres Island. Proceedings Of the International Symposium "Geomorphology and Sea" and the Meeting of the Geomorphological Commission of the Carphato-Balcan Countries, Mali Lošinj, Croatia.

Biondić, B., Ivičić, D, Kapelj, S. & S. Mesić 1995. Hydrogeology of Vrana Lake on Cres Island, Croatia. First Croatian Geological Congress, Opatija 18-21. october 1995, *Proceedings 1*: 95-100. (in Croatian with English abstract)

Bonacci, O. 1993. The Vrana Lake Hydrology (Island of Cres - Croatia). *Water Res. Bull.*, 407-414.

Cecconi, A. 1940. Il regine idraulico del lago di Vrana. *Annali dei Lavori Pubbulici*, Roma.

Gavazzi, A. 1902. Water Temperature of Transit Layer of the Vrana Lake. *JAZU, Rad 151*, Zagreb. (in Croatian)

Herak, M. 1986. A Concept of Geotectonics of the Dinarides. *Acta Geologica*, 16/1, 1-42, Zagreb.

Herak, M. 1991. Dinarides - Mobilistic View of the Genesis and structure. *Acta Geologica*, 21/2, 35-117, JAZU, Zagreb.(in Croatian with English abstract)

Hertelendi,E., Veres, M., Futo, I. & J. Hakl 1994. Environmental Isotope Study of Karst Systems. *COST 65 action "Hydrogeological Aspects of Groundwater Protection in Karstic Areas"*, Interim report.

Hertelendi, E., Svigor E., Rank, D. & I. Futo, 1995. Isotope Investigation of Lake Vrana and Springs in the Kvarner Area. First Croatian Geological Congress, Opatija 18-21. october 1995, *Proceedings 1*: 201-205.

Lerman, A. 1979. Lakes - Chemistry, Geology, Physics. Springer-Verlag. New York-Heidelberg-Berlin.

Lorenz, R. 1859. Der Vrana See. *Petermans Georg. Mitt., 1*, Gotha.

Mayer, E. 1873. Der Vrana See auf der Insel Cherso. *Mitt. Georg. Ges.,16,* Wien.

Nümann, W. 1949. Beitrage zur Hydrographie des Vrana - Sees (Insel Cherso), insbesonders untersuchungen über die organische sowie anorganische Phosohor und Stickstoffverbindungen. *Nova Thalassia.*

Parkhurst, D.L., Thorstenson, D.C. & L.N. Plummer, 1990. PHREEQE - A computer program for geochemical calculations, USGS, *Water-Resources Investigations Report 80-96*, Reston, Virginia.

Plummer, L.N., Prestemon, E.C. & D.L. Parkhurst 1994. An interactive code (NETPATH) for modelling net geochemical reactions along a flow path, Version 2.0 , USGS, *Water-Resources Investigations Report 94-4169*, Reston, Virginia.

Petrik, M. 1957. Hydrology regime of the lake Vrana. *Carsus Iugoslaviae,1,*Zagreb.

Petrik, M. 1960. A Contribution to the Limnology of the Vrana Lake. *Carsus Iugoslaviae, 2, 105-192, JAZU*, Zagreb.

Poljak, J. 1947. Hydrogeology of the Vrana Lake on Cres Island. Professional report, Zagreb. (in Croatian).

Ožanić, N. & J. Rubinić 1994. Analysis of the Hydrological Regime of the Lake Vransko jezero on the Island of Cres.*Hrvatske vode*, 2/8, 535-543, Zagreb.(in Croatian with English abstract)

Rubinić, J. & Ožanić 1992. Hydrological Characteristics of the Vrana Lake on the Island of Cres. Građevinar, 44, Zagreb. (in Croatian with English abstract)

Šegota, T. 1983. The History of the Adriatic Sea. Priroda 72/4. Pg. 106-107. Zagreb.

Tracer Hydrology 97, Kranjc (ed.) © 1997 Balkema, Rotterdam, ISBN 90 5410 875 4

Field tracer tests for the simulation of pollutant dispersion in the Doller river (France)

O. François & P. Calmels
Commissariat à l'Energie Atomique, Section d'Applications des Traceurs, Grenoble, France

F. Merheb
Burgeap S.A., Oberhausbergen, France

ABSTRACT : Tracer studies were carried out on the Doller river (Haut-Rhin, France) for the management of any accidental pollution, with a focus on ease of understanding and use rather than numerical models consideration. Rhodamine WT was used as the tracer and was injected into the river as an input pulse. Tracer concentrations were measured with time from water samples taken at several downstream locations. These experiments were conducted at three representative flow regimes of the river, whose water flow rates range from the low water to high water.

A first analysis of field data allows the presentation of the water residence time and the maximum concentration of the tracer versus the distance from the injection point and the water flow rate. With these results, a methodology, based on the interpolation technique, to estimate the travel time and the maximum concentration of the pollutant along the river for a given flow rate and a given injection point is presented. If the pollution point-source is unknown, but concentrations of pollutant versus time are measured downstream, the methodology enables estimation of the location of the pollution point-source.

When the tracer data is analyzed using the dispersion theory, transfer velocities of the water and dispersion coefficients are determined. Empirical relationships between velocity, dispersion coefficient and the distance from injection point and the water flow rate may be established.

1 INTRODUCTION

Tracer studies were carried out on a part of the Doller river, situated in the Haut-Rhin (France), for the management of any continuous pollution or accidental spills. For the protection of water supply downstream, the arrival time of the pollutant wave has to be evaluated as fast as possible.

Numerical models might be developed and form the basis of an alarm model such as the one proposed by Leibundgut et al. (1993). However, when such models are not practical (not available, not adapted or affordable to the water supply controllers and so on), the preference for defining the potential of pollution and the residence times has to be given over to easier understood and useful methodology, such as the interpolation method presented by Guizerix et al. (1970) and applied in this paper.

2 TRACER EXPERIMENTS

Rhodamine WT was used as the tracer, as a high proof, conservative, non toxic and non expensive tracer (these criteria have been presented for Rhodamine WT by Rubin et al., 1995).

The tracer was injected into the river as an input pulse, at a defined upstream cross-section S_0. Assuming homogeneous distribution of tracer concentration throughout any river cross-section, the tracer concentrations C were measured with time t, from water samples taken automatically at several downstream locations S_x (4, 9.9, 16.5 and 18 km far from the injection point), according to the « one point measurement technique ».

These experiments were conducted at three representative flow regimes of the river, whose water flow rates Q range from the low water to high water. The three flow rates measured at the reference station (0.8, 4.6 and 10.6 m^3/s) are assumed to be representative of a constant flow rate of the river section under consideration in this study. The main results of these tracer experiments are in Calmels et al. (1995).

3 DATA ANALYSIS FOR A PULSE INJECTION AT THE CROSS-SECTION S_0

A first analysis of the field data, as C(t) function of distance S_x and discharge Q, allows the presentation of the water residence time distribution and the maximum concentration of the tracer versus S_x and Q. The water residence time distribution h(t) is defined by $h(t) = C(t)/\int_0^{+\infty}C(t)dt$. Introducing the repartition function of h(t), $F(t) = \int_0^t h(t)dt$, the arrival time t_a, and the final time t_f, are defined as the quartiles 0.1 and 0.9 of F(t). The modal time t_{mod}, is the time at which C(t) or h(t) reaches the maximum ; the mean time t_{mean}, is defined by

$$\int_0^{+\infty} t\,C(t)dt \bigg/ \int_0^{+\infty} C(t)dt = \int_0^{+\infty} t\,h(t)dt .$$

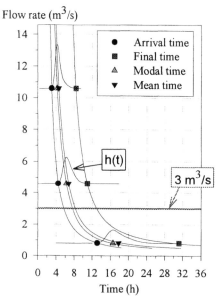

Figure 1 : Estimation of the temporal parameters at the measurement cross-section $S_{9.9}$ for the non experimental water flow rate of 3 m³/s

For a given experimental water flow rate (or cross-section), the plot of these temporal parameters versus S_x (or Q) leads to the estimation by numerical interpolation of these parameters for any cross-section of the river downstream from the injection point (or any water flow rate in the range of the experimental flow rates). Figure 1 shows the evolution of the temporal parameters with the flow rates (residence time distribution) at the measurements cross-section 9.9 km from the injection point, and how to define these parameters at any water flow rate, as an example 3 m³.s⁻¹.

According to these estimations, Figure 2 shows the evolution of the temporal parameters versus the distance from the injection point for the new water flow rate of 3 m³.s⁻¹.

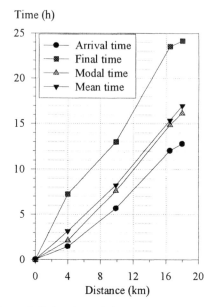

Figure 2 Estimation of the temporal parameters at the non-experimental water flow rate of 3 m³/s for any cross-section downstream from S_0

A similar analysis of the maximum concentration as a function of S_x and Q leads to the estimation of the maximum concentration at any cross-section and at any water flow rate, due to a pulse injection at S_0, as shown on Figures 3 and 4.

The shapes of C(t) or h(t) at any cross-section and for any water flow rate are interpolated from the plot of the temporal parameters (residence time distribution) versus the flow rate at a specific cross-section (Figure 1). They will be refined by introducing other temporal parameters corresponding to other quartiles of F(t), or with 3D interpolation on the graph.

The results of such tracer experiments would also allow the calibration of transfer and dispersion models of a tracer or a pollutant into the river when available.

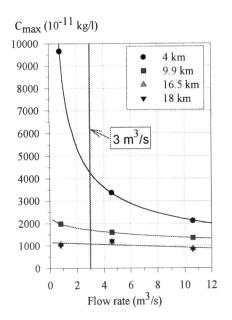

Figure 3. Estimation of the maximum concentration at the measurement cross-sections for the non-experimental water flow rate of 3 m³/s.

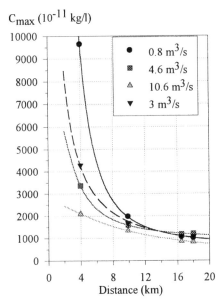

Figure 4. Estimation of the maximum concentration at the non-experimental water flow rate of 3 m³/s for any cross-section downstream from S_0.

4 OTHER INJECTION FUNCTION AND INJECTION POINT

The function C(t) for any other injection function is the result of the convolution of h(t), the residence time function corresponding to a pulse injection, with E(t), the entry (or injection) function :

$$C(t) = h(t) * E(t)$$

The previous figures allow the estimation of the travel time and the maximum concentration of the pollutant along the river for a given flow rate Q and a given pulse injection at the cross-section S_x. In this way, the temporal parameters at a downstream cross-section S_x will be defined after deconvolution of the residence time distributions $h_x(t)$ and $h_y(t)$, which are estimated by the method presented above, relative to the pulse injection at section S_0 (Figure 5). The maximum concentration is evaluated through the ratio of the maximum and the average value of h(t), and the average concentration which is defined by $\bar{C} = M t_{x,y}/Q$, with M the injected concentration mass and $t_{x,y}$ the travel time equal to t_f-t_a.

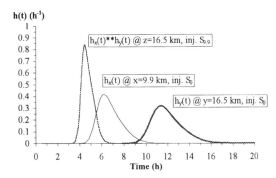

Figure 5. Determination of the residence time distribution, by deconvolution, at section $S_{16.5}$ resulting from a pulse injection at section $S_{9.9}$, for the experimental flow rate of 4.6 m³/s

If the pollution point-source S_x is unknown, but concentrations of pollutant versus time are measured downstream at S_y, the shape of the observed function (piston-dispersion like function with the estimation of $t_{x,y}$ or function with a plateau for the maximum concentration leading to the time for this constant regime to be established $t_{x,y}$) defines the type of injection function (pulse or Heaviside function). On the plot of the temporal parameters versus the distance at the given water flow rate, the time $t_{y,0}$ relative to the pulse injection at S_0 is defined and the new injection section S_x is the one for which the time $t_{x,0} = t_{y,0} - t_{x,y}$.

5 PISTON-DISPERSION APPROACH

When the tracer data is analyzed using the conventional dispersion theory (presented in Runkel & Bencala, 1995), transfer velocities and dispersion coefficients are determined. The well-known analytical solution used to define the transfer velocity U and the dispersion coefficient D is given by:

$$C(x,t) = \frac{M}{S} \frac{1}{\sqrt{4\pi Dt}} \exp\left[-\frac{(x-Ut)^2}{4Dt}\right] \quad (1)$$

where M is the mass of the contaminant spill and x is the distance in the downstream direction.

To each measurement data (x,Q), values of (U,D) are defined through the best fit of the analytical function C(x,t) with the experimental data curve. The variations of the calculated velocity U with the distance from injection point x and the water flow rate Q are such that the following empirical relationship is derived by regression analysis on the log transformed variables U and Q:

$$U = A(x) Q^{n(x)} \quad (2)$$

where the best fit U is such that n(x) has to be a linear function of x and A(x) an inverse of a linear function of x :

n(x) = 0.35 + 0.0176 x and A(x) = 1/(3.4 + 0.21 x)

In the same way, the empirical relationship between the dispersion coefficient and x and Q is given by :

$$D = a(x) - b(x) e^{-Q} \quad (3)$$

where the best adjustment for D is such that a(x) and b(x) have to be linear functions of 1/x :

a(x) = 32 + 216/x and b(x) = 69 + 288/x

Morales-Juberias et al. (1994), have presented a similar study but they have only considered one control station based downstream, which leads to determination of the following relationships between dispersion, velocity and flow rate :

$$U = 0.75 \ Q^{0.46}$$
$$D = 3.97 \ Q^{1.06}$$

Considering our experimental data, we could have defined a similar relationship for U, by imposing constant values of A and n in the determination of U as a function of x and Q :

$$U = 0.17 \ Q^{0.53}$$

This relationship will lead to a less satisfactory fit for experimental data of this study.

Hopkinson and Vallino (1995) have obtained, through tracer distribution study, the following equations to describe the dispersion and the velocity (they defined the cross sectional area S = Q/U, as a function of the distance x) :

$$S = \alpha_0 + \alpha_1 x^3 \text{ or } U = Q/S = Q/(\alpha_0 + \alpha_1 x^3)$$
$$D = \beta(x) \ Q^{0.55}$$

where $\beta(x)$ is inverse of a linear function of x.

These empirical relationships and the analytical solution (1) might be useful for a first estimation of the residence time distribution function resulting from a spill at any point located in the part of the river under consideration, although the experimental data indicate that equation (1) is not always satisfactory to describe the results of the tracer experiments.

6 CONCLUSION

The methodology presented in this paper, based on the interpolation technique from the different tracer experiments conducted at different flow regimes is one of the easiest techniques to use as an alarm model, where numerical models can not be considered. However, when dealing with transfer and dispersion models, results of these tracer experiments are of great interest for the calibration of these models.

These tracer experimental data are sometimes analyzed with conventional dispersion theory, which allows the determination of velocity and dispersion parameters as functions of water flow rate and distance from the injection point. These empirical relationships are then used in the concentration equation (1) to define the residence time distributions.

However, as the approximation of the water residence time distribution by the analytical solution of the advection-dispersion equation is not always satisfactory, it is advised to use the interpolation technique rather than this analytical approximation.

Acknowledgment : This work was supported by the Conseil Général du Haut-Rhin, the Direction Régionale de l'Environnement de l'Alsace, and the Agence de l'Eau Rhin-Meuse.

REFERENCES

Calmels, P., O. François & D. Getto 1995. Etude des vitesses de transfert des pollutions dans les rivières du Haut-Rhin. Réalisation d'opérations de traçages sur la Thur et la Doller. Synthèse des résultats. SAT/RAP/95.37/PC/CR, CEA Grenoble France

Guizerix, J., Y. Emsellem, P. Corrompt, R. Margrita, G. de Marsily, B. Gaillard, F. Rigot Muller, W. Kaufman and J. Max 1970, Méthodes des trcaeurs pour la détermination à priori de la propagation de substances polluantes dans le réseau hydrographique et à posteriori pour la localisation d'un point de rejet inconnu. In : *Nuclear techniques in environmental pollution*. Proc. Symp. Salzburg. IAEA-SM-142a/38. 583-602

Hopkinson & J. Vallino 1995, The relationship between man's activities in watersheds and rivers and patterns of estuarine community metabolism. Estuaries 18, 598-621.

Leibundgut, C., U. Speidel & H. Wiesner 1993, Transport processes in rivers investigated by tracer experiments. In : *Tracer in hydrology* (Proc. Yokohama Symp., 1993), ed. Peters N.E. et al. 211-217. IAHS Publ. no. 215.

Morales-Juberias, T., P. Zafra, M. Olazar, I. Antigüedad & I. Arrate 1994, Modeling solute transport in river channels using dispersion theory. *Environmental hydrology* 24, 275-280.

Rubin, N., M. Hitchcock, M. Streetly, M. Al Faihani & S. Kotoub 1995. The use of fluorescent dyes as tracers in a study of artificial recharge in northern Qatar. In : *Applications of tracers in arid zone hydrology* (Proc. Vienna Symp., 1994), ed. Adar, E.M. & Ch. Leibundgut. 67-78. IAHS Publ. no. 232.

Runkel, R.L. & K.E. Bencala 1995, Transport of reacting solutes in rivers and streams. In : *Environmental Hydrology*, ed. V.P. Singh, chap. 5, 137-164.

Tracer Hydrology 97, Kranjc (ed.)© 1997 Balkema, Rotterdam, ISBN 90 5410 875 4

Restoration of tritium inflow dynamics in regulated river systems

D. I. Gudkov & M. I. Kuz'menko
Department of Radioecology, Institute of Hydrobiology, Kiev, Ukraine

ABSTRACT: This paper presents the results of five-years long (1992-96) tritium monitoring in water objects of Ukraine. Tritium belongs to major polluters from emissions and discharges of NPPs, equipped with WWER reactors (analogous to PWR). When tritium distribution is studied in conditions, which make it difficult to carry out frequent and costly expeditions, it is beneficial to carry out these studies, using regulated water bodies with low rates of water exchange. The Dnieper river belongs to these systems, being the third major European river with its artificial cascade of six water reservoirs. Due to construction of the chain of hydroelectric stations, major water reservoirs were made with time of water exchange up to one and a half years (in years of low water). Therefore, the Dnieper river (taking into account its overall length and ten WWER units within the river basin) is a unique system with extremely low rate of water exchange. Leaving behind negative environmental consequences of the river regulation, one can surely admit, that this very feature is extremely favorable for tritium inflow studies. Due to the fact, that physical and chemical properties of tritium oxide and common water are practically the same, the tritium inflow dynamics remains unchanged downstream a contamination source, located in the upper part of the river. Therefore, it is enough to carry out one field expedition annually to take water samples within all the water area of the Dnieper river and its water reservoirs in order to restore the tritium inflow dynamics within the period of one - one and a half years. The above feature has allowed us to identify major increase of tritium inflow (more than six-time increase) in 1994 into the Dnieper and the river water reservoirs, to restore the dynamics of the inflow and to specify possible source of the increase.

1 INTRODUCTION

In case of nuclear power plants are located within basins of major rivers, it is often necessary to assess NPPs' share in radioactive contamination of these running water bodies.

Tritium is one of the major radioactive components of gaseous and liquid discharges of nuclear power plants, equipped with reactors use either boiling water or pressurized water as a coolant and moderator for fast neutrons. The first type includes BWR reactors; the second type are PWR reactors and their WWER-440 and WWER-1000 analogues (the letter forms 80% of the Ukrainian overall reactor capacity).

The characteristic peculiarity of the tritium is that it, being analogous to hydrogen in its chemical properties, is not trapped on the waste-processing facilities and as tritium oxide comes to the environment with NPP discharges. Tritium is assumed to be long-living, biologically active radionuclide and is capable to contaminate biosphere not only locally (in close proximity to a source) but also regionally and even globally.

Studies of tritium behavior in biological objects demonstrate its bio-accumulative pattern, its relative biological effects in some respects are higher than effects of some other radionuclides. These and other characteristic features of tritium allow one to classify it as one of the most potent radionuclides, causing multi-problem impacts on different organs and systems of living organisms.

When tritium distribution is studied in conditions, which make it difficult to carry out frequent and costly expeditions, it is beneficial to

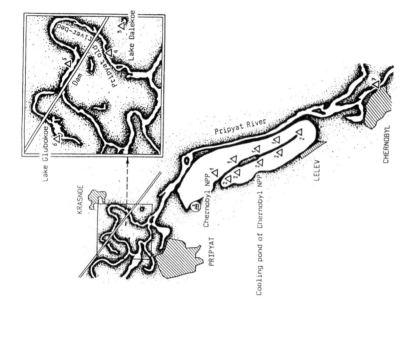

Figure 1. The catchment area of Dnieper River reservoirs with nuclear power plants

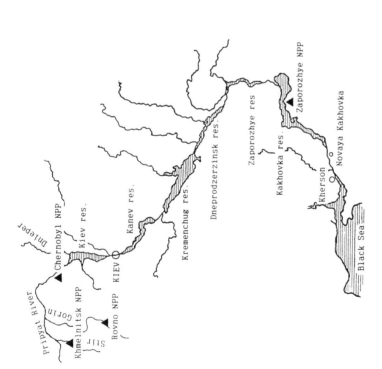

Figure 2. Sampling sites in the Chernobyl exclusion zone (numbers see in Table 1)

carry out these studies, using regulated water bodies with low rates of water exchange. The Dnieper river belongs to these systems, being the third major European river with its artificial cascade of six water reservoirs. Due to construction of the chain of hydroelectric stations, major water reservoirs were made with time of water exchange up to one and a half years (in years of low water). Therefore, the Dnieper river (taking into account its overall length and ten WWER units within the river basin) is a unique system with extremely low rate of water exchange. Leaving behind negative environmental consequences of the river regulation, one can surely admit, that this very feature is extremely favorable for tritium inflow studies. Due to the fact, that physical and chemical properties of tritium oxide and common water are practically the same, the tritium inflow dynamics remains unchanged downstream a contamination source, located in the upper part of the river. Therefore, it is enough to carry out one-two field expedition annually to take water samples within all the water area of the Dnieper river and its water 8reservoirs in order to restore the tritium inflow dynamics within the period of one - one and a half years.

Our studies were conducted: 1 - to study tritium contents in the basin of Dnieper water reservoirs; 2 - to identify tritium release sources for the Dnieper and the Dnieper cascade water reservoirs; 3 - to reconstruct time series of tritium inflow into the Dnieper watershed area; 4 - to assess tritium discharge with Dnieper water into the Black sea.

2 MATERIALS AND METHODS

Tritium content were determined in the Dnieper River upstream, in six reservoirs at Kiev, Kanev, Kremenchug, Dneprodzerzinsk, Zaporozhye, Kakhovka, and in the river downstream near Novaya Kakhovka and Kherson, as well as water objects of Chernobyl exclusion zone. Samples were taken from the scientific ship during the expedition period of the Institute of Hydrobiology in 1992-96. Sampling was carried out by Ruttner bathometer from under surface and near-bottom water levels in different locations of the Dnieper river and water reservoirs. After sampling, the samples were filtered to remove particulate matter and wereplaced into sealed 250 ml glass bottles. Before a measurement, water was distilled.

Tritium content was determined by the scintillating alpha-beta-spectrometer "Quantulus-1220" (LKB-Wallac, Sweden). The test samples were prepared using OptiPhase HiSafe-3 scintillation mixture at an 8:12 ratio between a sample and the scintillator. The measurements were carried out in 20 ml polyethylene vials. The spectrometer effectively validation was made by reference radioactive solution of tritiated water (average specific activity deviation not greater than 3%) and by Het-Trick (LKB-Wallac) standartization set with a deviation not greater than 1%. The minimum activity detectable for tritium determination mas 5.1 and 1.1 Bq/l at measurement times of 100 and 500 minutes, respectively. Here the deviation was not greater than 5-9%.

3 RESULTS

3.1 *Dnieper River reservoirs*

The Department of Radioecology of the Institute of Hydrobiology carries out studies of tritium content in water of open water objects at the territory of Ukraine (including regions of nuclear power plants' sites) from 1992. The most detailed tritium activity studies were carried out in the Dnieper river basin. The Dnieper is the major waterway of the country, the Dnieper basin includes sites of four from five Ukrainian NPPs (Figure 1).

Analyse of the data shows that in the summer of 1992 the highest (maximum) levels were observed for the upper part of the Dnieper (8.8 Bq/l), in the Kiev reservoir (10.6 Bq/l) and in the lower part of the Kakhovka reservoir (7.3 Bq/l). Average tritium contents in these location also show the largest values with respect to other reservoirs; 6.8 Bq/l in the upper Dnieper, 6.3 Bq/l in the Kiev reservoir, and 5.7 Bq/l in the lower part of the Kakhovka reservoir. Other reservoirs of the Dnieper cascade and the river downstream had lower tritium concentrations. In particular, the mean tritium content (Bq/l) was 4.4 in the Kanev reservoir, 3.7 in Kremenchug, 2.8 in Dneprodzerzhinsk and 4.4 in the Zaporozhye reservoir. The upper section of the Kakhovka reservoir (above the Zaporozhye NPP beam) had a mean tritium content of 3.9 Bq/l, with 5.5 Bq/l in the lower part of the Dnieper. Thus, in 1992, tritium content tended to decrease downstream but then increased near the Kakhovka reservoir. The tritium levels then increased almost 2-fold downstream from the Zaporozhye NPP beam.

A high tritium content in the Kiev reservoir might be attributed to the impact of the Rovno and Khmelnitsk NPPs, which are equipped with WWER reactors and location on the River Stir and River Gorin, respectively. These rivers flow into the River Pripyat that discharges into the Kiev reservoir (Figure 1). Besides that, one must also take into account that, after Chernobyl NPP disaster in 1986, substantial amounts of tritium were released into the environment (Salonen 1987, Florkovski et al. 1988, Katrich 1990), so we cannot rule out the possibility that some residual tritium (from the post-accident rainfall) located on the large Dnieper catchment, might come to the reservoir. All these factors appear to explain an increased tritium content in the Kiev reservoir water within the period of the study, particularly in its upstream section.

Higher tritium levels (with respect to other water reservoirs) in the downstream section of Kakhovka reservoir appear to be caused by the operation of the Zaporozhye NPP, located nearby (the NPP has the highest output capacity among all Ukrainian NPPs and equipped with five WWER-1000 reactors in 1992). The NPP operations are accompanied by substantial tritium release to the environment. This assumption is confirmed also by the fact that tritium concentration in the Zaporozhye NPP cooling pond (separated from the Kakhovka reservoir mainstream by an embankment) was almost 100 times higher than in the Kakhovka reservoir mainstream near Zaporozhye NPP site, and in 1993 the tritium concentration was one a half times higher than in 1992 (Gudkov 1993, Gudkov & Kuzmenko 1995, Gudkov et al. 1995).

Data for the summer of 1993 show ralatively uniform distribution patterns for tritium in the Dnieper and its reservoirs. In particular, mean values were 4.5 Bq/l in the upstream Dnieper and 5.4 Bq/l in the Kiev reservoir, slightly lower than 1992 averages for these location, and other values were generally similar. In the Dnieper and the cascade of water reservoirs considered together, the overall mean tritium concentration was 4.8 Bq/l in 1992 and 4.9 Bq/l in 1993.

Average tritium content in the downstream part of the Dnieper did not exceed 5.0 Bq/l. Notwithstanding this relatively uniform distribution of the radionuclide of interest in the Dnieper in 1993, maximal levels were observed (as well as in previous years) in Kiev water reservoir and in downstream part of Kakhovka water reservoir.

Samples of 1994 are of particular interest, in that year tritium activity in water reservoirs substantially exceeded the levels observed in 1992 and 1993, in particular, in the upstream part of the Dnieper - 2.8 - 4.3 Bq/l; in Kiev reservoir - 4.2-36.2 Bq/l; in Kanev reservoir - 21.9-59.4 Bq/l; in Kremenchug reservoir - 16.7-52.0 Bq/l; in Dneprodzerzhynsk reservoir - 21.5-33.4 Bq/l; in Zaporozhye reservoir - 8.5-26.6 Bq/l and in Kakhovka reservoir - 17.1-40.0 Bq/l.

As for the results of the samples, collected in May 1994 in the Pripyat river and in nearby water bodies (the issue will be described later in more details), we can see, that tritium levels in these water bodies (in Pripyat river, in particular) exceeded 120 Bq/l. More earlier and more later samplings were made in October 1993 and in July 1994. Therefore, relatively prolonged tritium inflow (as supported by tritium content in the Dnieper water reservoirs) could occur within that period of time.

Due to the fact, that samples in the Dnieper were collected in September 1994, to that time tritium levels had reduced to the levels of previous years, relatively high tritium contents were observed only near the dam. This distribution pattern is in good correspondence with water run time for Kiev water reservoir, provided that increased tritium inflow continued up to July 1994. Taking into account, that total Dnieper water discharge contains about 30% of Pripyat water, tritium levels in the Pripyat river could reach 170 - 180 Bq/l in some periods of 1994.

Thus, in 1994 increase of tritium content was registered in the Dnieper river and in Dnieper water reservoirs. The levels exceeded more than five times tritium contents in these water bodies in 1992 and 1993.

Two nuclear power plants equipped with PWR reactors are located within the Pripyat river watershed area (Rovensky and Khmelnitsky NPPs). These NPPs are prone to high release of tritium into the environment. According to our data, submitted by researchers from radioecology laboratory of Ukrainian Research Centre of Radiation Medicine (Acad. Med. Sci. of Ukraine), tritium content in the Stir river (the tributary of the Pripyat river, Povensky NPP is located at the river) reached about 400 Bq/l in November 1993. This level is substantially higher, than tritium levels in discharge water of the river, registered earlier (Los' et al. 1991, Buzinny et al. 1992). The content of tritium in water of open The period of increased tritium activity in the Stir river coincides with the time period for the assumed increase of tritium inflow into the Dnieper river and the Dnieper water reservoirs. Therefore, Rovensky NPP is the most probable source of increased tritium content in the Pripyat river with subsequent increase of tritium contents in the Dnieper water reservoirs. However, one cannot also exclude possible emissions and discharges of Khmelnitsky NPP (in that period there were no tritium measurement in water bodies nearby the NPP site).

In 1996 tritium contents in the Dnieper cascade water reservoirs, both in the upstream and the downstream ones, were registered within the range from 2.9 to 5.5 Bq/l; the contents did not show substantial fluctuations.

3.2 Water objects of Chernobyl exclusion zone

As it was discussed above, tritium release into the environment due to operations of nuclear power plants, is mainly associated with NPPs equipped with light water reactors. These reactors are very common both in Ukraine and abroad.

The Chernobyl NPP is equipped with RBMK (graphite moderated, light water cooled, channel type reactor) that show relatively minor tritium emissions to the environment. However, after the nuclear accident of 1986, substantial amount of tritium (besides a wide range of other radionuclides) has been released to the biosphere (Salonen 1987, Florkovski et al. 1988, Katrich 1990). Due to the above considerations, we have carried out determination of tritium content in the water objects located within 30-km exclusion zone of Chernobyl NPP since 1993 (Figure 2).

The studies show, that the radionuclide activity in the water objects in 1993 show characteristic concentration within the range from 4.1 to 11.0 Bq/l. Maximal values were determined for the cooling pond of Chernobyl NPP (Table 1). These concentrations do not exceed levels characteristic to water objects of Ukraine. Some minor increase of tritium concentration in the cooing pond of NPP shows that tritium discharges and emissions from Chernobyl NPP do not exceed mean weighted values for RBMK reactors (Kozlov 1991).

However, in May, 1994 we have registered substantial increase (more than ten fold) of tritium contents in all the water objects studied, located within 30-km exclusion zone. Maximal values (more than 120 Bq/l) were determined for the Pripyat river (the tributary of the Dnieper). The earlier and the later samplings were made in October, 1993 and June 1994, respectively, therefore, the substantial long term release of tritium (this assumption is proved by tritium content in the Dnieper river, downstream from the Pripyat influx) might be occur in this period of time. Due to the fact, that samples of Dnieper water were taken in September, 1994, tritium content in the river downstream (near and far) remain still high enough. This pattern of distribution is in good concordance with timings of water migration in the Dnieper (Novikov & Timchenko 1992), assuming that the tritium release was continuing up to July, 1994. Taking into account, that the overall Dnieper flow contains 30% of Pripyat water, we might assume that at some periods in 1994 tritium concentrations in the Pripyat river were as high as 170-180 Bq/l. Notwithstanding such a substantial increase of tritium activity in water objects within 30-km exclusion zone of Chernobyl NPP in May, 1994, two months later these tritium concentration declined down to 1993 levels. The cooling pond was an exception, where tritium content remained constant (17.2-25.5 Bq/l). This might be attributed to greater amount of water in the cooling pond and to more slow water exchange rate, comparing with other water objects studied, where water exchange was substantially faster. In November, 1994 tritium contents in all water objects continued to decrease, reaching 1993 levels at the end of the month.

Chernobyl NPP is not assumed to be the source of tritium increase in the water objects studied, because, notwithstanding the fact, that water supply to the cooling pond of Chernobyl NPP is made from the Pripyat river, the radionuclide concentration in the cooling pond was 1.5-2 times less than its content in the mainstream river. At the same time higher concentrations of tritium were found in the water objects located upstream from the Chernobyl NPP.

Taking into account, that Rovno and Khmelnitsk NPPs (equipped with WWER reactor) are located within the Pripyat river basin (Figure 2), one may presume, that one of these nuclear power plants was be a source of tritium increase in the water objects. Moreover, we have information, that in January, 1994 tritium concentration of more than 400 Bq/l were found in water from the tributary of Pripyat - Stir river (the river, Rovno NPP is located at).

Similar increase, however, a little less marked, was observed also in April 1995. Fortunately enough, next samples were collected two months later, as in 1994. In this case, the same

Table 1. Tritium content in water objects of Chernobyl NPP exclusion zone, Bq/l

No	Distance from Chernobyl NPP, km	Water objects	May-June 1993	October 1993	May 1994	July 1994	November 1994
1	2.9-7.8	Cooling pond of Chernobyl NPP ("cold" part)	8.6±0.4	8.9±0.4	84.0±4.2	25.5±1.2	7.8±0.3
2	3.2-9.6	Cooling pond of Chernobyl NPP ("warm" part)	11.0±0.5	8.5±0.4	66.6±3.3	17.2±0.8	6.5±0.3
3	3.9	Lake Dalekoe	x	5.8±0.2	61.2±3.0	7.4±0.3	4.5±0.1
4	4.8	Pripyat old river-bed (before the dam)	x	4.5±0.1	66.7±3.3	5.8±0.2	5.3±0.2
5	6.3	Pripyat old river-bed (after the dam)	x	4.1±0.1	94.2±4.7	6.2±0.2	3.6±0.1
6	7.0	Lake Glubokoe	x	4.2±0.1	75.0±3.7	7.9±0.3	5.4±0.2
7	16.0	Pripyat River (near from the Chernobyl)	1.5±0.1	7.3±0.3	120.4±6.0	9.7±0.4	5.7±0.2
8	29.0	Pripyat River (mouth)	x	5.3±0.2	x	5.9±0.3	x
		Average		6.1±0.2	81.2±3.8	10.7±0.4	5.5±0.2

No	Distance from Chernobyl NPP, km	Water objects	April 1995	June 1995	October 1995	July 1996	November 1996
1	2.9-7.8	Cooling pond of Chernobyl NPP ("cold" part)	11.5±0.9	12.0±1.0	9.31±0.9	5.6±0.6	5.8±0.6
2	3.2-9.6	Cooling pond of Chernobyl NPP ("warm" part)	16.6±1.2	11.8±0.9	4.9±0.5	5.4±0.5	7.0±0.7
3	3.9	Lake Dalekoe	29.8±1.8	5.8±0.6	7.6±0.7	5.7±0.6	2.9±0.3
4	4.8	Pripyat old river-bed (before the dam)	16.0±1.1	7.5±0.8	5.3±0.5	3.3±0.4	4.4±0.4
5	6.3	Pripyat old river-bed (after the dam)	15.0±1.1	4.8±0.5	6.8±0.6	5.1±0.5	5.2±0.5
6	7.0	Lake Glubokoe	14.9±1.0	7.6±0.7	7.6±0.7	3.3±0.4	4.6±0.4
7	16.0	Pripyat River (near from the Chernobyl)	10.2±0.8	4.1±0.4	8.4±0.8	5.1±0.5	4.2±0.4
8	29.0	Pripyat River (mouth)	x	x	x	x	x
		Average	16.3±1.1	7.7±0.7	7.1±0.6	4.8±0.5	4.9±0.5

Note: x - n.a.; No - see Figure 2.

changes of the radionuclide distribution patterns in the water bodies under study were observed, similarly to 1994, specific activity of tritium in practically all plain water bodies of the Pripyat river reduced down to its background levels. Cooling pond of Chernobyl NPP was the exception, its tritium activity, due to more low rate of water exchange, remained near 12 Bq/l. This feature confirms the assumption, that these water bodies are interconnected through ground water. The results demonstrate fluctuating pattern of tritium discharges by nuclear power plants, equipped with PWR nuclear reactors (located within the Pripyat river basin upstream from Chernobyl NPP).

Within next months of 1995 and within 1996 average tritium levels in water bodies of Chernobyl NPP 30-km exclusion zone steadily decreased (from 16.3 Bq/l in April 1995 to 4.8 Bq/l in July 1996). In November 1996 average tritium content in the water bodies under study remained at the level of 4.9 Bq/l (Table 1).

3.3 *Tritium discharge with Dnieper water into the Black Sea*

Zaporozhye NPP is the last major assumed artificial source, causing tritium enrichment of the Dnieper water, therefore, we assessed tritium discharge with the Dnieper water into the Black sea, based on tritium levels in the Southern part of Kakhovka water reservoir (downstream from the NPP) and accounting for the rate of water flow in the water reservoir.

Based on the data of flow rate through the Kakhovka hydro-station for 1992 - 1994 (the data were submitted by the State Committee of Ukraine on Water and Meteorology) and on average tritium levels in the part of Kakhovka water reservoir (the water from this part was to come into the Black sea within several months), we have assessed the radionuclide discharge (Table 2).

More high tritium discharge (1.5 times higher relatively to 1992) in 1993 at the same annual average tritium levels might be attributed mainly to particular features of hydrological patterns of the Dnieper, rather than to some increase of tritium inflow from the watershed area. This reason cannot explain the data for 1994, when average tritium contents were five times higher than in previous years of studies

In 1996, based on measurements of tritium levels in samples, collected monthly near Kherson and on the data of flow rate through the Kakhovka hydro-station, we have assessed monthly tritium discharge into the Black sea with the Dnieper water (Table 3).

Total amount of tritium discharge into the Black sea for 10 months of study reached 106.8·10^12 Bq.

Anticipated annual value will not exceed 150 - 160·10^12 Bq. Therefore, in 1996 tritium discharge into the Black sea with the Dnieper water corresponds to levels of 1992 and 1993.

Table 2. Tritium discharge with Dnieper water into the Black Sea, 10^{12} Bq

Months	1992	1993	1994
January	11.2±0.6	12.6±0.6	72.9±3.6
February	11.6±0.6	13.6±0.7	120.4±6.0
March	12.4±0.6	12.8±0.6	137.4±6.6
April	14.8±0.7	22.7±1.1	174.0±8.9
May	18.2±0.9	24.0±1.2	206.7±10.3
June	10.0±0.5	8.8±0.4	129.3±6.5
July	8.0±0.4	6.7±0.3	79.1±3.9
August	7.1±0.4	9.2±0.5	31.9±1.5
September	7.4±0.4	17.7±0.9	33.0±1.6
October	7.5±0.4	20.1±1.0	38.1±1.8
November	6.8±0.3	23.8±1.2	46.7±2.3
December	22.5±1.1	18.1±0.9	78.0±3.9
In all	137.6±6.9	190.1±9.5	1147.5±50.6

Table 3. Tritium discharge with Dnieper water into the Black Sea in January - October, 1996, 10^{12} Bq

Month	Tritium activity in water near from the Kherson, Bq/l	Flow rate through Kakhovka hydroelectric station, m³/sec	Tritium discharge into the Black Sea, 10^{12} Bq
January	4.5±0.4	965	11.3±1.1
February	5.5±0.6	1149	16.4±1.6
March	5.4±0.5	1075	15.1±1.5
April	5.3±0.5	1000	13.7±1.4
May	4.5±0.4	696	8.1±0.8
June	4.2±0.4	978	10.6±1.0
July	3.8±0.4	736	7.2±0.7
August	2.9±0.3	596	4.5±0.5
September	3.0±0.3	727	5.7±0.6
October	4.1±0.4	1335	14.2±1.4
In all			106.8±10.6

4 CONCLUSIONS

As a result of our research, conducted within 1992 - 1996, we have studied tritium distribution in both natural and artificial water bodies at the territory of Ukraine. The radionuclide levels in these water bodies were determined mainly by the global sources of tritium inflow (consequence of nuclear tests) and by emissions and discharges of nuclear power plants. The data for the period are in good conformity with results of other Ukrainian and foreign researchers on tritium activity in other water bodies.

Using the Dnieper water reservoirs as a test object, we have demonstrated increased share of regional sources of tritium inflow into water bodies (nuclear power plants, equipped with light water reactors - LWR). Due to irregular emissions and discharges of the radionuclide by NPPs, at some periods the share of global sources of tritium was negligible. This is confirmed by increased tritium activity in the Dnieper water reservoirs, observed in 1994, in that year the radionuclide concentration exceeded more than five times the levels registered in 1992 and 1993.

Based on the analysis of the data available, we presume, that one of nuclear power plants located within the Pripyat watershed area and equipped with LWR reactors, was the source of increased tritium contents in the Dnieper cascade of water reservoirs (either Rovensky, or Khmelnitsky NPP).

The results of our study demonstrate time series of tritium contents in surface water bodies, located at the territory of Ukraine, allowing one to differentiate them with respect of impact of artificial sources of discharges and emissions into the environment.

The studies have shown, that in 1993 tritium levels in water bodies of the 30-km exclusion zone of Chernobyl NPP did not exceed average levels in other water bodies of Ukraine. The concentration increase due to the Chernobyl accident of 1986 was, apparently, of short-term pattern (this is confirmed by rapid decrease of the radionuclide activity in the water bodies of the exclusion zone after its more than 10-fold increase in May 1994). Obviously, the contamination of 1986 was of aerosol pattern, and, consequently, was more large-scale. It was this pattern, that caused so heavy tritium fallout at large territories of the Pripyat and the Dnieper watershed areas. As a result, tritium inflow into these water systems was of more prolonged pattern. However, accounting for high rate of water exchange in major water bodies of the Pripyat plain within the Chernobyl NPP impact zone, the bulk of the radionuclide was washed out and was discharged with the Pripyat water into Kiev water reservoir at first several months after the accident.

The studies show, that it is necessary to set monitoring sites for sampling of water and fallout nearby sites of Rovno, Khmelnitsk and Zaporozhye NPPs to carry out regional tritium monitoring. The monitoring will allow to identify tritium sources and to assess levels of tritium inflow into the watershed area of the Dnieper and the Dnieper water reservoirs.

REFERENCES

Buzinny, M.G., A.V. Zelensky et al. 1992. Tritium monitoring in water of Ukraine in 1989-91. In A.E.Romanenko (ed.), *Chernobyl NPP accident*: 254-270. Kiev: USCRM (in Russian).

Denisova, A.I, V.M. Timchenko et al. 1989. *Hydrology and hydrochemistry of Dnieper and its water reservoirs*. Kiev: Naukova Dumka (in Russian).

Florkovski, T., T. Kuc & K. Rozanski 1988. Influence of the Chernobyl accident on the natural levels of tritium and radiocarbon. *Appl. Radiat. and Isotop.* 39(1):77-79.

Gudkov, D.I. 1993. Tritium in the Dnieper reservoirs and in the cooling water-body of Zaporozhye atomic power plant. *Proc. Annual Meeting Eur. Soc. Radiation Biology*, Stockholm, 10-14 June, 1993:P12:10. Stockholm: Stockholm University.

Gudkov, D.I. & M.I. Kuzmenko 1995. The study of Ukrainian NPPs impact on tritium distribution in the Dnieper River waters. *Proceedings of International Symposium on Environmental Impact of Radioactive Releases*, Vienna, 8-12 May 1995:301-302. Vienna: IAEA.

Gudkov, D.I., A.D. Andreev & M.I. Kuzmenko 1995. Tritium monitoring in the regions of Ukrainian NPP location and evaluation of Zaporozhye NPP impact on tritium distribution in the Dnieper River. *Revista de la Sociedad Nuclear Espanola.* 30(9):99.

Katrich, I. Yu. 1990. Tritium in natural waters after the Chernobyl accident. *Meteorologia i Gidrologia.* 5:92-97 (in Russian).

Kozlov, B.I. 1991. *Reference book on radiation safety*. Moscow: Energoatomizdat (in Russian).

Los', I.P., M.G. Buzinny & A.V. Zelensky 1991. The content of tritium in water of open reservoirs and sources of drinking water of some regions of Ukraine. *Vestnik AMN SSSR.* 8:54-56 (in Russian).

Novikov, B.I. & V.M. Timchenko 1992. The hydrological conditions of radionuclides migration in the Dnieper water reservoirs. *Vodnie Resursi.* 1:95-102 (in Russian).

Salonen, L. 1987. Carbon-14 and tritium in air in Finland after the Chernobyl accident. *Radiochim. Acta.* 41(4):146-148.

Tracer Hydrology 97, Kranjc (ed.) © 1997 Balkema, Rotterdam, ISBN 90 5410 875 4

Transport of Caesium radionuclides and Strontium-90 in the Dnieper water reservoirs within 10 year period after the Chernobyl nuclear accident

A. E. Kaglyan, V. G. Klenus, M. A. Fomovsky, Yu. B. Nabivanets, O. I. Nasvit & V. V. Belyaev
Department of Radioecology, Institute of Hydrobiology, National Academy of Sciences, Kiev, Ukraine

ABSTRACT: Authors propose for the discussion the results of ten year long studies on changes in concentrations of Cs radionuclides and ^{90}Sr in components of the Dnieper water reservoirs (water reservoirs of the Dnieper cascade is more than 1000 km long), Prypyat' river (Chernobyl'), plain water bodies of the Chernobyl NPP (ChNPP) 30-km exclusion zone. The materials include dynamics on caesium radionuclides and ^{90}Sr contents in water, suspensions and water plants. Interrelations between water-soluble forms of caesium radionuclides and ^{90}Sr and their suspensions-adsorbed ones are discussed.

1. INTRODUCTION

The Chernobyl NPP accident has sharply aggravated the radiological situation in the main water stream of Ukraine - the Dnieper river and the cascade of Dnieper water reservoirs. Our institute, from the first days of the accident, controls the radiological situation in both the cascade of Dnieper water reservoirs and in other major freshwater objects of Ukraine. The results of our studies were published in several papers (Кленус et al. 1991, 1994, Kuz'menko et al. 1992, Романенко et al. 1992) .

2. RESULTS AND DISCUSSION

At May 1986 we have identified ^{141}Ce, ^{144}Ce, ^{131}J, ^{103}Ru, ^{106}Ru, ^{140}Ba, ^{134}Cs, ^{137}Cs, ^{95}Zr, ^{95}Nb, ^{144}Pr and ^{89,90}Sr in water of Kiev and Kanev water reservoirs. Their concentrations in water varied from 0.04 to 2294 Bq/l (Романенко et al. 1992). Major radioactive contamination of water was determined mainly by short-living isotopes, e.g. the share of ^{131}J, in the total water radioactivity reached 80 - 90% .

Beginning from June 1986 the content of radionuclides in the water of these two upstream water reservoirs decreased, at Autumn 1986 the water radioactivity in the reservoirs was determined only by ^{90}Sr, ^{134}Cs and ^{137}Cs. Their concentrations in the water of Kiev water reservoir vary from 0.03 to 1 Bq/l (Kanev water reservoir: 0.01 - 0.16; Kremenchug water reservoir: 0.04 - 0.89; Dneprodzerzhinsk water reservoir: 0.02 - 0.52; Zaporozhie water reservoir: 0.01 - 0.07; Kakhovka water reservoir: 0.01 - 0.07 Bq/l) (Романенко et al. 1992) .

The dynamic picture of caesium radionuclides and ^{90}Sr concentrations in water is represented at Figures 1.

For these graphs we used only September-October data for water (period 1986 - 95). As Figure 1 shows, ^{90}Sr contents in water of Dnieper water reservoirs had stabilized, beginning from 1990 (except the Kiev water reservoir data for 1991).

Contemporary radio-environmental situation in the Dnieper water reservoirs is described by the data obtained in September-October, 1994. At these months water contents of ^{134}Cs+^{137}Cs vary in Kiev water reservoir from 0.24 to 0.40 Bq/l (Kanev water reservoir: 0.11 - 0.278; Kremenchug water reservoir: 0.042 - 0.084; Dneprodzerzhinsk water reservoir: 0.079 - 0.23; Zaporozhie water reservoir: 0.035 - 0.368; Kakhovka water reservoir: 0.044 - 0.11 Bq/l). Concentrations of ^{90}Sr in water vary in Kiev water reservoir from 0.02 to 0.33 Bq/l (Kanev water reservoir: 0.052 - 0.15; Kremenchug water reservoir: 0.025 - 0.11; Dneprodzerzhinsk water reservoir: 0.024 - 0.12; Zaporozhie water reservoir: 0.09 - 0.24; Kakhovka water reservoir: 0.003 - 0.11 Bq/l).

Calculated averages for concentrations of water-soluble forms of ^{90}Sr and ^{134}Cs+^{137}Cs in the water reservoirs for the period (except river mouth parts) demonstrate, that water concentrations of ^{90}Sr in Kiev, Zaporozhie and Dneprodzerzhinsk water

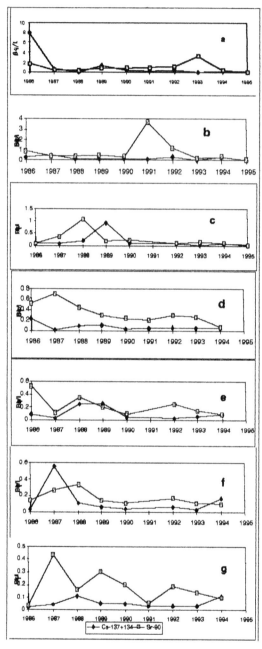

Figure 1.Dynamics of the radionuclide content in water of the Prypyat' river (Chernobyl) (a) and Dnieper reservoirs : Kiev res. (b), Kanev res. (c), Kremenchyg res. (d), Dneprodzer-zhinsk res. (e), Zaporozhie res. (f), Kakhovka res. (g).

reservoirs were higher than ^{90}Sr water concentrations in other water reservoirs. This is associated with the fact, that ^{90}Sr content in the Dnieper cascade water reservoirs is formed by the Prypyat' river drain (contamination due to the river watershed, located in the Chernobyl NPP 30 km exclusion zone). According to the data available, maximal concentrations of ^{90}Sr in the Prypyat' river are registered at the periods of Spring flood. When ^{90}Sr migrate downstream the cascade of water reservoirs, its concentrations do not change substantially, so its relatively high concentrations in Dneprodzerzhinsk and Zaporozhie water reservoirs are attributed to moving contamination front of Spring Prypyat' waters. The time period of water transport in the Dnieper reservoirs cascade is about year (water transport time is from Kiev res. - 43 days, Kanev and Dneprodzerzhinsk res. - 19, Kremenchug res. - 107, Zaporozhie res. - 23, Kakhovka res. - 166).

Average water concentrations of caesium radionuclides decrease downstream the cascade from 0.196 Bq/l (Kiev water reservoir) to 0.115 Bq/l (Kremenchug water reservoir), slightly increasing to 0.143 Bq/l (Dneprodzerzhinsk and Zaporozhie water reservoirs). Minimal contents were registered in Kakhovka water reservoir (0.07 Bq/l). The extremal content of radionuclides in water's reservoir (1991-1996) are: 1) radionuclides Cs: Kiev res. 0.015Bq/l (min Sep.1995) and 0.41 (max Aug.1994); Kaniv res. 0.009 Bq/l (min Sep.1995) and 0.59 (max Mar.1994); Kremenchug res. 0.038 Bq/l (min July 1992) and 0.084 Bq/l (max Sep.1994); Dneprodzerzhinsk res. 0.014Bq/l (min Aug.1993) and 0.12Bq/l (max Sep.1994); Zaporozhie res. 0.025Bq/l (min Aug.1993) and 0.37Bq/l (max Sep.1994); Kakhovka res. 0.008Bq/l (min Oct.1996) and 0.11Bq/l (max Sep.1994); 2) ^{90}Sr: 0.007Bq/l (min Sep.1995) and 1.21Bq/l (max July 1992); 0.018Bq/l (min Sep.1995) and 0.85 (max Apr.1994); 0.024Bq/l (min Sep.1994) and 0.62Bq/l (max July 1992); 0.025Bq/l (min Sep.1994) and 1.12Bq/l (max July 1992); 0.091Bq/l (min Sep.1994) and 0.3Bq/l (max July 1992); 0.012Bq/l (min Sep.1994) and 0.31Bq/l (max Sep.1991) respectively.

Due to the fact, that major water inflow into a cascade reservoir is associated with water discharge through the dam of the upstream one, coastal inflows, industrial and household wastewater from urbanized areas are practically of no importance on radio-environmental situation in the water reservoirs.

Because suspensions play substantial role in distribution and migration of radionuclides in water ecosystems, it is necessary to note, that major sources of sediment inflow into Dnieper water reservoirs include both allochthonous ones (inflow with river drain) and autochthonous ones (internal sources due to coastal erosion, products of plankton and water plants). Radionuclides are readily adsorbed at suspensions, caesium-137 accumulates very heavy at suspensions, in particular. At the first post-accident days of 1986 we registered (besides ^{137}Cs) also ^{141}Ce, ^{144}Ce, ^{131}J, ^{103}Ru, ^{106}Ru, ^{140}Ba, ^{140}La, ^{134}Cs, ^{90}Sr, ^{95}Zr, ^{95}Nb, ^{132}Te in quantities up to $5.55*10^8$ Bq/kg dry matter. Beginning from October, 1988 major contaminants at suspensions included radioactive caesium and ^{90}Sr, moreover, their contents practically did not change since October, 1986 (Романенко et al., 1992).

We calculated parameters of probable accumulation (accumulation factors K_n) of radioactive caesium and ^{90}Sr by suspensions from the environment (from ambient water in this case). Results show, that K_n values for ^{134}Cs+^{137}Cs varied from 10300 to 296300 in 1987 - 1988, at the same time K_n values for ^{90}Sr varied from 130 to 73400 . In September 1994 these values varied from 3300 to 20300 for caesium isotopes and from 100 to 82000 for ^{90}Sr. Maximal values of caesium accumulation factor were observed in Kiev water reservoir at the time of Spring flows, when there was drainage from heavily contaminated territories of the Chernobyl NPP 30 km exclusion zone with minor content of organic matter. Similar effects were observed for ^{90}Sr at places with intensive phitioplanctone growth.

If one compares ratios of water-soluble and suspension-adsorbed radionuclides, it can be seen, that major parts of both radioactive caesium and strontium-90 are in water-soluble form (from 43.42 to 98.40% of ^{134}Cs+^{137}Cs and from 56.36 to 99.87% of ^{90}Sr contents in water cross-section). At the same time, ^{90}Sr is known to exist preferably in water-soluble forms and its suspension-bonded share does not exceed, generally, 9% from its overall content. Only at periods of intensive phitoplanctone growth, due to its more high content in algae concentrations, this value might grow up to 40%. Majority of radionuclides is known to undergo rapid transfer to bottom deposits after their inflow into a water object.

Due to the consequences of the Chernobyl NPP accident, the bulk of radioactive fallout was concentrated within the 30 km exclusion zone at the nearest watershed area of Prypyat' river. The radionuclides were incorporated into both abiotic and biotic components of aqueous ecosystems. There is some risk that these substances might be transported with Spring flows into these water bodies and Prypyat' river and then into the downstream cascade of water reservoirs. Accounting for these factors we have studied the most contaminated water bodies of the exclusion zone: the Chernobyl NPP cooling pond

(water area about 22420000 m^2), Krasnenian branch of Prypyat' (water area about 1200000 m^2), closed lakes - Glyboke lake (water area about 69688 m^2), Daleke lake (water area about 3750 m^2). Prypyat' river was selected as a reference water body (near Chernobyl city).

Once entering the water ecosystems of the 30 km exclusion zone, the radionuclides from the accident fallout undergo distribution between the above components. At this stage bottom sediments play a major role, because due to their high absorption capacity, they can concentrate and deposit major part of radionuclides entering a water body, excluding the radionuclides from biological chains (however, only partly).

In 1994 - 1996 the most high concentrations of radioactive caesium were observed in October 1995 in 0 - 5 cm layer of bottom sediments from Glyboke lake (133000 Bq/kg dry matter), the most high concentrations of strontium-90 were observed at the same time in 0-5 cm layer of sludge from Daleke lake (15091 Bq/kg). Minimal concentrations of radioactive caesium and strontium-90 were observed in sandy bottom sediments from Prypyat' river (161 Bq/kg and 3.41 Bq/kg, respectively). Analyzing the cross-section distribution we have determined that maximal radionuclide levels (especially for ^{90}Sr) are located within 1 - 3 cm layer of bottom sediments. It might be attributed to accumulation by living hydrobionts and their further sedimentation.

As Table 1 demonstrates, maximal concentrations of caesium radionuclides and strontium-90 were observed in Glyboke lake and Daleke lake, being closed water bodies (Fig. 3). Minimal concentrations of these radionuclides were observed in water and suspensions from Prypyat' river (near Chernobyl city).

Depending on diverse factors: rainfall intensity, Spring flows intensity, changing solubility of "hot" particles (these particles are fairly common in the water bodies studied), desorbtion from soils, radionuclide concentrations in water of these water bodies increases and decreases, remaining however, near relatively high levels (Fig. 2). While analyzing radionuclide distribution in the water cross-sections between soluble forms and suspensions -adsorbed ones (Table 1) we have to note, that, similarly to the Dnieper cascade patterns, caesium radionuclides are almost equally distributed between water and suspensions, while ^{90}Sr is mainly concentrated in water.

Having radionuclide concentrations in water and suspensions, water area and depth of a water body, one can assess amounts of caesium radionuclides and ^{90}Sr in water cross-sections of the

137

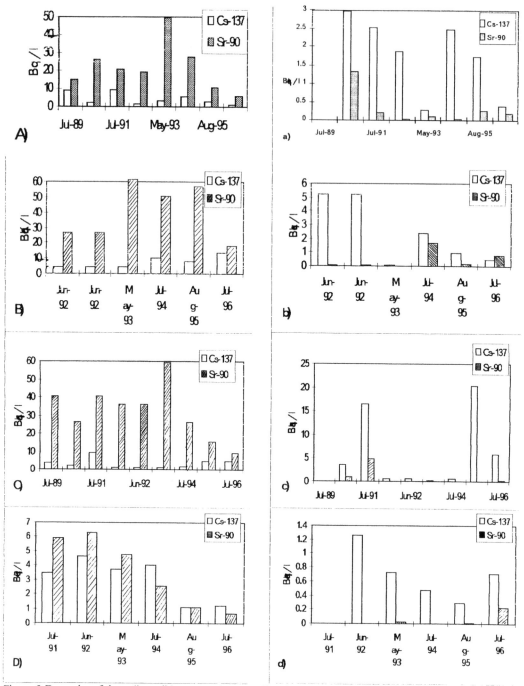

Figure 2. Dynamics of the radionuclide content in water (first letter) and suspension (small letter) of the Krasnenian branch (former river-bed of Prypyat') (A,a), Glyboke Lake (B,b), Daleke Lake (C,c), Cooling Pond of ChNPP (D,d).

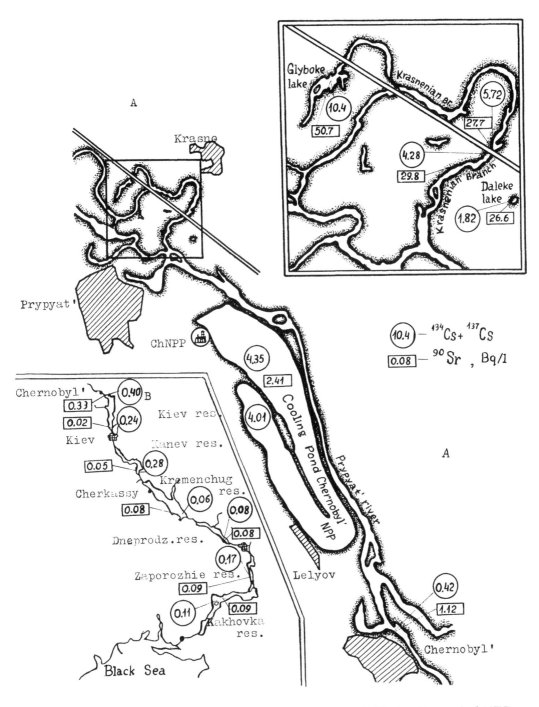

Fig.3. Distribution of radionuclides in water of 30-km.Chernobyl'NPP (ChNPP) exclusion zone water bodies in July 1994 (A) and Dnieper reservoirs in August-September 1994 (B).

Table 1. Radionuclides content in water and suspensions of the plain water bodies of the Prypyat' river of the Chernobyl' NPP 30km exlusion zone.

Date	$^{137}Cs + ^{134}Cs$				^{90}Sr			
	Water	Suspensions		Sorbed on suspensions	Water	Suspensions		Sorbed on suspensions
	Bq/l	Bq/l	Bq/kg	%	Bq/l	Bq/l	Bq/kg	%
Cooling pond ChNPP								
May-94	1.70	0.440	149000	20.56	1.68	0.020	5800	1.18
Jul-94	4.01	0.474	191268	20.56	2.53	0.001	283	0.04
Nov-94	2.22	0.980	572000	30.63	0.89	0.001	394	0.11
Apr-95	1.02	0.009	82510	0.87	0.98	0.009	4036	0.91
Aug-95	1.11	0.290	147577	20.71	1.08	0.014	7597	1.28
Oct-95	0.95	0.290	94852	23.39	1.10	0.020	6672	1.79
Jul-96	1.21	0.710	77106	36.98	0.65	0.230	25389	26.14
Aug-96	0.66	0.213	901876	24.40	0.61	0.008	35671	1.36
Krasnenian branch								
May-94	2.52	0.584	148000	18.81	18.41	0.218	55200	1.17
Jul-94	5.72	2.466	311507	30.13	27.70	0.031	3951	0.11
Nov-94	1.39	6.650	1392560	82.71	12.70	0.010	2270	0.08
Apr-95	1.27	0.003	730	0.24	8.86	0.001	292	0.01
Aug-95	2.92	1.730	57525	37.24	10.98	0.250	8172	2.23
Oct-95	1.18	0.877	243570	42.6	9.38	0.127	35131	1.33
Jul-96	1.01	0.390	27309	27.86	6.04	0.190	13106	3.05
Aug-96	1.20	0.650	100764	35.14	5.16	0.110	16583	2.09
Glyboke Lake								
May-94	7.99	1.410	694000	15.00	47.85	0.526	159636	1.09
Jul-94	10.40	2.378	281633	20.56	50.70	1.640	194345	3.13
Nov-94	1.03	1.090	247500	51.42	25.40	0.010	2260	0.04
Apr-95	5.45	0.290	7960	5.05	21.70	0.008	1833	0.04
Aug-95	8.07	0.383	59101	4.53	57.16	0.800	39938	1.38
Oct-95	8.68	1.690	32878	16.29	60.60	1.250	94829	2.02
Jul-96	13.98	0.514	76321	3.55	18.27	0.810	119882	4.25
Aug-96	3.13	2.506	378363	44.46	19.11	0.300	45728	1.55
Daleke Lake								
May-94	2.02	0.655	176949	24.48	32.22	0.056	15229	0.17
Nov-94	43.70	10.600	6660000	19.52	19.22	0.005	2810	0.03
Apr-95	25.99	0.026	8464	0.10	13.33	0.003	746	0.02
Aug-95	4.94	20.400	3916984	80.51	15.84	0.004	814	0.03
Oct-95	3.53	3.020		46.11	13.77	5.300		27.79
Jul-96	4.93	5.980	323468	54.81	9.43	0.240	13184	2.48
Aug-96	5.63	9.660	558467	63.18	8.26	0.470	27433	5.38
Prypyat' river (Chernobyl)								
May-94	0.16	0.137	15600	46.18	0.57	0.015	1740	2.62
Nov-94	0.21	0.700	36200	76.92	0.50	0.019	997	3.66
Apr-95	0.19	0.020	2248	9.52	0.19	0.003	327	1.55
Aug-95	0.06	0.043	2909	43.88	0.21	0.004	267	1.90
Oct-95	0.05	0.044	3770	49.44	0.16	0.037	32130	18.78
Jul-96	0.07	0.074	5788	51.39	0.22	0.006	443	2.65
Aug-96	0.05	0.040	4644	42.38	0.11	0.002	191	1.43

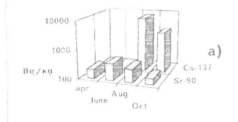

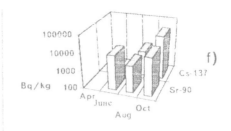

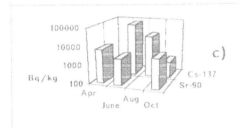

Figure 4. Seasonal variation of ^{137}Cs and ^{90}Sr content in water plants from some water bodies of the ChNPP 30-km exclusion zone (1995); a - reed (*Phragmites australis*, cooling pond of ChNPP), b - (*Stratiotes aloides*, Glyboke Lake), c - reed mace (*Typha angustifolia L.*, Glyboke Lake), d - Glyceria (*Glyceria aquatica*, Glyboke Lake), e - yellow pond-lily (*Nuphar lutea Smith*, Daleke Lake), f - horwort (*Ceratophyllum demersum L.*, Krasnanian branch of Prypyat')

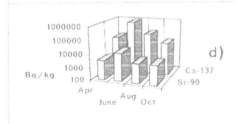

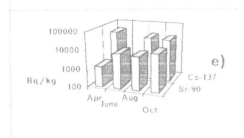

water body.

Our estimates show that amounts of radioactive caesium in Głyboke lake, Daleke lake, Krasnenian branch , ChNPP cooling pond reached: 1.51 10^9 - 4.04 10^9, 4.09 10^7 - 1.23 10^8, 2.52 10^9 6.72 10^9, 6.46 10^{10} - 3.87 10^{11} Bq, respectively (by July 1996). The same estimates might be made for strontium - strontium depo in Glyboke lake was - 1.99 10^9 - 5.32 10^9 Bq, in Daleke lake - 3.63 10^7 - 1.09 10^8 Bq, in Krasnenian branch 1.12 10^{10} - 2.99 10^{10} Bq, in ChNPP cooling pond - 2.96 10^{10} - 1.78 10^{11} Bq.

We have also estimated accumulation factors for the radionuclides under study (K_n) for sediments from the water bodies studied. Within the period of study (from 1994 to 1996) accumulation factors for radioactive caesium and ^{90}Sr were for Głyboke lake, Daleke lake, Krasnenian branch of Prypyat' river, Ch NPP cooling pond: 1460 - 240260 and 80 - 6560; 320 - 792910 and 50 - 3320; 570 - 206420 and 30 - 3210; 63700 - 1370000 and 110 - 58500, respectively.

Hydrobionts play major role in radionuclide accumulation. It is not worth to ignore biological factor, because it is also connected with internal ecosystem processes of radionuclide distribution and with their transfer from the exclusion zone by "biological path". As Fig. 4 shows, radionuclide content in water plants is relatively high, especially in Głyboke lake and Daleke lake. We have to note, that content of radioactive caesium and ^{90}Sr in fish depend on several factors, including: its age, length, nutrients, environment, etc..

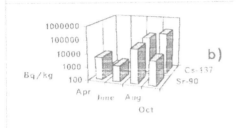

One pike (*Esox lucius*) from Daleke lake (age 4 years, weight 1.5 kg) shows the following radionuclide patterns: muscle tissue - caesium radionuclide content - 7429 Bq/kg (native weight), ^{90}Sr - 226 Bq/ kg; bone tissue - caesium - 4778 Bq/kg, ^{90}Sr - 3541 Bq/kg. Muscle tissue of a pike (0.75 kg of weight) from Prypyat' river shows: 1873 and 69 Bq/kg, muscle tissue of a bream (*Abramis brama*) (1 kg of weight) shows: 63 and 5.6 Bq/kg, muscle tissue of a sheatfish (*Silurus glanis*) (3 kg of weight) shows: 30 and 2.7 Bq/kg. Muscle tissue of breams from Ch NPP cooling pond shows 4-10 KBq/kg and 34.3 Bq/kg, respectively, and muscle tissue of sheatfishs 5-40 KBq/kg ^{137}Cs.

CONCLUSIONS

Radioactive contamination of biotic and abiotic components from the exclusion zone water bodies is a direct mirror pattern of complex radiological situation of the exclusion zone itself. For decades ahead the situation will be determined by the most environmentally hazardous radionuclides - ^{90}Sr, ^{137}Cs and trans-uranium elements. Due to high levels of radioactive contamination Ch NPP cooling pond, Glyboke lake, Daleke lake, Krasnianski branch of Prypyat' river are in fact unique radiological research test sites. (Fig.4)

Concentrations of radionuclides in different ecosystem components of the Dnieper cascade water reservoirs are substantially higher than before the accident. This also demands continuous radiological monitoring of the Dnieper cascade water reservoirs. In the case of major Spring flows and heavy rainfall there is a possibility for radioactive contamination of the exclusion zone water bodies, and the Dnieper cascade water reservoirs, through the Prypyat' water discharge.

REFERENCES

Кленус В.Г., Кузьменко М.И., Матвиенко Л.П. и др. 1991. Содержание радиону-клидов в воде и взвесях р.Днепра и его водохранилищ. *Гидробиол. ж.*4(27):82-87.

Кленус В.Г., Матвієнко Л.П., Каглян О.Є. 1994. Розподіл радіонуклідів по основних компонентам екосистем деяких водойм лівобережної заплави р.Прип'яті. *Доповіді Академії наук України* 1: 118-120.

Euzmenko M.I., Klenus V.G., Kaglyan A.E. et all, 1992. Transport of radionuclides in the Dnieper reservoirs after the Chernobyl disaster, *Tracer Hydrolody - Proc 6-th Int. Symp. on Water tracing, Karlsruhe,21-26 Sept.1992*: 115-118. Rotterdam: Balkema.

Романенко В.Д., Кузьменко М.И., Евтушенко Н.Ю. и др. 1992. *Радиоактивное и химическое загрязнение Днепра и его водохранилищ после аварии на Чернобыльской АЭС*. Киев:Наук. думка

Tracer Hydrology 97, Kranjc (ed.) © 1997 Balkema, Rotterdam, ISBN 90 5410 875 4

'In situ' determination of the dilution efficiency of submarine sewage outfall's with the help of fluorescent dye tracers

J. Roldão, J. Pecly & L. Leal
Laboratório de Traçadores, COPPE/UFRJ, Rio de Janeiro, Brazil

ABSTRACT: Fluorescent dye tracers were used to label industrial and domestic sewage introduced in the sea by submarine outfalls aiming to evaluate the effluent dilution pattern in the sea water. Uranine and Amidorhodamine G were simultaneously injected in the outfall's pipeline during 6 to 8 hours by a continuous injecting device and detected at different depths in the receiving coastal water. In order to obtain sufficiently detailed information to later plot the plume of sewage dilution, a large number of samples attached to theirs respective coordinates must be taken in the field. Computer programs to assist the navigation of the monitoring boat and to analyze the obtained data in the laboratory are essential. This paper provides indications on how fluorescent dye tracers can lead to a better understanding of the effluent dilution in the sea with answers that are also valuable for water quality modeling.

1 INTRODUCTION

For the past two decades several sewage submarine outfalls have been put in operation in Brazil, some are being planned and others are under construction. As the dilution efficiency of operating submarine sewage outfalls is quite difficult to evaluate, the demand for modern tools to do it has grown. Mathematical modeling can provide a first approach, but still some doubts remain. This paper shows how tracer techniques can be used to provide additional information for a better understanding about the performance of such important hydraulic structure.

Although there are several field experiments using drogues aiming to evaluate the effluent dispersion in the receiving body, the results obtained are valid only to evaluate the transport in the upper layer, which is strongly dependent on the wind stress. Also, oceanographic measurements have shown different current patterns between the upper and lower layers of the receiving body.

The need of artificial tracers appears as a consequence of the limitations of using parameters like turbidity, D.O., pH as "pseudo-environmental" tracers. Tests performed with samples of the effluent of a submarine outfall of a paper plant showed that the cellulose production from Eucalylipus tree did not produce enough amount of natural fluorescent

substances like Lignin Sulfonates which could be used as a tracer, as previously suggested (Wilander et al. 1974). The use artificial radioactive tracer is also not recommended, due to the large amount of water to be labeled, the high degree of dilution in the sea and the requirement of rather long injection time. In most cases the need of having a suitable artificial tracer to label the effluent makes possible quantitative monitoring after strong dilution with the receiving ocean water, which is fulfilled by fluorescent dye tracers. A proper device is necessary to guarantee a steady injection of the tracer solution in the effluent flowing in the pipeline and also a well equipped monitoring boat in the sea. The tracer injection must last for some hours in order to allow the systems to reach near "steady-state conditions".

This work gives some details about the methodology which is being developed by the Laboratório de Traçadores aiming the "*in situ*" determination of the dilution degree, in terms of dilution factor, when a submarine outfall discharges domestic or industrial effluent at the bottom of the sea.

In general the effluent is labeled by simultaneous injection of two well-known fluorescent dye tracers: Uranine (Acid Yellow 73) and Amidorhodamine G (Acid Red 50) diluted in fresh water. The field detection of these tracers is performed by portable fluorometers.

Typically, a submarine pipeline in Brazil has a diameter between 1 to 3 m, 1,000 to 3,000 m in length and the diffusers zone is placed in sites where the local water depth ranges from 15 to 30 m. The sewage flow rate ranges from 1 to 6 m3/s.

The range of minimum dilution of the liquid effluent in the sea can vary from 1/30 to 1/1,000 or dilution factors between 30 and 1,000.

In addition to the valuable information on the real dilution capacity of submarine outfalls, a tracer measurement offers as a "by-product" the precise determination of the effluent flow rate.

2 METHODOLOGY

2.1 *Tracer Selection and Injection*

Laboratory tests carried out with samples of industrial and domestic effluents of different origins have shown that strongly diluted Uranine and particularly Amidorhodamine G are stable enough, after being treated with chloroform, to withstand bacteriological degradation.

In order to label an effluent which is continuously released into the sea, the tracer solution introduced in the pipeline should be also continuously injected at a well known concentration and flow rate. A simple apparatus that fulfills these needs in general incorporates a large dilution barrel (200 l) with a mechanical stirrer and a peristaltic pump. The injection rate is around 1 liter/min and the tracer concentration lies between 5 and 10 %.

The mean tracer concentration of the labeled effluent ($\overline{C}$) is obtained when a proper sampling program is established at a point sufficiently away from the tracer injection site.

The definition of "dilution factor" is given by:

$$S = \frac{\text{mean tracer concentration in the effluent}}{\text{tracer concentration in the sea}}$$

This relation synthesizes the basic idea of using tracers to label the effluent in order to measure its dilution in the sea. The injection and sampling sites at the onshore pipeline, as well as the submerged pipeline and diffusers of the sewage outfall are shown in the following sketch.

The start of the injection is normally planned to be at the beginning of the ebb tide, for this condition supposingly leads to the worst case of shore contamination. During the scheduling of the field work, a delay must be allowed to take into account the transit time of the tracer between the injection site and the end part of the submerged pipeline where the diffusers are installed.

In general, a total time of around 1 to 2 hours is allowed before starting monitoring the tracer plume in the sea, in order to reach a "near steady" dilution pattern.

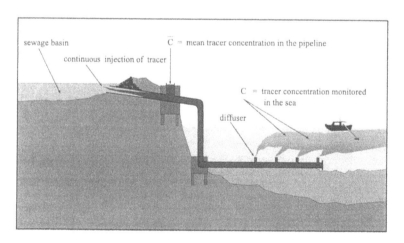

Figure 1.   Injection, sampling and detection of dye tracer in the sea.

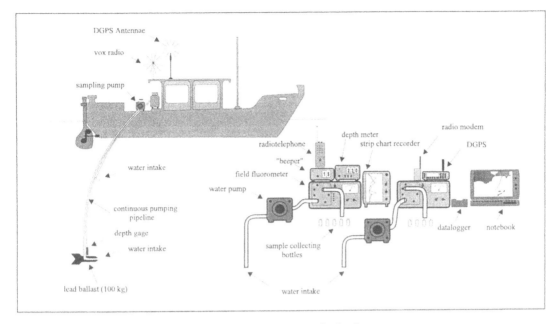

Figure 2. Equipment installed aboard the fluorescent tracer monitoring boat.

2.2 *Tracer Monitoring in the Sea*

The dynamic measurement of the tracer concentrations in the sea in order to determine the dilution factor (S) requires a boat equipped as shown in Figure 2.

In order to reach a better understanding of the effluent plume in the sea, the horizontal tracer concentration distribution is simultaneously measured at 2 or 3 different depths.

The simultaneous detection of tracer concentration at different depths provides a better knowledge of the vertical tracer distribution which is very important in most cases, and crucial in situations of stratified water columns. When the information of dilution at depths very close to the surface is needed, a compromise between this and the prevention of air suction by the circulation pumps installed aboard the monitoring boat must be made. In general the determination of the tracer concentration in the sea is made with the help of field fluorometers. The field procedures adopted by the Laboratório de Traçadores involve also a simultaneous sampling program of the sea water from each monitoring depth at every 15 sec.

The "*in situ*" reading of the tracer concentration by fluorometers is in general accepted as sufficient to obtain all the relevant information about the dilution. The additional procedure of taking samples, not only increases the detection efficiency (which decreases the involved tracer amount) but also makes possible to "take home" the tracer plume that can be in many cases very attractive, especially in case of equipment failure.

Each set of tracer detection comprises a 12 VDC circulation pump driven by a 56 Ah automotive battery which pumps sea water to a portable field fluorometer Turner model 10 equipped with a flow through measuring cell. Samples are taken at the outlet of the 1/2" diameter connecting polyethylene hose. The measured fluorescence values are recorded by a two channel stripchart recorder. The sampling hoses are kept at the selected depths by a 100 kg lead ballast whose depth is monitored by a pressure gage.

The navigation in the sea must be performed by a positioning system in order to know the exact location of each sampling point since the samples are taken by a moving boat. Some years ago the usual positioning system in the field works was comprised by 3 theodolites. Later this system was replaced by a Motorola positioning equipment Mini Ranger III and more recently by a DGPS (Differential Global Positioning System). All of them provide the necessary accuracy but the DGPS brings several additional advantages. The positioning system used nowadays employs a real time DGPS, a charting software integrated with a datalogger. Variables such as time, tracer concentration, present monitoring depth and water temperature are continuously logged and plotted on the notebook screen. The whole system is fully integrated and results in a valuable navigation aid. A simple device ("beeper") generates audio signals for synchronizing the time for water sampling and the time to acquire the boat position by the DGPS. The sampling time, coordinates, tracer concentration at different depths, depth of the lead ballast and water temperature are acquired in a datalogger and saved in the notebook. The tracer concentrations at two depths are still plotted, as a matter of security, by a stripchart recorder which is still being used. The sampling time is also marked on the chart paper by an event marker.

The field activities must be carefully planned due to the number of crucial equipment installed aboard the monitoring boat (in most of cases a small fishing boat). The planning includes a lot of checking and control procedures to minimize the possibilities of serious mistakes or failures that could completely jeopardize this "real time" field experiment.

The course of the monitoring boat zigzags over the plume path. In general the navigation lines (crossings) are closer at the near field, in order to assure that the highest concentrations (critical dilution factor) will be reliably detected.

Samples are taken along the navigation line every 15 sec at the outlet of the 1/2" plastic pumping hoses. Position corrections are made to compensate for the delay caused by the transit time of the water along the plastic pumping hoses.

Water samples are stored in 60 ml dark polyethylene bottles that are previously numbered. Every 10 or 15 samples, the samples numbers (all depths) and the instant of acquisition (time of the day) are also used to label the event marked in the stripchart recorder and on previously prepared forms to have a "checkpoint" control of the huge number of data produced after one day of field activities. The water samples from different depths are put in proper boxes to prevent photodecay and to make easier its transport to the laboratory. A typical campaign, lasting around 6 hours, generates something like 2,000 - 3,000 samples to be later analyzed in the laboratory.

An independent team aboard a second boat, which is moored in several sites, should take care of collecting samples to determine the vertical distribution of tracer concentration. These vertical profiles are very important to check the vertical mixing condition of the tracer/effluent plume. This boat is also used to support the measurements of the vertical profiles for temperature, turbidity, pH, D.O., conductivity, salinity, wind speed / direction and air temperature. The appropriate mooring locations of the second boat are decided by the monitoring boat based on the concentration detected during its navigation.

Water samples collected in the field during the vertical profiling (Figure 3) and the horizontal monitoring (Figure 4) are taken to the laboratory where their fluorescence and concentration are measured by the Synchronous Scanning (Behrens 1971). This method requires spectrofluorometers that are more expensive and fragile, and therefore are not recommended for *in situ* analysis.

The result of a vertical tracer distribution and other parameters measured at fixed point of the tracer plume are shown in Figure 3.

The same vertical concentration distribution pattern for both tracers confirms the strong stratification of the water column between 16 m e 22 m which generally occurs at Ipanema's sewage submarine outfall during summer conditions.

This strong stratification, caused by temperature effects, reduces significantly the possibility for vertical dilution, and has a clear correlation with vertical profiles of temperature, D.O. (affected by the presence of the effluent) and pH. This figure shows also how difficult it is to perform a field experiment when a mixing takes place in such a narrow and deep water layer.

In order to reach a better understanding of the horizontal dilution field of the effluent in the sea it is necessary to plot the horizontal tracer distribution maps at each monitoring depth.

As the complete set of data to be taken in account is large, the use of computers is welcome, if not the only choice. The computer analysis program employed in this kind of work is based on the geographic coordinates and concentration of each sampling point. It then defines a new regular grid (in general 30 m x 30 m) where dilution factors are calculated by proper interpolation. This is the set used for dilution factor isolines map.

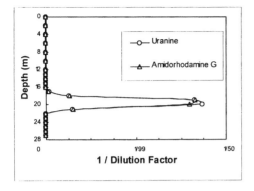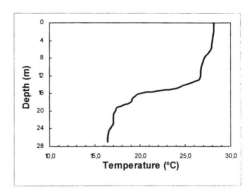

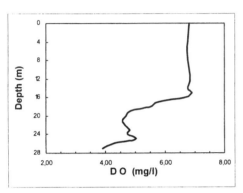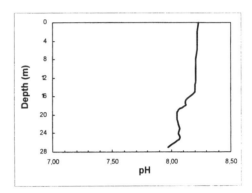

Figure 3. Vertical distribution of tracer concentration, temperature, DO and pH - Summer Campaign at Ipanema's sewage submarine outfall near the diffusers zone

The final shape (mostly the smoothness of the drawn isolines) is dependent of the proper choice for the values of the internal parameters.

The plot of the dilution factor isolines not only gives a better idea of the effluent/tracer path, but also contributes to a proper evaluation of the relative importance of any dilution factor range when its circumscribed area can be measured. An example of such an evaluation is seen in Table 1, where

"punctual" minimum dilution factors are presented as well as the associated area by the smaller dilution factors. The data was measured in field campaigns at two important outfalls in Brazil (domestic and industrial effluent).

From an engineering point-of-view the "minimum dilution factors" are less meaningful than the dilution factors that define a significant contamination area.

Table 1. Summary of data and critical dilution factors measured during seasonal campaigns using fluorescent dye tracer at two sewage submarine sewage outfalls in Brazil

Location of the outfall (type of effluent)	Campaign	Tide	Pipeline		Sea			
			Flowrate (m^3/h)	Mean conc. (ppb)	Minimum dilution factor	Minimum dilution isoline	Associated area (km^2)	Associated area (km^2)
Aracruz (industrial)	Spring	Flood	9,700	337	40	140	N. M. [2]	0.180 [6]
	Summer	Flood	9,530	352	94	140	0.026 [2]	0.002 [6]
Ipanema (domestic)	Summer	Flood	22,300	303	28	50	0.180 [17]	0.31 [21]
	Summer	Ebb	21,600	174	33	50	0.000 [17]	0.03 [21]

Notes : The values in brackets denote monitoring depth.   N.M. = not monitored

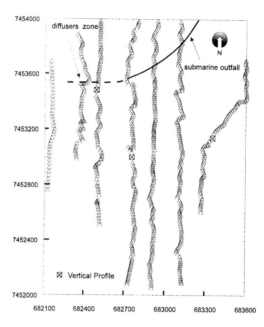

Figure 4. Navigation lines and sampling points in the Summer Campaign at Ipanema's outfall

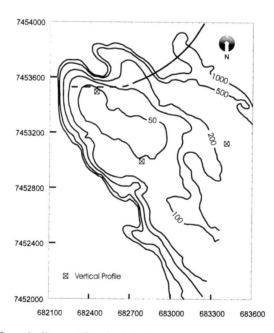

Figure 5. Tracer dilution factor isolines at 17 m depth in the Summer Campaign at Ipanema's outfall

3 CONCLUSIONS

The adopted fluorescent dye tracer methodology has proved to be very helpful to evaluate the *in situ* dilution patterns and factors of industrial and domestic effluents released in the sea. The horizontal dynamic tracer monitoring at several different depths and the vertical distribution of tracer concentration at selected sites allowed to measure the critical dilution factors and its affected areas in different seasonal situations, even when the tracer/effluent is trapped in narrow and deep water layers.

Tracer experiments should be performed in parallel with oceanographic campaigns in order to supply mathematical models with reliable and control data.

REFERENCES

Wilander, A., Kvarnäs, H., & Lindell, T. - A modified fluorometric method for measurement of Lignin Sulfonates and its *in situ* application in natural waters, *Water Research*, 1974, 1037-1045.

Behrens, H. - Untersuchungen zum Quantitativen Nachweis von Fluorezenzfarbstoffen bei ihrerer Anwendung als Hydrologische Markierungsstoffe - *Geologica Bavaria*, 1971, **64**, 120-131.

Roldão. J. et alli, - Monitoring the dilution capacity of the Aracruz Celulose S.A. submarine sewage outfall, seasonal and final reports ET-150525 (in Portuguese), Rio de Janeiro, 1994.

Roldão. J. et alli, - Evaluation the dilution efficiency of the Ipanema's sewage submarine outfall, summer report ET-150663 (in Portuguese), Rio de Janeiro, 1996.

ACKNOWLEDGMENTS

The authors wish to thank the financial support from the Brazilian Research Council (CNPq) and the sewage treatment companies which have made the field activities possible.

Unsaturated zone

Tracer Hydrology 97, Kranjc (ed.) © 1997 Balkema, Rotterdam, ISBN 90 5410 875 4

Tracer investigations in the unsaturated zone under different cultivation types in a mountainous catchment area

T. Harum & J. Fank
Institute of Hydrogeology and Geothermics, Joanneum Research, Graz, Austria

W. Stichler
GSF-Institute of Hydrology, Oberschleißheim, Germany

ABSTRACT: Since August 1995 three observation stations of infiltration water have been in operation under different cultivation types in a small cristalline catchment area S of the Alps. In May 1996 a combined tracing and irrigation experiment has been carried out using water from the Northern Calcareous Alps. This water is characterized by a different chemistry comparing to the natural background in the cristalline catchment and different values of the stable isotope ^{18}O. As artifical tracers Lithium and Bromide were used. The first results of the tracing experiment indicate important differences between the landuse systems forest, pasture and arable land concerning water movement and solute transport and allow the comparison of the transport behaviour of the natural and artificial tracers used.

1 DESCRIPTION OF THE TEST SITE

The area under investigation is a well delimited small catchment named Höhenhansl (surface 0.4 km², mean altitude 963 m a.s.l.), which is situated S of the alpine chain in the sub-alpine mountain ranges of the Steirisches Randgebirge (Eastern Styria, Austria). It is a tributary catchment of the Pöllau drainage basin (57.5 km²). The catchment is pilote zone in the frame of the project AGREAUALP (Agri-environmental measures and water quality in mountain catchments) financed by the European Commission, the Austrian Ministry for Science, Traffic and Culture and the Government of the federal province of Styria.

The drainage basin consists primarily of less permeable schists and gneiss (fissured aquifer) covered by a thick weathered sandy surface layer (porous aquifer). The typical alpine extensive landuse is characterized by alternating grassland and pasture, arable land and dominating forest (more than 70 % of the surface, s. fig. 1). The anthropogenic influence is primarily limited to dunging, pasturage and forestry. The mean annual precipitation of this region is in the range of 900 mm with an evapotranspiration of approximately 39 % of the precipitation.

The hydrological conditions are described in Bergmann et al. (1996), Gutknecht (1996), Fank et al. (1996, 1997), Fank & Harum (1997), Harum & Fank (1992) and Zojer et al. (1996a, 1996b). The catchment area is drained by a small creek, which is flowing in the outcropping gneiss. Therefore it can be assumed, that the total discharge is measured at

the outlet. Due to the thick weathered layer the hydrographs at the outlet of the small catchment area are characterized by a relatively low portion of surface flow during flood events.

The position of the hydrological measuring stations is visible in the landuse map (fig. 1). The total runoff from the small catchment area, which can be considered as a natural lysimeter, is measured continuously at a gauging station together with water temperature and electrical conductivity. The same parameters are measured at the main spring. The different types of impact from landuse on groundwater recharge are being detected at lysimeter stations under forest, pasture and arable land with crop rotation (outside of the catchment area, fig. 1) and detailed hydrological, hydrochemical and isotope investigations (c. fig. 2).

All lysimeter stations consist of a field lysimeter for the determination of the groundwater recharge, a profile with TDR-probes for measuring the time dependent water content of the soil in different depths, a profile with tensiometers to get information about the water tension of the soil in the same depths as the water content as well as the soil temperature. Suction cups in the same depths are used as well as field lysimeters to sample the infiltration water for hydrochemical and isotopical investigations. Additionally the amount of surface flow is being measured and sampled for analysis at small artificial catchments (surface 1 m²). At each lysimeter station the meteorological parameters precipitation amount, air temperature and humidity are being measured nearby.

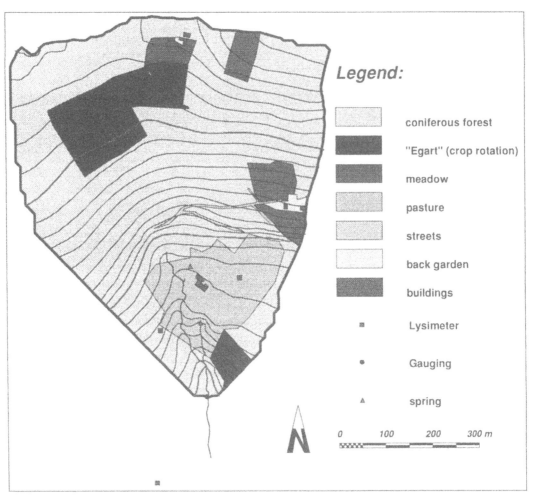

Figure 1. Distribution of landuse systems and measuring network in the catchment area Höhenhansl.

All research stations are constructed at an embankment. From a pit in the embankment of the test fields the probes were drilled up to three meters under the different landuse areas, so that the normal landuse practices can be done in the field. The small field lysimeters with suction plates are installed in a greater depth than the highest part of the roots of the plants. Therefore water consumption by the plants in these depths is impossible and the amount of water measured represents the groundwater recharge at the measuring point.

2 THE COMBINED TRACING EXPERIMENT

In May 1996 a combined tracing experiment has been carried out at the three lysimeter stations with the following main goals:

2.1 Objectives

- Determination of the maximum infiltration velocities under disturbed (lysimeter) and undisturbed (suction cups) soil conditions.
- Verification of the infiltration area of the lysimeters for the exacter determination of the amount of groundwater recharge under different landuse systems.
- Comparison of solute transport and water movement in the unsaturated zone under different typical alpine landuse systems.
- Comparison of the mobility of the artificial and natural tracers used and of their transport behaviour with those of contaminants from fertilization.
- Verification of model conceptions concerning water movement and solute transport in the unsaturated zone.

2.2 Boundary conditions and results with natural tracers

The soil types at the lysimeter stations can be characterized as typical weathered layer from cristalline rocks by loamy sands with sometimes intercalated blocks.

In fig. 2 the daily values of the soil tension measured by tensiometers in the lowest depths under the three landuse systems are compared with the daily precipitation amount (a) and the contents of the stable isotope Oxygene-18 in precipitation (b) and the suction cups in different depths under the three landuse systems (c,d,e).

The observation time before the tracer injections on May 2nd ,1996 can be characterized as period with unusually high summer and winter precipitation. Even in the summer period with high water consumption of the plants and evapotranspiration losses the matrix potential in the soil under meadow and arable land is low indicating processes of groundwater recharge by infiltration through the unsaturated

zone. Only under the forest a water deficit in the shallow soil can be recognized. The following long winter with thick snow cover from November 1995 to March 1996 is characterized by high water contents and nearly saturated soil conditions. The snow melt started at the end of March 1996 intensified by heavy rainfalls (70 mm/d) and high temperatures on the April 4th, 1996, which effected a high groundwater recharge in the catchment and consequently high discharges at the springs and the outlet.

The stable isotope [18]O represents a nearly ideal tracer allowing the determination of seasonal variations of solute transport in the unsaturated zone. The fluctuations in the precipitation (fig. 2b for the station meadow) are relatively high with approximately 22 ‰. Even in the shallowest suction cups a strong damping of the fluctuations can be recognized (for the station meadow 2.5 ‰ before snow melting, 7.5 ‰ including the snow melting period). With increasing depth no more damping is visible indicating the dominant piston flow in the unsaturated zone.

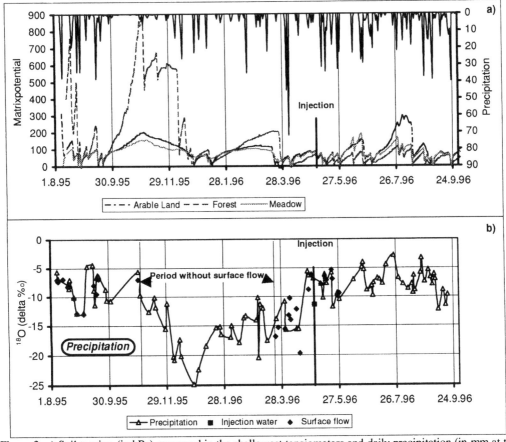

Figure 2. a) Soil tension (in hPa) measured in the shallowest tensiometers and daily precipitation (in mm at the station meadow. b) ^{18}O-contents in precipitation and surface flow (station meadow).

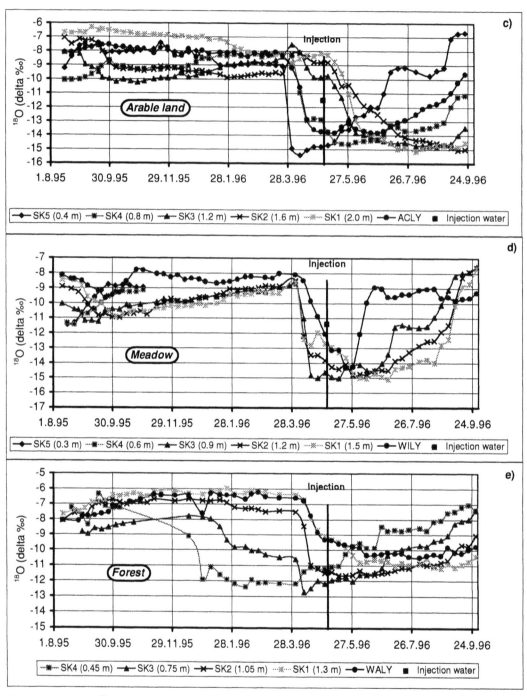

Figure 2. c), d) e) ^{18}O-contents of infiltration water sampled from the suction cups (SK) under undisturbed conditions in different depths and from the lysimeters (WILY, ACLY, WALY) under disturbed conditions under arable land (c), meadow (d) and forest (e).

During the snow melting period the soils were not frozen, therefore high amounts of melt water traced with low ^{18}O contents could infiltrate and initiated an important response in the infiltration water sampled in different depths. The different transport behaviour of ^{18}O is graphed in fig. 2c-e. It is clearly visible, that the breakthrough curves of ^{18}O show a nearly similar shape under arable land and meadow with a significantly higher retardation under arable land.

The highest heterogenities seem to be under the forest (fig. 2e). The breakthrough in the shallow suction cups (0.45 and 0.75 m) indicate earlier snow melting processes (milder microclimate) with a quick transport through macroporous systems intensified by the lower soil saturation before snow melt. The response to the input at the beginning of April is not so important comparing to the other landuse systems.

Fig. 3 shows the travel time of melted water computed from the beginning of the intensive snow melting (April 4th, 1996) to the minimum concentrations of ^{18}O in suction cups and lysimeters. In fig. 4 the maximum flow velocities calculated are being compared. It is clearly visible that the disturbed installation of the suction plates (lysimeters) effects a preferential flow through artificially created macroporous systems producing too high flow velocities.

The tracer breakthrough curves in the suction cups in different depths are well comparable between the three landuse systems taking under consideration the high inhomogenities and existing macroporous systems of the soils formed by the soil fauna, plant roots and freeze with preferential flow systems mainly under forest. This was indicated by dye tracing experiments with artificial irrigation in the unsaturated zone and following excavation of the soil (Theuretzbacher 1997). Even in greater depths (for example the suction cup in 2.0 m under arable land)

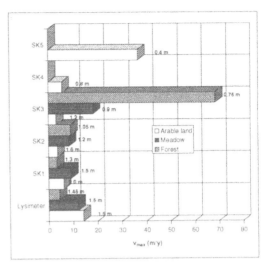

Figure 4. Maximum flow velocities in m/y in the test fields during the snow melting.

preferential flow systems with a quicker response can be detected (fig. 3).

The calculated maximum flow velocities vary in case of undisturbed flow conditions (suction cups) in greater depths between 3.1 and 6.1 m/y under arable land, 8.3 and 6.1 m/y under meadow and 3.4 and 8.9 m/y under forest. In shallow suction cups the velocities between 18.3 (meadow, less preferential flow), 36.5 m/y (arable land, disturbance by plugging) and 68.4 m/y (forest, disturbance by the root zone) indicate the macroporous systems in the shallow soil. A general decreasing of the flow velocity with increasing depth can be remarked.

The values correlate in a significant way with the groundwater recharge, which is much higher under meadow (310 mm/y) than under arable land with crop rotation (137 mm/y) and forest (130 mm/y; Fank et al. 1997). They are representative for the maximum infiltration velocities under nearly saturated conditions as they occur during periods of snow melting, i.e. the most important period of groundwater recharge in alpine regions.

2.3 Tracer injection by artificial irrigation

Due to the previous snow melting the injection of the tracers on May 2th,1996 happened in a period of nearly saturated soil conditions and maximum infiltration velocities. For the irrigation karst water from a high catchment area in the Northern Calcareous Alps (Hochschwab) was used. It can be characterized by a completely different chemistry concerning the major ions Calcium, Magnesium, Nitrate and Sulfate in comparison to the natural chemical background in

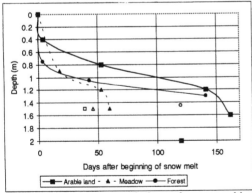

Figure 3. Maximum travel time of snow melting water under the three test fields. Filled signatures = suction cups (undisturbed flow conditions), non filled signatures = lysimeters (suction plates, disturbed flow conditions).

the cristalline catchment and different values of the stable isotope Oxygene-18 due the origin of the injection water in a higher catchment area with later snow melting (fig. 2). The chemical and isotope contents are summarized in table 1, where injection water is compared with the natural background in the lysimeter outflow before the tracer injection.

Table 1. Chemical composition (in ppm) and isotope content (in δ‰) of injection water and natural background (lysimeters under the three landuse systems.

	Injection water	Arable land	Meadow	Forest
Ca^{++}	82.40	21.84	29.51	19.47
Mg^{++}	18.40	3.62	6.81	9.35
Na^+	1.11	2.38	3.87	2.67
K^+	1.36	0.96	2.48	3.79
Cl^-	0.71	1.93	6.54	5.41
NO_3^-	4.64	59.94	45.70	21.55
SO_4^-	64.6	31.04	18.83	24.75
^{18}O	-11.39	-12.82	-9.03	-9.37

Additionally 2 kg of Lithiumbromide have been injected on each field. The tracer solutions (150-200 l) were injected by artificial irrigation on each test field, afterwards karst water from Hochschwab was irrigated. The total amount of irrigated water was at:

Meadow:	25.3 mm
Arable land:	18.2 mm
Forest:	15.6 mm

2.4 Flow velocities of migration water

The tracer breakthrough was measured at all suction cups and lysimeters in intervals of one week. The following tracers can be used for the evaluation (cp. table 1):
- Calcium, Magnesium, Nitrate and Sulfate with significant differences to the natural background.
- Lithium and Bromide with a background of 0.
- Oxygene-18 with significant signal but variations in background concentration due to snow melt.

The anion Bromide represents a very mobile conservative tracer, where effects of adsorption, retardation and chemical instability can be neglected. Therefore it can be accepted as reference tracer. The breakthrough curves for Bromide are plotted in fig. 5 for the different test fields. Important differences between the landuse systems can be observed, which confirm the results with natural tracers (chap. 2.2).

The concentrations of Bromide are much higher under arable land and forest due to the smaller area which was irrigated with the same amount of tracer.

The tracer experiment is not yet finished. But after an observation period of 5 months it is possible to draw first conclusions on the processes of solute transport under the different landuse systems.

All lysimeters show a quicker response comparing to the suction cups installed in similar depths (fig. 5). This effect confirms the results with natural tracers (^{18}O) and the existence of artificial macroporous systems due to the disturbed installation of the suction plates.

Most of the breakthrough curves from the suction cups (undisturbed flow conditions) are characterized by a first peak short time after the injection. It indicates that under all landuse systems preferential flow through macroporous systems mainly in the shallow zone of the soils exists.

Additionally it can be observed that under arable land and meadow Bromide reappeared in some deeper suction cups earlier than in the shallower ones (fig. 5). It proves the high heterogenity of the sandy weathered layer as indicated by the investigations of Theuretzbacher (1997).

The second peak in most of the breakthrough curves is due to the mobilisation of traced water by the heavy rainfalls in July 1996. In greater depths this peak did not yet occur during the observation period, therefore the sampling of infiltration water will continue.

For this reason it is only possible to evaluate the travel time of water traced with Bromide refering to the first appearance of the tracer (fig. 6).

The plot shows a rather quick response and high heterogenities under meadow followed by arable land and confirms that the maximum flow velocities depend mainly on the groundwater recharge which can be classified as follows. With the exception of the lysimeter under meadow the influence of the dis-

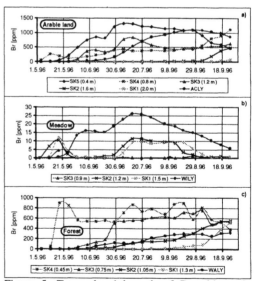

Figure 5. Tracer breakthrough of Bromide under arable land (a), meadow (b) and forest (c) in suction cups (SK) and lysimeters installed in a depth of 1.5 m (suction plates ACLY, WILY, WALY).

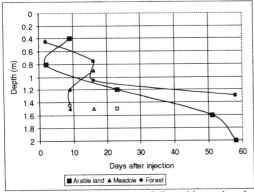

Figure 6. First appearance of Bromide under the three test fields. Filled signatures = suction cups (undisturbed flow conditions), non filled signatures = lysimeters (suction plates, disturbed flow conditions).

turbed installation with an artificially created preferential flow at the lysimeters is clearly visible.

2.5 Tracer comparison

In most of the suction cups and lysimeters the tracer passage is far away to be finished. A first comparison is possible at the shallowest suction cup under arable land. In fig. 7 the breakthrough curves normalized after the equations

$$C_{norm} = (C_{measured} - C_0)/C_i$$
$$C_{norm} = (C_i - C_0) / C_{max}$$

for the principal cations Calcium, Magnesium, Sodium, Potassium and Lithium and the reference tracer Bromide are plotted together with daily precipitation. The injected tracer mass for the different ions is listed in table 2.

Table 2. Irrigation at arable land: injected tracer mass in mg.

Ion	Tracer mass in mg
Ca^{++}	21667
Mg^{++}	4722
Na^+	304
K^+	197
Cl^-	201
NO_3^-	1104
SO_4^-	22489
Br^-	1840000
Li^+	160000

The normalized breakthrough curve of Lithium shows a clear retardation mainly concerning the first appearance and lower output comparing to the conservative tracer Bromide. The reason becomes clear observing the time-concentration curves of the other kations who indicate a typical tracer breakthrough

with a significantly higher output than Bromide. The phenomenon can be explained by adsorption of Lithium by the clay minerals of the soil and cation exchange against Sodium and Potassium (highest output), Calcium and Magnesium (partly breakthrough of the injected karst water, partly cation exchange).

The stable isotope Oxygene-18 shows until the second peak (produced by mobilisation of traced water due to an event of high precipitation) a similar breakthrough as the reference tracer Bromide (fig. 7). But in the following period contrary to Bromide the breakthrough of ^{18}O is characterized by increasing contents due to the mixing with enriched water from the high summer precipitation.

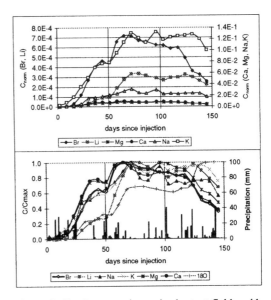

Figure 7. Tracing experiment in the test field arable land: normalized concentrations of selected kations and the stable isotope ^{18}O measured at the shallowest suction cup in a depth of 0.4 m.

3 CONCLUSIONS

The experiment is not finished and the observation of the measuring network will continue. Although it is to early for a final evaluation, calculation of the tracer output, modelling of the tracer breakthrough and comparison of the tracers, one can deduce first conclusions concerning the transport processes in the unsaturated zone under typical alpine landuse systems (forest, arable land and pasture) in the weathered layer of a cristalline catchment:

• The maximum flow velocities under nearly saturated soil conditions depend mainly on the groundwater recharge in the different landuse systems.

- The soils of the investigated landuse units are characterized mainly in the shallowest zone by systems of double porosity with significantly higher flow velocities in preferential flow systems. This effect is predominant under forest.
- The solute transport processes can be characterized by dominant piston flow with dispersive processes in macroporous systems mainly in the upper soil.
- From the tracers used Bromide and ^{18}O gave the best results concerning their mobility and the characterization of solute transport in the unsaturated zone. The Lithium breakthrough showed an important retardation. The cation was adsorbed by clay minerals in the weathered layer and exchanged against the other major ions. Their time-concentration curves show a breakthrough similar to Bromide.
- The relatively high residence times computed from the maximum flow velocities are of importance for future delimitations of protection zones for springs in comparable regions. They indicate, that often a delimitation following the 60-day-line is not only possible for pumped groundwater fields but also for springs.

REFERENCES

Bergmann, H., J. Fank, T. Harum, W. Papesch, D. Rank, G. Richtig & H. Zojer 1996. Abfluß-komponenten und Speichereigenschaften, Konzeptionen und Auswertemethoden. *Österr. Wasser- und Abfallwirtschaft*: 48 (1/2), 27-45, Wien.

Fank, J. & T. Harum 1994. Solute transport and water movement in the unsaturated zone of a gravel filled valley: tracer investigation under different cultivation types. *Future Groundwater Ressources at Risk (Proceedings of the Helsinki Conference), IAHS Publ.* no. 222: 341-354.

Fank, J. & T. Harum 1997. Stoffaustrag aus einem extensiv genutzten Einzugsgebiet. *Bericht der BAL über die 7. Gumpensteiner Lysimetertagung "Lysimeter und nachhaltige Landnutzung, 6.-8.4.1997, Gumpenstein: in press.*

Fank, J., T. Harum, H. Zojer, M. Gruber & H. Lang 1996. Agri-environmental measures and water quality in mountain catchments. Report Agreaualp (Austria), January 1995 - December 1995. *Unpubl. report, Inst. f. Hydrogeologie und Geothermie & Inst. f. Technologie und Regionalpolitik, Joanneum Research:* 79 p., Graz.

Fank, J., T. Harum, W. Poltnig, P. Saccon, M. Gruber, C. Habsburg-Lothringen, Ch. Ecker, W. Stichler, M. Eisenhut & H. Theuretzbacher 1997. Agri-environmental measures and water quality in mountain catchments. Report Agreaualp (Austria), January 1996 - December 1996. *Unpubl. report, Inst. f. Hydrogeologie und Geothermie & Inst. f. Technologie und Regionalpolitik, Joanneum Research:* Graz.

Gutknecht, D. 1996. Das interdisziplinäre Forschungsprojekt "Kleine Einzugsgebiete. *Österr. Wasser- und Abfallwirtschaft*: 48 (1/2), 1-5, Wien.

Harum, T. & J. Fank 1992. Hydrograph separation by means of natural tracers. In: Hötzl, H. & A. Werner [Editors]: *Tracer Hydrology. Proceedings of the 6th International Symposium on Water Tracing, Karlsruhe,* 21-26 September 1992, 143-148, Rotterdam: Balkema.

Theuretzbacher, H. 1997. Die Inhomogenität natürlicher Böden - dargestellt am Kleineinzugsgebiet Höhenhansl / Pöllau. *Bericht der BAL über die 7. Gumpensteiner Lysimetertagung "Lysimeter und nachhaltige Landnutzung,* 6.-8.4.1997, Gumpenstein: in press.

Zojer, H., H. Bergmann, J. Fank, T. Harum, W. Kollmann & G. Richtig 1996a. Charakterisierung des hydrologischen Versuchsgebietes Pöllau. *Österr. Wasser- und Abfallwirtschaft*, 48 (1/2), 5-14, Wien.

Zojer, H., J. Fank, T. Harum, W. Papesch, D. Rank 1996b. Erfahrungen mit dem Einsatz von Umwelttracern. *Österr. Wasser- und Abfallwirtschaft*, 48 (5/6), 145-156, Wien.

Tracer Hydrology 97, Kranjc (ed.)© 1997 Balkema, Rotterdam, ISBN 90 5410 875 4

Behaviour and comparison of dissolved silica and oxygen-18 as natural tracers during snowmelt

A.C. Hildebrand, M. Lindenlaub & Ch. Leibundgut
Institut für Hydrologie der Albert-Ludwigs-Universität Freiburg im Breisgau, Germany

ABSTRACT: This paper presents the results of a hydrograph separation during snow melt using the stable isotope oxygen-18 and dissolved silica as a hydrochemical tracer. The study was performed in the Brugga catchment (40 km²) and two subcatchments (15.9 and 6.4 km²) located in the Black Forest Mountains, southwest Germany.

An emphasis is laid on the characteristic behaviour of oxygen-18 and dissolved silica as tracers during spring flood. A comparison of the results obtained with both tracers indicates that hydrograph separation based upon dissolved silica generally tends to overestimate baseflow fraction. Furthermore, the lowest content of dissolved silica strongly corresponds with maximum discharge, ^{18}O reacts delayed. This might be a result of different flow pathways which have distinct effects (solution effects in contrast to transmission effects) on the behaviour of the two tracers.

An increasing gap between the two calculated baseflow fractions is interpreted as an increasingly filled interflow reservoir. Furthermore it is stated, that silica as a natural tracer is more suitable to show small and detailed effects on a progression line.

1 INTRODUCTION

During the last decades, hydrograph separation by natural tracers, has become a powerful tool for the study of runoff generation processes. However, by applying this method, the use of stable isotopes is limited to time periods with noticeable differences in isotopic content between system input and pre-event reservoirs. In addition, by the use of isotopes at least three different components can be detected, in respect to the two available isotope tracers oxygen-18 (^{18}O) and deuterium. Consequently, this leads to the need for hydrochemical tracers which do not show the limitations mentioned above.

However, with the application of hydrochemicals as natural tracers, other problems arise. Especially the question of conservative or non-conservative behaviour appears to be one of the main problems. Using stable isotopes the system input is strongly connected to the precipitation input, in contrast the hydrochemical tracers have their main source within the system and are partially independent from the system.

During snow melt it should be possible to consider runoff generation as a simplified two-component-system with one base component originating from groundwater and the second

component from melt water contribution. Under the assumption that within this particular hydrological condition, the two used tracers label the same reservoirs (snow cover and pre-event reservoir), a comparison of the results obtained with dissolved silica and 18O gives a deep insight into their characteristic behaviour.

2 STUDY SITE

The study was performed in the Brugga watershed (40 km²) located in the Black Forest Mountains, southwest Germany. The main catchment comprises of the smaller subcatchments of St. Wilhelmer Talbach (14 km²) and Buselbach (8 km²) as illustrated in figure 1.

The topography of the catchment shows an alpine character with steep slopes and narrow valley floors. The elevation for the main catchment ranges from 435 m to 1493 m above sea level. Runoff gauges for St. Wilhelmer Talbach and Buselbach subcatchments are situated at 633 m and 830 m ASL, respectively.

Underlying bedrock consists of gneiss and granite covered by soils and debris of changing depth, with a maximum of 10 meters at the foot of the slopes.

The catchment consists of three distinct

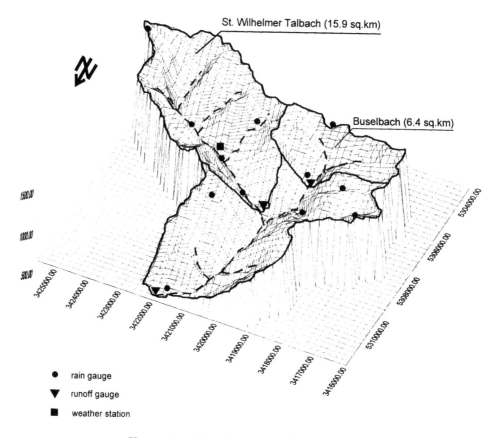

St. Wilhelmer Talbach (15.9 sq.km)

Buselbach (6.4 sq.km)

● rain gauge

▼ runoff gauge

■ weather station

Figure 1. The Brugga catchment and the subcatchments.

topographic units. Steep slopes cover 75% of the area, besides that high plains occupy 20% and narrow valley floors 5% of the remaining area. The main valleys are typically V-shaped, with relicts of glacial tills on the slopes and valley floors.

The mean annual temperature is between 3.2 and 7.5°C. Mean annual precipitation is approximately 1750 mm, generating a mean annual discharge of about 1250 mm. Maximum runoff occurs usually in spring following snow melt. The catchment is mostly forested (75%) and about 23% of the basin is used as grassland and less than 2% is sealed. In the upper part of the catchment the mean annual days with a snowcover reaches up to 177 with an average maximum depth of 175 cm.

3 METHODS

The hydrograph separation in this study was carried out according to the two-component mixing model, described by Pinder & Jones (1969), Sklash (1990),

Hooper & Shoemaker (1986) and others. The method is based on the assumption that the contributions of the reservoirs in the stream at any time can be calculated by solving the mass balance equations for the water and tracer fluxes in the stream, provided that the initial reservoir concentrations and stream water concentrations are known.

Snow cover was assumed to be free of silica. The isotopic content was determined from vertical snow cores, taken with cylindrical snow tubes with an inner diameter of 10.1 cm. Snow sampling was undertaken at nine stations within the catchment during the melt event. Snow was melted and the received, well-mixed meltwater was analysed on its isotopic content. The meltwater therefore represents a vertical mixture of the isotopic content of the snowpack at each station. There was a fast and complete depletion of the snow cover during the event. Therefore the isotopic content of meltwater was assumed to be equal to that of the snowpack.

Tracer concentrations of the baseflow components

were determined from stream samples collected during the preceding recession periods. For the main catchment tracer concentration before and after the event showed only very small variations. Therefore baseflow was considered to have an approximately constant value during the event.

For the smaller subcatchments the post-event concentrations varied significantly from the pre-event values. For calculating baseflow concentration during the event, a linear interpolation was made, according to Hooper & Shoemaker (1986), which was described as 'piecewise linear'.

Stream water sampling before and during the event was carried out with automatic samplers every three hours. The high sampling frequency made it possible to calculate hydrograph seperation precisely for both the rising and falling limb of the hydrograph. The ^{18}O contents were determined using the ^{18}O equilibration method with a Finnigan MAT mass spectrometer. For the silica contents a photometric method was used (blue silicomolybdate complex), following Deutsche Einheitsverfahren (DEV/DIN 38405 D 21).

The term „silica" in this text is used to describe the silica content, given in ppm Si, not the dissolved reactive silicate (SiO_2).

4 RESULTS AND DISCUSSION

4.1 Hydrograph separation in the main catchment

The calculated hydrograph separations for the Brugga catchment are illustrated in Figures 2 and 3. Part (a) of each shows the meltwater input, derived from the water equivalent of the snow cover and the input from precipitation. The baseflow hydrograph, calculated either using silica or ^{18}O as tracers, is shown below in part (b), in addition to the stream hydrograph. Part (c) shows the tracer concentrations used for the separations. Part (d) of the figure illustrates the calculated fractions of baseflow. This graph provides most of the information about the processes that influence the runoff generation during the event.

Comparing both figures (2 and 3) shows that the baseflow hydrographs are similar in shape to the runoff hydrographs. This suggests that the behaviour of silica during the event can be looked on as being quasiconservative.

Using silica, the calculated fraction of baseflow during the event is never below 0.68. In the separation with ^{18}O, a baseflow fraction of more than 0.5 is calculated for the event. This means that at least half of the stream water during the event is generated by reservoirs containing pre-event water. These reservoirs are considered to be located in the lower debris cover and the fissured hard rock aquifer (Lindenlaub et al. 1997). Similar to many other

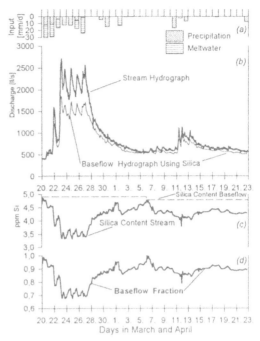

Figure 2. Hydrograph separation for the Brugga catchment with Silica. (a) Input from meltwater and precipitation, (b) stream hydrograph and calculated baseflow hydrograph, (c) silica variation in stream and baseflow component, (d) calculated fractions of baseflow.

studies, the reaction time of these reservoirs does not significantly differ from those of the direct components (e. g. precipitation or melt water).

In detail, the fraction of baseflow separated with silica shows a sudden decrease and increase with similar rates. The shape of the curve in Figure 2(d) between March 22 - 28 equals an inverted equilateral trapezium. Furthermore the minimal concentration of silica appears at exactly the same moment as maximum discharge. In contrast to this behaviour, the reaction of the isopic content is delayed and the minimal content of ^{18}O appears 24 hours after the minimum of silica concentration. The fraction of baseflow separated with ^{18}O in Figure 3(d) between March 22 - 28 therefore has a more parabolic shape than the one calculated with silica during the same period. This might be a result of different flow pathways which have distinct effects (solution effects in contrast to transmission effects) on the behaviour of the two tracers.

From March 28 onwards, the fraction of baseflow calculated with silica shows a smooth rising progression with a significant sharp bend (Figure 2(d)), that has no counterpart in total runoff behaviour. This sharp bend also appears, but stands

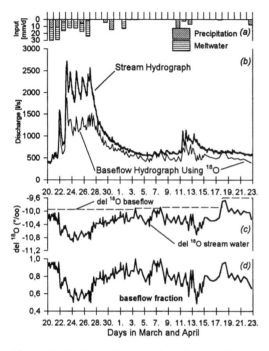

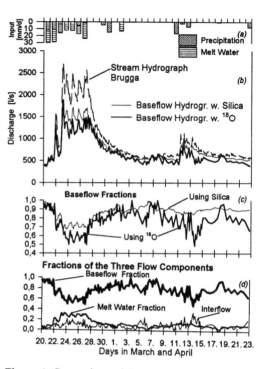

Figure 3. Hydrograph separation for the Brugga catchment using [18]O. (a), (b), (d) see Figure 2, (c) [18]O content of the stream and baseflow component.

Figure 4. Comparison of the hydrograph separations with both tracers for the Brugga catchment. (a) Input from meltwater and precipitation, (b) stream hydrograph and two baseflow hydrographs, (c) calculated baseflow fractions, (d) fractions of the flow components introducing a third interflow component.

out less clearly in the separation with [18]O. This is due to the more erratic proportion of the two components as indicated by the separation using [18]O. This corresponds strictly to earlier results reported by Hooper & Shoemaker (1986) and therefore appears to be a typical characteristic of different tracer behaviour and not limited to a distinct watershed. This 'bias' has an amplitude of 0.15‰, which exceeds the measuring uncertainty of the mass spectrometer. The sharp bend indicates the existance of more than two reservoir outflows with distinct tracer concentrations. A behaviour that can only be explained by assuming different flow pathways for one single component or by providing a system with more than two runoff components.

During April 1 - 3, a clear decrease in the amount of baseflow appears in both separations. This might originate from a small amount of precipitation which fell as snow on April 1st. This light snow cover melted within 2 or 3 days. A comparison of both figures incorporating the „bias"- effect in isotope behaviour, demonstrates that dissolved silica as a natural tracer is more suitable for showing such small and detailed effects on a progression line.

A comparison of the total amount of the baseflow fractions with both tracers indicates that the one based upon dissolved silica generally tends to overestimate baseflow fraction (Figure 4). This is

also in accordance with other studies carried out before (Maulé & Stein 1990, Hooper & Shoemaker 1986). An explanation for this distinguished tracer behaviour is a possible effect within flow paths. The existence of a fast interflow component (maybe via macropores) is assumed, consisting of event water and having its isotopic content. However, during the flow of the water through upper soil layers, an enrichment with dissolved silica is probable (Kennedy 1971 and Wels et al. 1991). This consequently leads to a general overestimation of the baseflow component.

Figure 4 shows the direct comparison of the two baseflow hydrographs with both tracers. Two main effects stand out and support the above mentioned theory of the additional interflow component:

1. The high resolution time-series during the second half of the investigation period indicates a fast reaction of stream isotopic content during small rainfall events (in contrast to dissolved silica).

2. During the event, an increasing gap occurs between the two baseflow fractions with each tracer.

This might be interpreted as the increasing 'filling up' of a possible interflow reservoir.

All the observations made in the main catchment were also found to be valid for the subcatchments.

4.2 Hydrograph Separation in the Subcatchments

Results of the hydrograph separations performed for the St. Wilhelmer Talbach and the Buselbach subcatchments are shown in Figure 5. For both subcatchments, the two groundwater hydrographs calculated with silica and isotopes lie closer together during the melt event compared to the main catchment. A predominant synchronus behaviour is evident for the first part of the event until March 26, especially for the Buselbach subcatchment (Figure 5(d)). For the second part of the event an increasing gap between both calculated base components arises in the same way as was observed for the main catchment. In respect to the explanations given in the earlier section of this paper this means a lack of the interflow component for the first six days of the event. Two different explanations for this behaviour can be given. The first one deals with an icy crust located at the interface between soils and snow cover. Both subcatchments are located at the upper parts of the main catchment with appreciable colder microclimates. Therefore the thawing of the soils may occur a few days after the melting of the snow cover has started.

The second explanation for the smaller interflow component of the subcatchments relates to the different morphology of the subcatchments compared to the main catchment. Refering to catchment area, the V-shaped valley of the St. Wilhelmer Talbach has the highest proportion of steep slopes compared with valley floors. Maybe the valley floors, that are assumed to have higher storage volumes than the debris cover of the slopes, play the most important role in generating the interflow.

Beside the climatic and morphologic differences, the used methods were not found to be affected by the transition from the 6.4 km^2 Buselbach subcatchment to the 40 km^2 Brugga watershed.

5 Conclusions

The comparison of the two used tracers shows a general overestimation of the baseflow fraction using dissolved silica. It also indicates an increasing gap with time between the fractions of groundwater either calculated with silica or stable isotopes. An interflow component affecting both tracers in a different way is postulated to be the reason for this distinct behaviour. This interflow consists of event water, which is already enriched with silica. Therefore dissolved silica cannot be considered to be a truly conservative tracer. Results of HOOPER &

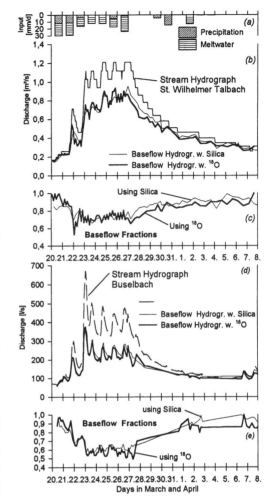

Figure 5. Comparison of the hydrograph separations with both tracers for the subcatchments. (a) Input from meltwater and precipitation, (b) stream hydrograph and two baseflow hydrographs for St. Wilhelmer Talbach, (c) calculated baseflow fractions for St. Wilhelmer Talbach, (d) stream hydrograph and two baseflow hydrographs for the Buselbach, (e) calculated baseflow fractions for the Buselbach.

SHOEMAKER (1986) concerning a more 'erratic' tracer behaviour of stable isotopes compared to dissolved silica were confirmed.

Beside the above described details in tracer behaviour, the comparison between the subcatchments and the main catchment does not show many qualitative differences. The transition in scale from 6.4 km² to 40 km² has not affected the methological aspects. It is expected that hydrograph separation with natural tracers can also be performed in even larger catchments.

ACKNOWLEDGEMENTS

We thank the Landesanstalt für Umweltschutz, Karlsruhe for providing discharge data for the St. Wilhelmer Talbach. We also would like to thank the students of the Institut für Hydrologie Freiburg, who helped collect water samples and assisted during various parts of the investigation.

REFERENCES

Hooper et al. 1990. Modelling Streamwater Chemistry as a Mixture of Soilwater End-Members - An Application to the Panola Mountain Catchment, Georgia, U.S.A. *Journal of Hydrology*, 116. Amsterdam: Elsevier Science Publishers B.V.; 321-343.

Kennedy, V.C. 1971. Silica variation in streamwater with time and discharge. *Adv. Chem. Ser.* 106.

Lindenlaub, M., Ch. Leibundgut, J. Mehlhorn & S. Uhlenbrook 1997. Interactions of hard rock aquifers and debris cover for runoff generation. Submitted for *IAHS Publication* of the 5th Scientific Assembly, Rabat, Morocco.

Maule, C. P. & J. Stein 1990. Hydrologic flow path definition and partitioning of spring meltwater. *Water Resources Research*, 26, 2959-2970.

Pinder, G.F. & J.F. Jones 1969. Determination of the groundwater component of peak discharge from the chemistry of total runoff. *Water Resources Research*, 15, 329-339.

Sklash, M. G. & R. N. Farvolden 1979. The Role of Groundwater in Strom Runoff. *Journal of Hydrology*, 43. Amsterdam: Elsevier Science Publishers B.V.; 45-65.

Wels, C., R.J. Cornett und B.D. Lazerte 1991. Hydrograph Separation: A Comparison of Geochemical and Isotopopic Tracers. *Journal of Hydrology*, 122. Amsterdam: Elsevier Science Publishers B.V.; 253-274.

Tracer Hydrology 97, Kranjc (ed.) © 1997 Balkema, Rotterdam, ISBN 90 5410 875 4

Water tracing tests in vadose zone

Janja Kogovšek
Karst Research Institute ZRC SAZU, Postojna, Slovenia

ABSTRACT: By water tracing tests in vadose zone we were searching the connection of defined points at the karst surface with its interior and also the spreading, the velocity of the percolation at given different permeable directions, the intensity and residence time of washing out the induced soluble stuff and the dynamics of rinsing. Cave systems provide an excellent opportunity for intercepting water in vadose zone. Seven water tracings above Planinska, Pivka and Postojnska Jama were carried out. The tracer was injected at the surface, in most cases directly into rocky base and watered; in one case only the tracer was spilt on soil without added water. Sampling points underground were from 40 to 100 m below the surface.

In addition to directions and spreading of water travel we recognized different ways of infiltration through single, variously permeable conduits. In all cases we stated relatively fast water percolation through permeable conduits and rather slower one through wider network of less permeable conduits. How quickly the tracer is washed out of the rocky mass depends mostly on the quantity, distribution and intensity of the rainfall that follows. To determine the returned tracer is made difficult by numerous sampling points where it is difficult to continuously measure a discharge.

1. INTRODUCTION

By water tracing tests we were primarily searching the connection of karst surface with its interior and velocity of percolation. But, detailed observations of the trickle discharge and different appearance of tracers underground, showed considerable differences between single trickles and drippings that have a connection with a specific point at the surface. In most cases the water tracing tests were carried out by adding a lot of water to tracers. Such a tracing experiment simulates the conditions appearing at road accidents by lorries transporting fluids. Several such accidents occurred in last years on the roads of Slovenia and as spilt harmful substances drained underground very quickly, no interventions were possible.

This is why it is useful to study the dynamics of substances spreading when bigger amount of fluids is spilt at the surface; but also to study the percolation and transport of substances from the karst surface in naturally wet conditions. Similar dynamics may be expected in the case of waste disposal sites.

2. THE RESULTS OF WATER TRACING TESTS

In 1977 the first water tracing test was carried out through 100 m thick unsaturated zone. At that time dye was injected as a solution of Uranin in a doline, about 15 m across, above Planinska jama and washed by 14 m^3 of water. Three weeks before and after the injection there was no rainfall worth mentioning. Most of tracer and added water drained through one point with velocity of 1.4 cm/s. In first 30 hours more than 60% of added water containing more than 20% of injected Uranin drained through this point. Positive but much weaker results were obtained at two other trickles; at the first one, discharge increased and tracer appeared, and at the second one less distinctive appearance of tracer and no change in discharge were noticed (Kogovšek & Habič 1981).

Similar tracing test was achieved from an old sink of waste water above Postojnska Jama which is not located in a doline. More then 2 weeks before there was no rainfall, but 3 days before 30 mm of rain fell, which did not cause the increase of discharge. First 0.5 m^3 of water was poured down the sink and then a solution of 60 g of Uranin was added, left to drain and later flushed with 5.5 m³ of water. The Uranin was detected in Postojnska Jama over a relatively small area of elliptical shape with a small axis of 30 m and a large axis of 40 m, at seven trickles. Already 75 minutes after the injection (it lasted 55 minutes due to slow run-off) we noticed a sharp increase of discharge and appearance of peak Uranin concentration in the most permeable trickle. Taking into account the time required for the injection, the flow velocity was 1.3 cm/s, and if we include the time for watering, the velocity was 2.2 cm/s. Rain came after 16 days and before this time only 3% of added water and 1.1% of injected Uranin reappeared in the cave. The longest residence time of tracer was three years; each of the periods of increasing dye concentration corresponds with a rainfall event.

One day later Uranin and a small discharge increase were detected in the nearby, rather thin trickle. At all the other points where Uranin appeared, its concentrations were considerably lower than those at trickle I; tracer appeared after rain which obviously squeezed it through badly permeable conduits with great resistence and slow water and mass exchange; the velocity was from 0.023 to 0.0007 cm/s (Kogovšek 1995).

2.1. Water Tracing Test in Pivka Jama

At the surface above Pivka Jama is a tourist camping; the lavatories have been built in one of nearby dolines. The waste waters are supposed to flow into a water treatment plant. But our analyses in the cave evidenced that the sewage system is not tight enough and that some water leaks into the cave (Fig. 1). We carried out several water tracing tests from washrooms in the above mentioned doline; samples were taken 40 m deeper in the initial part of Pivka Jama.

Water tracing by Uranin injection directly to grass covered surface

So far this was the only water tracing test when the solution of tracer was poured directly over the grass surface. We estimate the thickness of soil from 0.5 to 1 m and it was artificially accumulated. A good week before the injection there was no rain. The trickles in the cave had minimal discharge, some of them even dried up. The second and the third day after the injection 7 mm and 6 mm of rain fell; one must take into account also a diminished inflow from the washrooms. First of all Uranin appeared in

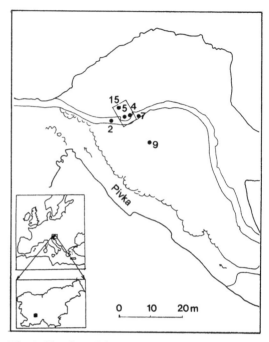

Fig. 1. Situation of the observed trickles (2, 4, 5, 7, 9 and15*) in Pivka jama 40 m below the surface.

the most permeable Trickle 15* (after 2 days), in a trickle that drops to the pathway in the cave, and one day later in Trickle 2. The initial increase of concentration was

followed by the second increase of concentration after rain when also discharges slightly increased. At that time Uranin was detected in poorly permeable Trickle 7. Such washing of tracer probably occurred after every rain. It means that soil, even in case of smaller rain does not importantly buffer the substances; the same was evidenced at road accidents. The experiment conveys some idea for future tests.

Water Tracing on June 19, 1992

On June 19, 1992 we injected a solution of 31 g of Uranin (by mistake instead of Rhodamin) into washrooms in a doline directly into sink. By the day of injection 107 mm of rain fell in June; after the injection there was no rain worth mentionning for two weeks. Before the injection started we poured 900 l of water and after the injection within 30 minutes we added 4.3 m³ of water and again, after a break of 15 minutes another 3.9 m³ of water. However, a minor part of water from the washroom flow into the cave also.

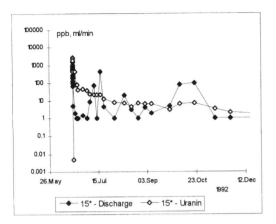

Fig.2. Tracing experiment in June 1992: Uranin breakthrough and discharge curves at trickle 15*.

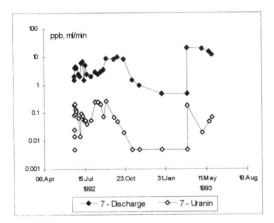

Fig.3. Tracing experiment in June 1992: Uranin breakthrough and discharge curves at trickle 7.

After one hour already, discharge increased and the tracer was noticed at Trickles 2, 4, 5 and 15*. In Figure 2 a sharp increase of Uranin concentration towards maximum (v_{dom}= 2.2 cm/s) is clearly seen at Trickle 15*, and its slow decrease that lasted up to October when heavy rain occurred (450 mm). Similar curves were determined at Trickles 4 (v_{dom}=0.9 cm/s) and 5 (v_{dom}=0.6 cm/s). Autumn rain resulted in intensive flushing of Uranin from wider background of trickles 15* and 5; the decrease of Uranin concentration in Trickle 4 indicates that its background is relatively narrow. We presume that at the same time also Uranin, that was previously injected directly over grass surface, appeared, but its share at the beginning was relatively low.

Essentially different was reaction of Trickles 2 and 7 (Fig. 3); at Trickle 9 the lowest concentration of

Uranin was found out and with substantial delay, after heavy rain when discharge only slightly increased. After injection when discharge noticeably increased, Uranin was substantially washed out through the Trickles 2 and 7 (velocity in respect of peak Uranin concentration v_{dom}=0.02 cm/s) but its concentration never expanded beyond 1 mg/m³. The rinsing of trickles background continued after every heavy rain in October 1992 and March 1993 and even later. We presume that impact of Uranin, previously injected over the grass surface played an important role.

By the end of June, 10 days after the injection, when there was almost no rain, 4% of added water and only one percent of tracer appeared in the cave. Major part of water with tracer flowed into water treatment plant and one part remained in the background of trickles; for a long time, after every rain, it was washed out (Figs. 2 and 3).

Combined water tracing test on May 17, 1994

On May 17, 1994 we poured into chosen sinks in washrooms 1.4 g of Uranin, 1 g of Rhodamin and 1 kg of NaCl; in 40 minutes we watered these tracers by 4.7 m³ of water. In May, before the injection, 75 mm of rain fell and after injection 44 mm in the following three days. Next rain came on May 24 and 27.

Half an hour after watering the tracers, discharge in the most permeable Trickle 15* increased and after next 15 minutes also less permeable Trickles 4 and 7 reacted. Discharge of Trickle 9 did not react earlier than on May 19 after rain when water appeared also in otherwise dry Trickles 2 and 5. Delayed and poor

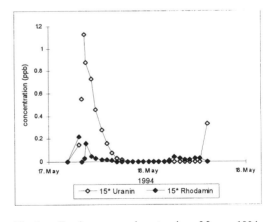

Fig.4. Tracing experiment in May 1994: Breakthrough curves of Uranin and Rhodamin at trickle 15*.

reactions of discharge increase indicated that background of trickles was not water-filled; it might be explained, that in spite of abundant rain, evapotranspiration, due to vegetation at the surface, used a lot of water.

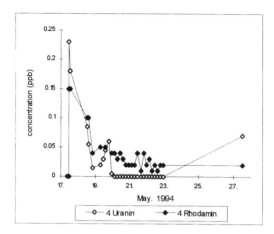

Fig.5. Tracing experiment on May 1994: Breakthrough curves of Uranin and Rhodamin at trickle 4.

The most distinctive and the first arrival of Uranin and occurred at Trickle 15[*] (v_{dom}=0.5 cm/s), followed by Trickle 4 (v_{dom}=0.67 cm/s), where the concentrations of both tracers reached lower values (Figs. 4 and 5). When discharge increased after rain, Uranin, and less distinctive also Rhodamin, appeared at Trickles 2 and 5. At Trickles 7 and 9 the appearance of tracers, specially Uranin, was less distinctive. Different appearance of Uranin and Rhodamin at single points in the cave is probably due to fact, that tracers were injected at different points, although only few metres apart.

Water tracing on June 16, 1994

After water tracing in May there were 240 mm of rain until the injection on June 16; already in the first week of June the tracer concentrations dropped under the limit of detection. In such water-filled background 2.6 g of Uranin was injected into external sink and watered by 3 m³ of water. In next days it was drizzling (20 mm) but some water probably came from washrooms, as camping site was open.

Again the peak discharge and Uranin concentration appeared in the most permeable Trickle 15[*] (v_{dom}=0.83 cm/s). Trickles 4 (increase of discharge and Uranin) and 7 (appearance of Uranin only) reacted by temporal delay (v_{dom}=0.3 cm/s). Discharge

at Trickle 2 slightly increased yet we did not detect any Uranin.

In this case we watered tracer by smaller quantity of water in well fed background of trickles. Similar fast reaction yet by slower decrease of Uranin concentration was noticed at Trickle 15[*] during water tracing test on June 1992, but at that time we used 10-times bigger amount of Uranin which was watered by 3-times more water.

Water tracing on June 26, 1995

Water tracing after one year followed by injecting 1 g of Uranin and 1 kg of NaCl into sinks in washrooms. In the week before 70 mm of rain fell and in the first half of June another 125 mm. The background of trickles was relatively well water-filled. The tracers were watered by 3 m³. The first week after injection was dry and in following three days 46 mm of rain fell.

As expected, Uranin appeared after half an hour after the discharge increase and in following 15 minutes the concentration of chloride levels and specific electric conductivity at Trickle 15[*] increased also (Figs. 6 and 7). In respect to peak Uranin concentration the percolation velocity was 1.2 cm/s, in respect to NaCl it was 0.95 cm/s.

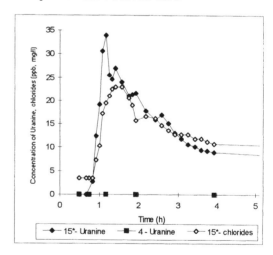

Fig.6. Tracing experiment in June 1995: Uranin and NaCl breakthrough curves at trickle 15*.

At the same time Uranin in Trickle 4 was detected but in essentially lower concentration. The next day we detected increased concentration of chloride level and during one week increased conductivity (SEC).

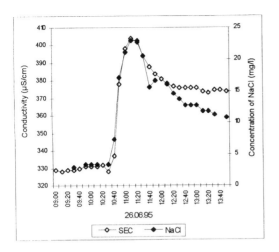

Fig.7. Tracing experiment in June 1995: parallel merasurements of specific el. conductivity and NaCl concentration at trickle 15*.

After rain the Uranin concentration at Trickle 2 was at the limit of detection; at Trickles 7 and 9 we detected Uranin only once so that this water tracing test can neither confirm nor exclude the connection.

Conclusions of water tracing in Pivka Jama

The water tracing results in the area of Pivka Jama show that only one, very permeable conduit leads the water from a doline through 40 m thick vadose zone, with velocity from 0.43 do 2.2 cm/s (controlled by hydrological conditions). We determined the velocities for less abundant Trickle 4 and slightly lower values for Trickles 2 and 5. All these trickles seasonally dry up. The slowest discharge (v = 0.02 cm/s) was assessed for Trickles 7 and 9 which are permanent (Table 1). Residence time of a tracer in the unsaturated zone of Pivka Jama was 18 months at the utmost (when the biggest amount of tracer was used (31 g) and the least (1 month only), when 1 g of

tracer was used. Water tracing in Postojnska Jama (injection of 60 g of Uranin) yielded still longer residence time - 3 years. Similar flow velocity through a vadose zone when tracers were placed to the rocky base beneath the soil and 20 mm of rain were simulated afterwards, was assessed for a nearby cave Črna Jama by V. Tropf (1995). Water tracing by a similar method in Yorkshire achieved by S.H.Bottrell and T.C.Atkinson assessed three components of storage in unsaturated zone with residence time of few days, 30-70 days and 150-170 days respectively. In one experiment, dye was still detectable after 75 months.

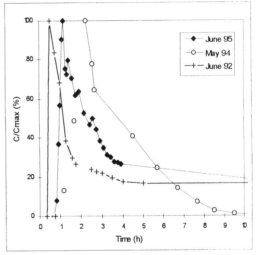

Fig.8. The comparison of breakthrough curves of Uranin at trickle 15* during different tracing experiments (June 1992, May 1994 and June 1995).

In Fig. 8 is shown the pattern of Uranin flow through the most permeable trickle 15* during the described water tracing tests controlled by different hydrologic conditions. Tracer through-flow at water tracing experiment in June 1992 and June 1995 is

Table 1

Tracing	Tracer	M	V	Vdom (cm/s) in trickles						Rain 2 weeks		L
experiment		g	m3	15*	4	2	5	7	9	before	after	month
June 1992	Uranin	31.4	9	2.2	0.9	0.44	0.6	0.02	0.02	0	0	18
May 1994	Uranin	1.4	4.7	0.5	0.67					75	105	0.75
	Rhodamin	1		0.43	0.67							0.75
June 1994	Uranin	2.6	3	0.83	0.3			0.3		135	20	3
June 1995	Uranin	1	3	1.2						117	47	1
	NaCl	1000		0.95								1

M - quantity of injected tracer, V - volum of added water, L - duration of tracer breakthrough (most of it)

relatively similar. In both cases the vadose zone was water-filled during the injection; yet in June 1992 we used 30-times more of tracer and it was waterd by 3-times bigger amount of water thus yielding faster appearance of tracer and steeper curve of Uranin break-through. At tracing in May 94 the vadose zone was not water-filled (important evapotranspiration) and Uranin appeared after longer residence time giving less steep curve of a dye break-through.

3. CONCLUSIONS

The experiments showed the mode of percolation of a tracer through the vadose zone in case of injection using additional watering. In all cases it was shown that water flows from a defined point at the surface via a series of more or less permeable fissures. The percolation through the unsaturated zone may be roughly divided into three components: rapid through-flow through most permeable conduits, flow velocity from 0.5 to 2 cm/s; slower velocity of some hundredth cm/s; and the slowest velocity below 0.001 cm/s. The trickles through which the water comes through the most permeable conduits dry up and their discharge vary the most. Slower percolation is provided by permanently filled fissures, trickles and droppings are permanent and discharge varies slightly.

Dye residence time in the unsaturated zone is mostly controlled by the quantity of injected tracer and amount of rain that follows. We assessed times from 3 weeks to 3 years.

If one wishes to recognize a detailed picture of flow patterns through vadose zone, by both more and less permeable conduits, 10 times greater amount of tracer must be used than in case when only the most permeable conduits from karst surface into its interior are to be assessed.

These first statements may be used as a base for following researches. Understanding the transport of substances through the unsaturated zone proves to be very important at planning the protection of karst water resources.

REFERENCES

Bottrell, S.H. & T.C.Atkinson, 1992. Tracer study and storage in the unsaturated zone of a karstic limestone aquifer. Tracer hydrology. Proc. 6th Int.Symps. on Water Tracing. H.Hötzl & A.Werner (eds), 207-211. Rotterdam: Balkema.

Kogovšek, J. & P.Habič, 1981. The study of vertical water percolation in the case of Planina and Postojna Cave. Acta carsologica, 9 (1980), 129-148, Ljubljana.

Kogovšek, J., 1995. The Surface Above Postojnska jama and its Relation with the Cave. The Case of Kristalni rov. Grotte Turistiche e Monitoraggio Ambientale, Simposio Internazionale. A.A.Cigna (ed), 29-39, Frabosa Soprana.

Tropf, V., 1995: Hydrologische und hydro-chemische Untersuchungen zur Perkolation in der ungesattigten Zone der Črna jama (Schwarze Hohle). Diplomarbeit, 105 p., Karlsruhe.

Tracer Hydrology 97, Kranjc (ed.)© 1997 Balkema, Rotterdam, ISBN 90 5410 875 4

Conceptual runoff model for small catchments in the crystalline border mountains of Styria, as developed from isotopic investigations of single hydrological events

D. Rank & W. Papesch
Österreichisches Forschungs- und Prüfzentrum Arsenal, Vienna, Austria

ABSTRACT: Isotope-hydrological investigations in small catchments in the crystalline Styrian border mountains (Eastern Alps, Austria) during storm events showed that a two-component runoff model (direct runoff, base flow) cannot sufficiently explain the course of the runoff isotope data during the events. At the very least, a "third" runoff component must exist, which cannot be attributed to either the direct runoff of precipitation water, or to the base flow before the event. Isotope data indicate that this third component consists of precipitation water, which is stored for a few weeks or months in the unsaturated zone (weathered material), whereas the base flow before a storm event consists mainly of water from the joint aquifer, with a mean residence time of about ten years. It may be concluded that the increase of the discharge during storms is primarily due to the pressing out of this third component from the previously unsaturated zone by infiltrating precipitation water. Direct runoff plays a minor role; its influence could not be detected during most of the precipitation events investigated.

1 INTRODUCTION

In two neighboring test areas in the Styrian border mountains, isotope methods were utilized to investigate the composition of runoff from small catchments and springs following storms (Papesch, Rank 1996). The weathered layer which covers the slopes of the crystalline hills is between one and several meters thick. The Pöllau test area includes several micro-catchments on the slopes of the Pöllau Basin (Zojer et al. 1996). The investigations concentrated on the micro-catchment "Höhenhansl" (0.43 km²), in which anthropogenic influence is insignificant. The "Ringkogel", the slopes of which are the source of numerous small springs (discharge ≤ 1 l/s), was included into the investigations as a second test area (Figure 1, Rank, Häusler, in press).

In addition to a quantitative approach to the course of hydrological events, isotopic investigations also provide insight into the age structure of waters (for the fundamentals, see Moser, Rauert 1980). Such knowledge helps us to draw conclusions on the storage processes in hydrological systems and on the composition of the runoff; that is, on the relative shares of base flow, direct runoff following storm events, and interflow. Distinguishing the components of runoff is an important basis for the hydrological assessment of utilizable water reserves in a particular region.

A prerequisite for successful isotope investigations of single hydrological events are variations in the isotope ratios of the input - the precipitation. The deviation of isotope ratios for single precipitation events from the average yearly course depends on the origin of the humid air masses, as well as on the respective climatic conditions during the precipitation events.

Figure 2 depicts the variations in the ^{18}O content of the daily precipitation at the "Heiling" Station in the Pöllau test area, during the period June/July 1991. For example, significant precipitation events in mid-July occurred consecutively, and had ^{18}O contents, which differed considerably from one another. These events provided useful input signals, which could be followed within the hydrological system.

Even within one and the same event, changes in the isotope ratios can be observed over time, which can cover a range in the magnitude of the seasonal variations of the monthly mean values (Rank 1991). Assuming an adequate depth of precipitation, the suitability of a single precipitation event for isotope-hydrological investigations increases with the deviation of the isotope ratios in the precipitation water from the mean values in the hydrological system under consideration. As a result of seasonal variations in the ^{18}O content of precipitation, the greatest deviations of the ^{18}O content of single precipitation events from the mean value of the

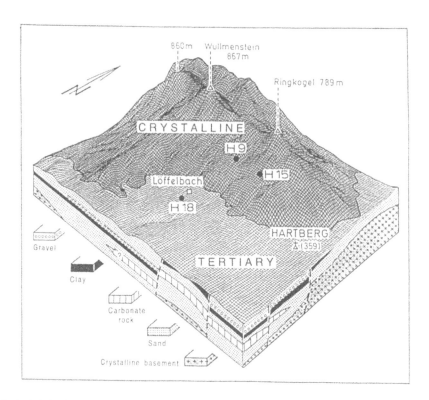

Figure 1. Geological situation in the Ringkogel test area. The shists and gneisses of the Ringkogel crystalline are plunging under the Tertiary series towards the south. The Tertiary is formed by a series of sands, limestones, mudstones and gravels that is quickly changing, both horizontally and vertically.

system can be expected in winter and in summer. In the areas under investigation, precipitation during the winter generally takes the form of snow, and does not directly reach runoff. Mid-summer is therefore the most favorable period for research.

2 SEASONAL VARIATIONS IN THE ISOTOPE RATIOS IN RUNOFF AND THE MEAN RESIDENCE TIME OF THE WATERS

During the period June 1991 to June 1992, the annual course of the isotope ratios in the runoff from the partial catchments in the Pöllau test area was recorded. In addition, some data is also available from 1990. Two samplings, which should indicate the effect of precipitation events on the composition of runoff, were taken after storm events (22-23 July 1991 and 21 November 1991).

Illustrating the results of the monthly samplings, Figure 3 depicts the course of the isotope ratios in the runoff water from two micro catchments. In one case, (1.1.1.1, 0.74 km^2), the influence of the two precipitation events can be seen somewhat more distinctly, while in the Höhenhansl catchment

(1.1.2.2, 0.43 km^2), the influence is relatively insignificant. Aside from precipitation events, the isotope ratios display only small seasonal deviations, which suggest a longer residence time of the base-flow water. This assumption has been confirmed by an evaluation of the ^{3}H data. According to the exponential model, mean residence times for the base flow are between 6 and 13 years. Such a residence time is relatively high for slopes, supporting the conclusion that the movement of ground water takes place at least partially in solid rock (joint aquifer).

Even when the fluctuations in the isotope ratios during the events of 21-23 July can be attributed approximately to the direct influence of the precipitation water - with a simultaneous increase in the base flow (Figure 3), the same conclusion is not possible for the event of 21 November. Although the precipitation in this event exhibited a low ^{18}O content (Figure 5), the ^{18}O content in the runoff increased significantly. This observation leads to the conclusion that in addition to the base flow (in the state it is in before the precipitation event) and to the share of precipitation water which could possibly run off directly, there must be at least one

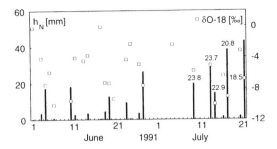

Figure 2. Pöllau test area: Precipitation depth and [18]O content of daily precipitation at Heiling, including some ^{3}H values (TU) for several events during June/July 1991.

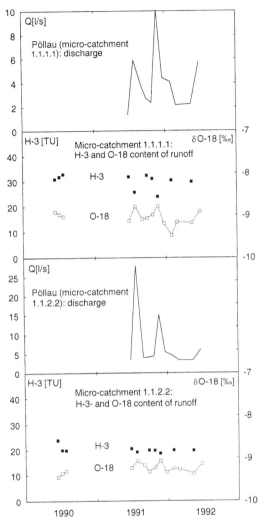

Figure 3. Pöllau test area, micro catchments 1.1.1.1 and 1.1.2.2 (Höhenhansl): Seasonal variations of ^{3}H and [18]O content of runoff water and discharge data.

additional runoff component, which is responsible for the ^{18}O-maximum. The springs in the neighboring Ringkogel region provided the basis for a more exact investigation of this event.

In the Ringkogel area near Hartberg, samples were taken for isotope analysis on 21 November 1991, following severe storms. Discharge from the springs was higher than usual (Rank, Häusler, in press). In the springs of the crystalline zone, which primarily exude fissure water, there occurred a significant, short term rise in the [18]O content (Figure 4, springs H9 and H15), while the springs in the Tertiary zone exhibited no variations in [18]O content. The mean residence time of the base-flow water of the springs ranged between 7 and 13 years, similar to that in the Pöllau test area.

Although the higher [18]O content in the Ringkogel springs on 21 November initially suggested that the runoff contained a high amount of water from summer precipitation, which had been stored for several months, the results of the ^{3}H investigations placed this assumption in question. On November 21st, a common feature of all fracture springs was a significant ^{3}H minimum (decrease at H9 from 36 to 28 TU, at H15 from 28 to 20 TU), while spring H18 in the Tertiary zone indicated no change in either [18]O content, or in ^{3}H content. Summer precipitation, characterized by the seasonal ^{3}H maximum, cannot therefore be responsible for the raised [18]O content on November 21st.

A more precise explanation of the composition of the discharge water on 21 November was provided by a thorough analysis of the input - the precipitation. From the Ringkogel test area, neither precipitation and/or discharge data, nor precipitation samples for isotope measurements, were available. Therefore, attempts were made to reconstruct the hydrological situation in November 1991, utilizing the measurement data from the neighboring Pöllau Basin (Figure 5). Even when the data cannot be directly transferred and applied, they do permit qualitative statements.

On November 4th, the region experienced significant storms, (45 mm of precipitation in the Pöllau test area), with an [18]O content unusually high for this season (Figure 5), and a low ^{3}H content, which corresponded to the usual seasonal values (approximately 5 TU). The precipitation which fell immediately prior to the sampling had, in contrast, a low [18]O content, and can therefore be eliminated as a possible cause for the [18]O maximum in the spring water on 21 November. This precipitation can therefore have had little, or no, influence on the runoff of 21 November.

175

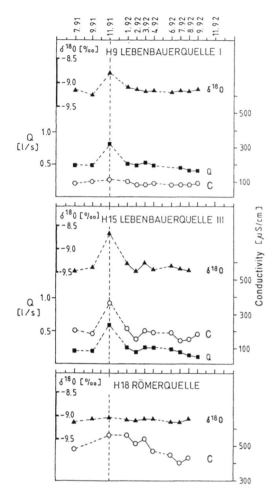

Figure 4. Ringkogel test area: [18]O content, conductivity (C), and discharge (Q) of two fracture springs in the Ringkogel area (H9, H15) and one spring in the Tertiary zone (H18). The fracture springs show a quick, significant increase of the [18]O content after the storms in November 1991, which cannot be attributed to the direct runoff of precipitation water.

in their respective conductivities. For H9, the value is lower, and hardly changes, even during the storm period, while H15, with a higher initial value, exhibits a clear maximum in November 1991. The cause of this behavior could be the influence of agriculture on the weathered layer in the drainage area from H15. Corresponding to the conceptual model in Section 4, during the storm period, the longer path of transport for the infiltrating precipitation water in the weathered layer probably led to a rise in conductivity.

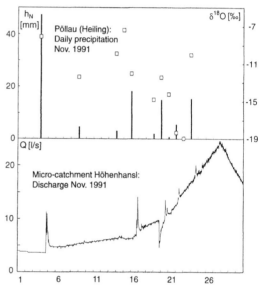

Figure 5. Pöllau test area: Daily precipitation depth, $\delta^{18}O$ values of daily precipitation and discharge (Höhenhansl) in November 1991. The discharge minimum on 19 November is probably due to anthropogenic influence.

3 INVESTIGATIONS DURING PRECIPITATION EVENTS IN THE MICRO-CATCHMENT HÖHENHANSL

3.1 The precipitation event of 18 August 1991

For a preliminary experiment in the Höhenhansl area, precipitation water was collected manually in a makeshift funnel every half hour. Runoff samples were attained by means of commercially available automatic samplers. The isotope ratios in the runoff changed distinctly at the start of the precipitation (Figure. 6). However, the [18]O content did not rise drastically enough to permit the additional runoff to be attributed solely to the precipitation. We could assume from the course of the ^{3}H content that the

Apparently, the infiltrating precipitation water from November 4th was primarily stored in the unsaturated zone, raising the moisture content of the soil, but not contributing significantly in the short term to a rise in the spring discharge (Figure 5). Later on, such water can be mobilized by further storms, quickly reaching springs, and/or small creeks.

The conductivity of the spring water during the storm period proved not to be very indicative (Figure 4). While the springs H9 and H15 behave similarly, with respect to their isotope ratios and discharge, significant differences can be observed

additional runoff was not coming from an increased discharge of water, corresponding to the base flow before the event. The ^{3}H content of the runoff, which was significantly higher than the ^{3}H content of the precipitation water, could only be explained through the role of an additional runoff component, having a relatively high ^{3}H content and, in comparison to the precipitation, a somewhat lower [18]O content. The most likely candidate is precipitation water (summer precipitation), which infiltrated the ground several weeks earlier.

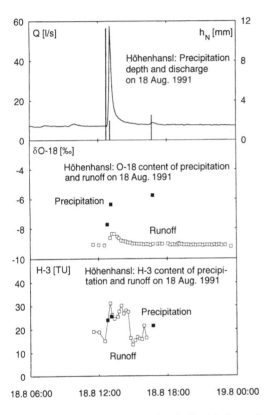

Figure 6. Micro-catchment Höhenhansl: Precipitation and discharge on 18 November 1991; [18]O and ^{3}H content of precipitation and runoff.

These considerations led to the recognition that the base flow changed not only in amount during the hydrological event, but also according to the origin of water. In addition to the water of the base flow before the event, which has a mean residence time of approximately 6 years, at least one additional component with a relatively short residence time must be involved. In this case, a two-component model (base flow, direct runoff of precipitation) cannot sufficiently describe the runoff composition.

According to this result, the most logical step is to split the "base flow" in the area under investigation into two components: on the one hand, the real base flow (with a mean residence time between 6 and 13 years), and secondly, a short term component (with a mean residence time somewhere between several days and a few months). In this case, the real base flow remains more or less constant during the precipitation event and during discharge, while the share of the short term component rises quickly and resides just as rapidly, supporting the assumption that there must be intermediate storage in the unsaturated zone. The rapid rise provokes the question as to the mobilization of these waters through the transmission of pressure from the infiltrating precipitation water via so-called "soil-air cushions".

3.2 The precipitation event of 11 June 1993

The summer of 1993 witnessed the utilization of an automatic collector, which we had developed ourselves, for the investigation of precipitation events (Papesch, Rank 1996). Common to all events is an $\delta^{18}O$ value for the base flow ranging between -9.1 and -9.2 ‰.

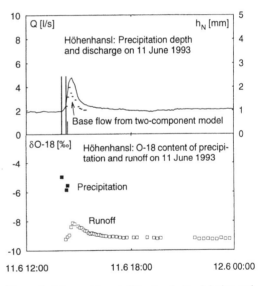

Figure 7. Micro-catchment Höhenhansl: Precipitation and discharge on 11 June 1993, including the base flow calculated on the basis of a two-component model (base flow and direct runoff); [18]O content of precipitation and runoff.

During this event, in which the $\delta^{18}O$ content of the precipitation was approximately 4 ‰ above that of the base flow (indicating a strong input signal), the [18]O content of the runoff displayed a

177

maximum, as a result of the precipitation (Figure 7). However, the "base flow" must also have risen sharply during the event, as otherwise, the maximum would have been even more pronounced. The ^{3}H data provide, in this case, no information as to a possible change in the composition of the "base flow" during the event, due to the fact that, by chance, the ^{3}H contents of precipitation and discharge (both approximately 18 TU) do not differ significantly from one another.

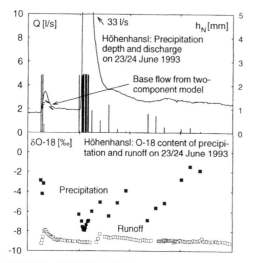

Figure 8. Micro-catchment Höhenhansl: Precipitation and discharge on 23-24 June 1993, including the base flow calculated on the basis of a two-component model (base flow and direct runoff); ^{18}O content of precipitation and runoff.

3.3 The precipitation event of 23-24 June 1993

The course of the isotope ratios during the first - smaller - discharge peak (Figure 8), can be interpreted in a similar manner as in the events of 18 August 1991 and 6 June 1993: during the event, the base flow rises, and the ^{3}H content implies that a change takes place in the composition of the "base flow". During the considerably higher discharge peak (up to 33 l/s, with 2 l/s base flow before the event), attributable to the storm which set in four hours later, the ^{18}O content initially does not change at all. This means that during the greatest extent of the discharge peak, the runoff contains no notable quantities of directly discharging precipitation water. After the discharge peak has fallen off, the ^{18}O content begins to rise, which could suggest that some precipitation water runs off directly. From the ^{3}H values, we can assume that a large portion of the "base flow" is of an origin

other than the water of the base flow before the event.

Also in this case, the question is raised as to which role, in addition to moisture content in the unsaturated zone, is played by the transmission of pressure via soil-air cushions. Apparently, it is through this process that the influence of the infiltrating precipitation water can first enable a comprehensive pressing out of the waters stored underground.

3.4 The precipitation event of 29 September 1993

The precipitation activity during this event was distributed over a period of twelve hours (Figure 9). Subsequently, the discharge exhibited no distinct peaks. The ^{18}O content of the runoff clearly illustrated that precipitation water which runs off directly plays no role in the rise of the discharge to a level which is twice the initial value. The difference between the ^{18}O signal in precipitation and base flow amounts to as much as 6‰, so that if precipitation does account for a portion of the discharge, its presence would have to be clearly evident. The ^{3}H data also do not exhibit a direct influence of precipitation. The increase in the discharge during the event can therefore be attributed exclusively to the "base flow". Furthermore, this result permits the conclusion that a larger portion of the precipitation water is stored in the unsaturated zone.

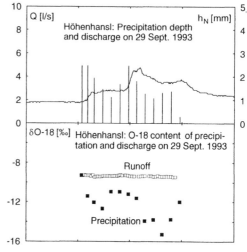

Figure 9. Micro-catchment Höhenhansl: Precipitation and discharge on 29 September 1993; ^{18}O and ^{3}H content of precipitation and runoff.

4 CONCEPTUAL MODEL OF RUNOFF GENERATION DURING STORMS IN THE CRYSTALLINE BORDER MOUNTAINS OF EASTERN STYRIA

The findings of the isotope investigations in the Pöllau Basin and in the neighboring Ringkogel-region point towards a conceptual hydrogeological model of runoff generation during storms in the area of the crystalline border mountains of eastern Styria. The conceptual model which follows is developed upon the example of the reaction of the Ringkogel springs to the precipitation event of 20-21 November 1991 (Rank, Häusler, in press).

An essential share of the water discharged on 21 November 1991, from the Ringkogel springs (fracture springs), following the storms of the previous day, could be attributed to the storm of 4 November 1991. Following a brief storage period, the discharge to the springs (which usually exit near the border between the joint aquifer and the weathered layer) of this precipitation water, on which the most recent precipitation had no noticeable effect, apparently did not occur via the joint aquifer system (mean residence time of the water approximately 10 years), but rather via the otherwise unsaturated zone of the weathered layer.

Normally, the weathered layer is unsaturated (Figure 10, "Low precipitation rate"). The infiltrating precipitation of 4 November is stored primarily in the unsaturated zone, and increases the moisture content of the soil. It does not, however, contribute significantly in the short term to a rise in the discharge from the springs (Figure 5). Under the influence of additional storms (16 and 20 November), a saturated zone developed above the joint aquifer, because the joint aquifer was either full, or respectively, the infiltrating precipitation water could not seep into the joint aquifer quickly enough (Figure 10, "Storms"). Following the slope, within a short time, this water reaches the spring area, and here, mixes with the water of the base flow.

The fact that the runoff of 21 November contained no notable amounts of water from the most recent precipitation event leads to the conclusion that precipitation waters from consecutive storms infiltrate the soil as fronts (Figure 10). Between the fronts, "soil-air cushions" are formed temporarily, which on the one hand prevent the short term mixing of the waters from consecutive storms; while on the other hand, the formation of a saturated zone above the fissured water system is encouraged through pressure transmission, and the water is pressed out towards the springs. The pressure

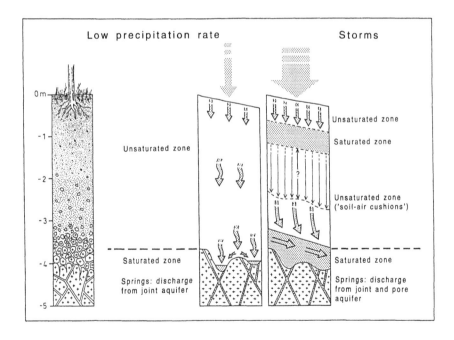

Figure 10. Conceptual hydrogeologic model for the generation of runoff during storm events in the crystalline border mountains of Styria. Normally, the infiltrating water enters the joint aquifer. Provided that the soil moisture is high enough, a saturated zone will be formed above the joint aquifer during storms, and young water will quickly reach the springs and creeks. Thereby, pressure transmission via "soil-air cushions" obviously plays a certain role, also.

transmission through the soil-air cushions also has the effect that during storm events, the discharge from springs and creeks rises sharply with no notable delay, and a discharge peak develops (Figures 6, 7, and 8). Laboratory investigations have confirmed that soil-air cushions can assume such a pressure transmission function (Bergmann et al. 1996).

The finding that discharge peaks during precipitation events are created essentially through the pressing out of water stored underground, and are not primarily attributable to the direct runoff of precipitation water, leads to the conclusion that precipitation water infiltrates the soil on a greater scale than originally assumed.

By the time of the next measurement series in January 1992, the isotope ratios of the majority of the Ringkogel springs had returned to the base-flow values; that is, the springs were again being fed primarily from the joint aquifer (Figure 4).

The results of the observations of this event in the Ringkogel test area, as well as the results of the investigational observations in the Pöllau test area, both lead to the conclusion that following precipitation events, waters are pressed out, which cannot be attributed to either the base flow before the event (primarily fissure water with a higher residence time), or to the actual precipitation itself. The fact that the measurement series at the Ringkogel springs by chance took place at a particularly favorable time (greatly varying isotope ratios for consecutive precipitation events) did however enable the origin of this "third" runoff component from the normally unsaturated zone to be determined. Due to the lack of local precipitation data, no quantitative statements as to composition are possible.

5 CONCLUSIONS

Isotope investigations of single precipitation events have brought progress to the understanding of runoff dynamics and storage behavior in a region with pore and joint aquifers. When enough care is taken in performance, and the sampling and registration in the field are well looked after, isotope measurements can be an excellent method of separating the runoff components during hydrological events. Short term variations in the isotope ratios in precipitation are used as input signals. The course of the isotope ratios in runoff provides information on the share of base flow and direct runoff of precipitation water, as well as on the role played by other runoff components, and thereby, on the storage processes in hydrological systems. The most important conclusion of the results is that a two-component model - base flow and the direct runoff

of precipitation water - does not suffice to describe the actual process in the generation of runoff, and that the water movement, respectively the storage of water in the underground (above all in the unsaturated zone) must be incorporated into the conceptual model. The influence of the moisture content in the unsaturated zone, of the time between consecutive precipitation events, and of pressure transmission via soil-air cushions is clearly evident.

This research has been supported by the Austrian Academy of Sciences.

REFERENCES

Bergmann, H., J.Fank, T.Harum, W.Papesch, D.Rank, G.Richtig & H.Zojer 1996. Abflußkomponenten und Speichereigenschaften, Konzeptionen und Auswertemethoden. *Österreichische Wasser- und Abfallwirtschaft* 48: 27-45.
Moser, H. & W.Rauert 1980. *Isotopenmethoden in der Hydrologie*. Berlin: Borntraeger.
Papesch W. & D.Rank 1996. Isotopenuntersuchungen zur Erfassung der Wasserspeicherung und der Abflußvorgänge (Teilprojekt zu: Erfassung der Abflußvorgänge in kleinen natürlichen Einzugsgebieten). Report 90.16/2, Wien: Bundesforschungs- und Prüfzentrum Arsenal.
Rank, D. & H.Häusler. Zur Abflußentstehung nach Starkregenereignissen im kristallinen Randgebirge der Oststeiermark. *Mitt.österr.geol.Ges.* (in press).
Rank, D. 1991. "Umweltisotope" - Fortschritte in Forschung und Anwendung. *Mitt.österr.geol. Ges.* 83: 91-108. Wien.
Zojer, H., H.Bergmann, J.Fank, T.Harum, W.Kollmann & G.Richtig 1996. Charakterisierung des hydrologischen Versuchsgebietes Pöllau. *Österreichische Wasser- und Abfallwirtschaft* 48: 5-14.

Tracer Hydrology 97, Kranjc (ed.) © 1997 Balkema, Rotterdam, ISBN 90 5410 875 4

Investigation of preferential flow in the unsaturated zone using artificial tracers

S. Uhlenbrook & Ch. Leibundgut
Institute of Hydrology, University of Freiburg, Germany

ABSTRACT:

Different tracer methods were used to investigate the generation of runoff components in the Loechernbach catchment (1.7 km^2) in the Southwest of Germany. This paper describes a tracer experiment in the unsaturated zone, using the artificial tracers uranine and bromide. It was carried out on a crop field, which is drained by an artificial drainage system. The flow through the macro- and micropore systems was evaluated quantitatively by using a non-destructive method.

For a simulated heavy rainstorm the mean residence time of water in the unsaturated zone (depth 1.2 m) was 2.5 hours. Different tracer transport models (deterministic and probabilistic models) were compared. These models take only one flow system in the unsaturated zone into consideration. In a second step, a new method was generated for the simulation of the tracer transport in the unsaturated zone. The method distinguishes between the macro- and micropore system. Very good results were obtained when modeling the flow in the macropores with the Single Fissure Dispersion Model (SFDM) and modeling the flow in the matrix of the soil with the Convection Dispersion Model (CDM). The SFDM was developed to describe the flow through fissures of saturated hard rock aquifers. This inverse validation of the SFDM for the macropore flow in the unsaturated zone shows its further potential.

1 INTRODUCTION

In soils, containing macropores, water and dissolved chemicals can move preferentially, bypassing most of the soil matrix. This phenomenon has an important influence on different hydrological processes, such as infiltration, groundwater recharge and evapotranspiration. The importance of the preferential flow for the runoff generation processes has been widely discussed (e.g. Mosley 1982, McDonnell 1991, Hornberger et al., 1991, Buttle 1994) and will be enlarged by this contribution.

Beside the influence of the preferential flow phenomena on the quantitative hydrology, they have a great influence on the qualitative hydrology. They allow agricultural chemicals, applied to the surface, to move rapidly through the root zone, thereby contributing significantly to groundwater contamination (e.g. Jabro et al. 1994, Everts & Kanwar 1994, Flury et al. 1994). By bypassing most of the soil matrix, the buffer effect of the unsaturated zone, overlying an aquifer, is reduced.

The major formation processes of macropores are known and have been well described. The macropore system can be created by animal burrows, for example earthworms can produce a very efficient pore system (Omoti & Wild 1979b). Plant roots can also provide a permeable macropore system, which is often interconnected over long horizontal distances (Mosley 1982, Mikovari et al. 1995). Cracks formed by shrinkage or other fissures, which are formed by weathering of the matrix, may produce a pore system, depending on the water content of the soil (Beven & Germann 1982). Macropores can also be formed by subsurface erosion (soil pipes). Pores, which are developed by agricultural activities (plowing, trenching etc.), also belong to the group of macropores.

Although the distinction between two pore systems (micro- and macropores) is simple, it is a practical way to explain the observed fast and slow flow processes in soils. Different approaches describe the water and solute transport in soils, containing macropores. The Richard's equation can not be used, because the water movement in the macropores is very often turbulent, heterogeneous and anisotropic. Therefore, every application of the Richard's equation must fail, e.g. a Darcy-based model, (Beven & Germann 1982, Chen & Wagenet 1992, Demuth & Hiltpold 1994). One way to model the water flow in the unsaturated zone is to use the double porosity approach, wherein two distinct flow systems with interactions between each other are modeled (Chen & Wagenet 1992). However, it is problematic to determine the properties of each flow system, as well as the empirical interaction term. Germann & Beven (1985) suggested a kinematic wave concept to describe the flow within macropores, but it is difficult to evaluate the friction of the water. Some attempts were made to model the macropore flow hydraulically (Scotter 1978), but the immense need of parameters made this concept not feasible.

Even if the spatial distribution of the flow pattern in the unsaturated zone is very variable (Demuth & Hiltpold 1994, Flury et al. 1995), for process studies and for the validation of transport models, tracer experiments on small plots are a practical method to explore the movement of water and solutes in the unsaturated zone in situ (Mosley 1982, Hornberger et al. 1992, Flury et al. 1995, Mikovari et al. 1995). To investigate flow pathways and residence times of water and solutes in the unsaturated zone under field conditions, a tracer experiment with artificial tracers was conducted in this study. The main interest is focused on the generation of fast runoff components in the unsaturated zone. It was possible to show, how much the macropore flow in agricultural soils can contribute to the formation of fast runoff components.

2 STUDY SITE

The study was conducted in the small Löchernbach catchment (1.7 km^2) within the Hydrological Experimental Basins East-Kaiserstuhl in the Southwest of Germany. The underlying bedrocks are mafic, volcanic rocks, which are widely covered by a thick (up to 50 m) loess stratum. 90 % of the basin are covered by loess. The mean annual precipitation is 700 mm, the evapotranspiration in this relatively temperate region (mean annual temperature 10° C) can reach up to 600 mm.

The Loechernbach catchment was altered by the consolidations in 1969/70 and 1974/76. These actions changed the morphology of the basin completely, and had micro-climatic, pedological, hydrochemical and hydrological impacts. Large terraces were built, where nowadays mainly wine is grown.

77 % of the total catchment are used for agriculture; 6.1 % of the area is sealed. The large terraces are very often equipped with an artificial drainage system, so that around 20 % of the total area is drained by this system.

The soils within this anthropogenically altered area are mainly young, hardly developed loess soils. The base capacity is very high, and the soils are well structured in general. The mean grain size distribution is: 85 % silt, 8 % sand and 7 % clay. A broad population of earthworms can be found even in the intensively used farmlands. The average effective porosity amounts to 20 % (Luft 1981). At some sites evidences for episodic wetness are found.

An investigation of runoff behavior showed a clear change in runoff characteristics in the altered catchment Loechernbach compared to an unchanged, small-terraced catchment. The rate of peak discharge became twelve times higher and the base flow was significantly reduced (Luft 1980, Bucher & Demuth 1985). For the generation of fast runoff components it is supposed, that the sealed areas and the artificial drainage systems are important. So far it was unknown to which extend the artificial drainage system in the dense loess soils can contribute to the runoff generation of floods.

3 METHODS

3.1 Tracer experiments

To find out the system properties of a soil compartment, it is adequate to carry out a tracer experiment with artificial tracers under defined boundary conditions. Tracer experiments are black box experiments. To determine the system function of an unknown system, it is necessary to know the system input and output. This means in the case of a tracer experiment, that the system output has to be measured, which is in general the concentration as a function of time. The system is applied with a defined input, during stationary conditions.

For the application of the tracers, a so-called „double-frame system" was constructed, designed on the basis of a double-ring infiltrometer. The double-frame system was composed of two rectangular frames (17 and 8 m^2) fitted into one another (Figure 1). Using this system, the tracers, which were applied on the inner frame, are not lost to the surrounded unsaturated zone, due to capillary suction (Mehlhorn et al. 1995).

With the application of the tracers, a heavy rainstorm was simulated by artificial sprinkling (20 mm precipitation in 90 minutes, including 6 mm sprinkling before and 10 mm after the tracer injec-

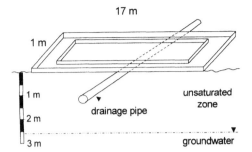

17 m

1 m

1 m

drainage pipe

unsaturated zone

2 m

3 m

groundwater

Figure 1: Schematic sketch of the test plot.

tion). A rainstorm in this order occurs statistically five times a year. For the investigation of the production of fast runoff components, it is important to irrigate with realistic amounts of water. Ponded infiltration experiments, e.g. with a ring-infiltrometer, should be avoided.

Two tracers were injected on the inner frame: 460 g of the fluorescence dye uranine and 12 kg sodium bromide. To estimate the amount of tracer to be injected (system input), the equation of Leibundgut & Wernli (1982) was used. The concentration of the traced water was measured at different outlets of the drainage system (system output). Additionally, the discharge was measured at the outlets, with a volumetric method, in order to calculate the tracer recovery rate.

For the interpretation of the results it is important to recognize the following facts: First, the actual landuse was the cultivation of sweet corn after the harvest. The yearly plowing did not happen before the experiment was executed. This means that the supposed macropore network was not destroyed in its vertical continuity. Second, at the day before the experiment started, the soil had field capacity, because the soil had been irrigated two days in advance. This way defined initial moisture conditions would have been established. Unfortunately, in the night before the experiment was carried out, an additional event of 20.3 mm happened. That means, that the soil was still unsaturated to the depth of 1.20 m, but the water content was higher than field capacity. It has to be taken into consideration, that the upper soil was almost saturated, which probably has an influence on the macropore flow mechanism (Mosley 1982, Germann 1986).

3.2 Modeling of tracer tests in soils

Three different models were used to simulate the breakthrough curves of the tracer experiment:

In the domain of ground- and surface water the classic *Convection-Dispersion Model* (CDM) is widely used (e.g. Maloszewski & Zuber 1984). The one-dimensional CDM describes the solute transport in porous media as a variation of different velocities in a deterministic approach. The CDM is based on Darcy's law. In the case of one-dimensional steady-state flow and an instantaneous injection of a conservative tracer, an analytical solution of the transport equation is available:

$$C(t) = \frac{M}{Qt_0} \cdot \frac{1}{\sqrt{4\pi Pe^{-1}(t/t_0)^3}} \cdot \exp(-\frac{(1-t/t_0)^2}{4t/t_0 Pe}) \quad (1)$$

where:

$C(t)$	=	tracer concentration [mg/m^3]
t	=	time [s]
M	=	mass of tracer injected [mg]
Q	=	discharge [m^3/s]
t_0	=	mean transit time of water [s]
Pe	=	Peclet-number [-]

To fit the observed concentration to the simulated concentration, two parameters have to be determined: The mean residence time and the Peclet-number, in which the longitudinal dispersivity and the flow distance is included (Maloszewski 1994).

As a second model the *Single Fissure Dispersion Model* (SFDM) after Maloszewski (1994) was applied. It describes the transport of solutes in saturated fissured aquifers, mainly in hard rock aquifers. The model assumes, that the water transport takes place within the fissures, but that solutes can move into the matrix or back to the fissures by diffusion. The water in the matrix is assumed to be stagnant. How long a chemical is retained in the matrix depends on the concentration gradient between the matrix and the fissures. This gradient controls the diffusion of the solutes, and thereby the tailing of the breakthrough curve.

In the case of one-dimensional steady-state flow and an instantaneous injection of a conservative tracer, an analytical solution of the transport equation is also available:

$$C_f(t) = \frac{aM}{2\pi Q} \cdot \sqrt{Pet_0} \cdot$$
$$\int_0^t \exp(-\frac{Pe(t_0-u)^2}{4ut_0} - \frac{a^2u^2}{t-u}) \frac{du}{\sqrt{u(t-u)^3}} \quad (2)$$

where:

$C_f(t)$ = tracer concentration in the fissure [mg/m³]
t = time [s]
M = mass of tracer injected [mg]
Q = discharge [m³/s]
t_0 = mean transit time of water [s]
Pe = Peclet-number [-]
a = diffusion parameter [s$^{-1/2}$]

To fit the simulated concentrations to the observed concentrations, three parameters have to be determined: The mean residence time of the water, a dispersion value (Peclet-number) and a diffusion parameter (depending on the tracer, the porosity of the matrix and the width of the fissures). Another assumption of the SFDM is the approximation of the water transport in the fissures by a single fissure. A detailed description of the model is given in Maloszewski (1994).

The *Transfer Function Model* (TFM) after Jury (1982) was chosen as a third model. This model does not describe the flow processes in a deterministic way, like the CDM or the SFDM, but in a stochastic way. The TFM does not make any assumptions concerning the flow processes, except that the velocity distribution in the unsaturated zone may be described as log-normal. This assumption was made by the developer of the model after evaluating several tracer tests. The system of the unsaturated zone transforms an input function into an output function (tracer breakthrough). The transfer function is the function of the log-normal distribution:

$$f(x) = \frac{1}{\sqrt{2\pi}\sigma} \cdot \frac{1}{x} \cdot \exp(-(\ln x - \mu)^2 / 2\sigma^2)$$ (3)

where:

μ = mean value
σ = standard deviation

The transfer function can be fitted to the observed data with two parameters: The mean value and the standard deviation of the distribution function. The mean residence time and the mean velocity of the water can be calculated with these parameters (Jury et al., 1986).

4 RESULTS AND DISCUSSION

4.1 Results of the tracer experiment

During the sprinkling overland flow was observed at some sites. Local ponding occurred, where microto-

pographic depressions started to be filled. It was possible to measure tracer concentrations at one outlet of the drainage system for the following 45 days. The most interesting part of the breakthrough curve is the period of the first 12 hours after the application (Figure 2), where 90 % of the tracers were recovered. The discharge at the outlet during this time was constant.

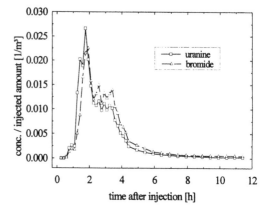

Figure 2: Breakthrough curves of uranine and bromide. To compare both tracers the concentrations were standardized to the injected amount.

Both breakthrough curves have a similar shape with a small „pre-peak" (around 1 hour after the application), a definite „main peak" (almost 2 hours after the application) and a „shoulder" at the decreasing limb of the concentration before a considerable tailing starts (4 hours after the application). Only one hour after the application, relatively high concentrations were measured. That means, the tracers were transported very fast through the unsaturated zone (1.2 m) and the following drainage pipe (350 m). The total recovery rate amounted to 12.5 % for uranine and 13.4 % for the non-sorptive bromide. This shows, that only a part of the 17 m² double frame system was influenced by the drainage pipe.

From Figure 1 it can clearly be seen, that the uranine concentrations increase and decrease earlier than the bromide concentrations. Also all the calculated velocities of uranine were higher than the velocities of bromide. These results are contrary to other investigations, where bromide was described as the fastest tracer (Jacob & Loeffelhardt 1991).

4.2 Modeling the results assuming a single flow system in the unsaturated zone

The breakthrough curve was modeled with the three

different transport models, described above (Figure 3). The following results and discussions refer to the observed concentrations of uranine because of its higher sensitivity and its better reproducible analysis.

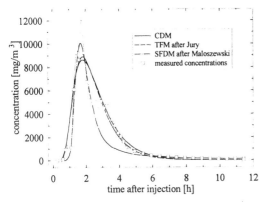

Figure 3: Comparison of the simulated concentrations of three different models with the observed concentrations, assuming a single flow system in the unsaturated zone.

It is noticeable that the results of the TFM are very similar to the results obtained by the CDM. The simulation with the SFDM shows a higher maximum concentration and a more pronounced tailing after six hours.

Table 1: Results of the simulation with the CDM, SFDM and TFM.

	CDM[1]	SFDM[2]	TFM
mean residence time	155.7 min	98.2 min	150.2 min
dispersion parameter	Pe = 8.26	Pe = 45.1 a = 0.0537	μ = 4.909 σ = 0.454

[1] assuming a total flow distance of 350 m
[2] molecular diffusion value $4.5*10^{-5}$ m²/s; fissure width 2 mm

The visible conformity of the CDM and the TFM resulted in almost the same mean residence time. The residence time of the water calculated with the SFDM is shorter, because of the diffusion of the tracer into the matrix. The SFDM is the best model to reproduce the tailing of the breakthrough curve. However, the results of the simulations, assuming a single flow system in the unsaturated zone, are not satisfying. None of the models can simulate the specific shape of the breakthrough curve. This leads to the conclusion, that distinct flow systems were active.

4.3 Modeling the results assuming several flow systems in the unsaturated zone

4.3.1 Using the CDM in the unsaturated zone

If a soil has an effective macropore system beside the micropore system, which is represented by the soil matrix, at least two distinct pore systems can be identified. Assuming the distinct pore systems have no or only minor hydraulic interactions (water flow between the macropores and the soil matrix), the transport of the tracers in each flow system can be modeled individually. The superposition of the simulated breakthrough curves leads to the total simulated concentration. Figure 3 illustrates the results of the simulation with the CDM, assuming three flow systems.

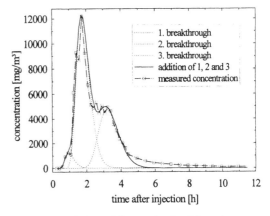

Figure 4: Simulation of the uranine breakthrough curve with the CDM, assuming three distinct flow systems.

The conformity of the simulated and the observed curves is very good. The „pre-peak", the „main peak" and the „concentration shoulder" at the decreasing limb are well reproduced. About 4.5 hours after the application a striking discontinuity in the curve of the observed concentrations is visible. Then the simulation obviously underestimates the observed concentrations.

It seems to be easy to generate an arbitrary number of breakthrough curves to represent the observed concentrations. However, it is important to find a reasonable interpretation of the physical processes. In this case, the first breakthrough seems to be generated by a single long-lasting crack, which transported the applied water very fast to the drainage system. The second (main) peak in the breakthrough curve was likely produced by the general macropore system. This pore system transported the main portion of the applied water. Consequently, the last breakthrough curve simulates the tracer transport through the soil matrix. This flow system was responsible for the shoulder of the breakthrough curve.

Darcy-based hydraulic calculations for the third breakthrough curve resulted in a hydraulic conductivity of $2.1*10^{-5}$ m/s for the matrix flow system (assuming a vertical flow direction and an effective porosity of 20 %). This seems to be a reasonable value, concerning the grain size distribution described above. The conductivities, which were estimated for the saturated zone by pumping tests and tracer experiments in equal loess soils, are in the same order of magnitude (Luft 1981).

The higher concentrations of bromide compared with uranine at the „shoulder" of the breakthrough curve (Figure 2) also indicate a transport through the matrix. Bromide is a conservative (nonsorptive) tracer, while uranine was adsorbed in the matrix in batch experiments (Uhlenbrook 1995). The lower concentrations of uranine compared to bromide were probably caused by the sorption of uranine in the micropore system of the loess matrix.

4.3.2 Using the SFDM coupled with the CDM in the unsaturated zone

The tracer transport in the macropores of a soil is assumed to be similar to tracer transport in the fissures of a hard rock aquifer. Therefore, the SFDM was used to simulate the macropore flow (Figure 5). The matrix flow was still simulated with the CDM. The first pre-peak was neglected, because it seams to be a local particularity.

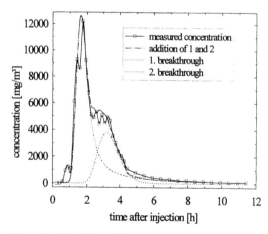

Figure 5: Simulation of the uranine breakthrough curve with the SFDM coupled with the CDM, assuming two flow systems.

This simulation gives an almost ideal reproduction of the observed concentrations. Even the tailing of the breakthrough curve is well simulated. The mean residence time of the water in the macropore system and the corresponding drainage system is 92.5 min; that of the micropore system of the matrix is 202.2 min. For the application of the SFDM the same parameters for the fissure width and for the molecular diffusion were used as indicated in table 1. The matrix porosity was assumed to be 40 %. These values are reasonable and are in the order of magnitude, described in the literature.

4.3.3 Discussion of the results assuming several flow systems

First of all, it should be mentioned that the physically-based models (CDM and SFDM) describe tracer transport through a saturated medium. During the investigation, the soil was not saturated. The CDM assumes Darcy-flow in the soil, which can not be expected in natural soils, at least not for soils containing macropores (Beven & Germann 1982, Demuth & Hiltpold 1994). However, it seems to be an adequate possibility to apply the CDM for the flow through the soil matrix. The pores, which are small enough to have field capacity, were saturated during the time of the investigation.

Following the type of modeling described above, two or more flow systems are modeled in a parallel way. In nature water exchange exists between the flow systems, for example suction of water from macropores to finer pores of the matrix. These interactions were neglected so far. It has to take into consideration, that the diffusion gradient, which controls tracer transport between the fissures and the matrix, was disturbed. The SFDM assumes stagnant water in the matrix. This modeling method suppose a Darcy-flow in the matrix. This means for the tracer transport, that at the beginning the tracer transport to the matrix was overestimated, because of a reduced diffusion gradient. After the main part of the macropore flow (approximately 2.5 hours after the injection), the diffusion of the tracer into the macropore system was underestimated, because of an increased diffusion gradient, caused by the tracer transport in the matrix. Future investigations have to consider modeling these interactions between the flow systems.

Nevertheless, under certain assumptions and interpreting that macropores are similar to fissures of a hard rock aquifer, the linking of the SFDM and the CDM seems to be a good possibility for physically-based modeling of tracer transport in soils. This is a basis to generate a new two-domain transport model for the unsaturated zone.

5 CONCLUSIONS

It was possible to show the importance of macropore flow for the runoff generation process in a small, anthropogenically influenced catchment. The tracer experiment showed, that preferential flow can dominate the transport of solutes. This is contrary to other investigations (e.g. Lindström & Rhode 1992), where the importance of macropore flow was not detectable. More than 60 % of the applied water was transported through the macropore system. Compared to other investigations is this a high but reasonable rate (Mosley 1982, Demuth & Hiltpold 1993). With this type of experiment it is possible to evaluate the preferential flow mechanism of an unsaturated soil quantitatively. Furthermore, it is a nondestructive method, with no impairment, caused by the installation of devices.

Both applied tracers were suitable for the application in experiments in the unsaturated zone. The fluorescence dye uranine is slightly sorptive, which was shown with batch experiments (Uhlenbrook 1995) and with the results of the field experiment. However, the use of uranine is still advisable because of the very easy and sensitive analysis and its harmless toxicology.

Very good results were obtained by simulating the flow in the macropore system with the SFDM and parallel the flow in the micropore system with the CDM. The combination of these two models is a basis to generate a new model for the solute transport in unsaturated medias as a two domain concept. Future investigations should focus on modeling of the interactions between the distinct pore systems.

Beside the influence of the preferential flow phenomenon on the generation of fast runoff components, the macroporosity has ecological and economic consequences. Fertilizer, pesticides etc. may be transported quickly below the root zone. From there they will be washed out to the drainage system, where they pollute stream ecosystems; or the pollutants could penetrate deeper and contribute to groundwater contamination. This has to be taken into consideration even in these dense loess soils, where the preferred landuse is agriculture.

ACKNOWLEDGEMENTS

The authors wish to thank J. Mehlhorn for his version of the SFDM and for many fruitful discussions. Thanks are also due to W. Ruf for many helpful comments on an earlier draft of the manuscript and for very useful input at various stages of the research. V. Armbruster and M. Weiler are thanked for their critical reviews. Last but not least, the authors are grateful to all the students for their collaboration by the fieldwork. The research was financially supported by the „Förderverein Hydrologie" of the University of Freiburg.

REFERENCES:

Beven K. J., Germann P., 1982: Macropores and water flow in soils; *Water Resources Research*, Volume 18, 5, 1311-1325.

Bucher B., Demuth S., 1985: Vergleichende Wasserbilanz eines flurbereinigten und eines nicht flurbereinigten Einzugsgebietes im Ostkaiserstuhl für den Zeitraum 1977 - 1980; *Deutsche Gewässerkundliche Mitteilungen*, 29, H1, 1-4.

Buttle J. M., 1994: Isotope hydrograph separations and rapid delivery of pre-event water from drainage basins; *Progress in Physical Geography*, 18, 1, 16-41.

Chen C., Wagenet R. J, 1992: Simulation of water and chemicals in macropore soils. Part 1: Representation of the epuivalent macropore influence and its effect on soil water flow; *Journal of Hydrology*, 130, 105-126.

Cui Y. F., Demuth S., Leibundgut CH., 1995: Runoff separation using tracer; *Proceedings of the Solution 95*, Edmonton.

Demuth N., Hiltpold A., 1993: "Preferential flow": Eine Übersicht über den heutigen Kenntnisstand; *Zeitschrift Pflanzenernährung Bodenkunde*, 156, 479-484.

Everts C. J., Kanwar R. S., 1994: Evaluation of Rhodamine WT as an adsorbed tracer in an agricultural soil, *Journal of Hydrology*, 153, 53-70.

Flury M., Flühler H., Jury W. A., Leuenberger J., 1994: Susceptibility of soils to preferential flow of water: A field study; *Water Resources Research*, Volume 30, 7, 1945-1954.

Germann P. F., 1986: Rapid drainage response to precipitation; *Hydrological Processes*, 1, 1-13.

Germann P. F., Beven K. J., 1985: Kinematic wave approximation to infiltration into soils with sorbing macropores; *Water Resources Research*, 21, 990-996.

Hornberger G. M., Germann P. F., Beven K. J., 1991: Throughflow and solute transport in an isolated sloping block in a forested catchment; *Journal of Hydrology*, 124, 81-99.

Jabro J. D., Lotse E. G., Fritton D. D., Baker D. E., 1994: Estimation of preferential movement of bromide tracer under field conditions; *Journal of Hydrology*, 156, 61-71.

Jacob H., Löffelhardt P., 1991: Vergleich von ^{3}H, Cl⁻, Br⁻ und Li⁺ als Tracer für die Wasserbewegung und den Stofftransport in der ungesättigten Bodenzone; *Steirische Beiträge zur Hydrogeologie*, 42, 151-164.

Jury W. A., 1982: Simulation of solute transport using a transfer function model; *Water Resources Research*, Volume 18, 2, 363-368.

Jury W. A., Sposito G., White R., 1986: A transfer function model of solute transport through soil. 1. Fundamental concepts; *Water Resources Research*, Volume 22, 2, 243-247.

Leibundgut Ch., Wernli H. R., 1982: Zur Frage der Einspeisemengenberechnung für Fluoreszenztracer; *Beiträge zur Geologie der Schweiz - Hydrologie*, Band 28, 1, 119-130.

Leibundgut Ch., Cui Y. F., 1994: Ganglinienseparation in kleinen Einzugsgebieten; *Beiträge zur Hydrologie der Schweiz*, Gedenkschrift M. Keller, Hydrologie kleiner Einzugsgebiete, 35, 96-104.

Lindström G., Rohde A., 1992: Transit times of water in soil lysimeters from modeling of ^{18}O; *Water, Air and Soil Pollution*, 65, 83-100.

Luft G., 1980: Abfluß und Retention im Löß dargestellt am Beispiel des hydrologischen Versuchsgebietes Rippach, Ostkaiserstuhl; Beiträge zur Hydrologie, Sonderheft 1, Verlag Beiträge zur Hydrologie, Kirchzarten.

Luft G., 1981: Zur Schätzung der Parameter Abstandsgeschwindigkeit und Dispersionskoeffizienten aus Makierversuchen mit Uranin in schluffigen Aquiferen; *Deutsche Gewässerkundliche Mitteilungen*, 25, H1, 12-18.

Maloszewski P., 1994: Mathematical modeling of tracer experiments in fissured aquifers; *Freiburger Schriften zur Hydrologie*, Band 2, University of Freiburg.

Maloszewski P., Zuber A., 1984: Interpretation of artificial and environmental tracers in fissured rocks with a porous matrix; in: *Isotope Hydrology 1983*, 635-651, IAEA, Vienna.

McDonnell J., 1990: A rationale for old water discharge through macropores in a steep, humid catchment; *Water Resources Research*, 26, 2821-2832.

Mehlhorn J., Leibundgut CH., Rogg H., 1995: Determination of the flow and transport parameters of the unsaturated zone using dye tracers; *IAHS Publication No. 229*, 183-192.

Mikovari A., Peter C., Leibundgut CH., 1995: Investigation of preferential flow using tracer techniques; *IAHS Publication No. 229*, 87-98.

Mosley M. P., 1982: Subsurface flow velocities through selected forest soils, south island, New Zealand; *Journal of Hydrology*, 55, 65-92.

Omoti U., Wild A., 1979a: Use of fluorescent dyes to mark the pathways of solute movement through soils under leaching conditions: 1. Laboratory experiments; *Soil Science*, Volume 128, 1, 8-33.

Omoti U., Wild A., 1979b: Use of fluorescent dyes to mark the pathways of solute movement through soils under leaching conditions: 2. Field experiments; *Soil Science*, Volume 128, 2, 98-104.

Scotter D. R., 1978: Preferential solute movement through larger soil voids, 1, Some computations using simple theory; *Australian Journal of Soil Resources*, 16, 257-267.

Uhlenbrook S., 1995: Investigation on fast runoff components; M. of. Sc. Thesis, *University of Freiburg, Institute of Hydrology*, Germany (written in German).

Tracer Hydrology 97, Kranjc (ed.) © 1997 Balkema, Rotterdam, ISBN 90 5410 875 4

In-situ tracer investigations on the water balance of an alternative surface covering system on a sanitary landfill

R. Zischak & H. Hötzl

Department of Applied Geology, Hydrogeology Section, University of Karlsruhe, Germany

ABSTRACT: For a surface sealing system on a sanitary landfill in SW-Germany a long term water balance has been calculated by measurements of nearly 3 years. The results show interesting phenomenia in the hydraulic flow behavoir of different layers. These problems could be evaluate by two tracer experiments in the sequence of the layers under unsaturated conditions. The existence of short circuit flows or any other preferentiell flow have been neglected by the results of the tracer experiments. The mass balance and the shape of the breakthrough curves document the functionability of the sealing system especially the quality of the mineral clay liner. Furtheron the influence of temperature on the outflows and the effectiveness of the capillary barrier could be reproduced by the tracer experiments. Also the results of soil properties from laboratorium measurements for unsaturated conditions have been evaluated by the results of the tracer breakthroughs.

1. INTRODUCTION

In Karlsuhe (SW-Germany) groundwater contaminations were detected downstream of the main sanitary landfill since the mid 80-ties. The owner decided to develope a remediation program to protect the intensively used aquifer from forthgoing contamination.

Beside several measures such as installing of a hydraulic drainage system and a waste water collecting and conditioning plant the building up of an effective surface covering system was the most important step. Hence the landfill was positioned in a formerly gravel pit within the groundwater fluctation zone without any basic sealing system.

Since the beginning of the dumping in the late 50-ties

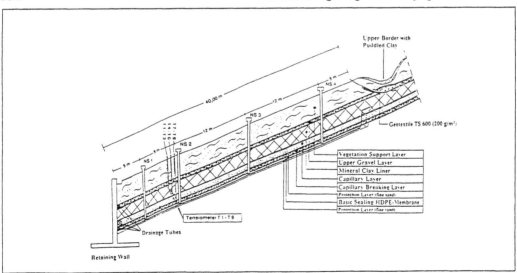

Figure 1: Cross section of the lysimeter

the height of the waste deposit grew up to 60 meter above the surrounding ground level, therefore the angles of slope have increased up to 23 degrees (1:2.3) in average. The regular surface sealing system required by the Technical Instruction for Domestic Waste (TA Siedlungsabfall, TASi 1993) guarantees not enough slope stability at angles steeper than 21° (1:2.5) (Burkhardt & Egloffstein 1994). Thus the community of Karlsruhe has been forced to install an alternate system which must be proofed for equivalence to the regular system. This improvement is mainly based on a water balance.

The alternate system, so called „Enforced mineral clay liner with an underlying capillary barrier", shows vertically from top to the bottom the following sequences (Fig. 1):

- Vegetation Support Layer (1 meter)
- Upper Gravel Layer (0.15 meter)
- Compacted mineral clay liner (0.6 meter out of 3 lifts each 0.2 meter)
- Capillary Layer (0.3 meter)
- Capillary Breaking Layer (0.15 meter)

This system has been installed 1993, since then frequently measurements on the water balance have been carried out by the Hydrogeology Section of the Department of Applied Geology at the University of Karlsruhe (Zischak & Hötzl 1994, 1996a+b).

2. METHODOLOGICAL APPROACH

2.1 Equipment

A lysimeter (10 x 40 meter, Fig.1) has been built up for the evaluation of the water balance to furnish proof the equivalence of the sealing system. The position of the relevant instruments installed is depick in Figure 1.

A precipitation recorder, positioned equal to the ground level, and two rainfall sampler, one downslope and on the top of the lysimeter serve for the basic input data of the water balance equation.

At the retaining wall downslope 4 different outflows are continiously sampled by drainage tubes and measured by gage recorders, while the soil water storage is registered with neutron logging each week at 4 probes allover the lysimeter area. Furthermore with a tensiometer profile we have been able to watch out the hydraulic gradients.

2.2 Results

Table 1 shows the long term water balance. According to the outflow from the capillary breaking layer the amount is equal to the input into the waste deposit. It results that only 0.6 % of the total

Table 1: Long Term Water Balance

14.11.93 - 30.06.96	[mm]	[% of P]
Precipitation	2,281.57	
$Q_{Surface}$	10.55	0.5
$Q_{Upper\ Gravel\ Layer}$	373.06	16.4
$Q_{Capillary\ Layer}$	176.40	7.7
$Q_{Capillary\ Break.\ Layer}$	13.28	0.6
SumQ	573.3	25.1
$S_{Soil\ Water}$	-84.0	-----
ET = P - SumQ - S_{SW}	1,708.2	74.9
ETP_{Haude}	2,1.98.2	96.3

precipitation would percolate through the surface sealing system. Related to the total area of the landfill (22 ha) only 1,056 m³ would infiltrate into the landfill per year at an average precipitation height of 800 mm/a and produce new waste water (Zischak & Hötzl 1994, 1996a+b). Taking this balance into account, results in a high functionability of the alternate system and an execellent effectiveness of the sealing system. Additionally the alternate system save about 6.4 millions DM compared to the regular one.

In the test-site the surface flow seems to be too low for the existing steep angles but the very early cultivation of the vegetation cover prevented a remarkable surface flow on the one hand and a high evapotranspiration rate had been produced on the other hand. Furthermore the interflow near the surface has been not drained together with the surface flow but with the outflow of Upper Gravel Layer. Egloffstein et al. (1995) pointed out using the model HELP 3.0 that intensive cultivation results in a very low surface flow about 0.1-0.4 % of total precipitation.

About 8 % of total precipitation infiltrates into the capillary barrier and creates the outflow of the capillary layer and the capillary breaking layer.

The difference between potential evapotranspiration after Haude (Tab. 1) and from measurements calculated actual evapotranspiration seems to be quite less so the reliability of the lysimeter itself is given.

Also the effectiveness of the capillary barrier shown in Table 2 reflects the high quality standard of the sealing system. Mean effectiveness is about 95 %, however the variation ranges from 98 down to 57 % (Tab. 2).

Table 2: Quality of the Sealing System in 1994 (Comparison of the measured discharges)

	Precipitation	Capillary Layer	Capillary Breaking Layer	Effectiveness of the Capillary Barrier	Effectiveness of the Sealing System
	[mm]	[mm]	[mm]	[%]	[%]
Nov'93	23.60	1.61	0.24	86.98	98.98
Dec'93	152.10	7.30	0.37	95.19	99.76
Jan'94	61.59	8.91	0.27	97.06	99.56
Feb'94	49.74	7.01	0.24	96.75	99.53
Mar'94	71.17	8.43	0.24	97.22	99.66
Apr'94	60.79	8.40	0.26	97.00	99.57
May'94	79.72	6.49	0.16	97.62	99.80
Jun'94	78.47	4.94	0.23	95.48	99.70
Jul'94	52.36	4.21	0.44	90.47	99.15
Aug'94	83.15	3.00	0.81	78.64	99.02
Sep'94	107.53	1.49	0.78	65.64	99.27
Oct'94	38.42	1.83	0.60	75.41	98.45
Nov'94	42.34	1.91	0.45	80.87	98.93
Dec'94	79.16	1.51	1.13	57.20	98.57

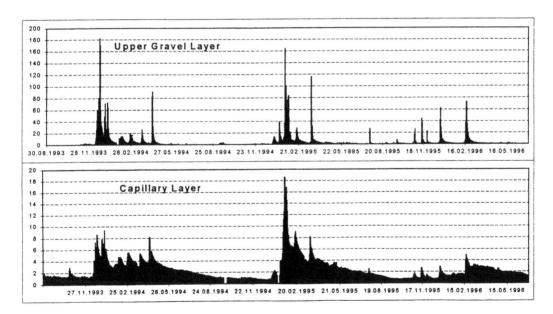

Figure 2: Outflow curves and discharges in l/h

2.3 Open Questions

One open question has ever been the seepage water rate under the compacted mineral clay liner. The clay liner have been built up in 3 lifts while in the same time saturated hydraulic conductivity proofed by several tests seems to be around 5×10^{-10} m/s. The resulting calculated percolation rate should be 2.1 % at 800 mm yearly precipitation. Fig. 2 shows the correlated outflow curves of upper gravel layer and capillary layer. Obviously a synchronous flow behavior between the two curves exists although the highly compacted mineral clay liner separates both layers. In the winter period there is a discrete reaction of the capillary layer to every peak of the upper gravel layer with a phase displacement of 6 to 8 hours (Fig. 2). Close to this phenomenon first thought conforms to short circuits in the area of the retaining wall (s. Fig. 1). This problem has been proofed by the tracer experiments (Section 3) .

The other problem is given by the effectiveness of the capillary barrier (Tab. 2). The outflow of capillary breaking layer increases during summer period while the capillary layer outflow decreases, so the sealing property of the capillary barrier system slopes down to 57 % at minimum (Wohnlich et al. 1996, Zischak & Hötzl 1994, 1996a+b).

3. TRACER TESTS

3.1 Objects of Investigation

Both phenomenia discussed above were investigated by tracer tests in the Upper Gravel Layer and in the Capillary Layer. Furtheron hydraulic properties of unsaturated and nearly saturated conditions should be pointed out for these layers, so a combined tracer experiment has been carried out (Watter 1994).

By tracing water of Upper Gravel Layer it has been possible to detect short circuit flows and also to quantify them. Due to the general setup of the experiments, we have also the possibility to trace the clay liner and furtheron the Capillary Layer.

3.2 Description of Tracer Input

Three different flourescent dyes have been injected via boreholes ($\varnothing$ 32 mm) directly into the layers of the covering system in different level and distances to the outflow tubes (Tab. 3). Uranine and pyranine were used for the high compacted mineral clay liner and eosine for the capillary layer consisting of homogenous very fine sand. While within the input probes the dye solution was rejected by overflowing the upper gravel layer was also indicated with fluorescent dyes. Each dye has been injected in a row of boreholes across the lysimeter.

Table 3: Listing of Tracer Tests

Dye	Mass	Level/Layer	Distance
Uranine	99 g	Up. Gravel Layer	11 m
Pyranine	47 g	Up. Gravel Layer	23 m
Uranine	142 g	Clay Liner (Top)	11 m
Pyranine	91 g	Clay Liner (Bottom)	23 m
Eosine	68 g	Cap. Layer (Top)	35 m
Naphthionate	125 g	Cap. Layer (Mid)	5 m

The samples at the output systems were taken by automatic sampling systems for nearly 4 months and later on by activated carbon charcoals. In this period there was no response of eosine in the capillary layer outflow so a second tracer experiment has been carried out with sodium-naphthionate at a shorter flow distance to the retaining wall (s. Fig. 1) and another level within the layer (Tab. 3). The injection of naphthionate has been made via 3 boreholes.

The water samples have been analysed with a synchronous scan fluorimeter (Beherens 1971) also the carbon samples after desorption.

3.3 Results of Tracer Tests

The breakthrough curves of Uranine and Pyranine in Upper Gravel layer are dominated by singular quite sharp peaks within the first 20 hours after injection. The pyranine curves seems to have a broader peak related to longer flow distance and therefore greater dispersivity (Fig. 3a).

During the first 500 hours after injection no response of pyranine and uranine was found in the outflow of the capillary layer (Fig. 3b). Then first uranine has been detected, lateron pyranine was detectable on a higher concentration because organic compounds have been overlying the low pyranine-peaks in the fluorimeter scanning (Fig. 3b). The measured and calculated results are shown in each textbox within the two figures. Comparing the concentrations in the upper gravel layer to them in the capillary layer related to the flow rates, it results that only 0.3 to 3.6 % of total outflow of the capillary layer belongs to short circuit flows out of the upper gravel layer. The relative portion seems to be proportional to the height of the flow rate in the gravel layer.

In respect to the water balance, this means no significant change for the total discharge underneath

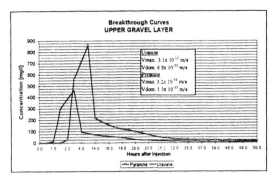

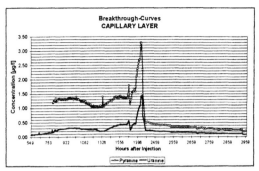

Fig. 3a: Breakthroughs in the Upper Gravel Layer

Fig. 3b: Breakthroughs in the Capillary Layer

the compacted mineral clay liner. These results have been evaluated also by hydraulic tests (Schnell 1996), which defined the portion with 0.4 to 4.2 %. This demonstrates the reliability, both in quantity and quality, of tracer experiments to detect macropore flow, short circuit flow or other preferentiell flow.

During a period of nearly 100 days no eosine was found in the outflow of the capillary layer. Due to the lower unsaturated hydraulic conductivity and the variable water content in the vertical profile of the capillary layer the distance has been too long and the retardation at the fine sand matrix has been too strong. Thus a second tracer experiment with sodium-naphthionate has been carried out 5 meter from the retaining wall (Tab. 3).

The resulting breakthrough curves are shown in Fig. 4a+b. Different to the other tracer experiments also an tracer output outof the capillary breaking layer has been detected (Fig. 4b). The mass balance of recharges yields to 0.2 % of tracer passes the interface of the capillary barrier and infiltrates into the capillary breaking layer.

The broad peak of the outflow curve of the capillary breaking layer reflects the increasing outflow compared to the decreasing outflow of the capillary layer in the summer period. That means during heating up the soil matrix in summer more seepage water will pass the interface of the capillary barrier.

Kämpf & v.d. Hude (1995) and Wohnlich et al. (1996) also described this physical phenomenon as a result of changing flow conditions in a capillary barrier caused by temperature effects. The higher soil temperature in summer penetrating from the surface produce a higher vapour flow downward into the capillary breaking layer. Basicly the temperature also reduces the viscosity of soil water. Thus the effectiveness of the capillary barrier decreases in summer period (Tab. 2).

Eosine, injected in the first tracer experiment, occured in the activated carbon charcoals after 370 days for the very first time (Fig. 5). The failure of this tracing has been caused by the location of injection. The dye was injected in the upper very low saturated part of the capillary layer in the top region of the lysimeter (Tab. 3). Thus the active cross section for the unsaturated flow is very narrow and we have had a long period of retardation.

The results of uranine and pyranine in the charcoals seems to be quite complex. The curves of extinction in the charcoals shape like the Figure 5 in

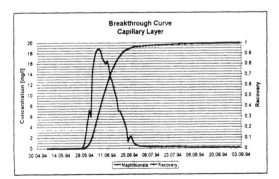

Fig. 4a: Breakthrough in the Capillary Layer

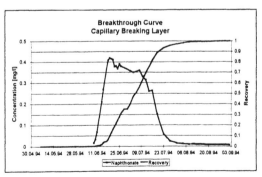

Fig 4b: Breakthrough in the Capillary Break. Layer

quality . All the time over nearly 3 years we found portions of both fluorescent dyes in the outflow of the capillary layer. Then in the summer of 1996 the scale units of extinction rose up significantly.

The measured distribution over time looks like a tracer breakthrough curve.

Taking the mean velocity out of the tracer experiment with sodium-naphthionate into account (Fig. 4a+b), we could estimate a travel time in the capillary layer. The leaving time should remain for the transport through the mineral clay liner. The results caculated with 1 to 5x 10^{-9} m/s correlate quite well with the Darcy-velocity of 5x10^{-10} m/s and a effective porosity of 5 %. Therefore we can postulate the missing of any macropore or preferentiell flow in the mineral clay liner (Zischak & Hötzl 1994, 1996a+b).

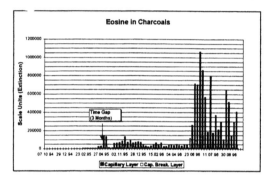

Figure 5: Results of Eosine in the Charcoals

4. CONCLUSIONS AND OUTLOOK

Tracer experiments have been a useful instrument to verify and confirm the investigations in the testsite of the sanitary landfill Karlsruhe-West (SW-Germany).

Even for unsaturated conditions tracer experiments evaluate the unsaturated hydraulic properties due to the laboratorium examinations. The most time exactly quantifying is not possible but the description of the quality of the flow processes works quite good.

The experiments results in no significant evidence of short circuit flow at the retaining wall and no evidence for any kind of preferentiell flow through the mineral clay liner. Therefore another process causes the synchronous flow behavior of the upper gravel layer and the capillary layer (Fig. 2). The hydraulic tests and the laboratorium experiments confirm the process characteristic and also the phase displacement of 6 to 8 hours. The estimation and evaluation of the transform function is still going on and will be finished in 1997.

REFERENCES:

Bronstert, A., Merkel, U. & Zischak, R. (1996): Flow dynamics in a hillside area with a landfill cover.- in Wasser und Boden Vol. **48/12**: 38-54; Hamburg, Berlin.

Burkhardt, G. & Egloffstein, Th. (1994): Ausführungsvarianten von Oberflächenabdichtungssystemen und Hinweise zu deren Auswahl.- in Schriftenreihe der Angew. Geologie Karlsruhe, **34**: 59-101; Karlsruhe (Eigenverlag).

Behrens; H. (19971): Untersuchungen zum quantitativen Nachweis von Fluoreszensfarbstoffen bei ihrer Anwendung als hydrologische Markierungsstoffe.- Geologica Bavaria, **64**: 120-131; München.

Egloffstein, Th., Burkhardt, G. & Heidrich, A. (1995): Wasserhaushaltsbetrachtungen bei Oberflächenabdichtungen und -abdeckungen.- in Egloffstein & Burkhardt (1995): Oberflächenabdichtungen für Deponien und Altlasten - Abdichtung oder Abdeckung? - Schriftenreihe der Angew. Geologie Karlsruhe, **37**: 12/1- 12/53; Karlsruhe (Eigenverlag).

v.d. Hude, N., Kämpf, M. & Montenegro, H. (1995): Capillary barriers- State of science/Practical application.- in Egloffstein & Burkhardt (1995): Oberflächenabdichtungen für Deponien und Altlasten - Abdichtung oder Abdeckung? - Schriftenreihe der Angew. Geologie Karlsruhe, 37: 6/1- 6/29; Karlsruhe (Eigenverlag).

Kämpf, M. & v.d. Hude, N. (1995): Transport Phanomenia in Capillary Barriers: Influence of Temperature on Flow Processes.- in CISA: Proceedings Sardinia 95, Fifth International Landfill Symposium Vol II: 565-576; Cagliari

Schnell, K. (1996): Hydraulic investigations of the surface covering system in the lysimeter of the sanitary landfill Karlsruhe-West under the aspect of interface effects.- Studywork at Department of Applied Geology, University Karlsruhe: 112 p.; Karlsruhe [unpublished].

TA Siedlungsabfall (1993): Technische Anleitung zur Verwertung, Behandlung und sonstigen Entsorgung von Siedlungsabfällen.- Bundesanzeiger: 117 S.; Köln.

Watter, U. (1994): Investigations on functionability of the surface covering system of HMD Karlsruhe-West.- Studywork at Department of Applied Geology, University Karlsruhe: 86 p.; Karlsruhe [unpublished].

Wohnlich, S., Bauer, E., Franke, C., Hötzl, H. & Zischak, R. (1996): Die Kapillarsperre als alternative Oberflächenabdichtung.-TABASARAN (Hrsg.):Vertieferseminar Zeitgemäße Deponietechnik 1996 vom 12.-13.03.96; Veröffentlichung in Vorbereitung; Stuttgart.

Zischak, R. & Hötzl, H. (1994): Results of the testsite for the combined capillary barrier on the waste deposite Karlsruhe-West.- in Schr. Angew. Geol. Karlsruhe 34: 161-180; Karlsruhe.

Zischak, R. & Hötzl, H. (1996a): Mineral clay liner with an underlying capillary barrier - Testsite HMD Karlsruhe-West -.- in Schr. Angew. Geol. Karlsruhe 45: 13/1-13/22; Karlsruhe.

Zischak, R. & Hötzl, H. (1996b): Capillary Barriers for Surface Sealing of Sanitary Landfills.- Oral presentation H22D-7 at AGU Fall Meeting 1996 publ. as EOS, Transactions, AGU Vol. 77, No. 46, Washington.

Aquifer

Tracer Hydrology 97, Kranjc (ed.)© 1997 Balkema, Rotterdam, ISBN 90 5410 875 4

Karstic sources in Malatya Province, east of Turkey

I. Atalay
Department of Geography, Buca Faculty of Education, Dokuz Eylül University, İzmir, Turkey

ABSTRACT: Turkey has karstic terrains due to the fact that limestones which had formed during the all geological times are widespead. Most of them occur along the southern part of Turkey. These areas are also rich in karstic sources or springs some of which feed some rivers and obtain the drinking water of settlements and irrigation water for agricultural lands. One of the most important karstic source areas is found in the vicinity of Malatya Province, located in the eastern part of Turkey. Here there are a lot of karstic sources. The big one named Pınarbaşı (Turkish: head of the source) gives the major part of the drinking and irrigation water of the Malatya Province and its discharge is more than 10 m^3 s^{-1}. The major part of the Malatya agricultural land totalling more than 50 km^2 area is irrigated by the Pınarbaşı spring.

INTRODUCTION

As it is known, human life in the karstic land mostly depends on the existence of the karstic springs and/or sources. Such springs determine the settlements and economic activities of inhabitants especially in the arid and semi-arid regions. Indeed some rivers which are mostly fed by karstic springs and springs give adequate water both for irrigation and drinking. For example, karstic springs mostly emerging on the edge of Taurus Mountain belt provide both irrigation and drinking water of Malatya city and the other settlements living totalling 500 000 population.

Malatya Basin is under the continental arid climate. Mean annual precipitation is over 300 mm, minimum is under 200 mm. This figure increases on the mountainous areas. A major part of the precipitition falls during the winter and spring. Severe drought prevails during summer season. Agricultural activities of the Malatya Plain, in which arid climate prevails have been done thanks to karstic sources. Especially apricot which is famous in the world grows in the Malatya Plain.

GEOLOGIC and GEOMORPHIC PROPERTIES

Malatya and its surroundings area is located within the Taurus Mountains belt and Malatya Plain occur at an altitude of 850-1000 m on the plateau surfaces in the northern section of the Taurus Mountains. Bey Mountain which is the main branch of the Taurus Mountains abruply rises on the southern edge of the Malatya Plain. The highest peaks exceed 2500 m. Relative altitude between Malatya Plain and Bey Mountain is over 1500 m.

The foundation of Bey Mountain is generally made up of Paleozoic epi-metamorphic schists such as phyillite and clayey schists. Gneiss and mica schists belonging to Lower Paleozoic are outcropped in the southern part of the mountains.

The upper part of the Bey Mountain is composed of crystallized and cracked limestones belonging to upper Paleozoic. Cretaceous clayey limestones and marl are common in the lower part of the mountains and lies discordantly on the Paleozoic terrain. Eocene clayey and hard limestones, inclined as a monoclinic structure towards the north, mostly occur in the norhern part of the mountains.

Bey Mountain mass was uplifted and partly folded during the Alpine orogenic period and then vertical tectonic movement occurred so that Malatya Basin was collapsed via the fault lines extending northern and western part of the mountains.

Rivers which had been set up on the upland areas have cut deeply their own valleys according the Malatya Basin level. Pre-Neogene basement associated with schists were outcropped along the deeply incised valley on the montainous areas. Malatya Basin which was filled with fluvio-limnic materials is deeply cut by the streams joining the Euphrates River. And plateau appearance come into scene.

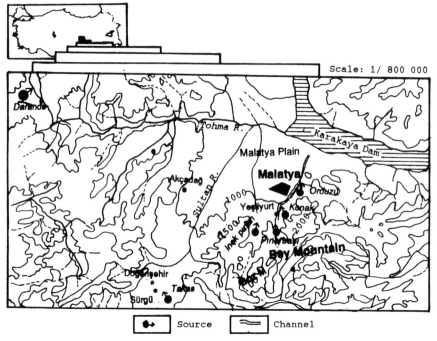

Figure 1: Location and topographic map showing karstic sources in the vicinity of Malatya Province.

Karstification process has begun at least at the end of the Mesozoic so that some streams originating from the upland areas have been shifted so that partly underground drainage system has been formed.

KARSTIC SPRINGS

As a general rule, karstic springs are found on the egdes of the mountains and contact lines in places where between limestone and impervious layers composed of schists, clayey limestone and marly strata which are exposed along the deeply cut valley.

Main karstic springs and their formations are as follows:

Pınarbaşı Spring

This karstic spring emerges at an altitude of c. 1200 m in the bottom of the valley named Kozluk along the contact line between clayey schist and the paleozoic limestone. Paleozoic limestones covering the Bey Mountain are the main supplier of the spring (Fig. 2). Indeed, underground flows originating northern sections of the of limestones collect along the valley in the Bey Mountain. In addition to this, the water derived from the southern slopes of this mountain supplies the water of Lake Abdulharap.

Pınarbaşı spring's discharge changing with precipitation is more than 10 m^3 s^{-1}. Maximum attains 20 m^3 s^{-1}. The water of this spring also works hydroelectric power plant which is established at Kapuluk place, 5 km north of Pınarbaşı. Pınarbaşı spring's water flow throught the cannel from emerging point to near Euphrates river in order to realize both irrigation and drinking water of settlements such as Gündüzbey, Yeşilyurt, Yakınca, Tecde, Malatya, Battalgazi, etc.

Takas (Sürgü) Spring

It appears at an elevation of c. 1000 m SE of Sürgü Town at the edge of western part of Boz Mountain. This spring feeds considerably the Sultan Stream which is the main tributary of Euphrates in Malatya Basin.

Darende Spring

This spring is among the main karstic spring of study area with a mean flow of 10 m^3 s^{-1} It exits at the valley bottom at an elevation of c. 1100 m. This spring is in exsurgence characters. That is deeply incised karstic valley collects seepage water through the karstic rocks and feeds considerably the Tohma Stream which is the main tributary of Euphrates in the Malatya Plain.

200

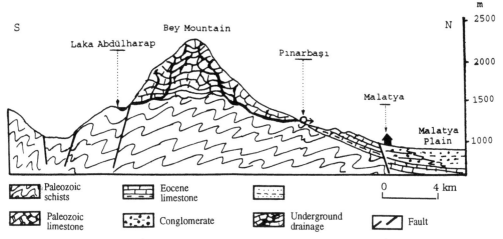

Figure 2: Cross-section of Bey Mountain and the formation of Pınarbaşı karstic source.

Konak Spring

This spring is one of the another karstic sources of Malatya Province. It appears at the bottom of the valley in places where the contact line between limestone and marly layer is exposed at the edge of Bey Mountain (Figure 3). Its flow is about 3 m³ s⁻¹.

appears on the impervious clayey limestone in the karstic depression like doline. Seepage water derived from the eastern part of the Bey Mountain feeds this spring.

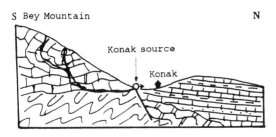

Figure 3: The formation of Konak Spring.

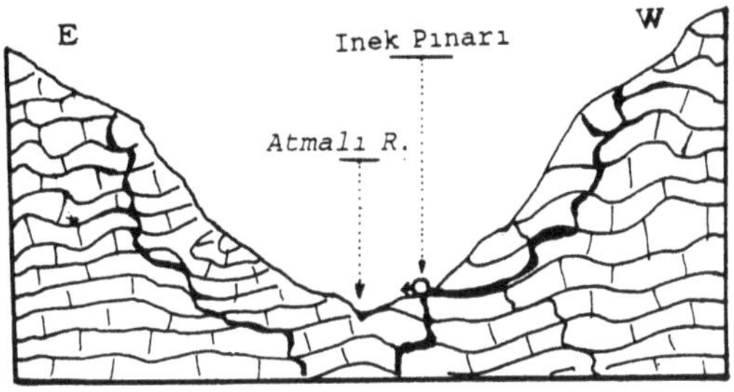

Figure 4: The formation of Inek Pınarı Spring.

Inek Pınarı (Cattle spring)

It emerges in the lower slope of valley and it is one of the main supplier of Atmalı streams (Figure 4). The flow of this source changes considerably according to the amount of precipitation and changing of groundwater flow. Its flow increases when underground karstic hole is plugged by fine transported material. For this reason this source can be termed as a exsurgence type. Maximum flow is more than 6 m³ s⁻¹.

Orduzu Spring

It is found 2 km in the eastern part of Malatya. It

REFERENCES

Atalay, I. 1994. *Geography of Turkey*. 480 p. Ege Univ. Press, Izmir, Turkey.

Atalay, I. 1995. Pedogenesis and ecology of karstic lands in Turkey. *Acta Carsologica*, XXIV: 53-67.

Atalay, I. 1996. Karstification and karstic lands in Turkey. *Karren Landforms* (Ed. by J. J. Fornos, A. Gines), Universitat de les Illes Balears, Palma.

Jennings, J.N. 1985. *Karst geomorphology*, Oxford, 293p, Basil Blackwell, U.K.

Tracer Hydrology 97, Kranjc (ed.) © 1997 Balkema, Rotterdam, ISBN 90 5410 875 4

A negative dye tracing in the Grintovec massif (Kamnik Alps)

Philippe Audra
CAGEP-URA, Aix-en-Provence Groupe de Valorisation de l'Environnement (GVE) & URA D1476 du CNRS, Université de Nice-Sophia-Antipolis, France

Abstract : a tracing with fluoresceine was undertaken in August 1996 in the Grintovec massif (Kamnik Alps, Slovenia). All the samples taken in the Kamniška Bistrica spring were negative. This unexpected result is analysed.

The Kamnik Alps constitute a crest line along the Austrian border, 30 km north of Ljubljana city. They culminate at the Grintovec (2558 m asl.). The studied area (Veliki Podi) is a glacio-karstic amphitheatre, which is located on the eastern slopes of the Grintovec, between 1900 and 2300 m asl. The rocks are triasic limestones *(Dachsteinkalk)* affected by a strong pendage (around $40°$, but with important variations). Belonging to overthrust sheets inclined to the north, these layers are affected by an intense fracturation.

The discovery of a 200 m deep shaft, in a virgin speleological area, have incited us to make a dye tracing. According to the pendage, the large Kamniška Bistrica spring (1,5 m^3 / s at low waters, more than 5 m^3/s during storms high waters), located 4 km away just at the foot of the clints (600 m asl.), seemed to drain this part of the massif.

The fluoresceine injection (1 kg dissolved in 2l alcohol) had been realized in this shaft (Brezno pod Koglom, 2040 m asl.) after a storm, the 8. August 1996.

The emergence had been watched over with an automatic sampler during one week, with a sampling frequency of 2 hours the two first days and 4 hours the following days. Moreover, a temporary emergence located upstream (690 m asl.), and probably acting as an overflow of the main spring, had also been sampled.

The analyses were made in France with a fluorimeter with a precision for dilution of 10^{-10} kg/l. None of the analysed samples (about 60) showed any trace of fluoresceine.

This negative result suggests several hypothesis about the happening of the fluoresceine and the structure of the karstic aquifer :

- Taking into account the sampling frequency and the accuracy of the fluorimeter, it is highly improbable that the passage of the tracer could not have been detected. Moreover, knowing that we are dealing with a mountain karst characterised by a high altitude gradient where the discharges are very rapid (more than 100 m/h), that the aquifer was saturated and that two storms happening after the injection should have expelled the tracer, it is also unlikely that the fluoresceine could have been stored in the aquifer, and that its restitution could have occurred after the watch period. We must recognise that the tracer had reappeared somewhere else, in one of the surrounding valleys.

- In the north, three valleys flank the massif. In the NE, the Logar valley, where the Savinja spring pours out (1300 m asl.), 3 km away. In the N, the Vellach valley, located in Austria. In the NW, the abrupt slopes of the Grintovec. A tracer restitution in one of these valleys is unlikely, according to the pendage sloping to the south and to the moderate altitudinal gradient.

- Finally, in the western part the Kokra river flows, at the foot of the Grintovec. A direct link with springs along the river would be improbable, according to the presence of an aquiclude. However, south of the Veliki Podi, the pendage slopes to the west. It is then possible to envisage an oblique flow under the Veliki Podi, in direction of the south beyond the Kokrsko Sedlo, which then could attain the area where the pendage slopes to the west and finally joins the Kokra valley. However, we do not know of any important spring in this place. Such an eventuallity is of course highly hypothetical, but it could explain this negative dye tracing.

So, even if we could not finally settle for a negative dye tracing, it is nevertheless acceptable to admit that the Veliki Podi area does not belong to the Kamniška Bistrica catchment area. A hypothetical outlet in the Kokra valley would need to be taken into account. This could mean that this large valley edging the massif had known a complex evolution, allowing a conqueror drainage to be installed over the neighbouring karstic basins. Such a hypothesis would be interesting to check in the future.

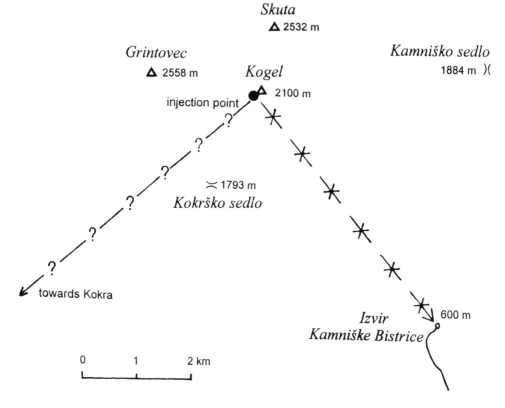

Figure 1: Sketch of the tracing under Kogel

This mission was supported by the Slovene Art and Sciences Academy (SAZU), in the frame of the collaboration with the French Centre national de la recherche scientifique (CNRS). I am gratefull to J. Biju-Duval for the analyses of the samples.

Tracer Hydrology 97, Kranjc (ed.)© 1997 Balkema, Rotterdam, ISBN 90 5410 875 4

Ground water flow in the fresh water lens of northern Guam

W. L. Barner
ICF Kaiser Engineers, Inc., Pittsburgh, Pa., USA

ABSTRACT: Recent hydrogeologic investigations on the northern portion of the island of Guam, Mariana Islands, have provided interesting data regarding expectations of rapid (conduit) flow within the fresh water lens. Evidence from cave exploration and boreholes drilled inland and near the coast suggest karstification has occurred between the phreatic and vadose zone at a depth of approximately 150 meters (500 feet) below ground surface, and within the transition zone between the fresh and salt water interface, near sea level. Tracing results, and non-flashy responses on water levels in wells, suggests ground water movement is representative of macro-porous-media flow, not conduit flow that is commonly assumed in most karst aquifers. Groundwater flow is further complicated by lateral and vertical variations of reef facies, partial or complete diagentic removal of primary porosity. Horizontal and vertical groundwater flow in the aquifer converges along the coastal area, and discharges as resurgences, or submarine springs. In contrast, flow within the vadose zone is rapid, and the direction of movement is highly unpredictable.

INTRODUCTION

During the past decade, site characterization investigations on the island of Guam have increased as a result of requirements for regulatory compliance. Since the liberation of Guam from the Japanese in 1944, sanitary and industrial wastes have been disposed in various trenches, borrow pits, quarries, and dolines. Potential ground water degradation may result if hazardous substances, in the form of leachate, are released from these disposal areas. To address this concern, geologic and hydrogeologic investigations resulted in the installation and sampling of ground water monitoring wells to address this concern. As part of these investigations, a dye trace study was performed to determine which monitoring wells and borehole locations are downgradient from and could intercept a release from the main landfill area. The results of the dye trace are being used to identify ground water flow direction, flow velocity, appropriate borehole locations to be converted into monitoring wells, to confirm existing monitoring wells are capable of detecting releases from the landfill area, and for locating new monitoring wells. These investigations are part of ongoing environmental investigations by ICF Technology, Inc. under contract to the U.S. Air Force.[1]

PHYSIOGRAPHIC SETTING

The island of Guam is the southern-most island in the Mariana Island chain, located approximately 13° 27' North latitude and 144° 47' East longitude (Figure 1). Guam is approximately 2,500 km (1,550 miles) south of Japan and approximately 5,300 km (3,300 miles) southwest of the Hawaii Islands. The Mariana Islands are a complex island-seamount system, divisible geographically, tectonically, and chronologically into island arcs (Siegrist and Randall, 1992); an older frontal arc (middle Eocene; 43 million years before present, [mybp]) which includes the larger islands of Guam, Rota, Tinian, and Saipan, plus two smaller uninhabited islands. A younger arc (early Pleistocene; 1.3 mybp) of active seamounts and islands lies to the west and north of the older arc.

Guam is approximately 48 km (30 miles) in length running north to south, ranging in width east to west from approximately 6.4 km (4 miles) near the center of the island to 13 km (8 miles) in the northern part of the island. Depositional environments of Guam represents a limestone-veneered peak on a nearly submerged volcanic ridge between the Philippine Sea to the west and the Pacific Ocean basin to the east

---

[1] USAF Contract No. F33615-90-D-4010.

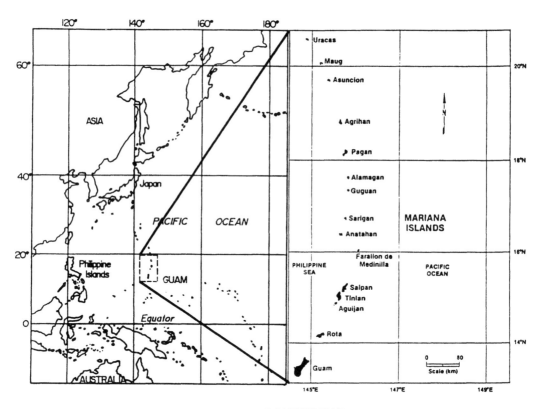

FIGURE 1 - LOCATION MAP

(Cloud, 1951). The elevations on the northern plateau region within the Air Force properties range from about 60 meters (200 feet) to 180 meters (800 feet) above sea level. The surface of the limestone plateau is interrupted by two volcanic peaks, Mount Santa Rosa and Mataguac Hill, with elevations of 252 meters (826 feet) and 192 meters (630 feet),respectively, above sea level.

The island is subequally divided into the northern limestone plateau and the higher volcanic hills to the south. The northern limestone plateau rises to approximately 200 meters (660 feet) above sea level and is covered with thick vegetation consisting of limestone forest, mixed natural and exotic forest, and mixed natural and exotic shrubs (ICF Technology, 1994; Raulerson and Rinehart, 1991). The southern half of the island rises to a maximum elevation of 378 meters (1,240 feet) above sea level and the volcanic hills support mainly sword grass and small shrubs.

GEOLOGY OF NORTHERN GUAM

The geology underlying the northern plateau of Guam consists of two primary limestone reef deposits overlying volcanic rocks (Figure 2). The Barrigada (late Miocene to early Pliocene; 11.2 to 3.4 m.y.b.p.)is usually white, chalky, fine grained texture composed of a foraminiferal-algal wackestone distinguished by sparse, but locally abundant accumulations of coral molds consisting of *Porites* and *Astreapora*. Foraminefers consist of *Operculina, Cycloclypeus*, and *Gypsina*, which are criteria for recognition (Tracey, et al.,1959). The maximum thickness of the formation is unknown, but has been presumed to be greater than 180 m (540 feet) based on similar lithology and fossil assemblage from borehole drilling. The Barrigada is interpreted as being a submarine carbonate bank formed at depth of approximately 200 m (600 feet) below sea level. The peripheral slopes, and the abundance of coral and molluscan remains, indicate bank shoaling during late Miocene or early Pliocene time, allowing reef formation on the underlying volcanic (Alutom) formation (Tracey, et al., 1959; Siegrist and Randall, 1992).

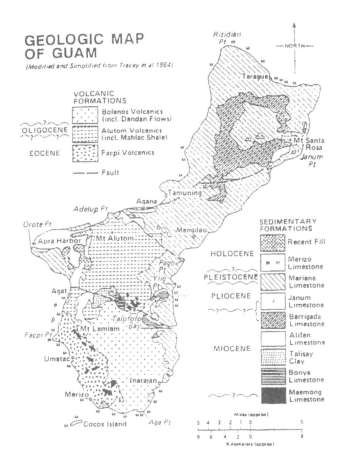

FIGURE 2 - GENERALIZED GEOLOGIC MAP OF GUAM (SIMPLIFIED AND MODIFIED FROM TRACEY ET AL. (1964) AND REAGAN & MEIJER (1984) (FROM SIEGRIST, H.G., AND R.H. RANDALL, 1992)

Overlying the Barrigada Limestone is the younger Mariana Limestone (late Pliocene to Pleistocene time; 3.4 to 1.6 m.y.b.p.)which comprises most of the surface area of the northern plateau and onlaps the Barrigada Limestone as a vertical and transgressional facies, changing from a deep to a shallow water depositional sequence.

The main body of the Mariana consist of four reef-associated facies which include the reef facies, the fore-reef facies, the detrital facies, and the molluscan facies. The reef facies are massive, generally compact, porous and cavernous, white coralline and coralgal boundstones, coarse grainstones, packstones, and wackestones of reef origin, and are made up of mostly corals consisting of the following genera: *Acropora, Favia, Goniastrea, Leptoria, Platygyra, Pocillopora, Porites, Stylophora, Symphyllia,* and *Turbinara.*

Gastropods and bivalves also occur as molds and casts (Tracey, et al., 1959; Siegrist and Randall, 1992). The detrital facies are friable to well cemented, coarse to fine-grained, generally porous and cavernous, white detrital coralliferous rudstone, micitic wackestone and mudstones mostly of lagoonal origin. Fossil assemblages include benthic foraminerfer, corals, mollusks, oysters, turritellid, and *Halimeda.* The molluscan facies are fine-grained, white to tan detrital limestone of lagoon origin containing casts and molds of mollusks. The fore-reef facies consist of well-bedded, friable to indurated, white foraminifera packstones and wackestones deposited as a fore-reef and (Tracey, et al, 1964; and Siegrist and Randall, 1992).

GEOMORPHOLOGY

The limestone formations have undergone geomorphic

processes that have created karst features consisting of dolines, caves, and cenotes. Numerous dolines are abundant in northern Guam. These features have been formed through the chemical dissolution of limestone by weak carbonic acid derived from the atmosphere and soils (Palmer, 1991). Continual chemical dissolution through time allows for this dissolution to penetrate the limestone bedrock to deeper depths. In addition to the surface and near surface dissolution of the bedrock, additional, and probably the most important dissolution of the limestone occurs at the interface between the two chemically contrasting ground waters. Solutional aggressiveness can be enhanced or renewed by mixing waters of contrasting chemistry (Esteban and Wilson, ed., Fritz, et al.,1993; Palmer, 1991; and Bögli, 1980).

GROUND WATER

The Barrigada and Mariana Limestone formations are the primary water bearing aquifers and drinking water supply on Guam. The ground water found within the aquifer is commonly referred to as the Northern Guam Lens (NGL). The NGL has been designated as a sole-source aquifer by the EPA (Barrett Consulting Group, 1992; Siegrist and Randall, 1992). The NGL is recharged by rainfall which generally exceeds 200 cm (80 inches) annually on the northern portion of the island (Mink, 1976). Selected geochemical parameters and tritium concentrations were evaluated by Mink and Lau (1977) in a study conducted in 1976 to more fully understand the hydrologic cycle on island and to identify prospects for additional water yield. Concentrations in rainwater have been monitored and act as an excellent tracer to determine how recently water has been in contact with the atmosphere and were used in this study to ascertain the rapidity at which infiltrating rainfall would recharge the aquifer. Tritium analysis of ground water samples were compared to rain water and a ground water age date of "about five or less years old" (circa 1972) was determined by their study.

The ground water exhibits a relatively flat hydraulic gradient resulting from the high permeable limestones, with the elevation of the water table ranges from sea level at the coastal areas to approximately 1.8 meters (6 feet) above sea level . Throughout northern Guam, fresh ground water floats on seawater in approximate buoyant equilibrium, which in combination of the effect of the dynamics of flow of the fresh water, results in a body of water with parabolic surfaces at both the fresh water-vadose and the fresh water-sea water interface (Figure 3). This equilibrium is referred to as the Ghyben-Herzberg model (Ward, et at., 1965). Based on this model, for every one unit of water above sea level, 39 units of fresh water will occur below sea level.

Based on the measured water levels, the calculated theoretical lens thickness can be in excess of 60 meters (approximately 200 feet), assuming the existence of a sharp boundary or interface between the fresh and sea water. Observed fresh water thickness on Guam, based on geophysical logs (fluid conductivity) is approximately 37 meters (120 feet). The observed boundary, however, is usually diffuse in nature because of hydrodynamic dispersion induced by movements of the interface from tidal changes, seasonal differences in recharge rates, and withdrawals of fresh water by pumping (Mink, 1976). This diffuse zone of brackish water between the salt and fresh water is referred to as the transition zone, and the thickness of this zone is dependent on the dynamics of flow and potential for mixing with the fresh water portion of the lens. The transition zone varies in thickness and has been measured to range from 6 to 15 meters (20 to 50 feet respectively). The depth to the water table across the northern plateau generally approximates sea level, and ground water exists as seeps and submarine springs along the coast.

DISCUSSION

Geologic and hydrogeologic investigations included water level measurements in existing monitoring wells and borehole drilling around the landfill area. The resulting water levels indicated a relatively flat ground water gradient sloping toward the northern coastal area. Borehole drilling consisted of placing boreholes in sinkhole areas and in assumed downgradient positions from the landfill complex. In addition, a natural potential survey along the northern beach area was performed to place boreholes on natural potential anomalies (Lange and Barner, 1994). Geophysical and borehole video logging of the boreholes suggest that most of the caverns, vugs, and dissolution porosity occurs at the vadose/phreatic interface and at the fresh water/salt water interface. This was also confirmed from drilling logs. The size of the dissolution porosity zones and caverns range from as little as a foot to several ten's of feet. The caves discovered along the northern plateau margin can vary in height from 1-2 meters (3 to 6 feet) to approximately 12-15 meters (40 to 50 feet). The caves generally have small entrances (about a meter wide) and either drop several meters to the main cave floor, or have sloping entry ways before opening into the main cavern which can range in horizontal distance of up to 60 meters (200 feet) and widths up to 45 meters (150 feet). The caves usually terminated at a sharply sloping wall into a sump containing fresh water. In many casing, speleothems were observed below the fresh water elevation in the cave pools, evidence the former water table level was lower during previous sea level stillstands.

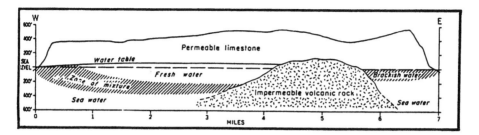

Figure 3 Cross-section of the freshwater lens in limestone in northern Guam (From Ward, et al., 1965).

Caves in Guam appear to have formed from the dissolution of limestone resulting from the mixing of two contrasting water types (fresh and saline water), which occurs within the mixing zone of the fresh and saline ground water, as discussed within the ground water section that follows, and at the vadose/phreatic interface. The emerged caves are found along the margins of the limestone plateau and were found to discovered along the same topographic expression, in this case along terraces, indicating former sea level stillstands. Along the margins of the island, the fresh water lens is thin in comparison to the lens thickness inland. This thinning of the lens, and the influences from tides and storm surges, create a chemically aggressive mixture that produces dissolution porosity and caves. Drilling of several deep boreholes inland and along the coastal areas identified cavernous zones and dissolution porosity grouped at several different elevations or horizons which correspond to emerged caves and also support provide evidence of former sea level stillstands. This phenomena has been documented in the literature from the Yucatan Peninsula (Back et al., 1984), in the Bahamian blue holes (Smart et al., 1988), and the flank margin caves of the Bahamas (Mylroie et al.,1995).

A dye trace was initiated to determine ground water flow direction and estimated flow velocities within the fresh water lens and to simulate releases from a landfill areas. The tracer study monitored over 70 locations that included existing wells and boreholes around the landfill area, boreholes along the beach area drilled on natural potential anomalies, and in caves that contained fresh water sumps, discovered along the base of the northern cliff face of Tarague Beach. One month of background results were collected prior to the dye placement to establish background fluorescence of the aquifer. Background fluorescent conditions of the aquifer and dye standards were used to compare data during the evaluation of sample analsis.

Three dyes were used consisting of; rhodamine WT, Uranine, and Phorwhite BBH Pure. The dyes were placed in boreholes drilled around the existing landfill area and were at depths of 15 and 60 meters (50 and 200 feet) below ground surface. The third injection was directly into the ground water, approximately 137 meters (450 feet) below land surface. Samples collected from these sampling locations consisted of activated carbon and undyed cotton samplers.

The dye trace program lasted 15 months and was terminated with positive results (Figure 4) indicating two types of flow; gravity flow or drainage through the vadose zone and saturated flow. Flow velocities were calculated assuming straight line flow path directions. Because of the porous nature of the limestone, the numerous caverns and other karst features located, and previous tritium studies showing relative young water, it was anticipated that ground water movement would be very rapid. However, saturated flow was not as rapid as originally anticipated, and the combined vadose and saturated flow near the landfill area indicated more rapid movement in the vadose zone prior to reaching the water table. Flow within the ground water system is essentially advective flow along the hydraulic gradient and is indicative of diffuse transport with the average flow velocity calculated to range from 6 meters (20 feet) per day to 11 meters (36 feet) per day.

The vadose system appears to act as a rapid flow system with the water draining in multi-directions and at different inclinations. This is understandable due to the deposition and stratigraphic relationship of the reef limestone, and fracturing and faulting of the bedrock associated with island arc tectonics. When the water flows through the vadose zone, velocity is abruptly reduced when the water table is encountered. Travel time for the combined vadose and saturated flow ranged from 91 meters (300 feet) per day. The combined flow regime includes approximate 76 meters (250 feet) of vertical movement through the vadose

209

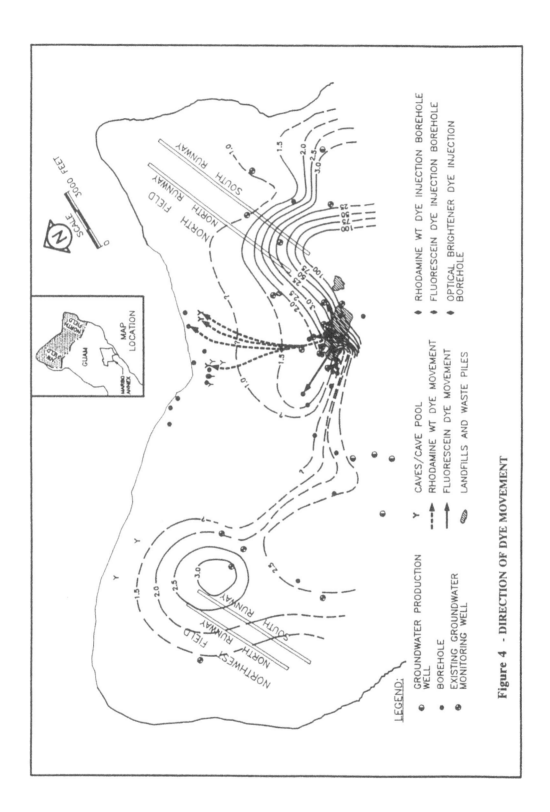

Figure 4 - DIRECTION OF DYE MOVEMENT

LEGEND:

- ● GROUNDWATER PRODUCTION WELL
- • BOREHOLE
- ⊖ EXISTING GROUNDWATER MONITORING WELL

- Y CAVES/CAVE POOL
- RHODAMINE WT DYE MOVEMENT
- FLUORESCEIN DYE MOVEMENT
- LANDFILLS AND WASTE PILES

- ◆ RHODAMINE WT DYE INJECTION BOREHOLE
- ◆ FLUORESCEIN DYE INJECTION BOREHOLE
- ◆ OPTICAL BRIGHTENER DYE INJECTION BOREHOLE

SCALE

0   3000 FEET

MAP LOCATION

N

210

zone to the water table before the saturated flow condition predominates.

In addition to the two flow components mentioned above, dye samples placed deeper within the lens, that were positive for dyes, indicate that ground-water flow within the lens is both horizontal and vertical. Dye arrival times in the deeper samples were similar to those near the water table. However, the vertical flow gradient may be greater than the horizontal flow gradient, because of the extra travel distance required for the dye to migrate both downward and laterally within the lens to reach the deep sites.

The conceptual model for ground-water flow within the NGL. Rain falling on the surface percolates into the soils and limestone bedrock. The upper portion of the limestone is highly corroded and consists of a vast network of dissolution porosity capable of storing large volumes of water (epikarst). This zone continues to develop as new water is introduced to the system and dissolution of the limestone continues. The epikarst water is gradually released through other, smaller pathways to the underlying vadose zone and to the aquifer as diffuse recharge. Discrete concentrated runoff occurs only where there are enlarged joints, fractures, fault zones, and surface depressions that can direct surface runoff to the aquifer. As the groundwater continues to travel to the aquifer, the flow can be altered, by encountering interconnecting fractures, dissolution cavities, or lithologic changes.

Upon reaching the aquifer, the influx of new meteoric water displaces water in the aquifer causing downward (vertical) flow as was as along the subtle hydraulic gradient. Because of the highly permeable limestone, water levels do not respond rapidly by this influx of water. The rate and direction of flow can be altered within the aquifer by the occurrence of preferential pathways. Flow velocities within the phreatic zone mimic a macro-porous-media flow velocity averaging 6 to 15 m per day. Flow within the vadose zone is similar, but can be very rapid if discrete, highly permeated zones are encountered.

These findings and the limestone aquifer of Guam is similar to other recent limestone systems such as provided in the literature for the Yucatan, Florida (Beck, 1986), and the Bahamas (Mylroie, 1995; Mylroie et al., 1995). These aquifer systems are characteristic of young limestone formations, relatively flat ground water gradients, and highly diffuse permeability.

CONCLUSION

The geology of northern Guam is characterized by porous reef deposited limestones on a active volcanic seamount. The two primary limestone formations; the Barrigada and Mariana Limestones, contain the primary drinking water source for the island of Guam. Geologic and hydrogeologic investigations conducted on the northern portion of the island demonstrate that water moves through the vadose zone rapidly, but is then slowed when reaching the water table. Flow is then along the subtle hydraulic gradient and is several times less rapid than flow in the vadose zone. Ground water flow direction in the vadose zone is unpredictable because of the thick sequence of the reef limestone the water drains through under geologic structural influence before reaching the water table. Ground water flows along the hydraulic gradient and exits at seeps and possible submarine springs located along the coastal regions of the island.

DISCLAIMER

The views, opinions, and conclusions are those of the authors' and does not constitute endorsement by the Air Force.

REFERENCES

Back, W., Henshaw, B.B., Herman, J.S., and Van Driel J.N., 1986. Differential Dissolution of a Pleistocene Reef in the Ground-Water Mixing Zone of Coastal Yucatan, Mexico *Geology* 14: 137-140.

Barrett Consulting Group, 1992. *Water Facilities Master Plan Update*. Prepared for the Public Utility Agency of Guam and Government of Guam, February, 1992, 308 p. + Appendices.

Bögli, Alfred, 1980, edited by June C. Schmid. *Karst Hydrology and Physical Speleology*. Springer-Verlag, New York, 270 p., plus 12 plates.

Cloud, P.E., Jr., 1951. *Reconnaissance Geology of Guam and Problems of Water Supply and Fuel Storage*. U.S. Geological Survey, Military Geology Branch for Intelligence Division, Office of the Engineer, General Headquarters, Far East Command, 50 p.

Esteban, M., and Wilson, J.L., 1992 eds., Richard Fritz, James L. Wilson, and Donald A. Yurewicz. Paleokarst Related Hydrocarbon Reservoirs, *SEPM Core Workshop No. 18*, New Orleans, April 25, 1993, 275 p.

ICF Technology, Incorporated, 1994. *Work Plan Addendum to Operable Unit 6 for Operable Unit 2, Andersen Air Force Base, Guam, Final Report* prepared for Andersen Air Force Base, March, 1994,

1st Revision, October 24, 1994, 210 p., + 1 plate and appendices.

Lange, A.L., and Barner, W.L., 1995. Application of the Natural Electric Field for Detecting Karst Conduits on Guam. *Fifth Multidisciplinary Conference on Sinkholes and the Engineering and Environmental Impacts of Karst*. Gatlinburg, Tennessee, April, 1994.

Mink, J.F., 1976. *Groundwater Resources of Guam: Occurrences and Development, Technical Report No. 1*, Water Resources Research Center, University of Guam, September, 1976, 275 p.

Mink, J. F. and Lau, S. L., 1977. *Groundwater Analysis by Tritium Technique: A Preliminary Evaluation*. University of Guam Technical Report 2, 25 p.

Mylroie, John E., Carew, James L., Edward Frank F., Panuska, Bruce E., Taggart, Bruce E., Troester, Joseph W., and Carrasquillo, Ramon, 1995. Comparison of Flank Margin Cave Development: in Boardman, M. R., ed., *Proceedings of the Seventh Symposium on the Geology of the Bahamas: San Salvador Island, Bahamian Field Station*, 62pp.

Palmer, Arthur N., 1991. Origin and Morphology of Limestone Caves. *Geological Society of America Bulletin*, 103: No. 1, January, pp. 1-21.

Raulerson, Lynn and Rinehart, Agnes, 1991. *Trees and Shrubs of the Northern Mariana Islands*. Coastal Resources Management Office of the Governor, Commonwealth of the Northern Mariana Islands, Saipan, Northern Mariana Islands, 120 p.

Siegrist, H.G., Jr., and Randall, R.H., 1992. Carbonate Geology of Guam: Summary and Field Trip Guide, *7th International Coral Reef Symposium, Guam, U.S.A.*, June 18-20, 1992. University of Guam, Water and Energy Research Institute of the Western Pacific, 39 p., + 1 plate.

Smart, Peter L., Dawans, J.M., Whitaker, F., 1988. Carbonate dissolution in a modern mixing zone, *Nature* 335: 811-813.

Tracey, J.I. Jr., Stensland, C.H., Doan, D. B., May, H.G., Schlanger, S.O., and Stark, J. T., 1959. *Military Geology of Guam, Mariana Islands: Part I Description of Terrain and Environment; Part II Engineering Aspects of Geology and Soils.* Engineer Intelligence Dossier Strategic Study: Marinas, Subfile 19 Analysis of the Natural Environment. Prepared under the direction of the Chief Engineers,

U.S. Army by the Intelligence Division, Office of the Engineer, Headquarters United States Army Pacific with personnel of the United States Geological Survey. Ward, P.E., Hoffard, S.H., Davis, D.A., 1965. Geology and Hydrology of Guam, Mariana Islands, U.S.G.S. Professional Paper 403-H, 29 p.

Tracer Hydrology 97, Kranjc (ed.) © 1997 Balkema, Rotterdam, ISBN 90 5410 875 4

Experiences in monitoring Timavo River (Classical Karst)

F. Cucchi, F. Giorgetti & E. Marinetti
Department of Geological, Environmental and Marine Sciences, University of Trieste, Italy

A. Kranjc
Karst Research Institute, ZRC-SAZU, Postojna, Slovenia

Introduction

The Reka-Timavo underground karst system is one of the main karst phenomena of Mediterranean region which roused the interest of scientist from the second century AD up to present.

The karst system develops in a 50 km long, 12 km wide and several thousands meters thick carbonatic massif, with a general SE-NW direction (see fig. 1). The massif consists in a rather undulated platform with elevation head going from 450 m (in Skocjan area) to 50 m (in Isonzo river alluvial plane). The carbonate complex (part of the Placer's "Komen Platform") is composed of several lithological units. The oldest terms of the carbonatic succession is Lower Cretaceous limestone (about 300 m thick) subtly bedded with a medium-high karstification level and high degree of permeability.

This is overlaid by a dolomite unit of Albian-Cenomanian age with a 300 to 600 m thickness. Dolostone is poorly stratified and karstified and show a relative permeability degree lower than that of the underlying and overlying limestones. Succession follows with a thick sequence (300-1000 m) of Upper Cretaceous massive pure limestone. This is compact and fractured in great blocks and pervasively karstified both in depth and on the surface and represent the most important karst unit of the hydrogeologic system.

Stratified limestone unit of Paleocene-lower Eocene age follows, thinly bedded, fractured and partly karstified (40 to 450 m thickness).

The platform is characterized by a wide anticline with a NW-SE axis, complicated by a series of subparallel folds and some faults. The main ones are parallel to the main structure, other minor perpendicularly intersect the structure.

The Reka-Timavo underground stream develops in a NE direction for at least 40 km (as the crow flies), according to a hipsographic curve which shows more than 5% slope during the first 4 km (from the 323 m above sea-level in Skocjan Cave entrance, to the 88 m above sea level in Kacna Cave). The following 11 km of stream shows a 0.7% slope (from Kacna Cave to Trebiciano Cave) and the rest, until the San Giovanni Springs, a 0,05% slope only (Cucchi & Forti, 1981; see fig. 2).

The Classical Karst aquifer has been investigated from the roman epoch up to the present day with vast but partial studies. At present the collection, completion and comparison of old and new, important hydrological data has commenced. This to define in a more accurate way the hydrogeology and the hydrodynamics of the karst aquifer and to be preliminary remark to the computation of its intrinsic vulnerability. The karst network consists of a principal important developed channel dendritic pattern drainage of waters coming from the karstified fractures system. The saturated zone is completely karstified as far as 100 m under sea level. Unsatured one thick from few to 350 meters is deeply karstified and characterized with large and deep dolines and more then 5000 caves (some halls large 600 cubic meter, large galleries some telemeters long and pits more then 100 meters deep).

The data initially collected confirm the importance of continuous measurement in karst waters study. The Timavo hypogean river indeed is a very complex aquifer with unpredictable behavior: our work illustrate first results and preliminary comments.

Used instruments

Since April 1994 three stations of continuous measurement of level, conductivity and temperature of Karst Timavo waters are active (see Fig. 1). The three gauges are located at the entrance of Skocjan Cave (Slovenija), in the Lindner hall of Trebiciano Cave (Italy, about 20 km as the crow flies between sinkhole and springs) and near the S. Giovanni di Duino Springs (Italy). At the Skocjan Cave the instrument has been located 300 meters far from the entrance right to the big gully, 280 meters above sea-level; at the Linder hall has been located near the active siphon, 11 meters above sea-level. At the Timavo Springs the temperature and conducibility gauge has been installed into the water board laboratory; the water level meter has been located into the Pozzo dei Colombi Cave, 1 meter above sea-level (see fig. 2). All sensors had been carefully tested before use.

The large change of water level (more than 70m in the Linder hall) and the violence of flow caused many problems at the bottom of the Trebiciano and in the Škocjan caves. Special impermeable containers have been planned and built (Fig.3), they have been welded to the cave walls and the connection cables has been protected.

Temperature and conducibility gauges are adjustments of THERMOS DATA-D.A.S. Co., a digital recorder connected through a cable line with the sensors. After the first experiments it was necessary to design and built a special container able to support hydrostatic pressures up to 10 Atm. Figure 3 shows the complete device. The technical specification of Thermos Data Sensors are: temperature: range -25°+100°; accuracy (between 0°-70°) ±0.2°C; resolution 0.1°C. Conductivity range is 100-1000 µS/cm; accuracy ±2µS/cm; resolution 0.1µS/cm.

The water level meter is designed and built by MECCATRONICA-Trieste and is composed by the meter measuring the pressure variation and the digital equipment for memorizing the recorded data at fixed intervals. This two components are located into a steel pipe and provided with own power supplies. Figure 4 shows the complete device. The gauge weighs the variation of the water level, measuring the variation of the total hydrostatic and atmospheric pressure. The tecnical specification of the water level meter renges between 0 and 20m; accuracy 8cm; resolution 0.1cm.

The gauge weight the variation of the water level, measuring the variation of the total hydrostatic and atmospheric pressure. The technical specification of the water level meter are: range 0-20 m; accuracy 8 cm; resolution 0.1 cm.

Flows

Flows are daily analyzed near the sinkhole in Cerkvenikov Mlin (Slovenija) and in the springs; rainfall data come from Borgo Grotta Gigante (Italy) meteorological station. (See figs. 5,6,7). The year considered 1995 is characterized by high rainfall and therefore by water discharge above average. Rainfall has been 1740 mm (average is 1355 mm), June show the maximum with 235 mm (115 mm fall during one week). The Reka river average discharge has been almost 15 cu.m/s (average is 9 cu.m/s); maximum discharge has been 113 cu.m/s recorded in march, 3rd. The spring average annual discharge has been 42 cu.m/s (average is lower then 30 cu.m/s) with six floods higher than 70 cu.m/s, two higher than 110 cu.m/s (in March and in December) and a maximum flood of 141 cu.m/s.

The water level in the Trebiciano Cave is terribly changeable: normally is arranged between 12 and 14 m above sea-level, during maximum floods has reached more then 70 m above sea-level. Since the instrument had been calibrated to gauge water level differences below 20 m four times has been out of range four times(see fig.8). In the Pozzo dei Colombi Cave the minimum water level recorded has been 1.55 m above sea-level, the maximum one 5.58 m in March (see Fig. 9).

Crest velocity during the highest floods (rate of flow at the spring higher than 90 cu m/sec) are very variable: from the sinkhole to the Trebiciano Cave time range from 10-12 to 24-30 hours; from Trebiciano to springs from 4-5 to 20-24 hours and they do not seem to depend from rain rate.

Temperature and conductivity

The annual temperature excursion at the Skocjan Cave entrance ranges from 0°C to 23°C (see fig. 10). In the Trebiciano Cave phreatic waters (after at least 30 km of hypogean presumed stream) goes from 6,4°C to 14,0°C; at the springs from 7,8°C to 15,8°C. During floods the temperatures excursion in Trebiciano Cave range less than 4° and the effect goes on for more or less ten days (see fig. 11). To the springs the temperature variations are lower: only 1° (see Fig. 12). The atmospheric temperature recorded by Borgo Grotta Gigante meteorological station is represented in Fig. 13.

Conductivity decrease characterizes floods, but the river's regimen depends from the ratio between meteoric waters percolating in the hypogean aquifer (more rich in solute) and Reka River waters swallowed near Skocjan (they have low values of solute, but high values in suspended matters). See figures 14, 15, 16, taking into account that the conducibility values recorded, due to the peculiar feature of this gauges, are not reliable in absolute but must be read as general trend.

REFERENCES

Boegan E.; 1938: Il Timavo. Studio sull'idrografia carsica subaerea e sotterranea. Mem. Ist. Ital. Spel., Mem. 2°.

Cucchi F., Forti F.; 1981: La cattura del Timavo superiore a Vreme. Atti Mem. Comm. Grotte Boegan, Vol 21.

D'Ambrosi C., Mosetti F.; 1963: Alcune ricerche preliminari in merito a supposti legami di alimentazione fra il Timavo e l'Isonzo. Boll. Geof. Teor. Appl., Vol 5° n. 17.

Gemiti F.; 1994: Indagini idrochimiche alle risorgive del Timavo. Atti Mem. Comm. Grotte Boegan, Vol. 31.

Placer L.; 1981: Geologic structure of SW Slovenija. Geologija, Vol 24/1.

Work supported by G.N.A.C.I., Pubbl. No. 1033.

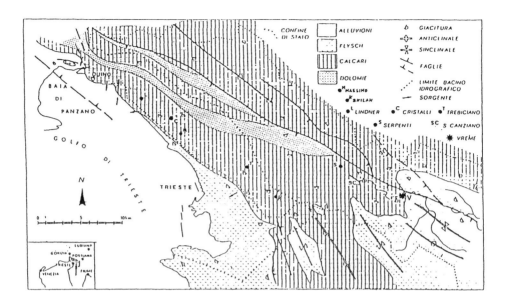

Figure 1: Geological sketch of the drainage basin of Timavo River.

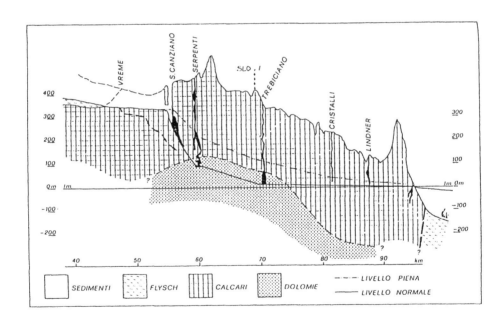

Figure 2: Section along the assumed Timavo stream.

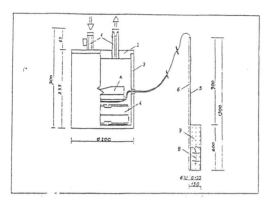

Figure 3: Termos Data equipment.

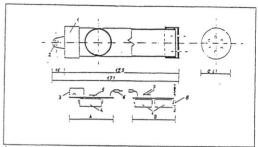

Figure 4: Meccatronica equipment.

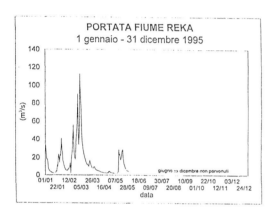

Figure 5: Reka River Discharge.

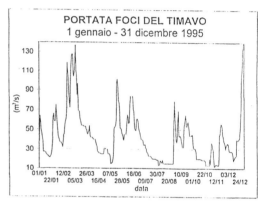

Figure 6: **Timavo River Spring Discharge.**

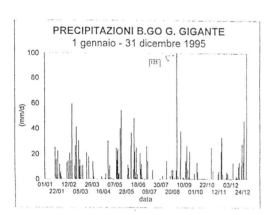

Figure 7: Rain fall recorded in Borgo Grotta
Gigante meteorological station.

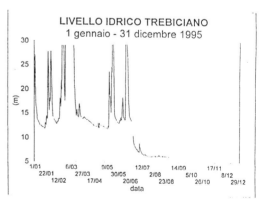

Figure 8: Water level recorded in Trebiciano Cave.

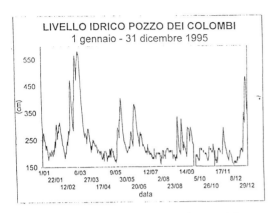

Figure 9: Water level recorded in Pozzo dei Colombi Cave.

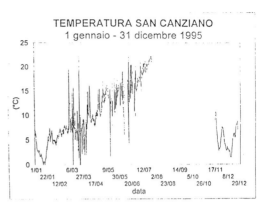

Figure 10: Water temperature recorded in Skocjan entrance.

Figure 11: Water temperature recorded in Trebiciano Cave.

Figure 12: Water temperature recorded in Timavo River Spring discharge.

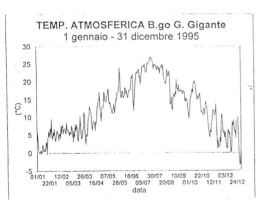

Figure 13: Air temperature recorded in Borgo Grotta Gigante meteorological station.

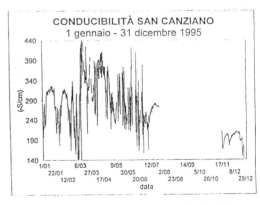

Figure 14 Water conductivity recorded in Skocian Cave entrance.

217

Figure 15: Water conductivity recorded in Trebiciano Cave.

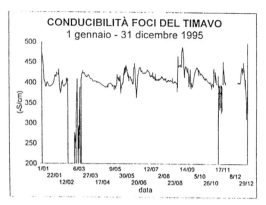

Figure 16: Water conductivity recorded in Timavo River Spring discharge.

Tracer Hydrology 97, Kranjc (ed.) © 1997 Balkema, Rotterdam, ISBN 90 5410 875 4

Tracer tests in the Joèu karstic system (Aran Valley, Central Pyrenees, NE Spain)

A. Freixes, M. Monterde & J. Ramoneda
Servei Geològic de Catalunya, Institut Cartogràfic de Catalunya, Barcelona, Spain

ABSTRACT: In the first part of the present paper a brief historical review of tracer tests performed in the Joèu karstic system is given. These tests have allowed an accurate delimitation of the Joèu system. The second part is about five quantitative tracer tests carried out in this karstic system between 1991 and 1993. The results demonstrate the existence of a well developed karstic network (structure), and significant differences in the behaviour of the system according to its hydraulic state (time of the hydrological cycle). These differences are observed in hydrodynamics (flow velocities, residence times, ...) as well as in hydrogeochemistry (mineralization, CO_2-H_2O-carbonate system, ...).

1 INTRODUCTION

The objectives of this work are: a) to delimit the Joèu karstic system, and b) to improve our understanding of the behaviour and structure of this system. In this regard, the tracer tests form an additional aspect of the research in hydrodynamics carried out in this system, and complement the studies in hydrogeothermics and hydrogeochemistry (Freixes et al. 1993 & in press).

2 THE JOÈU KARSTIC SYSTEM. PHYSICAL CHARACTERISTICS

The Joèu karstic system (Figure 1), which is the most outstanding karstic capture in the Pyrenees, drains through swallow holes the waters of the upper catchment area of the river Ésera (Mediterranean basin) towards the river Garona (Atlantic basin).

This system has a complex structure of a binary type (Mangin 1975, 1978 & 1994, Walliser 1977, Freixes et al. 1993 & in press). The impermeable part of the basin is made up of Late Hercynian granodiorites and Carboniferous metapellites and metapsamites, whereas the aquifer consists of Devonian metamorphic limestones with metapellites and metapsamites forming the top (Figure 2); the aquifer is, therefore, captive. This system is characterized by a significant surface drainage (Aigualluts, Barrancs, Escaleta and Renclusa rivers) and by a number of important swallow holes (Aigualluts, Renclusa, Hòro Barrancs, Hòro Nere, Hòro Aran, Mall Artiga, ...). This important surface drainage is characterized by a permanent behaviour because of ice melting from the

Aneto and Maladeta glaciers and because of the considerable snowfalls into the basin.

Table 1 summarizes other physical and hydrodynamical characteristics of the system. The marked discrepancy between the highest parts of system and the spring (approx 2000 m) should be pointed out. The relationship between the dynamic reserves and the total anual runoff means that we are dealing with a system with significant karst development (well developed karstic network) (Mangin 1975, Freixes et al. 1993).

Table 1. Physical and hydrodynamical features of the Joèu karstic system.

Catchment area (km^2)	27-31
Maximum altitude (m a.s.l.)	3404
Swallow holes altitude (m a.s.l.)	2000
Spring altitude (m a.s.l.)	1410
Mean altitude (m a.s.l.)	2541
Spring discharge (m^3/s):	
. mean	2.16
. maximum	11-12
. minimum	0.25
Dynamic reserves (hm^3)	3-5
Total anual runoff (hm^3)	60-90
Swallow holes discharge (m^3/s):	
. Aigualluts	0.05-5.0
. Renclusa	0.02-3.0
. Hòro Nere	0.01-0.5
. Hòro Barrancs	0.01-0.2
. Mall Artiga	0.01-0.2
. Hòro Aran	0.01-0.2

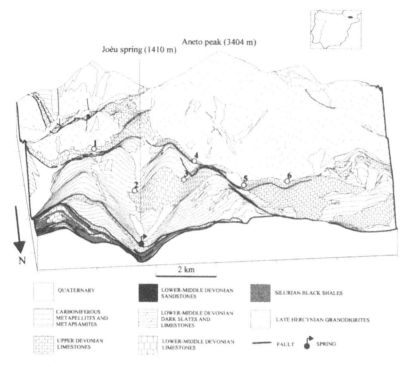

QUATERNARY

CARBONIFEROUS
METAPELLITES AND
METAPSAMITES

UPPER DEVONIAN
LIMESTONES

LOWER-MIDDLE DEVONIAN
SANDSTONES

LOWER-MIDDLE DEVONIAN
DARK SLATES AND
LIMESTONES

LOWER-MIDDLE DEVONIAN
LIMESTONES

SILURIAN BLACK SHALES

LATE HERCYNIAN GRANODIORITES

FAULT   SPRING

O Swallow Hole: 1 Hòro Nere 2 Mall Artiga 3 Hòro Aran 4 Hòro Barrancs 5 Aigualluts 6 Renclusa

Figure 1. Block diagram of the Joèu karstic system showing the relief, the surface geology, the main swallow holes and the Joèu spring.

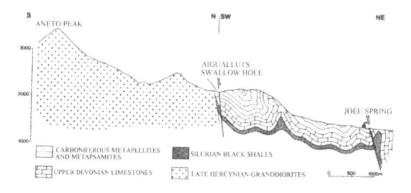

CARBONIFEROUS METAPELLITES
AND METAPSAMITES

SILURIAN BLACK SHALLS

UPPER DEVONIAN LIMESTONES

LATE HERCYNIAN GRANODIORITES

Figure 2. Geological cross-section of the Joèu karstic system.

3 LIMITS OF THE SYSTEM: TRACER TESTS IN THE LAST 150 YEARS

The first tracer tests performed in the Joeù system were focused on the Aigualluts swallow hole (Table 2). At the end of the XVIII century the fate of the large volume of water drained through this swallow hole was already a subject of debate. In this regard, it is worth noting that Ramond de Carbonnieres, in accordance with popular belief, already considered that the Aigualluts swallow hole was drained by the Joèu spring. During the XIX century a number of authors such as Joanne, Lambron, Lezat, Leymerie, Schrader and Reclus, supported this hypothesis. Joanne (1858) describes, without naming the author, a tracer test using sawdust which confirmed the connection between Aigualluts and the Joèu spring (in Faura 1916).

Belloc (1896) traced the Aigualluts swallow hole with an amount of fucsina that was not sufficient to be detected in the spring. Given that this supposedly "negative" result had been obtained by a reputed author, several scientists accepted this thesis despite the findings cited by Joanne. Nevertheless, Martel (1905) took issue with Belloc's conclusions (in Faura 1916 & 1926).

The connection between Aigualluts and the Joèu spring was definitively demonstrated by Casteret in 1931. On the 19th July he injected 60 kg of fluorescein which emerged in the Joèu spring after less than 24 hours (more than 40 km of the river Garona downstream was coloured) (Casteret 1931).

In 1947 and 1948 De Lizaur performed tracer tests with fluorescein in the Aigualluts and Renclusa swallow holes, confirming that the residence times were relatively short (De Lizaur 1958).

In the 1960s Vedruna and Leniger used fluorescein to demonstrate the hydraulic connection between other swallow holes, such as Hòro Nere, Hòro Aran, Mall Artiga and Hòro Barrancs, and the Joèu spring (Rijckborst 1967).

The most recent tracer tests have been performed in the 1990s by Freixes et al.. These tests were of a quantitative type and were carried out within a wider research context.

In order to delimit the basin of the Joèu system, a series of tracer tests were performed in the Alba karstic system and in the Pla d'Estanys, both next to Joèu (Table 2).

4 ROLE OF TRACER TESTS IN UNDERSTANDING THE BEHAVIOUR AND STRUCTURE OF THE JOËU KARSTIC SYSTEM

4.1 Methodology

Between 1991 and 1993 quantitative tracer tests were carried out at three swallow holes in the Joèu karstic system on the basis of a systems analysis approach: Aigualluts (3 tests), Renclusa (1 test) and Hòro Nere (1 test) (Figure 1).

Table 2. Tracer tests performed in the karstic systems of Joèu, Alba and Pla d'Estanys.

Swallow hole	Spring	Modal time (h)	Author (tracer test date)
Aigualluts	Joèu	?	Joanne (1858)
Aigualluts	Joèu	≈10	Casteret (Jul-1931)
Aigualluts	Joèu	12	De Lizaur (Jul-1947)
Aigualluts	Joèu	15.5	Freixes et al. (Aug-1991)
Aigualluts	Joèu	14	Freixes et al. (Jun-1992)
Aigualluts	Joèu	212	Freixes et al. (Feb-1993)
Renclusa	Joèu	13	De Lizaur (Jul-1947)
Renclusa	Joèu	5.2	De Lizaur (Aug-1948)
Renclusa	Joèu	17.2	De Lizaur (Aug-1948)
Renclusa	Joèu	13.8	Freixes et al. (Aug-1992)
Hòro Aran	Joèu	14	Leniger (1965)
Mall Artiga	Joèu	14	Leniger (1965)
Hòro Barrancs	Joèu	9	Leniger (1965)
Hòro Nere	Joèu	151	Vedruna (Sep-1961)
Hòro Nere	Joèu	72	? (1964)
Hòro Nere	Joèu	61.5	Freixes et al. (Nov-1992)
Lake Alba	Alba	5	Freixes & Garriga (Jul-1970)
Lake Alba	Alba	6	Freixes et al. (Jun-1992)
Forao Tancao	Alba	?	Monterde (Aug-1988)
Padernes	Alba	≈20	Monterde (Aug-1985)
Padernes	Alba	17	Freixes et al. (Aug-1992)
Pla d'Estanys	Hospital	20	Monterde et al. (Nov-1996)

Each test involved slug injections of a given amount of fluorescein into one of the swallow holes and the subsequent collection of samples at the Joèu spring (automatic sampler) in order to obtain the dye breakthrough curve (spectrofluorometer). Fluorocaptors of active carbon were simultaneously left at neighbouring springs (Artiga, Pomero, Gresilhon and Horno) in order to establish their possible hydraulic connection with the traced swallow holes and, therefore, with the Joèu karstic system.

Table 3. Quantitative results of the tracer tests performed in the Joèu karstic system between 1991 and 1993.

	Aigualluts			Renclusa	Hòro Nere
	Aug-1991	Jun-1992	Feb-1993	Aug-1992	Nov-1992
Swallow hole - spring lineal distance (m)	3700	3700	3700	4750	4100
Swallow hole - spring altitude difference (m)	660	660	660	705	630
Injected tracer (g)	1514	1530	229	400	2015
Swallow hole discharge (m³/s)	1.50	1.20	0.06	0.80	0.01
Spring mean discharge (m³/s)	2.75	2.98	0.25	2.34	1.20
Time of tracer arrival (h)	11	11	180	11	59
Maximum velocity (m/h)	336	336	21	432	69
Modal time (h)	15.5	14	212	14	61.5
Modal velocity (m/h)	238.7	264.3	17.5	339.3	66.7
Mean residence time (h)	16.2	16.2	243.4	17.0	69.1
Mean travelling velocity (m/h) Apparent velocity (m/h)	235 229	238 229	15.6 15.2	305 280	61.8 59.4
Maximum tracer concentration (mg/l)	2.4×10^{-2}	1.8×10^{-2}	1.8×10^{-3}	3.9×10^{-3}	1.5×10^{-2}
Mean tracer concentration (mg/l)	9.7×10^{-3}	4.8×10^{-3}	4.9×10^{-4}	2.0×10^{-3}	2.3×10^{-4}
Restitution factor (%)	90.0	65.5	43.8	53.8	44.2

During the tracer tests the Joèu spring was equipped with a datalogger for continuous recording of the water level. Measurements of discharge were carried out, by dilution gauging and by current meters, to obtain the water level - discharge relationship.

4.2 Results and discussion

4.2.1 Quantitative results of tracer tests

As none of the fluorocaptors installed at the Artiga, Pomero, Gresilhon and Horno springs showed any signs of having adsorbed fluorescein it seems that Joèu is the only spring in this karstic system.

The quantitative results of the tracer tests performed as well as the experimental conditions are given in Table 3.

The tracer tests at Aigualluts and Renclusa in summer were performed under conditions of relatively high water flow. Both the mean residence time (16-17 h) and the mean travelling velocity (230-305 m/h) indicate a high development of the drainage structures that connect these swallow holes with the spring. It is worth noting that the results of the two tracer tests at Aigualluts performed in summer (August 1991 and June 1992) are comparable, which is a sign that, under similar hydraulic conditions, the response of the aquifer to a particular event is constant. Likewise, the result of the tracer test at Aigualluts in winter is of especial interest as it demonstrates the influence of the time of the hydrological cycle on the behaviour of the system. It is worth noting that between the Aigualluts swallow hole and the Joèu spring in winter the mean travelling velocity is approx 14 times lower than in summer, which means a time lag of 8 days in order to obtain the maximum dye concentration.

The tracer test at Hòro Nere, despite being performed during the falling phase of the hydrogram, also shows a relatively good connection between the swallow hole and the spring.

The dye breakthrough curves, given as residence time distributions (Lepiller & Mondain 1986, Meus 1993) (Figures 3 and 4), have a tendency to be unimodal. In the summer tests appear some minor modes after the passage of most of the dye probably due to the daily discharge variations (snow melting). These tracer tests, excluding that at Aigualluts in winter, show how most of the dye emerges after approx 20-30 h (low dye dispersion) despite

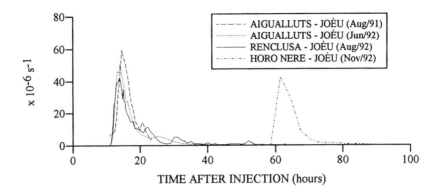

Figure 3. Residence time distributions (RTD) for tracer tests in summer and autumn.

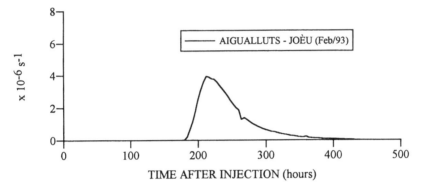

Figure 4. Residence time distribution (RTD) for the tracer test in winter.

differences in time of tracer arrival. This finding suggests that the connection between the swallow holes and the spring is constituted by well-developed drains that are very transmissive. On the other hand, the relatively high mean travelling velocity of the flow prevents the dye breakthrough curve from being significantly affected by the daily discharge variations, which is a characteristic feature in the summer months.

The dye breakthrough curve corresponding to the tracer test at Aigualluts in winter (Figure 4) shows a greater dye dispersion (most of the dye emerges in an interval of approx 150 h). This finding again highlights the role of hydrodynamics in the behaviour of the system.

4.2.2 Role in the chemical kinetics of water flow

It is worth noting how the different residence times of the water through the aquifer in winter and in summer affect the chemical characteristics of the water at the Joèu spring. In fact in winter when the residence times are longer the water that emerges is

more mineralized. This coincides with a saturation index (ΔpHc) and a pCO_2 significantly higher than in summer (t-test, p<0.05). Nevertheless, the values of pCO_2 are always very low (mean pCO_2 = 0.011 %, n = 68) and, in any case, the water that emerges is relatively close to equilibrium. This indicates that although most of the year the residence time through the aquifer is very short it is always sufficient to achieve a relative equilibrium in the CO_2-H_2O-carbonate system. This suggests that the chemical kinetics, in this case, is not substantially altered by the differences in the residence time. Hence, the lower mineralization and the lower pCO_2 in summer are attributed to the dilution caused by larger volumes of water, regardless of the residence time (Freixes et al. in press).

Similar aspects of chemical kinetics are demonstrated when comparing water chemistry at the entrance of the aquifer (Aigualluts swallow hole) with that at the Joèu spring (Table 5). However, the most striking finding during the passage of the water through the aquifer, whether summer or winter, is the increase in mineralization. Thus, at the Joèu spring there is a concentration of bicarbonates approx 2.4 -

2.5 times higher than that at Aigualluts, and also a concentration of calcium approx 2.6 - 2.8 times in excess of that at Aigualluts. This mineralization is achieved regardless of the residence time of the water through the aquifer.

However, although the water that enters Aigualluts is clearly undersaturated as regards calcite, it always reaches the Joèu spring with characteristics close to equilibrium.

In this case, although the residence time of the water through the aquifer is conditioned by the hydraulic state of the system, the kinetics of the reactions in the CO_2-H_2O-carbonate system is scarcely affected.

Table 4. Mean physico-chemical features of the Joèu spring in winter and summer (1988 - 1993).

	Winter	Summer
Conductivity (μS/cm at 20 °C)	92	49
Bicarbonates (mg/l)	56.1	29.3
Calcium (mg/l)	18.7	9.0
pCO_2 (%)	0.019	0.004
ΔpH calcite	-0.22	-0.35
n	16	15

Table 5. Mean physico-chemical features of the Aigualluts swallow hole and the Joèu spring at different times of the hydrological cycle (1992 - 1993).

	Winter		Summer	
	Aigualluts	Joèu	Aigualluts	Joèu
Conductivity (μS/cm at 20 °C)	43	91	21	51
Bicarbonates (mg/l)	22.4	57.0	13.0	31.5
Calcium (mg/l)	7.6	19.6	3.4	9.6
pCO_2 (%)	0.062	0.024	0.047	0.005
ΔpH calcite	-1.94	-0.29	-2.51	-0.38
n	4	4	5	5

5 ACKNOWLEDGEMENTS

The authors wish to express their gratitude to A.Mangin and D.d'Hulst (Laboratoire Souterrain de Moulis, CNRS) for their help in methodological aspects of the work, and to M.Estrada, M.Latasa and E.Bardalet (Institut Ciències del Mar, CSIC) for giving us facilities to use the spectrofluorometer of their laboratory.

6 BIBLIOGRAPHY

Bakalowicz, M. 1979. Contribution de la géochimie des eaux à la connaissance de l'aquifère karstique et la karstification. *Thèse de doctorat es Sciences naturelles, Universite Pierre et Marie Curie, Paris.*

Bakalowicz, M. 1992. Géochimie des eaux et flux de matières dissoutes. In. J.N.Salomon & R.Maire (eds), *Karst et Évolutions Climatiques. Hommage à Jean Nicod*:61-74. Presses Universitaires de Bordeaux.

Casteret, N. 1931. Le problème du Trou du Toro. Détermination des sources du rio Esera et de la Garonne Occidentale. *Bull. Soc. Hist. Nat. Toulouse* LXI:89-131.

De Lizaur, J. 1958. Estudio sobre las conexiones subterráneas de las cabeceras de lor ríos Esera y Garona. *Libro Jubilar (1848-1949), Instituto Geológico y Minero de España, Madrid.* 2:381-425.

Faura, M. 1916. Sobre hidrología subterránea en los Pirineos Centrales de Aragón y Cataluña. Supuesto origen de los Güells del Jueu. *Boletín de la Real Sociedad de Historia Natural* XVI:353-354.

Faura, M. 1926. Del Valle de Aran a los Montes Malditos. Expedición complementaria. Excursión C-3. *XIV Congreso Geológico Internacional (Madrid)*:165-181.

Freixes, A., M.Monterde & J.Ramoneda 1991. Respuestas hidrogeoquímicas y funcionamiento de los sistemas acuíferos de Güells de Jueu y Alguaire (Valle de Aran): resultados preliminares. *IV Congreso de Geoquimica de España, Soria*:591-601.

Freixes, A., M.Monterde & J.Ramoneda 1993. Hidrología de los sistemas karsticos del Valle de Aran (Pirineos, Catalunya). In H.J.Llanos, I.Antigüedad, I.Morell & A.Eraso (eds), *I Taller Internacional sobre Cuencas Experimentales en el Karst. Matanzas (Cuba), Abril 1992. Libro de Comunicaciones*:131-140. Castelló: Publicacions de la Universitat Jaume I.

Freixes, A., M.Monterde & J.Ramoneda in press. The karst development potential of the aquifers in the Val d'Aran (Catalonia). *Acta Geologica Hispanica.*

Lepiller, M. & P.H.Mondain 1986. Les traçages artificiels en hydrogéologie karstique. Mise en oeuvre et interprétation. *Hydrogéologie* 1:33-52.

Mangin, A. 1975. Contribution à l'étude hydrodynamique des aquifères karstiques. *Thèse doctorat es Sciences naturelles, Dijon, Annales Spéléologie France* 29(3):283-332; (4):495-601; 30(1):21-124.

Mangin, A. 1978. Le karst, entité physique, aborde par l'étude du système karstique. *Réunion AGSO - Colloque de Tarbes*, 21-37.

Mangin, A. 1994. Karst Hydrogeology. In J.Gibert, D.L.Danielopol & J.A.Stanford (eds), *Groundwater Ecology*: 43-67. San Diego: Academic Press.

Meus, P. 1993. Hydrogéologie d'un aquifère karstique dans les calcaires carbonifères (Néblon-Anthisnes, Belgique). Apport des traçages à la connaissance des milieux fissurés et karstiques. *Thèse de Docteur en Sciences, Université de l'Etat a Liege*.

Rijckborst, H. 1967. Hydrology of the Upper-Garonne basin (Valle de Aran, Spain). *Leidse Geologische Mededelingen* 40:1-74.

Walliser, B. 1977. *Systèmes et modèles. Introduction critique à l'analyse de systèmes*. Paris: Editions du Seuil.

Tracer Hydrology 97, Kranjc (ed.)© 1997 Balkema, Rotterdam, ISBN 90 5410 875 4

An example of geological investigation in a karst area in Lower Austria

G.Gangl
Verbundplan, Donaukraft Engineering, Vienna, Austria

For the design of a Hydro Power Plant at the Danube east of Vienna detailed geological investigations were performed 1981-1984. The hot springs of Bad Deutsch Altenburg at the eastern margin of the Vienna basin were of special interest. As the result of a drilling campaign a detailed 3-dimensional geologic model was constructed. Chemical and isotopic data of water samples allow a separation into different groups. Additional measurement of the temperature show where thermal water raises and is mixed with karstic water and water of the Danube. As chemical analyses have been continued, the results of the different groups are presented for 13 years of water sampling in a poster.

1. INTRODUCTION

The area of investigation is at the eastern edge of the inneralpine Viennese Basin. The edge is built up of mesozoic and elder rocks. The basin fill is of tertiary sands and shales, in the outer part of the basin are calcareous strata („Leithakalk"). The hot iodine-sulfur springs of Bad Deutsch Altenburg in Lower Austria have been used balneological for centuries. The water rises at the south-eastern roof of the carbonatic rocks of the Kirchenberg.

2. GEOLOGICAL MODEL

In the poster session photos of a geological model of the underground of Bad Deutsch Altenburg are presented. The main units are

- quaternary gravel
- sands and clay of tertiary age
- tertiary calcareous strata („Leithakalk")
- mesozoic limestone

3. HYDROGEOLOGICAL INVESTIGATIONS

Measurements in the raising thermal springs shows temperatures up to 28°C. Sections with isothermals are drawn from borehole measurements, where upwelling of water can be observed. Measurements of soil gases confirm the existance of a southeast-northwest striking fault normal to the Danube river.

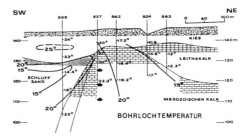

Fig. 2 Schematic geologic section crossingthe park of Bad Deutsch-Altenburg with temperature measurements in boreholes:The uppwelling of water can be observed inthe interior zone of the 20° C isotherm.

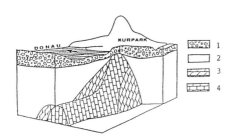

Fig. 1 Geological model of Bad Deutsch-Altenburg seen from the North:1. Gravel, 2. Sand and shale, 3. Tertiary calareous strata „Leithakalk", 5 Mesozoic limestone

The main results of chemical analysis of water samples are presented. Different groups of water can be distinguished. With the help of isotop concentration the age of the water can be determined:

- Thermal water of an age of more than 35 years due to Tritium measurements
- Very young water from carstified limestone
- Pore water in the quaternary and tertiary layers and
- Danube water

REFERENCES:

Gangl G. 1988.
    Geologische und hydrogeologische Voruntersuchungen zum Bau des Kraftwerkes Hainburg, Baugeologische Tage in Hüttenberg, *Mitteilungen für Baugeologie und Geomechanik, Band 1, 233-247, Technische Universität Wien*

Gangl G. 1990.
    Hydrogeologische Untersuchungen an den Heilquellen von Bad Deutsch Altenburg (Niederösterreich) im Rahmen der Vorarbeiten für das Donaukraftwerk Hainburg. *Österreichische Wasserwirtschaft, Jahrgang 42, 1-17*

Maurin V. 1988.
    Hydrologische Stellungnahme zur Frage der Beeinflussung der thermalen Mineralquellen in Bad Deutsch Altenburg durch den Bau der Kraftwersstufe Hainburg 1983, In M.Welan und K.Wedl (Hrg.): *Der Streit um Hainburg in Verwaltungs- und Gerichtsakten*, Akademie für Umwelt und Energie, Laxenburg

Tracer Hydrology 97, Kranjc (ed.) © 1997 Balkema, Rotterdam, ISBN 90 5410 875 4

Karst water tracing in some of the speleological features (caves and pits) in Dinaric karst area in Croatia

Mladen Garašić
Croatian Speleological Association, Civil Engineering Institute of Croatia & Department for Geotechnics, Zagreb, Croatia

ABSTRACT: In Croatian part of Dinaric karst which is "locus typicus" for karst in general, hydrogeological and speleohydrogeological researches are absolutely unreliable if appropriate methods of ground water tracing are not applied. Various methods were tested, but Na-fluorescein coloring method was and still is in major use. Many speleohydrogeological researches of speleological features and forms abundant in this region were made by use of this method. In Croatian Dinaric Karst area, which spreads on over 53% of Croatia, something like 8000 speleological features were registered and explored until now. Ground water was found in more than 35 % of that features, and permanent ground water is present in cca. 18% of all registrated caves and pits. This is very high percentage of appearance according to other karst areas in the world. Three separate areas of Croatian karst (Lika, Gorski kotar and Kordun regions) were selected (Fig.1) in order to present successful solutions of hydrogeological problems inside the speleological features by water tracing. If water tracing had not been made "in situ" probably the result interpretations would be different and inaccurate. These are areas of underground bifurcation's, and sometimes trifurcation's. In hydrogeological meaning areas that are mentioned here like Jezerane and speleological system of "Rokina bezdana" in Lika, area around Fužine in Gorski kotar and area around Rakovica with many cave systems and upper flow of Korana river in Kordun are especially interesting. In discovering of speleological features, directions and kinds of ground water flow, water tracing will certainly be of great help in the future. The suggestion is that water tracing in karst must be combined with speleological researches when ever possible.

1. INTRODUCTION

Speleology has very important role in karst hydrogeology. Measuring of absolute neotectonic movements combined with explanations of speleogenesis and resolving of many karstification processes is possible in speleological features. Tectonic elements, invisible or "masked" on the surface (bedding, folds, faults, nappes, etc.), are resolved and documented by these neotectonic movements. Connection of speleological features (caves and pits) and karst ground water is even bigger. Flows which are directly measured and observed under ground are easier to understand than those which are only perceptive as "swallow hole - spring" phenomenon. Great number of underground connections were discovered only by

Figure 1. Situation map of the Croatian Karst area /LEGEND: 1 = Rokina bezdana pit (Lika); 2 = Vrelo Cave and Fužine (Gorski Kotar); 3 = Panjkova and Muškinja Cave System (Kordun)/

injecting of special colors in swallow holes (ponors) and by registrating of color appearances in springs. The objective of speleohydrogeological researches that are conducted directly inside speleological features is comprehensive knowledge of speleological water channels, ground water flowing rules, amounts and directions of flows, chemical and bacteriological specialties of groundwater etc.

Areas with differences in lithostratigraphic characteristics of primer rocks, in hydrogeological zoning, in speleomorphological types and hydrogeological functions of speleological features were selected.

2. KARST AREA OF MALA KAPELA MOUN/ TAIN MASIFF, DREŽNICA , STAJNICA AND JEZERANE (LIKA) - ROKINA BEZDANA CAVE SYSTEM

This is the area of Central karst region (belt) which contains cca 850 different speleological features discovered until now.

2.1. SPELEOGEOLOGY

The Mala Kapela mountain area was geologically researched by I. Velić (1973), and geological mapping (Velić & all, 1980, 1982) discovered minor range of Jurassic layers than it was presumed by previous labors of J. Poljak.

2.1.1. Lithostratigraphy and tectonics

Jurassic layers (beds) are represented as carbonate facies - by dolomite and limestone exchanging. Those are well layered sediments in gray color and with many micro fossils. The layer thickness is estimated from 1300 to 1600 m. The Lower Cretaceous sediments in Velika and Mala Kapela are represented by uniformed carbonate strata series, mostly limestones (really dolomites), sedimented in very shallow sea. Average thickness is cca 1400 m (Velić, 1973).

Closer determination of tectonic relations in this region is prevented by well developed vegetation cover. Nevertheless it is possible to observe that area of Mala Kapela and south part of Velika Kapela were formed in synclinal structure. Lower Cretaceous layers appear in syncline limbs and upper Cretaceous sediments appear in nucleus. Limbs are slightly inclined and the biggest inclination angle is 35°. This area has remarkable tectonic disturbance, so kilometer long fault extending from Jasenak, over Drežnica and Zrnić to Crnačko polje and Jezerane can be seen at first sight. Fault which divides Juras-

sic and Cretaceous layers is parallel to the first one. Just on that fault disturbance and karstification of thick carbonate Jurassic and Cretaceous layers refer to presumption that big speleological systems could be expected here (deep pits and caves with water flows).

2.1.2. Speleomorphology

One of the most interesting speleological features of Croatian karst is Rokina bezdana. It is located between mountain massifs of Velika and Mala Kapela in Lika. This complex speleological feature has a vertical entrance (pit) 102 m deep, followed by horizontal cave channels, most of them rich with water. In the main channel measured water flow in minimum was 2 m^3/s, and in maximum was estimated to several hundreds of m^3/s. Underground river forms huge waterfalls and lakes, and diving was performed in cca 10 siphons. From 1975. endemic amphibians in this karst region (*Proteus anguinus*, Laur.) were discovered in Rokina bezdana. Due to very complex feature, 30 speleological actions (exploring) were undertaken in order to explore 1016 m of length and 127 m of depth. Researches were performed over thirty years and are not over yet.

2.1.3. Speleohydrogeology

According to classifications of hydrogeological functions of speleological features (Garašić, 1991b), Rokina bezdana is percolating speleological feature with steady underground flow. Water neither sinks in it, nor springs out, it just flows through.

Multiple measuring of underground river flow in area of the main rock debris have revealed that minimum flow is 2 m^3/s, and according to indicators, maximum could be estimated to several hundreds m^3/s. Ground water level rises for 10 to 30 m depending on the amount of precipitation in drainage basin. Underground river in Rokina bezdana is very powerful, forming waterfalls, torrents and underground lakes. All year long water has constant temperature of +5,3° C. Water chemical analyses (Garašić, 1976) made in Physical Medicine and Rehabilitation Institute of Faculty of Medicine University of Zagreb have revealed that type of Rokina bezdana water is calcium - bicarbonatic, with total mineralization of 406,769 mg/l. These results are characteristic for carbonate areas water, because the amount of calcium bicarbonate is dominant, while the amount of chloride is rather small. Concerning other elements, magnesium is higher, but its impact on water character is not significant. Based on this analysis this water can be used for drink. According

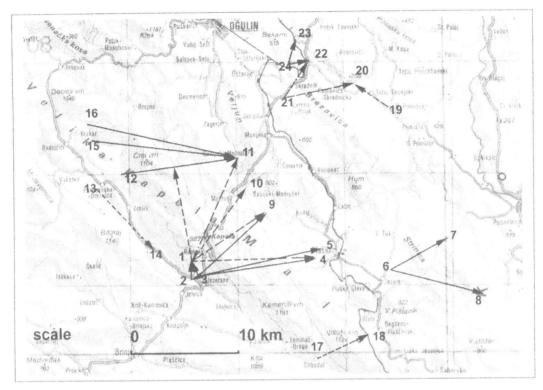

Figure 2. Results of Water tracing in Rokina bezdana (Mala Kapela, Lika) Karst area
(LEGEND: 1 = Rokina bezdana pit, 2 = Crnačka Cave , 3 = Jaruga swallow hole (ponor), 4 = Dretulja spring I, 5 = Spring Dretulja II, 6 = Swallow of Dretulja, 7 = Spring of Mrežnica, 8 = Savača Cave, 9 = Vrljika cave, 10 = Vrnjika cave II, 11 = Zagorska pec Cave, 12 = Ponor Sušik Cave, 13 = Swallow hole in Dreznica, 14 = Spring cave in Crnačko polje, 15 = Ponor in Krakarsko polje I, 16 = Ponor in Krakarsko polje II, 17 = Ponor spilja kod Glibodola, 18 = Velika pećina Cave, 19 = Ponors in Mrežnica river beds, 20 = Caves of Rudnica, 21 = Ponors of Zagorska Mrežnica, 22 = Kukača Cave, 23 = Tuonjčica Cave, 24 = Ponor Ambarac)

to bacteriological analysis made in 1984. the quality of water is excellent.

Crnačka cave has hydrogeological function of estavela (Garašić, 1992b) - in rain period is functioning like spring, and in arid period, with water sinking in it, functions like swallowing hole. Chemical analysis of the water from Crnačka cave showed the same results as water from Rokina bezdana (Garašić & Kovačević, 1992).

Total surveyed volume of Rokina bezdana channels is estimated to 400 000 m³. Paleohydrogeological as well as recent water flow had an intense impact on erosion and corrosion of the cave. From hydrogeological point of view it is interesting to understand where are the sources that collect the water to the underground river and what its farther direction is. We can be almost certain that it is connected to Crnačka cave estavela, which is located 1210 m to the south of Rokina bezdana entrance, and at the border of karst polje. The field's elevation is cca 100

m lower than Rokina bezdana entrance. The connection between two features was found for the rain period, when water springs from Crnačka cave and in Rokina bezdana water level rises for 10 to 30 m. This rising of ground water level was observed in both features, but just in the period when ground water level is high in general. In the arid period we speak of "table water" between Crnačka cave and unknown part of Rokina bezdana. This two features should be connected by channel which brings water to the Rokina bezdana system during the whole year, but up till now it has not been completely discovered. In Crnačka cave the siphon diving occurred in three places. Total length is 611 m, but it goes farther.

The karstification depth in the area of Rokina bezdana is estimated to several hundreds of meters, and vertical circulation zone from 150 to 200 m. It is followed by permanent horizontal (siphonic) circulation zone, which transports the water to Aegaean

(Black) Sea drainage basic. Future researches by use of water tracing in the rain period of the year will detect if there is connection between swallow hole Jaruga, Rokina bezdana, Crnačka cave, Dretulja spring, or on the other side towards swallow hole Sušik and drainage basin of spring - cave Zagorska Mrežnica. In hydrogeological sense the area of Mala and Velika Kapela represents very complex issue, and researches in those speleological features will certainly contribute to the general understanding of its problematics. The length of Rokina bezdana system is estimated to 10 or more kilometers, and it is just the question of time when it will be explored and researched completely by speleologists.

2.2. RESULTS OF WATER TRACING

Investigating the same water levels in Rokina bezdana and Crnačka cave is an interesting issue. During the flood on Crnačko polje (karst field), in the rain period, its water level (457,6 m a.s.l.) is similar to water level in Rokina bezdana (Jurak,

1984). Besides that, the color that was injected in Crnačka cave in the rain period, appeared in Rokina bezdana, but unfortunately it was impossible to define exactly the channel where it came from (lower parts of South channel). The researches were only qualitative. The appearance of the color in water flow was certainly detected, but its concentration was not measured. The speed of water flow based on measuring the appearance of color theoretical velocity was estimated to 0,6 cm/s. It confirms the presumption that there are big underground spaces filled with water.

The issue is where the water in Rokina bezdana comes from and where does it flow farther. Future systematic water tracing will give more certain answers. Nevertheless, we can be positive, according to many data, about connection between Rokina bezdana and Crnačka cave. The connection with swallow hole Jaruga in Stajničko polje is possible. If that underground connection exists, then it could be possible that water from Rokina bezdana flows to Dretulja spring. The connection between swallow

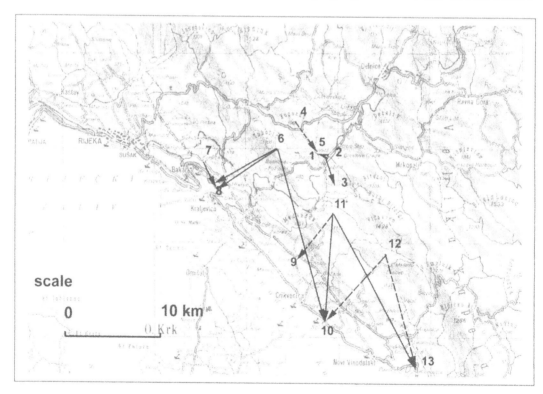

Figure 3. Results of Water tracing in Vrelo Cave System and Fužine Karst area (Gorski kotar)
(LEGEND: 1 = Vrelo Cave, 2 = Ponor on Vrata Cave, 3 = Potkoš Spring, 4 = Pits on Rogozno, 5 = Spring Cave of Ličanka, 6 = Ponors of Lepenica, 7 = Pit in Krasica, 8 = Vruljas in Bakarac, 9 = Pits in Sušik, 10 = Springs in Selce, 11 = Ponor Cave of Ličanka, 12 = Stupina jama pit, 13 = Spring near Klenovica)

hole Jaruga and Dretulja spring was determined by the color injecting (Baučić, 1965). There is also a possibility that water from Rokina bezdana flows northwest, towards Drežničko and Krakarsko polje, and farther towards Zagorska Mrežnica spring (in Zagorska peć cave) by certainly confirmed underground connections (Biondić & all, 1987; Turner, 1954a, 1954b, 1954c). What area collects the water for Rokina bezdana is even more difficult issue. Probably that area is water collecting zone region of the mountain massif Mala Kapela (Figure 2).

Analyzing all the facts, this hydrogeological system appears to be one of the most complicated systems in Croatian karst. The direct systematic coloring of the water flows inside of Rokina bezdana could resolve this problematic (Garašić & Kovačević, 1992).

3. AREA AROUND FUŽINE (GORSKI KOTAR)

This is the mountain area of Central Karst Belt (Gorski Kotar) where 450 different speleological features were explored until now.

3.1. SPELEOGEOLOGY

Interesting features and characteristics of Gorski Kotar and Fužine area have been attracting the geologists for many years now. The geologists paid attention to surface and ground water, underground karst spaces, minerals, and other valuable rocks (rocks for engineering, barite, clay, gravel etc.). A Short publications and reviewing maps with basic data about this area have been published during the first half of 19th century by Austro-Hungarian geologists. T. Kormos & V. Vogl (1913) had published the results of two-year researches of Fužine surroundings. They focused on kinds of rocks and their age history, and they paid less attention to structural relations. F.Koch (1925/26) had fundamentally reviewed surroundings of Fužine being the first one who mentioned the real possibility of accumulation in Ličanka river valley. His conclusion was based on existence of waterproof "carbonate shales and slates" what makes this accumulation possible. M. Tajder (1965) and M. Vragović & Lj. Golub (1969) were writing about rocks found in the south of Fužine, near Fužinski Benkovac and about their practical value. Those are andesite rocks which, according to other areas, made breakthrough to the surface on the break between central and upper Trias. I. Grimani & all (1973) had geologically described the area of Vrelo cave. M. Herak (1987) had described geological researches previous to HE "Nikola Tesla" construction. During the construction of the part of the highway Rijeka - Zagreb, extending from Oštrovica

to Vrata, M. Garašić had multiply described layers and made geological maps of Vrelo area focusing on geological engineering and hydrogeology (Garašić, 1991a,1992a,1996).

3.2. WATER TRACING RESULTS

The Vratarka brook, which sinks in Ponor and has an underground flow through the feature, ends as a siphon after 456 meters of waterfalls and lakes. Underground water connections between Ponor in Vrata and Potkoš spring in Lič polje (cca 1800 m away) had been detected by injecting fluorescein. It is also presumed that water from Ponor appears (Garašić, 1992b) in Vrelo cave (1650 m away). Theoretical flow direction following the contact of Triassic and Jurassic rocks, and partly (under Vrata tunnel) following Triassic and Paleozoic rocks, makes the connection with Potkoš spring very interesting. Theoretical apparent velocity of the underground flow towards Potkoš spring is estimated to 15 cm/s, which is rather quick, and that refers on speleological channels with concentrated water flow. Justified presumption is about water pollution in Potkoš caused by waste in Ponor near Vrata. There were no bacteriological analyses. Water pollution in Ponor comes from secondary speleological channels bringing water from areas near settlement Vrata (sanitary effluence, system of sewage). The main role in water transporting towards Vrelo cave, and towards Veliko vrelo spring in the same time, could have the fault contact of Jurassic and Triassic strata with Paleozoic rocks.

Water from Potkoš directs its flow through Lič polje till the swallow hole under Kobiljak, sinks (Biondić & all, 1980; Biondić & all, 1985), flows underground and reappears on Tepla spring near Novi Vinodolski (the distance from swallow hole is 18,5 km, apparent velocity 1.04 cm/s), on Vrulja (submerged under sea spring) Žrnovnica near Klenovica (the distance from the swallow hole is 19,3 km, apparent velocity 1,10 cm/s) and in Žminjac spring near Bakar (the distance from the swallow hole is 11,5 km, apparent velocity 0,61 cm/s) (Figure 3).

4. AREA OF RAKOVICA KRŠLJA AND UPPER FLOW OF KORANA RIVER (KORDUN)

This is the area of Inner Karst Belt (Kordun) where the number of explored speleological features is estimated to 650. There are several very long caves, longer than 10 kilometers.

Speleological research will be continued in the future (Panjkova and Muskinja Cave near Rakovica).

4.1. SPELEOGEOLOGY

The rocks where the most important cave system: Muškinja - Panjkova cave was formed, are mostly Upper Cretaceous and somewhere Lower Cretaceous carbonate sediments. That are mostly limestones and lime dolomites. From hydrogeologic point of view, these rocks are water permeable, especially because of their secondary porosity. Upper Cretaceous layers are in fault contact with Lower Cretaceous limestones. Those are good bedding lightly grey limestones with significant addition of organic detritus remaining from Ruddiest fragments. Dolomites are rear and appear in lens form (Korolija & all, 1981). Neogenic sediments are present in closed karst depressions (karst valleys and uvalas, sinking rivers Perlinac and Kršlja). Those are sandstones and clays, Miocene layers with *Congerious, Melanius, Melanopsidius* and *Fusarulusime* formed after Cretaceus with emersion phase which lasted till the beginning of Miocene. Fresh-water lakes were formed by decreasing of particular parts of land caused by neotectonic movements (which are still actual). Miocene sediments are primary transgresive on all older rocks. M. Garašić (1984) quotes the results of neotectonic measuring with contributions of some newer data. The velocity of tectonic blocks decreasing in eastern part of the cave system is in average 0,926 mm, which is relatively high. The Mašvina hill increasing in Neogene (Prelogović, 1976) was approximately 400 m. All those data confirms very intensive neotectonic activities (local decreasing and increasing of tectonic blocks) in Muškinja and Panjkova spilja cave system area. Beside hydrogeological conditions, those neotectonic activities are relevant factor of karst feature speleogenesis.

4.2. CAVE SYSTEM SPELEOMORPHOLOGY

By its morphology cave system Muškinja - Panjkova spilja is determined as branch, etage, several levels speleological features with several water flows. The swallow hole of Perlinac brook represents one entrance in the cave while the other one is dry. Channels are getting closer to the surface in several locations, and are collecting water from it. For example, in western bank of Kršlja brook, 150 - 200 m from the Muškinja cave entrance, the distance towards speleological channels, explored by diving in siphons is only 10 m. The next example is the distance between Kršlja brook active swallow sand hole and explored cave channel which is 34 m, or in Blatni chanal just 5 m. The interesting discovery is that several levels channels appear more often in eastern part of the system which is genetically older. The

diving in conducting system of the Muškinja main channel revealed chambers in several levels. The total length of the topographically surveyed channels is 12385 m, and more than 700 m of the cave system were discovered but not precise surveyed. Muškinja and Panjkova cave were connected by cave diving tectonics in several points, therefore it is not correct to talk about length of particular cave, but of the cave system. Of 12385 m system length, 11 km is filled with water flows. Certainly that is the biggest amount of accumulated ground water in one cave system, discovered by speleological exploring in Croatia.

In general, we can say that the basic directions of cave channels extending are west - east and southwest - northeast. But there are some aberrations from the data gathered in this statistical cloud, especially in carbonate rocks categorized as Lower Cretaceous. It is presumed that well trained speleological team would need at least a week to pass through all the already explored cave channels in Muškinja and Panjkova Cave system. The future explorers should count on that.

4.3. RESULTS OF WATER TRACING

The hydrogeological function of the cave system Muškinja - Panjkova is very complex but already has enough results of explorations that leads to general understanding. The whole system was formed in the water permeable rocks categorized as Lower Cretaceous and Upper Cretaceous. Those are well bedding limestones, intensively karstified from the surface, and underground (inverse karstification). Water is collected from Rakovačka valley and slopes of Mala kapela, 10 - 20 km west from the cave system. Water springs from the caves near Rakovica, flow through the field, sinks again in several caves (Švica, Durlić, Ponorac etc.). Underground flow goes through speleological features below Bliznica and Lipovača and appears again in spring-caves (Perlinac, Suvaja, Baričevac, Točak etc.). Farther on, water has a surface flow through Neogene and Miocene water-resistant layers, then sinks again (Ponor near Muškinja or Varićakova cave, Panjkova cave, Ponor Jovac etc.), on the contact with Cretaceous layers. Mašvina mountain dominates over that part of Kordun, and has a huge cave system beneath, collecting the water from Mašvina hills. This system also collects water from river Korana underground flow, after its sinking 9 km south east from the cave system. That refers to the existence of three different underground flow directions in the system at the same time. It was confirmed by long lasting speleogeological and hydrogeological researches. From

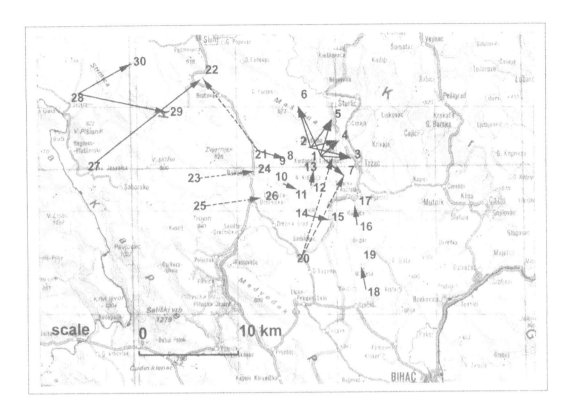

Figure 4. Results of water tracing in Panjkova and Muskinja Cave system Karst area (Kordun)
(LEGEND: 1 = Muškinja Cave, 2 = Panjkova Cave, 3 = Crno Vrelo Cave, 4 = Zecevac spring, 5 = Bijelci spring, 6 = Basare spring, 7 = Bukovac spring, 8 = Perlinac spring cave, 9 = Tocak spring cave, 10 = Ponorac cave, 11 = Suvaja cave, 12 = Baraceva pecina cave, 13 = Baričevac, 14 = Pits on Lipovača, 15 = Jankovica pecina cave, 16 = Swallow hole Gatica, 17 = Gatica Cave system, 18 = Šarićeva Cave system, 19 = Mala Peca spring, 20 = Ponors in Korana river (near old bridge), 21 =   Ponor Cave Svica, 22 = Spring of Slunjčica, 23 = Pits on Rakovačka uvala, 24 = Spring in Rakovica, 25 = Pits on Trovrh, 26 = Spring of Rakovački brook, 27 = Swallow of Lička Jasenica, 28 = Ponors of Dretulja, 29 = Suvača cave, 30 = Mrežnica river spring)

Muškinja - Panjkova cave system, after underground flow, water springs in several locations (spring-cave Crno vrelo, spring Zečevac, Basare, Bjelci) all in direction of northeast - east and influents in Korana river. Connections between all those water flows were proofed by water coloring and some were actually (physically) explored by speleologists (Ponorac - Suvaja).

Ground water found in the channel behind the Zeleni siphon comes to the system from two directions. First is from Poskokova jama, 39 m long, 4 m deep. That cave has occasional swallow function and was formed between two limestone layers which caused the wide channel forms, but in the same time is very low. It is impossible to pass through because of narrowness. Between those two features the air distance is 420 m. Second direction is from the sink in right bank of Kršlja brook. The air distance between

swallow holes and distal channel end behind Zeleni siphon is 725 m. Swallow holes had been widened in the very dry season, but it increased the length just for 11 m. Nevertheless, water coloring revealed the connection between those two locations. The color had been passing through unknown parts of the system for 8 hours and then appeared in Zeleni siphon. If we know the average water velocity in that area (10 cm/s), we can estimate the total length of unknown part to 2 km. The important fact is the existence of the water in the channel even in periods of Kršlja brook bed total dryness, and when there is no water sinking in Paukova jama. This refers to the fact that there must be another, unknown groundwater flow in that channel. This conclusion is confirmed by the constant water temperature that does not depend on surface temperatures. The water from that unknown source must be staying underground

235

for long period. The presumption is towards the water of Korana river that sinks in its bed near Drežnik-grad and Korana Bridge, but it has not been proved by water tracing yet. In cave channel behind Zeleni siphon in the time of diving, the water flow was cca 500 - 600 l/s. However, multiple flow estimations during rain period in the canal in front of Zeleni siphon were cca 2-3 m³/s. It refers to the important conducting hydrogeological role of the cave channel with Zeleni siphon in whole Muškinja and Panjkova cave system (Figure 4).

In whole cave system the amount of accumulated water has been estimated to 100 000 m³ during the arid period. The longest lake is 635 m long, with the biggest depth over 17 metres. Those are the imposant data. Water temperature in the channel was 8,9° C. Important is to mention several locations in that system where measured water temperatures were higher than it is common for Croatian karst (15,4°C in the channel under Jamarje). Those hypotherme could be caused by active neotectonic relations, but there have not been enough researches on that subject.

5. CONCLUSION

The great importance of water tracing was revealed in the area of Rokina bezdana cave system in Lika, speleological features around Fužine in Gorski Kotar and Muškinja - Panjkova cave system in Kordun, because such valuable results couldn't be overtaken by water coloring from the surface. The suggestion is to include water tracing in all researches of speleological features with water. On all locations coloring of the water will continue.

REFERENCES

Baučić, I.(1965): Hydrological characteristics of the Dinaric karst in Croatia with a special regards to the underground water connections. Naše jame, vol.7, br.1-2, str. 61-72, Ljubljana.

Biondić, B., Ivičić, D. & Viljevac, Ž. (1987): Mogućnost zaustavljanja vodnih valova u visokim dijelovima sliva Zagorske Mrežnice. Zbornik referata IX jugoslovenskog simpozijuma o hidrogeologiji i inženjerskoj geologiji, Priština, knjiga 1, str.33-39, Beograd.

Garašić, M.(1984): Neotektonske aktivnosti kao jedan od uzroka geneze i morfologije jednog od najvećih spiljskih sistema u Hrvatskoj. Deveti jugoslavenski speleološki kongres, Karlovac, Zbornik predavanja, str.457-465, Zagreb.

Garašić, M.(1991b): Morphological and Hydro geological Classification of Speleological structures (Caves and Pits) in the Croatian karst area. Geološki vjesnik, vol.44, str.289-300, Zagreb.

Garašić, M.(1992b): Ronjenje sifona u spilji Vrelo (Gorski kotar). Spelaeologia croatica, vol 3, str.78, Zagreb.

Garašić, M & Kovačević, T.(1992): Speleološki sustav Rokine bezdane u rješavanju hidrogeoloških odnosa u području Male i Velike Kapele. Spelaeologia Croatica, vol.3, str.15-22, Zagreb.

Grimani, I., Šušnjar, M. Bukovac, J., Milan, A., Nikler, L., Crnolatac, I., Šikić, D. & Blašković, I. (1973): Tumač za osnovnu geološku kartu, list Crikvenica, L33-102, str.1-47, Beograd.

Herak, M.(1987): Geološka karta Fužina i okolice. Istraživački radovi za izgradnju HE " Nikola Tesla". Monografija Fužine, str.214-220, Rijeka.

Jurak, V. (1984): Hidrogeološka interpretacija krškog porječja kao putokaz zs speleološka istraživanja. Deveti jugoslavenski speleološki kongres, Karlovac, Zbornik radova, str.237-249, Zagreb.

Koch, F.(1925): Tektonika i hidrografija u kršu. Glasnik Hrv. prir. društva, br.37-38, str.71-87, Zagreb.

Kormos, T. & Vogl, V. (1913): Weitere Daten zur Geologie der Umgebung von Fužine. Jahresber. Ung, geol. Reichsanst, (1912), str. 57-61, Budapest.

Korolija, B., Živaljević, T. & Šimunić, A.(1981): Osnovna geološka karta. Tumač za list Slunj, L33-104, str.1-47, Beograd.

Maucci, W.(1973): L'ipotezi dell'errosione inversa. Le Grotte d'Italia, vol.4, pp.4-54, Bologna.

Redenšek, V.(1950): Nova špilja "Vrelo" kod Fužina. Naše planine, vol.2, br.10-11, str.305-310, Zagreb.

Turner, S.(1954a): Bojenje Ćupić ponora kod Krakara. Fond stručne dokumentacije HMZ, br.9, Zagreb.

Turner, S.(1954b): Bojenje ponora Maravić kod Drežnice. Fond stručne dokumentacije HMZ, br. 10, Zagreb.

Turner, S.(1954c): Bojenje ponora u južnom dijelu Drežničkog polja (M-6), Fond stručne dokumnetacije HMZ, Zagreb.

Velić, I. (1973): Stratigrafija krednih naslaga graničnog područja Velike i male Kapele. Geološki vjesnik, vol.26, str. 93-109, Zagreb.

Velić, I. & Sokač, B. (1980): Osnovna geološka karta, 1:100000, list Ogulin, L 33-103, Beograd.

Velić, I., Šokač, B. & Šćavničar, B. (1982): Tumač za OGK, list Ogulin, L33-103, str.1-46, Beograd.

Zanoškar, D. & Garašić, M. (1992): Ponor kod Vrata i hidrogeologija bliže okolice Fužina. Spelaeologia Croatica, vol. 3, str.27-31, Zagreb.

Tracer Hydrology 97, Kranjc (ed.) © 1997 Balkema, Rotterdam, ISBN 90 5410 875 4

Isotopic and hydrochemical tracing for a Cambrian-Ordovician carbonate aquifer system of the semi-arid Datong area, China

Gu Wei-Zu
Nanjing Institute of Hydrology and Water Resources, People's Republic of China

Ye Gui-Jun, Lin Zeng-Ping, Chang Guang-Ye, Fu Rong-An & Fei Guang-Chan
National Administration of Coal Geology, Handan, People's Republic of China

Jing Zhi-mno & Zheng Ping-Sheng
Datong Coal Mining Administration, People's Republic of China

ABSTRACT: This study of groundwater in the South Mouthspring Cambrian-Ordovician carbonate aquifer system is undertaken isotopically and hydrochemically. The radiocarbon, tritium and stable- isotope data lead to the suggestions that groundwater in this system is resulted from mixing of paleo-groundwater with precipitation recharge under varying circulation paths and histories. The Cambrian-Ordovician carbonate outcrops form the main recharge area of this aquifer system with a recharge rate estimated from two methods up to 48 to 64 mm per year. The circulation environment of groundwater is distinguished into types of active, static and sluggish. It is further grouped according to its significant isotopic and hydrochemical features. On these bases, five sub-aquifer systems having their own recharge, circulation and discharge paths are identified.

1 INTRODUCTION

Water resources problem has been raised notably with the semi-arid Datong region, a main coal mine area of China, due to the shortage of water supply which is estimated at about 9×10^{4} m³/day towards the close of this century. It leads to focus on the exploitation of deep groundwater from Cambrian (ϵ) Ordovician(O) carbonate aquifer system within which the South Mouthspring ϵ-O aquifer system (SMϵO) seems the most important one for recent development. Water tracing was used as constraints on groundwater origin, recharge, regionalization and delimitation for this aquifer system which seems difficult to be resulted and quantified by standard hydrogeological techniques alone.

2 HYDROGEOLOGIC SETTING AND SAMPLING METHOD

The study area covers 1493 km² with main part of about 750 km² within which most deep boreholes are distributed for hydrogeologic exploration. It is mountainous on the Datong Sycline with highest elevation of about 1968 m a.s.l. and the lowermost of a river bed of about 1160 m a.s.l. The metamorphosed gneiss of Archean (Ar) is the oldest formation and the basal in this area, it outcrops along a large fault and makes up the east geohydrologic boundary of the aquifer system. (Fig.1). It is unconformablly overlain by various stratified deposits and formations of Paleozoic, Mesozoic and Cenozoic except that of Triassic Period. The Cambrian and Ordovician carbonate strata of marine deposits extend in much of the study area. It oucrops along the outcropping Ar at the east boundary and sporadically exposures within the southern part of study area. It dips generally to northwest with maxium depth below the ground surface of about 800 m. It varies in thickness from about 994 m near the midst of study area to few hundreds of meters by its south, west and disappears towards the north boundary. The ϵ-O strata are composed of thick sections of sandstone, dolomite and limestone interbedded with lime marl and shale. In downward order the succession consists of formations of the Middle and Lower Series of Ordovician Age with absence of its Upper series followed by the Upper, Middle and Lower Series of Cambrian Age. It behaves hydraulically as the primary water-bearing units of this aquifer system. This carbonate groundwater regime is complicated by numerous faults. All boreholes are finished in the Lower Series of Cambrian Strata and structured in the way that it is available only for the water from ϵ-O units.

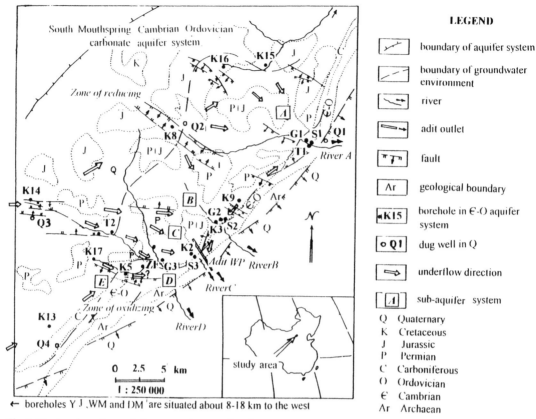

Fig.1. Regional geology of the research area showing the generalized geology, outcrops of Є-O carbonate strata, sampling boreholes, groundwater environments and sub-aquifer systems.

Precipitation was sampled monthly according to IAEA methods (IAEA, 1994) from 5 stations with altitude from 1050 m to 1530 m a.s.l. Surface water was sampled from 4 rivers originated in research area. Quaternary phreatic groundwater was sampled monthly from 7 wells. Groundwater of Є-O carbonate aquifer system was sampled monthly from 25 sites within research area in specially drilled project boreholes finished in depths of 450 m to 850 m below the ground surface with diameters of 110 mm to 500 mm. Special sampling equipments were designed for this purpose. Drainage from coal mine adits was sampled, too.

Sample sizes of 50 ml and 500 ml were used for ^{2}D, ^{18}O and T respectively while about 100 l was collected for preconcentrating carbonate in field for radiocarbon analysis using the Bag Method from South Africa (Verhagen B.Th.,1993). Another set of sample with size of 50 ml again was used for determination of uranium content in waters.

3 ORIGIN OF WATERS

The isotopic composition of local precipitation with 35 data sets from 1994 - 1995 fits a equation slightly different from that of WML (Craig,H., 1961)

$$\delta\,^2D = 7.409\ \delta\,^{18}O + 9.548 \tag{1}$$

Oxygen and hydrogen isotopes in groundwater of this SMЄO aquifer system with 61 data sets of 1994 - 1995 from 22 boreholes vary in a way that is distinguishable from that of meteoric water (Fig.2). The observed $\delta\,^{18}O$ values range from -8.12 ‰ to -12.56 ‰. For all data sets it fits

$$\delta\,^2D = 6.173\ \delta\,^{18}O - 10.656 \tag{2}$$

For mean values of all boreholes, it fits

$$\delta\,^2D = 6.449\ \delta\,^{18}O - 8.125 \tag{3}$$

It seems reasonably to suggest that the original water of

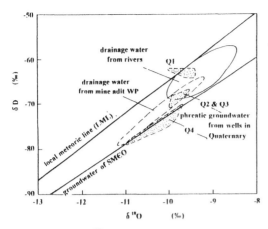

Fig.2. δD - δ ^{18}O diagram showing the local precipitation line, groundwater line of SMЄO aquifer system and the clusters of other waters.

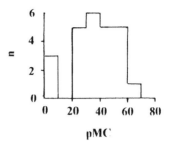

Fig.3. Frequency histogram of radiocarbon values in groundwater of SMЄO aquifer system.

this Є-O aquifer system before various mixing and underflow circulation has a mean value of δ ^{18}O about -17.12 ‰ and δ ^{2}D about -117.27 ‰. This isotopically very depleted water could not have been produced in the surface temperature regime recently. If the Dansgaard relation correlating δ ^{18}O and temperature (Dansgaard, W.,1964) is valid approximately for this water, then this δ ^{18}O depleted value should correspond to an annual mean surface temperature of about -2°C. It is inferred that this is the synsedimentary water of marine origin left in Є-O carbonate strata from marine deposits during the last glacial period. The data of δ ^{18}O and δ ^{2}D of present groundwater throughout this Є-O aquifer system comprise that this paleo-meteoric water has been recharged and mixed with modern meteoric water by percolation via various overlying Carboniferous, Permian and Mesozoic to Cenozoic sediments, and, via the outcropping Є-O rocks.

δ ^{18}O and δ ^{2}D data of 1994 - 1995 of phreatic groundwater from the dug wells finished in Quaternary strata show the different origins with three types as shown in Fig.2: (a) from precipitation, e.g., the Q1 which lies right on the LML with high ^{14}C contents up to 59 pMC and high tritium content up to 58 TU; (b) from the mixing of precipitation and Є-O groundwater, e.g., the Q2 - Q3 and (c) from Є-O carbonate aquifer, e.g., the Q4 which lies right on the SMЄO line, it is due to that this kind of wells is located near the outcrops of Є-O carbonate strata, then it accepts recharge from karst groundwater with low pMC of 33.9 and low TU of about 17.5.

Drainage water from rivers occupies clusters between LML and SMЄO groundwater line as shown in Fig.2. It is a mixing from precipitation with groundwater of Є-O aquifers.

The cluster of drainage water from the main out-let of an coal mine adit WP (Fig.2) shows that the Є O groundwater has contributed preferentially.

4 SOURCES OF GROUNDWATER OF SMЄO CARBONATE AQUIFER SYSTEM

The frequency histogram of radiocarbon values of the groundwater of this aquifer system (fig.3) also reveals a mixing mechanism approximately from recent meteoric infiltration water with high pMC and the paleo-groundwater with lowermost pMC. It is more obviously from the plot of TU versus δD as shown in Fig.4, groundwater in a variety of localities in this aquifer system is a mixing from precipitation water as the cluster of highest TU and variant δD resulted from different altitudes and seasons, with groundwater of very low TU and very depleted δD. It could be reasonably suggested that it composes of all the sources of karst water in this system from the inclusiveness of this plot, three end-members α, β and γ, as their regional sources could then be identified.

1. Source α This K15-like water containing lowermost contents of tritium of lower than 0.5 TU and lowermost radiocarbon concentration of about 4 pMC is the typical paleo-groundwater, the geological features of its Є-O producing units about 560 to 690 m below the ground surface are specified by nearly parallel formations as shown in Fig.5a.

2. Source β This T1/G3-like water containing highest content of tritium of about 40 TU and high pMC up to 60 is one kind of typical form of heavy recharge by precipitation. It is formed in inclined Є-O producing units of great gradients in exposed area without / with thin Quaternary coverages. Water change is enhanced by faults and, by inclined basal Ar as shown in Fig. 5b.

3. Source γ This S3-like water containing also highest content of tritium about 40 TU but moderate pMC of about 40 pMC is another kind of heavy recharge by precipitation due to its geological features specified by large thickness of exposed Ordovician formations as shown in Fig.5c.

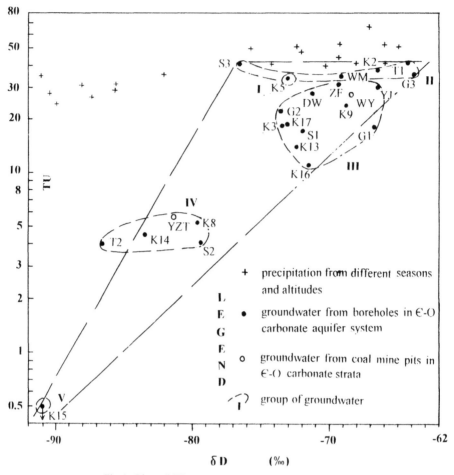

Fig.4. Plot of TU versus δD for precipitation and groundwater in SMЄO aquifer system.

It seems that the ongoing precipitation recharge for SMЄO carbonate aquifer system is happened mainly in the area with outcropping / semi-outcropping Є-O strata and is enhanced by faults. No evidence for such a recharge is found from groundwater in Є-O aquifers on which overlain with strata of various other geological series in large thickness.

It is found that the altitude effect of isotopic composition in precipitation is varied seasonally within this area. The change of $\delta^{18}O$ with altitude is -0.28 ‰ per +100 m , -0.34 ‰ per +100 m, -0.12 ‰ per 100 m and -0.05 ‰ per 100m for spring (April-May), summer (July - September), autumn (October - November) and winter (December - February) respectively. It seems that it is greater than the reported value of -0.3 ‰ per +100 m (Eriksson,

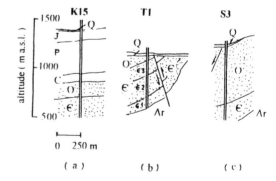

Fig.5. Geological features of three boreholes.

1983) in summer and smaller in other seasons. It implies that in addition to the different geological features, precipitation from different altitudes is another mechanism which makes the difference between sources β and γ.

5. ENVIRONMENT OF Є-O GROUNDWATER

Uranium tends to be a mobile element in aqueous environment, the concentration of dissolved uranium (UC) in SMЄO carbonate aquifer system shows a very wide range from 0.14 to 11.8 μg / l. The changing of dissolved uranium in deep carbonate aquifer is characterized by both environmental and kinetic factors (Cowart, J.B. and Osmond, J.K., 1974). The low UC related to very low radiocarbon concentration as shown in Fig.6 reveals that the groundwater is involved in a reducing environment with static or sluggish circulation while the higher

UC with moderate / high pMC implies an oxidizing environment with active groundwater circulation. This oxidizing environment is by and large coincide with the exposed Є-O area (Fig.1).

6. REGIONALIZATION OF GROUNDWATER IN THE SMЄO CARBONATE AQUIFER SYSTEM

For regionalization of groundwater for this aquifer system, the first thing is to identify the sub-aquifer systems. It is done from comparing pMC of groundwater at different sites and checked by hydrostatic field observed in boreholes. Five sub-aquifers, A to E, are identified as shown in Fig.1, it could be recognized by surrounding arrows of underflow.

The discharge of sub-aquifer A and D is towards the River A and D respectively while that of sub-aquifer B and C is towards Rivers B, C and the mine adit WP. However, no isotopic evidence for what waterbody the discharge of sub-aquifer E is towards, it is still not clear. Several exploitation boreholes and coal mine pits with small water quantity of pumping distributed in this area serve as their release outlets, too. It is interesting that groundwater in reducing environment with small TU and pMC,

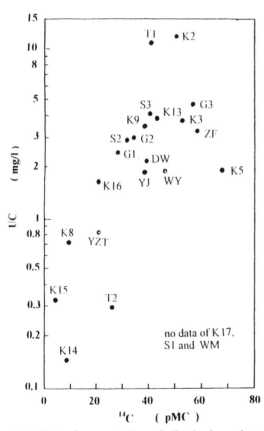

Fig.6 Plot of concentration of dissolved uranium versus that of radiocarbon in groundwater of SMЄO aquifer system.

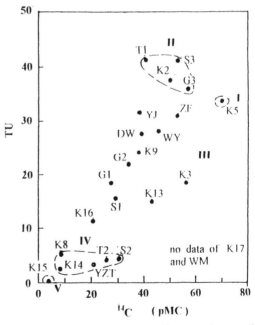

Fig.7. Plot of TU versus pMC in groundwater of SMЄO aquifer system.

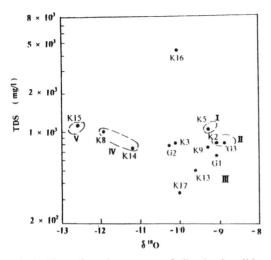

Fig.8. Plot of total contents of dissolved solids against $\delta^{18}O$ in groundwater of SMЄO aquifer system.

e.g., the K15 and K8 which occupy a large portion of the west part of SMЄO aquifer system takes a part into all sub-aquifers A to D.

Groundwater in SMЄO aquifer system is grouped from various isolines of its isotopic contents and, from comparing various isotopic and hydrochemical correlations as follows.

1. Mixing proportion: by TU versus δD (Fig.4);
2. Circulation and environment: by UC versus pMC (Fig.6);
3. History: by TU versus pMC (fig.7);
4. Hydrochemistry: by TDS versus $\delta^{18}O$ (Fig.8) and the Piper diagram (Fig.9).

Six groups are separated, grouping get from individual correlations is in coincidence generally except several in the Piper diagram due to the variety of hydrochemical background resulted from their geological environments. Group 1 of K5 belongs to a special sub-aquifer as mentioned above, groundwater here has a most active circulation environment in exposed area of Є-O carbonate rocks with faults and caves, highest pMC is detected in this group. Group 2 together with Group 1

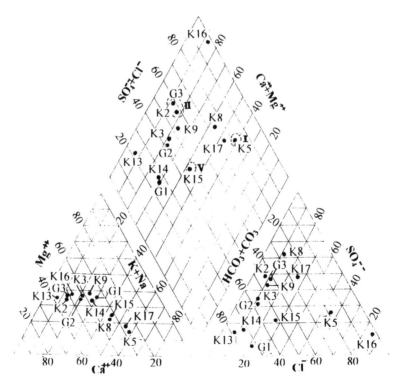

Fig.9. Piper diagram of major ion composition of groundwater in SMЄO aquifer system.

coincides with the outcropping Є-O region along the east boundary of study area. Group 5 is specialized from the paleo-groundwater of lowermost pMC. Group 5 and Group 4 occupy the region of reducing environment of Є-O groundwater.

7 ESTIMATION OF PRECIPITATION RECHARGE

The evidence of ongoing diffuse precipitation recharge for SMЄO carbonate aquifer system could be demonstrated in general from the plot of radiocarbon versus tritium as shown in Fig.7. Monthly variations of $\delta^{18}O$ in precipitation and in groundwater within a year are used for quantitative estimation of precipitation recharge. It is approximated as simple-harmonic. The recharge time T, time period for precipitation recharge reaches groundwater, is estimated from a differential equation established from the differences of phase-and-amplitude between two sine curves. T varies from 0.67 years for T1 to 38.6 years for K8. An estimation of precipitation recharge rate R is then resulted, it ranges from 16% for Є-O exposured area to about 1% for the area with deep buried Є-O formations.

As a comparison, another method from South Africa is used for estimation of R (Verhagen, B. Th., 1990). It is based on the assumption that our SMЄO aquifer system is a normally recharged dynamic groundwater basin, it could be reasonably realized from isotopic evidences as mentioned above. With initial radiocarbon of 70% to 80% for the carbonate aquifer system as discussed by Prof. Geyh (Geyh, M.A., 1972), from this method, R has the range of 12% to 15% for exposured Є-O area and 0.5% to 0.8% for the area with deep buried Є-O formations.

The annual mean precipitation of this area is about 400 mm, it follows that the precipitation recharge rates of some 48 to 64 mm/a for exposured Є-O area and 2 to 4 mm/a for area with deep buried Є-O strata could be excepted.

8 CONCLUSIONS

For this SMЄO carbonate aquifer system, the main conclusions that can be drawn from this study are:
1 Groundwater of this aquifer system is originated from the synsedimentary water of marin origin left in Є-O carbonate strata during the last glacial period. It has been mixed with meteoric water with varying circulation paths and histories.

2 Groundwater in this system is fossil only to the extent imposed by the regional geology, in a variety of localities in this system, it goes in the present climatological phase. The sources of groundwater in SMЄO aquifer system are both the paleo-groundwater and precipitation recharge. Precipitation recharge is in different versions formed from different altitudes and percolation paths via different formations of Є-O carbonate strata. The Є-O carbonate outcrops and semi-outcrops constitute the main recharge area of this system. An estimation of recharge rate of about 48 to 64 mm per year for outcropping area and 2 to 4 mm per year for area with deep-buried Є-O formations could be expected.

3 Two kinds of active circulation with oxidizing environment and static / sluggish circulation with reducing environment are separated for groundwater of this aquifer system. Five sub-aquifer systems are identified from varies isotopic and hydrochemical correlations. Every sub-aquifer system contains an outcropping Є-O area as its recharge area and, the river and/or mine adit as its discharge outlet. Groundwater in reducing environment with low tritium and radiocarbon contents takes part in every sub-aquifer system. Groundwater is also grouped according to its isotopic and hydrochemical features, every sub-aquifer consists of several groups of groundwater, sometimes the same group may be involved into several sub-aquifer systems.

ACKNOWLEDGMENTS

The authors wish to express their gratitude to Prof. Dr. Balthazar Th. Verhagen, Schonland Research Centre of Nuclear Sciences, South Africa, and Prof. Dr. Mebus A. Geyh, Niedersachsisches Landesamt für Bodenforschung, Germany, for their technical advise and sustained efforts in producing data. In particular, they wish to thank Prof. Dr. Gultekin Günay, Turkey, for his encouragement. Prof. Gu Wei-Zu is much indebted to a Dutch anonymous scientist who sponsor him as the member of IAH, it is of great benefit to his research.

REFERENCES

Cowart, J.B. and J.K.Osmond 1974.U-234 and U-238 in the Carrizo Sandstone aquifer of South Texas. In *Isotope Techniques in Ground Water Hydrology 1974*. Vienna : IAEA

Craig,H. 1961. Isotopic variations in meteoric waters. *Science* 133:1702

Dansgaard, W. 1964. Stable isotopes in precipitation. *Tellus* 16:436-468

Eriksson E. 1983. Stable isotopes and tritium in precipitation. In *Guidebook on Nuclear Techniques in Hydrology*. Vienna: IAEA

Geyh, M.A. 1972. On the determination of the initial ^{14}C content in groundwater. In T.A.Rafter and T.Grant-Taylor (eds.) *Radiocarbon Dating-Proc.8th Int. Conf.* Vienna: IAEA

IAEA. 1994.*Environmental Isotope Data No.10: World Durvey of Isotope Concentration in Precipitation (1988-1991)*. Vienna:IAEA

Verhagen B.Th.1993. The plastic bag method for collecting groundwater for radiocarbon analysis. Personal communication

Tracer Hydrology 97, Kranjc (ed.) © 1997 Balkema, Rotterdam, ISBN 90 5410 875 4

Analysis of fracture-induced water mixing by natural tracers at a granitic pluton (El Berrocal, Spain)

J. Guimerà & L. Martínez
Universitat Politècnica de Catalunya, Barcelona, Spain

P. Gómez
Centro de Investigaciones Energéticas Medioambientales y Tecnológicas, Madrid, Spain

ABSTRACT: We use hydrochemical data and natural isotopic tracers to study the groundwater flow regime, at the regional scales of a granitic pluton. The tracers involved in this study are tritium, deuterium and ^{18}O. Renewal times inferred by tritium content considering an exponential mixing flow model, allowed us to distinguish between the local discharge zones (shallow waters) from the regional ones, with renovation times above 1000 years. The exponential model was validated using stable isotope data. Most of the springs appear associated to highly conductive fractures in the middle-slope of the batholith. They display conspicuous differences in stable isotopes composition, tritium content as well as major and minor chemical elements. The contrast in hydraulic conductivity of the rock mass and the transmissivity of the fractures induces an important mixing of groundwater of different ages and origins. Hence, the "mixing model" which relies on geologic and hydraulic evidences explains hydrochemical and isotopic anomalies.

1. INTRODUCTION

Groundwater flow at regional scale usually depicts a homogeneous behavior. That also applies for low permeability fractured media (LPFM, also known as hard rocks). When detailed analysis is done, anomalies arise. That is, conspicuous differences exist in water composition and/or age of the springs, or distribution of hydraulic parameters. Usually, a rigorous study of geology, hydrochemistry and hydraulics of the rock mass, leads to a refinement of the conceptual model, thus explaining early anomalies. As regards groundwater age determination, Zuber and Motyka (1994) showed that matrix diffusion plays an important role, yet isotopes tend to be selectively retarded. The role of fractures in fluid processes such as mixing, has been evidenced in many geological media (Bredehoeft et al., 1992 among others).

In this work, we make use of geological information to explain anomalies in springs composition of a granitic batholith. For that, the regional picture is first introduced and following, the methodology. Then, we combine geology, hydraulics and hydrochemistry to explain spring anomalies.

1.1 Regional and project frames

The international El Berrocal project was an integrated exercise in geological, geochemical and hydrogeological characterization and had the aim of understanding and modeling the past and present-day migration processes that control the behavior and distribution of naturally ocurring radionuclide in a granitic environment. The objectives of the project were broadly focused on those processes which have relevance to safety assessments for radioactive waste disposal. However, El Berrocal was not under consideration as a potential repository site.

The El Berrocal site is located some 90 km SW of Madrid (Spain) in a granitic pluton at an altitude over 900 m.a.s.l. The surface expression of the pluton is roughly circular in shape, with an approximate diameter of 5 km (Figure 1). The pluton hosts a number of U mineralizations, at least two of which were considered economic and were mined until late 1960's. Most of the field and modeling efforts within the project were focused at site scale ($\cong$ 100 m x 100 m), around a sub-vertical 2 m-thick U-rich quartz vein.

Groundwater flow at regional scale, follows the shape of the pluton, being radial distributed when plan viewed. Recharge occurs sparsely along the

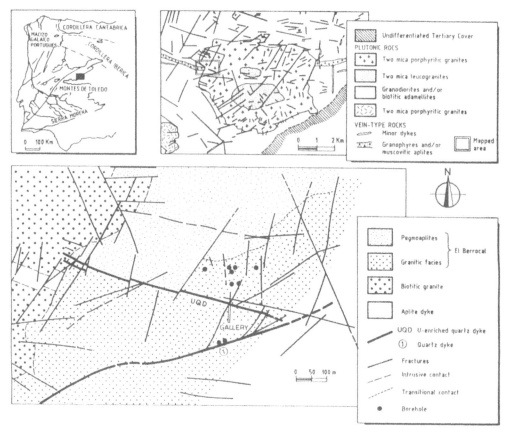

Figure 1. Location of El Berrocal site (after Marín et al., 1996)

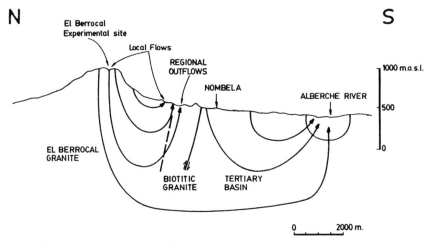

Figure 2. Groundwater flow scheme at regional scale of El Berrocal site. Flow lines are indicative yet no information is available in a major part of the cross section. Note the presence of ephemeral flows as local discharges on top of the pluton. This scheme is partially fed-back with information presented in this article.

surface expression of the rock mass, and exists together with outflows. However, high altitude zones are recharge-dominated while discharge mainly occurs along low altitude areas. Head distribution in boreholes up to 609 m deep confirmed that regional flow points downwards (Figure 2). For more detailed analysis at regional scale selected springs were sampled.

2. METHODOLOGY

Flowrates of springs were of the order of few ml/min when measurable. Complete chemical and isotope analyses where performed at most of them. Gómez et al. (1996) explain further details on field and lab procedures for sampling, analysis and criteria for representativeness. Such details are out of the scope of this paper. Samples were taken at permanent springs, and some of them were affected by seasonal variations. Comparison of winter-summer composition allowed to discriminate those springs which did not changed dramatically from those that were affected by climate conditions. Figure 3 shows the distribution of the springs, their chemical composition and a rough description of the local geology. Rain water samples were also available.

Data analysis was carried out by means of conventional multivariate analysis (Norasis, 1994) as a first approach. Such methodology is quite well known and widespread used (Seyhan et al., 1985; Siegel et al., 1991). Categories arose from the analysis were used for further geochemical calculations of heavy metal complexion and uranium geochemistry (Gómez et al., 1996). The most significant factors were those describing the evolution degree of groundwater (TDS) and the influence of pure carbonate phases. Both factors give close relationship for those springs located in top and the middle slope of the batholith, thus indicating a certain degree of continuity (evolution) from those springs located at high altitudes (recharge-dominated) with those at the middle-slope (recharge-discharge). However, samples taken at springs at low altitudes indicated highly evolved water, and did not correlated to the former.

Cluster analysis pointed to a groups of springs that appeared correlated in composition and location in the batholith. In general terms, they are following zones of rapid renewal (recharge-dominated) mixed recharge-discharge and discharge-dominated. Such subdivision will be revisited later in this article.

Conventional diagrams (Stiff polygons, Figure 3 and trilinear, Figure 4) were used to discriminate springs according to their chemical composition. Furthermore, stable isotopic composition was studied by means of meteoric line for rain water ($^{18}O/^{2}H$, Figure 5) and by relationship of $\delta^{18}O$ $^0/_{00}$ and altitude (Figure 6). A complete mixing model was used to correlate tritium content and probable age of water (Figure 7).

3. RESULTS

Deuterium-oxygen-18 local rain water line highly correlates with the world-wide one. Excess of deuterium (14) is comparable to those obtained for coastal Mediterranean areas, and show the effects of local storms, specially during autumn (Araguás, 1991). Meteoric line is then expressed by:

$$\delta D = 8 \delta^{18}O + 14 \qquad [1]$$

Yurtsever and Gat (1981) showed that $\delta^{18}O$ and δD of rainwater decrease within the altitude. Such variation depends on topography and local climate. Gradients of -0.15 to -0.5 $^0/_{00}$ per 100 m may be reached in the ^{18}O case. In the present case, rainwater samples at different altitudes were not available. However, we assimilated discharge points of rapid renovation as the uppermost limit of the samples at the area. Figure 6 shows that the gradient for El Berrocal pluton in some -0.5 $^0/_{00}$ in 100 m.

Springs 14, 18, 19 and 20, that show higher mineralization (Figure 3) and outflow at low altitudes, display the highest recharge altitudes. Provided that the analysis is correct and the composition unique, spring 14 shows a height above datum line actually not existing in the area. It is likely that either the altitude line does not apply to all the samples or that recharge altitude was different than present. We discuss such possibilities on next section.

Tritium content analyses showed a sparse distribution, both in terms of TU and location. According to the mixing model established for rainwater of the area (Plata, 1995) tritium contents above 15 TU are not expected (Figure 7). Likewise, tritium contents below 5 TU are indicative of long renewal times.

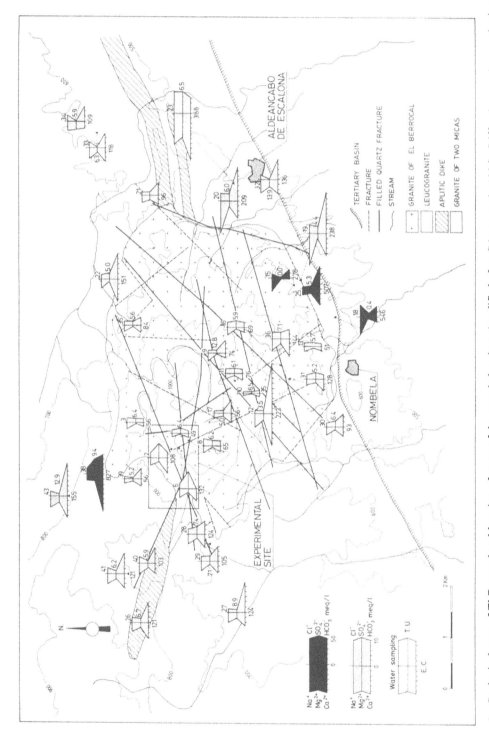

Figure 3. Geological map of El Berrocal and location of some of the sampled springs (modified after Gómez et al., 1996). Stiff polygons show major ions composition, Electrical conductivity (EC) and tritium content (TU) are also shown. Spring 14, not displayed in this figure appears very close to #15 with similar chemical composition and tritium content = 0 TU.

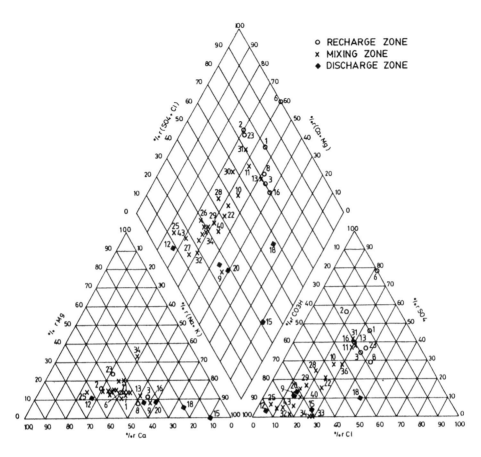

Figure 4. Trilinear (Piper-Hill-Langelier) diagram of springs. Symbols are distributed according to the "attributed" nature of each spring which depends on their position in the regional groundwater flow distribution (Figure 3). Springs 31, 11 and 13 are not located at high altitude areas (recharge-dominated) but display composition similar to them, close to the transition to mixing (discharge-recharge dominated) zone.

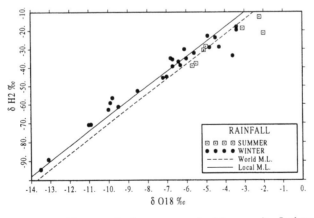

Figure 5. Local meteoric line making use of rain water samples taken on-site. It shows a good agreement with the world-wide one. Excess of δ^2H $^0/_{00}$ is 14.

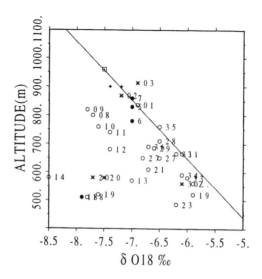

Figure 6. Change of altitude effect on δ^{18} O of spring samples. Note that springs located at the surrounds of the pluton (14, 18, 19, 20) fall well separated from the theoretical recharge altitude. The rest of springs discharge groundwater recharged up to 300 m above them. The sample above the line lacks of proper interpretation.

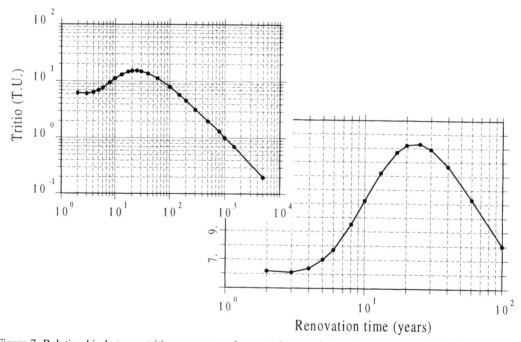

Renovation time (years)

Figure 7. Relationship between tritium content and expected renewal times for an exponential mixing model (Plata, 1995). Figure 3 shows the tritium content for most of the springs. Spring 14, which is located close to # 15 displayed 0 TU.

4. DISCUSSION

Based on stable isotopes, major ions concentration and tritium content, we classify springs in zones of rapid renewal, at recharge-dominated zones; mixed recharge-discharge and a third discharge-dominated group. We try to be consistent with the regional flow model scheme depicted in Figure 2 and explain the anomalous values.

Springs located at high altitude areas, depict tritium content values similar to present rainwater. Chemical composition is sometimes of sulfate type due to the dissolution of pyrite associated to quartz veins. TDB are usually low, as expressed by the small area of Stiff polygons. According to $\delta^{18}O$, recharge altitude in the order of 100 m above the outflow altitude.

Springs located at low altitudes and at material boundaries, depict high TDS, low tritium content (sometimes null) and recharge areas close to the uppermost altitude of the area. It is worth mentioning that spring 14, related to recharge altitude of 1200 m.a.s.l., higher than those present in the area ($\cong$ 1000 m.a.s.l.), displays zero tritium content. The expected age is older than 1000 years, although we should rely on carbon isotope data to be more accurate. It is likely that during the period that groundwater evolved from rainfall-infiltration to discharge, erosion of the higher altitudes of the rock mass have occurred. The high mountainous landscape together with the closely Tajo river basin and high energy deposits (Marín et al., 1996), support this hypothesis. Assuming no uplift movement of the pluton, a rough estimation of erosion rate would be: 200 m/$\geq$ 1000 yr gives $\leq$ 20 mm/year. Local estimates of erosion rate by means of classic geomorfological approaches (Fournier's climate factor) give erosion rates on the order of 2 mm/yr (Guerra and López-Vera, 1985). Thus, a broad agreement between the two figures exist. A recharge age of 5000 to 10000 years, would fall in a cold period of the area (BRGM, 1994) which agrees with the light - negative - content of $\delta^{18}O$.

Springs located at the middle-slope of the pluton deserve special analysis. Groundwater flow is driven by topography, although contrasts in permeability refract and mix flow lines. That would be the case of springs located along one fracture filled with quartz (#9, 10, 11 and 12, see Figure 3), where an apparent increase of the mineralization based on the electrical conductivity, would favor the hypothesis of fluid flow along the fracture. However, detailed analysis of tritium content and total mineralization shows that compositions are too heterogeneous to maintain such hypothesis. That part of the batholith is characterized by a relatively high number of crossed faults. Contrasts in permeability together with topography, favor mixing, thus resulting in a heterogeneous distribution of water composition. Figure 8 illustrate the conceptual model for such mixing process, where water of different depths (A, B, C) and different compositions are mixed by the presence of a high permeable fracture. An apparent contradiction arise from Figure 6, that shows that spring 12, of low tritium content has a difference of recharge-discharge altitude comparable to the rest of the springs of the area. While this should not favor the differences observed, we should keep in mind the heterogeneity of the rock mass. Not all flow paths occur within the same flow velocity; that is, infiltration on a certain area takes place preferentially along pervious fractures, but also through the rock mass. Indeed, Motyka and Zuber (1994) contended that matrix diffusion, upscales in the sense that it becomes more dominant within the volume of affected rock mass. Solutes are retarded by a factor of porosity: $(\phi_f + \phi_M)/\phi_f$. being the subscripts, f for fractures and M for rock matrix. Then, if matrix porosity is not negligible - which is not at regional scale - anomalies such as that of spring 12 are likely to occur. The more significant the difference between fracture and matrix porosity, the more important the anomalies. Motyka and Zuber (1994) showed that the discrepancies in groundwater dating in sedimentary aquifers are more significant, yet matrix displays high porosity values. Varni et al (1994) and Carrera et al (1995) provide further discussion on the effect of mixing processes on groundwater age.

5. CONCLUSIONS

Classical techniques of groundwater analysis, proved to be of great help even in highly heterogeneous media. The study of groundwater composition at regional scale (5 km x 5 km) was achieved by simple techniques of groundwater sampling, analysis and dating. The conceptual model thus produced is consistent with observations based on topography, hydraulic heads and ephemeral springs: groundwater recharge takes place mainly in areas of high altitude and discharge

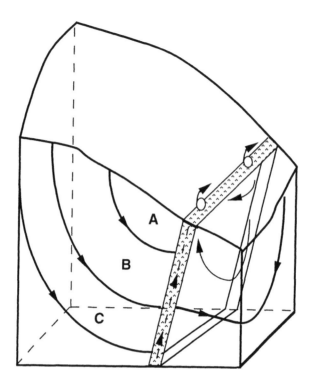

Figure 8. Change in groundwater composition formed by mixing of water of different origin A, B, C along and through a permeable fracture zone.

at lower boundaries of the system. Flow paths are not homogeneously distributed and a range of recharge-discharge probabilities exist due to the presence of pervious faults.

A contribution of this study, yet to be validated with more independent information, is the correlation of water age recharge altitude and inferred erosion rate.

It has been evidenced the role of slow flow paths, likely to occur though the rock mass, rather than pervious fractures. Numerical models simulating large scale flow, must take into account this contribution of the regional hydrogeology.

ACKNOWLEDGEMENTS

El Berrocal project was funded by ENRESA (Spain) and partially by the EC under contract FI2W/CT91/0080. Authors express their gratitude to those individuals that made possible field laboratory work. Special thanks are devoted to Pedro Rivas (CIEMAT, Spain) and J. Carrera (UPC, Spain) for encouraging the authors (and all participants) along the project. C. Marín (CIEMAT) contributed to the discussion of local Quaternary processes. Prof. E. Custodio (UPC, Spain) contributed to the discussion of stable isotopes interpretation out of the frame of the project.

REFERENCES

Araguás, L. (1991) Adquisición de los contenidos isotópicos (^{18}O y D) de las aguas subterráneas: variaciones en la atmósfera y en la zona no saturada del suelo. Ph. D. Dissertation, Univ. Autónoma. Madrid.

Bredehoeft, J.D.; Belitz, K. and Sharp-Hansen, S. (1992) The hydrodynamics of the Big Horn basin: the study of the role of faults. *AAPG Bull.* 76/4:530-546

BRGM (1994) Western Europe and Iberian peninsula from -120000 years to present. Glacial isostasy, paleogeography and paleotemperatures. ENRESA tech. rep. 05/94.

Carrera, J.; Varni, M. and Saaltink, M.W. (1995) Una ecuación en derivadas parciales para la edad del agua subterránea. Hidrogeología y Rec. Hidráulicos (XIX) 325-336. VI Simposio de Hidrogeología. Sevilla, Spain.

Gómez, P. and 13 more authors (1996) Hydrochemical and isotopic characterization of the groundwater from the El Berrocal site (Spain), Topical report 4, Vol. II, El Berrocal project, ENRESA, Spain.

Guerra, J. and López-Vera, F. (1985) Análisis y aplicación del factor climático de Fournier a la estimación de la erosión específica en la región de Madrid. *Cuad. Inv. Geograf. I Col. sobre Proc. Act. en Geomorf.* vol. XI (1-2):149-159

Marín, C.; Campos, R.; Pérez del Villar, L. and Pardillo, J. (1996) Geology of El Berrocal site (Spain). Topical report 1, Vol. I, El Berrocal project, ENRESA, Spain.

Norasis, M.J. (1994) *SPSS Advanced Statistics 6.1.* Spss Inc., USA.

Seyhan, E.; van de Griend, A.A. and Engelen, G.B. (1985) Multivariate analysis and interpretation of the hydrochemistry of a dolomitic reef aquifer, N. Italy. *Water Resour. Res.,* 21(7) 1010-1024.

Siegel, M.D.; Lambert, S.J. and Robinson, K.L. (1991) Hydrochemical studies of the Rustler Formation and related rocks at the WIPP area, SE New Mexico. *Sandia Nat. Lab. Rep.* SAND88-0196.

Plata, A. (1995) Tritio de origen termonuclear en las aguas de El Berrocal. CEDEX, 51-493-1-010, Madrid. (Unpub. tech. report).

Varni, M.; Saaltink, M.W. and Carrera, J. (1994) Aplicación de campos de velocidades continuas al cálculo de edades cinemáticas de las aguas subterráneas. *IAH Congress (Spanish Group), Análisis y evolución de la contaminación de las aguas subterráneas,* vol. 1, 279-294. Alcalá de Henares, Madrid.

Yurtsever, Y. and Gat, J.R. (1981) Atmospheric water. In "Stable isotope hydrology: D and [18]O in the water cycle". Vienna, IAEA, pp. 103-142.

Zuber, A. and J. Motyka (1994) Matrix porosity as the most important parameter of fissured rocks for solute transport at large scales. *J. Hydrol.,* 158:19-46

Tracer Hydrology 97, Kranjc (ed.) © 1997 Balkema, Rotterdam, ISBN 90 5410 875 4

Properties of underground water flow in karst area near Lunan in Yunnan Province, China

Janja Kogovšek & Metka Petrič
Karst Research Institute ZRC SAZU, Postojna, Slovenia

Hong Liu
Yunnan Institute of Geography, Kunming, People's Republic of China

ABSTRACT: To solve the problems of water shortage in the wider karst area of the Lunan Stone Forest in Yunnan Province, China several hydrogeological researches were planned. In this paper the results of a tracing test in the karst aquifer near the Tianshengan village are presented. Hydrogeological properties of the region were determined by existing geological map and by additional field observations. Basing on karst cave and passage explorations that are accessible from the surface we assumed the directions and basic properties of the underground water flow. Precipitations and recharge regime were evaluated by daily data of precipitations of the area and by discharges at underground gauge station from August 8, 1993 to August 21, 1996. Also the basic physico-chemical analyses of water were done for 16 sites and they indicated possible connections of observed waters. These analyses helped to select the injection and sampling points and to define the frequency of sampling at given sites. In July 19, 1996 at medium water level we injected in two different injection points 2 kg of Uranine and 100 kg of NaCl and started to sample at 9 points at an interval from 2 hours to up to once per day at the end of observations. Sampling and discharge measurements (every 2 hours) at gauging station Dakenyan enable to calculate the returned quantity of tracer at this point. The described water tracing test gave us some basic information regarding the properties of the underground water in the treated area but it also raised a series of interesting questions. Additional investigations are suggested to get some of the answers.

1 INTRODUCTION

The Institute of Geography from Kunming in Yunnan Province, China and the Karst Research Institute ZRC SAZU from Postojna, Slovenia have been working for two years already on the common project "Karst Phenomena's Preservation, Protection and Large Cave System's Exploration in Yunnan Province". One study polygon of this project is the karst area around the village Tianshengan, which is located in the north-eastern part of the Lunan county in Yunnan province (Fig. 1). In this predominantly agricultural area problems of water shortage are frequent. This typical karst aquifer has plenty of underground water, but its time distribution is disadvantageous. A rainy period from May to October is followed by a long dry period, in which the water reserves are often not large enough for the irrigation of the mainly rice fields. Therefore the agricultural production of the area is very low; problems with water supply for the villages are also frequent.

To resolve this problem, different surface and underground water accumulations are created. To the west and east of Tianshengan two reservoirs are already used for this purpose, but their capacities are too small. Therefore the possibility of constructing a new reservoir in this area has to be studied and some basic hydrogeological conditions must first be delineated.

As one of the research methods a tracing test was proposed in order to find out underground water flow characteristics in the studied karst aquifer. As part of the common Chinese-Slovene project this was done in July and August 1996 and the results are presented in this article. It was the first such experiment in this area, so our goal was predominantly to define the main underground flow and to get some basic information required for planning a more complex tracing and other tests in future.

2 NATURAL BACKGROUND

2.1 Geological conditions

The polygon for the tracing test covers of 25.5 km² and is located in a broader area of the Lunan Stone Forest (Fig. 1). The altitudes range between 1920 m in the east to 1750 m in the west. The relief undulates very gently. On the slopes of depressions and hills,

stone teeth are developed, from 0.5 to 5 m high. Also, pinnacles more than 20 m high can be seen in this area near the Dakenyan.

The eastern border of the studied karst area consists of Precambrian grey slate, sandstone and tuff of the Niutoushan group and white-grey quartz sandstones of the Doushantou group. Carbonate rocks of Carboniferous and Permian age in the central part of the area outcrop as narrow belts in the north - south direction. They dip towards the west

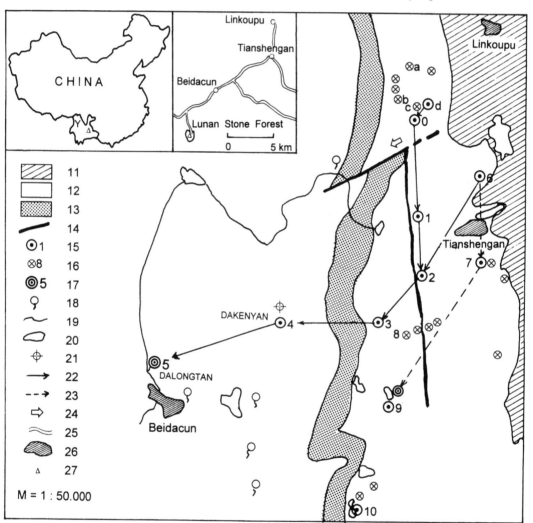

Fig.1: Hydrogeological map
11. Precambrian noncarbonate rocks, 12. Carboniferous and Permian carbonate rocks, 13. Lower Permian clastic rocks, 14. fault, 15. cave or shaft with underground water flow - a sampling point, 16. cave or shaft with underground water flow, 17. karst spring - a sampling point, 18. small spring, 19. surface stream, 20. lake or water reservoar, 21. precipitation station, 22. on the base of tracing test proved direction of underground water flow, 23. on the base of tracing test supposed direction of underground water flow, 24. possible direction of underground water flow, 25. road, 26. village, 27. Lunan Stone Forest

with a dip angle between 7° and 15°. The oldest are Lower Carboniferous white-grey limestones and oolitic limestones in the eastern part. Towards the west they concordantly pass to younger rocks. The most important for our research are well karstified Middle Carboniferous grey limestone and oolitic limestone and Upper Carboniferous grey and red limestone, where large underground channels have developed. Carboniferous and Permian carbonate rocks are separated by a 20-30 m thick layer of quartz sandstone and shale claystones. Also massive or thick-bedded limestones, dolomites and dolomitic limestones in the western side of the non-carbonate barrier are of the Lower Permian age. It was in this part that the stone forest developed. In the studied area two faults are important: Tianshengan fault in the N-S direction and the Shibanshou fault in the NE-SW direction. Several shafts and caves with underground flow, that were observed during the tracing test, are distributed along the Tianshengan fault. (Hydrological team of Yunnan Geological and Minerals Survey Bureau 1979, Yunnan and Guizhou team of Yunnan-Guizhou-Guangxi Petroleum Oil Exploration Company 1964)

2.2 Meteorological and hydrological properties

The climate of the studied area can be classified as subtropical with an average annual precipitation of 796 mm, average relative humidity of 75,3% and average annual temperature of 15.6° C (for the period 1980 - 1992). The amount of total annual rainfall varies considerably and for the observed period between 1980 and 1992 it ranges from 542 mm (1992) to 1066 mm (1991). Each year can be divided into dry and rainy season. The dry season from October to April has only 12 to 20% of total annual precipitation.

There is no surface drainage in the studied area and underground karst channels are well developed. They are not deep below the surface and are accessible through several shafts and caves. At different locations pumps are installed there. Water pumped from the underground stream is used mainly for the irrigation of fields.

The most important spring Dalongtan, is situated in the south-western part of the area at an altitude of 1810 m. As yet there are no exact data about the capacity of this spring. The discharges are measured in the underground karst channel at the Dakenyan pumping station (Fig. 1), which is situated 2.45 km north-eastern of the spring. From the measuring slat

which was installed in August 1993, the water levels are observed by a local farmer four times a day and from these values a mean daily discharge is calculated.

The precipitation station is also at Dakenyan and the daily precipitations were measured for the same period.

Daily values of precipitation and discharges between August 8th, 1993 and August, 21st 1996 are presented in figure 2. The maximum discharge is 6.28 m³/s and the average discharge 0.466 m³/s. In May 1996 the measuring profile in Dakenyan was empty, but it is hard to define the minimum discharge, because in the dry period it is largely controlled by the quantity pumped by different pumping stations. The maximum daily value of precipitation was 75 mm.

These characteristic values were calculated also for two separated hydrological years and great differences were noticed between them. For the hydrological year between May 1st 1994 and May 29th 1995 the total amount of rainfall was 1153 mm, maximum daily precipitation 75 mm, maximum discharge 3.48 m³/s and the average discharge 0.464 m³/s. From May 30th, 1995 to May 16th, 1996 the total amount of rainfall was 732 mm, maximum daily precipitation 45.6 mm, maximum discharge 6.28 m³/s and the average discharge 0.396 m³/s.

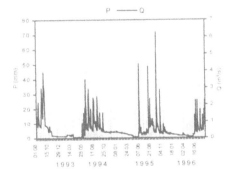

Figure 2: Daily precipitations and discharges at Dakenyan station

The tracing test was planned for July, 1996, so the characteristic values for this month were compared also. The maximum discharges were 2.92 m³/s in July, 1994 and 0.523 m³/s in July, 1995. Although we have the data only for a short period, a great variability of hydrologic conditions can be observed. On the basis of these values the broader interval of expected discharges in the time of tracing test has been predicted and used in constructing the plan of the experiment.

3 TRACING TEST

3.1 Hydrological conditions at the time of the tracing test

After the injection of tracer on July 19th, 1996, the discharges were measured at the Dakenyan station at shorter time intervals than usual: between July 19th and 25th every two hours, and from July 26th to 30th, every four hours. The values obtained are presented in figure 4, together with the concentration curves, and were used for the quantitative analysis of the experiment. At the time of injection the discharge was 0.783 m^3/s. No water for irrigation was pumped at the pumping stations in this area.

The characteristic discharge values for the time of tracer test between July 19th and August 18th, were compared with the values of the two previous years (Tab. 1). On the basis of this comparison the observed period in 1996 can be defined as average; only the extreme maximum discharge was lower than in previous years.

Table 1: Hydrological situation at the time of the tracing test

Time interval	Q_{min}	Q_{mean}	Q_{max}	P_{dmax}	P_{tot}
19.7.-18.8.1996	0.393	0.817	2.24	35.1	160.9
19.7.-18.8.1994	0.426	1.064	2.40	75	271.9
19.7.-18.8.1995	0.214	0.689	4.25	23.3	124.3

At the time of the tracer test daily precipitations were measured also. The total amount (P_{tot}) was 160.9 mm and the maximum daily value (P_{dmax}) 35.1 mm. Comparison with the same period of the years 1994 and 1995 shows the same picture as for the discharges. In 1995 the total amount and the maximum daily precipitation were lower, and in 1994 higher than in the observed period in 1996. Therefore we can characterize this 1996 year as an average without extreme values.

On July 3 1996 10.4 mm of rain fell. The next intensive rain occurred on July 16 in the evening and continued during the night to the next morning. In this time interval 56.6 mm of rain fell. Rain continued to fall the next morning but less heavy as there were only 11 mm of rain. The next rain followed on July 21 when 5.5 mm fell and on July 22 when 26.5 mm fell. Up to the end of July it rained three more times with a few mm of rain only. In the first three days of August 28.1 mm of rain fell, and from July 8 to 11 an additional 38.8 mm in total; heavier rain occurred on August 16 and 17 when 35 and 5.7 mm fell.

The rainfall in the middle of July increased the discharge of underground flow considerably and it reached 1.7 m^3/s at Point 4; the water transported huge amount of soil and was of an intense orange-brown colour. On July 19 when we decided to inject a tracer, the discharge was decreasing and was 0.783 m^3/s. It continued to decrease over the next few days and it reached the initial value not earlier than after the rain of July 22. By the end of August the discharge decreased slowly and after the rain it increased again to 1.23 m^3/s.

3.2 The tracing test

Inspection of the area where we planned the water tracing indicated possible injection points and springs and places where the appearance of the tracer would be probable. We measured temperature, SEC and pH and took blind samples and samples for determination of carbonate, calcium and magnesium levels.

On a basis of these first measurements and analyses we assumed that the water flows from Point d (Quighuadong) towards Point 0 (Yanshidong) and not in a direction towards Point c as had been supposed earlier (Fig. 1). The analyses suggested that water flows from the Point 0 towards Points 1 (Maoshuidong), 2 (Shihuiyao), 3 (Xiangshuidong) and 4 (Dakenyan) up to the Dalongtan spring at Point 5. We assumed that water at Points 6 (Wayaodong), 7 (Guanyindong) and the springs 9 (Changshuitang) and 10 (Xiniutang) had a different origin but we did not exclude the probability of connection.

Possible injection points could be points a, b, c and d; as the most favourable conditions existed at point d where discharge was approximately 30 l/s we chose that one. The second injection point was Point 6 which is a stream in the Wayaodong cave, having a discharge of 5 l/s.

3.2.1 Injection of tracers and water sampling

On July 19 1996 from 10.35 to 10.40 p.m. we injected 2 kg of Uranine that was dissolved in 50 l of water into a stream at the Point d. On the same day from 12.30 to 12.40 we also injected 100 kg of dissolved NaCl at Point 6.

Referring to the previous analyses of rainfall and discharge we made a sampling plan. Point 4 was sampled in most detail. For the first 7 days we sampled every 2 hours, later every 4 hours and then every 6 hours then every 12 hours and after 15 days once per day.

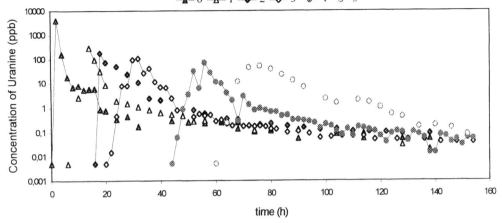

Figure 3: Breakthrough curves of Uranine at the points 0, 1, 2, 3, 4 and 5

At Points 0 and 3 we started sampling every 2 hours and at Point 1 and 2 every 4 hours. To define precisely the maximal concentrations of Uranine and the time when it appeared we should need slightly more frequent sampling at Points 1 and 2. We also sampled at Points 5, 7, 9 and 10.

3.2.2. *The results of Uranine tracing*

Uranine appeared at consecutive points 0, 1, 3, 4, and 5 as shown in figure 3. At other observed points 7, 9 and 10 Uranine was not detected and we may exclude the possibility of bad connection with these points as we used relatively great quantities of tracer.

Table 2. Air distance (d), time of tracer's travel in respect to maximal tracer's concentration (t) and dominant velocity (v) of Uranine and NaCl between observed points

Points	URANINE				NaCl		
	d m	t h	v m/h	v cm/s	t h	v m/h	v cm/s
d-0	500	0.6	830	23			
0-1	1800	11	164	4.5			
1-2	1050	8	131	3.6			
2-3	1250	16	78	2.2	16	78	2.2
3-4	1850	24	77	2.1	24	77	2.1
4-5	2450	20	123	3.4	22	111	3.1
6-2	2000				17	118	3.3
6-7	1500				13.5	111	3.1
6-9	4050				200	21	0.58
d-5	8900	80	112	3.1			

The flow velocity for single sections of central runoff towards the Dalongtan spring (5) were calculated from air distance and time when maximal

concentrations of Uranine appeared at single points (dominant velocities). Real velocities thus might be higher. Measured and calculated values are presented in the table 2.

The fastest water flow appeared in the initial part of its route between the Points d and 0; the slowest was between points 2 and 4 where it passes over impermeable rocks. After this section its velocity rises up to spring 5 again (Fig. 1). The average velocity between injection point d and the Dalongtan spring (5) is 3.1 cm/s. A tracing experiment in Wulichong subsurface drainage system in Mengzi County, Yunnan shows the average velocity between 0.6 and 49 cm/s. Water tracings in the Slovene karst (Habič et al. 1990, Novak 1991, Habič & Kogovšek 1992), when the tracer was injected directly into the water flow, indicated flow velocities between 0.2 and several cm/s, and in rare cases only the flow velocity reached values as high as 10 cm/s.

We studied more in detail the Uranine appearance at point 4 (Dakenyan) where every 2 hours we also measured discharge and daily rainfall. Figure 4 shows the breakthrough curve of Uranine and relative recovery. By the end of July the Uranine concentration decreased to 0.01 mg/m^3. Heavier rainfall at the beginning of August increased the drainage and washed out the remained Uranine. The Uranine concentration rose to 0.070 mg/m^3, yet it decreased by the middle of August to 0.01 mg/m^3. By the middle of August, namely one month since the injection, the returned quantity of Uranine was more than 1100 g, more than 55 %. Some of remaining tracer was probably washed off later after each rainfall but we did not continue sampling.

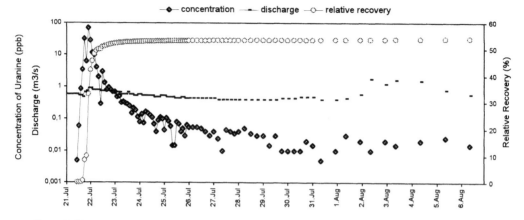

Figure 4: Discharge, breakthrough curve of Uranine and relative recovery of Uranine at Dakenyan (point 4)

3.2.3 *The results of tracing by NaCl*

As a second tracer the only one available was NaCl. Injection at point 6 and observations at the above-mentioned springs showed the water flow to be towards point 2 and towards spring 5 as already assessed by Uranine (Fig. 5). Dominant flow velocity up to point 2 was 3.3 cm/s; other velocities to the points 3, 4 and 5 were practically the same as those given by Uranine.

point 9 is very questionable as there the flow velocity would be only 0,6 cm/s. More frequent sampling or a greater amount of tracer, or a more suitable tracer would probably give clearer picture. It is most probable that water flows from point 6 in two directions, towards points 2 and 7; this gives a higher dilution effect and consequently lower concentrations which must be considered in future. These results indicate a need to repeat the water tracing from point 6 with a more suitable tracer.

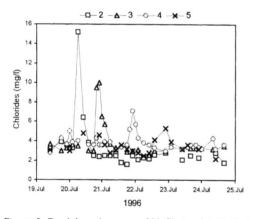

Figure 5: Breakthrough curves of NaCl at points 2, 3, 4 and 5.

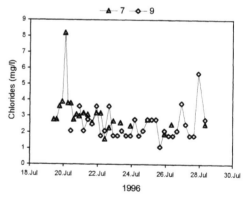

Figure 6: Breakthrough curves of NaCl at points 7 and 9.

At points 7 and 9 we sometimes noticed increased chloride levels (Fig. 6) but only on a few occasions so we cannot state that these points are connected. We may conclude that drainage from point 6 towards point 7 is very probable as the flow velocity obtained from the results would be 3.1 cm/s; drainage towards

No increase was noticed at point 10. Even from the physical measurements and chemical analyses obtained we cannot suggest possible connections which could be used for future tracing tests. But we noticed the very different reactions to rainfall of the discharges at springs 9 and 10.

4. CONCLUSIONS

In the studied area the underground karst channels are well developed not deep below the surface. They may be observed at various locations where shafts and caves reach the underground flow and they are proved by high flow velocity which was established by water tracing during medium high water level. When discharge is high, drainage may be still faster which is an important fact due to be considered for eventual pollution. The flow velocity slightly diminishes when water reaches the non-carbonate barrier. Stated velocities predominantly vary between 2.1 and 4.5 cm/s, the average velocity for the whole section is 3.1 cm/s, and they correspond to values given for underground waters of southern China by Yuan Daoxian et al. (1991) and Song Linhua et al. (1993): medium values of more mm/s and maximal from several cm/s to more ten cm/s.

In spite of well-developed underground passages, fast drainage and fact that we injected a tracer directly into the underground flow, the returned quantity of tracer is relatively low. Uranine adsorption into soil studied by laboratory experiments did not show any possibility of an important tracer adsorption. This is why we admit a possibility that a part of underground flow was not included in the sampling after tracing test. It seems probable that a part of water flows along the Shibanshou fault through a barrier of non-carbonate rocks and feeds smaller springs or superficial flow on its western side. The maximal decrease of tracer concentration was noticed between the points 0 and 1. Similar conclusion indicate the stated main directions of underground flow N-S and E-W and comparison with underground flow between points 3 and 4 (Fig. 1).

Comparing the discharges at points 4 and 5, which were due to changeable hydrologic conditions only estimated, we noticed considerable increase in discharge of the Dalongtan spring. We assume that the observed underground flow from NE direction is not the only feeding source of this spring.

As already mentioned, the described water tracing test gave us only some basic information regarding the properties of the underground water in the treated area but it also raised a series of interesting questions. Let us quote only some of them: does the water drain along the Shibanshou fault and if this is the case, what is its rate? What role does the flow passing points 6 and 7 play and where is its continuation? Is there any connection between springs at points 9 and 10 in spite of our negative results? The answers may be obtained by new investigations which are already planned by Chinese experts for the future.

REFERENCES

Yuan Daoxian, Zhu Dehao, Weng Jintao, Zhu Xuewen, Han Xingrui, Wang Xunyi, Cai Guihong, Zhu Yuanfeng, Cui Guangzhong & Deng Ziqiang 1991. *Karst of China*. Geological Publishing House: 1-224. Beijing.

Habič, P., J. Kogovšek, M. Bricelj & M. Zupan 1990. Dobličica Springs and their wider Karst Background. *Acta carsologica*, 19: 5-100. Ljubljana.

Habič, P. & J. Kogovšek 1992. Water Tracing in Krupa Karst Catchment, SE Slovenia. *Acta carsologica*, 21: 35-76. Ljubljana.

Hydrological team of Yunnan Geological and Minerals Survey Bureau 1979. *A General Survey Report on Farmland Irrigation Hydrogeology in Lunan Shilin Area*. Unpublished report.

Novak, D. 1991. Recent Tracing of Karstic Waters in Slovenia. *Geologija*, 33: 461-478. Ljubljana.

Song Linhua, He Yueben & Feng Yan 1993. Ground water tracing in Wulichong surface drainage system mengzi county, Yunnan province. *Proc. of 11th International Congress of Speleology*: 227-229. Beijing.

Yunnan and Guizhou team of Yunnan-Guizhou-Guangxi Petroleum Oil Exploration Company 1964. *The Synthetical Report on the Structures of Stoneforest in Yiliang County; Yunnan Province, China*. Unpublished report.

Tracer Hydrology 97, Kranjc (ed.) © 1997 Balkema, Rotterdam, ISBN 90 5410 875 4

Preliminary results of the submarine outfall survey near Piran (northern Adriatic sea)

V. Malačič & A. Vukovič
National Institute of Biology, Marine Biological Station Piran, Slovenia

ABSTRACT: During the winter period 1996-1997 two small-scale surveys of two adjacent submarine diffuser outfalls were conducted using a fine-scale CTD probe with mounted *in-situ* fluorometer. Chlorophyll-*a* was the most heterogeneously distributed parameter in the ambient sea, at all depths. Salinity was mostly homogeneous. In future studies, which involve fluorescent dye tracing, spatial distribution of chlorophyll-*a* around the diffusers needs to be considered.

1 INTRODUCTION

In the year 1976, a new sewage treatment plant commenced operation. It was designed to discharge sewage from the digestion tanks out into the northern Adriatic, about two miles offshore (Figure 1), through a submerged sewage pipe, of length about 3.5 km and of diameter 0.4 m. A maximum flow rate of 150 l/s was believed to be adequate in accommodating for excess loads (population of 50,000) during the tourist season. The submarine pipe terminates with a 109 m long diffuser which is comprised of 25 regularly spaced, alternating, side outfall orifices of diameter 0.1 m. The diffuser was installed 1m above the sea-floor.

After ten years of operation it was apparent that a second pipe was necessary. In 1987 a new pipe was placed alongside the existing pipe, separated by about 150 m at the ends. The pipe is of diameter 0.6 m, and its diffuser of length 185 m, is composed of two parts. The inner section is also of diameter 0.6 m, and has 11 side orifices of diameter 0.1 m, while the outer section is of diameter 0.4 m and contains 6 side orifices with an additional tapered-head opening. The diffuser was designed to give a mean, side outflow rate of 19 l/s (per orifice), which corresponds to a maximum pipe outflow rate of 380 l/s.

Studies carried out in 1974/75 and 1978/80 monitored the nutrients, plankton and benthic communities, and even faecal pollution indicators (Avčin et al. 1979, Malej 1980, Faganeli 1982). No significant environmental deterioration, due to sewage disposal, was observed. Further studies of sewage impact on water quality and the ecosystem in the Bay of Piran and the Gulf of Trieste have not been udertaken until recently.

2 METHODS

The diffuser vicinity was studied using a fine-scale CTD probe, designed by the University of Western Australia. Besides the standard parameters (temperature, conductivity and pressure), oxygen and *in-situ* fluorescence was also measured. The vertical profiles of these quantities were collected during the free-fall of the probe, at a speed of about 1m/s. The data was retrieved at a frequency of 50 Hz, providing a vertical resolution of 2.5 cm. An area of 0.4 miles x 0.4 miles around the diffusers was monitored in under a two hour period with a distance between stations of about 200 m.

Scuba-diving was necessary in order to locate the exact position of the diffusers. Underwater photographs of the sea floor ecology, surrounding the second diffuser, were examined.

3 SEA-FLOOR SURROUNDING THE DIFFUSER

At the beginning of October 1996, the sea-floor towards the end of the second diffuser was teeming with flora and fauna. In Figure 2, sewage emanating from the left-hand side orifice can be seen. A polychaet *Spirographis spalanzani*, close to the

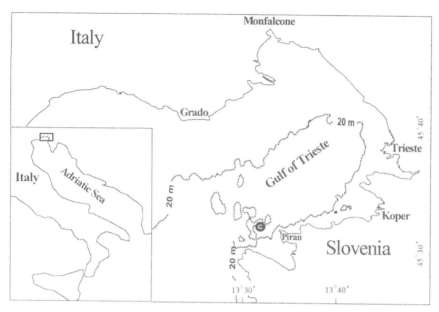

Figure 1. Map of sewage diffusers located offshore from the tip of Piran (thick circle).

Figure 2. The polychaet *Spirographis spalanzani* settles near the left-side diffuser orifice, from which sewage emerges.

orifice, is not disturbed by the sewage. Figures 3, 4 and 5 demonstrates abundant benthic biocenosis around the diffuser- an otherwise hard and barren detrital sea-floor.

4 DISTRIBUTION OF CHLOROPHYLL

On the 10th October 1996 the total sewage outflow rate (from 11:30 to 12:30) was 162.6 l/s. This is separated into two streams: 5/9 flows through the new submarine pipe, and 4/9 flows through the old.

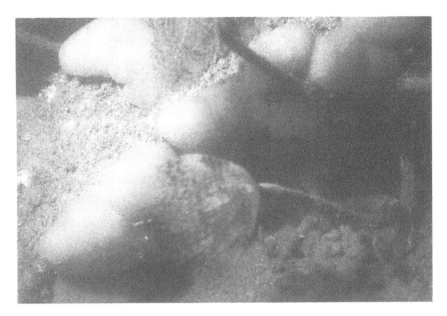

Figure 3. Marine organisms anchored to the diffuser wall, exploiting the nutrient-rich environment.

Figure 4. The sea urchin normally inhabits unpolluted areas; their settlement around the diffuser indicates that the water is unspoiled by sewage.

During the survey, the tidal outflow from the Gulf of Trieste was almost at a maximum. The tidal range was rather high (84 cm), giving a maximum outflow velocity between 6-8 cm/s. Superimposed on the tidal outflow was the wind-driven flow - the "burja" wind, which blew from ENE with a half-hour average speed of around 4 m/s. The wind intensity was at a declining phase; the previous night the speed was around 6 m/s. Salinity and temperature in the upper part of the water column, above 16 m depth, displayed homogeneity. Sea-water in this region was well mixed with the sewage (Figure 6).

265

Figure 5. The demersal fish *Scorpaena sp* revels on fauna which thrives in the diffuser vicinity.

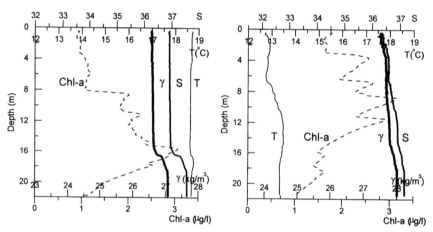

Figure 6. Vertical profiles of chlorophyll-*a*, temperature, salinity and density excess (γ). The left figure is the profile of the 10th October 1996 survey; the right figure is the profile of the 17th December 1996 survey. See Figure 7 for station locations.

Chlorophyll-*a*, however, was gradually increasing with depth, reaching a maximum at the bottom pycnocline (the vertical density jump). Horizontal variations of temperature, salinity, density, and chlorophyll-*a* were examined statistically at depths of 0.5m, 5m, 10m, 15m, and 20 m. It appears that temperature varied at most by 0.1 °C; salinity varied by 0.04, except at 15m where the variation reached a value of 0.4. Density excess varied horizontally by less than 0.1 kg/m^3, except at a depth of 15 m,

where the variation reached a peak value of 0.3 kg/m^3. On the contrary, the chlorophyll-*a* concentration varied greatly, with variations ranging between 1-2 μg/l. The horizontal and vertical variations of chlorophyll-*a* were of an order of magnitude equal to the mean value. Figure 7 illustrates that southwards of the diffusers, concentrations of chlorophyll-*a* are lower, falling to a value of zero.

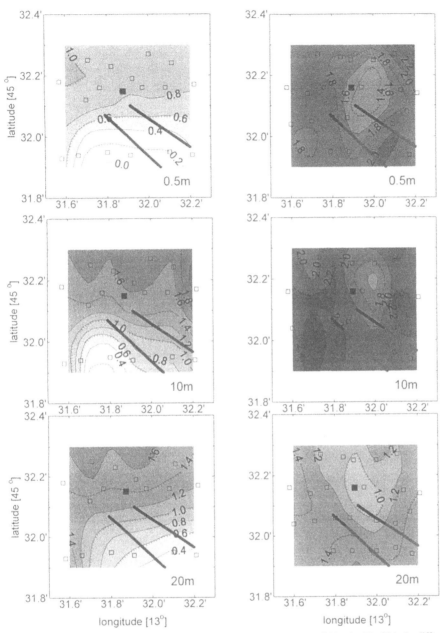

Figure 7. The distribution of chlorophyll-*a* (μg/l) at three levels (0.5m, 10 m and 20m depth) within the diffuser vicinity. Figures on the left are of the survey performed on 10th October 1996; figures on the right on 17th December 1996. Black lines are the diffusers; gray lines are the sewage pipes; empty rectangles are the profiling stations; full rectangle is the station in which the vertical profiling is described.

During the second survey, conducted on 17th December 1996, the sewage outflow rate was higher (207.5 l/s), tides were at a local minimum (zero tidal flow), and the tidal range was rather low (11 cm).

The weather was relatively calm on this day. The chlorophyll-*a* minimum was centered in front of the second diffuser (Figure 7). Concentrations were higher in the upper part of the water column (Figure

6), which was a little more stratified than in the previous survey. At the bottom pycnocline the density jump was less intense. The horizontal variations of chlorophyll-*a* were again the same order of magnitude as the mean value, while other quantities changed only by a few percent with respect to their mean values.

5 CONCLUSIONS

The two preliminary surveys of temperature, salinity and chlorophyll-*a*, in a weakly stratified environment, exhibit different horizontal distributions of chlorophyll-*a*, while the temperature and salinity (therefore also density) were almost horizontally homogeneous. Winds and tides which prevailed during the first survey, had a marked influence on chlorophyll-*a* distribution and growth. Mixing of the sea-water with sewage was enhanced, whilst the abundance of chlorophyll-*a* was low. Further studies are necessary to explain the chlorophyll-*a* horizontal distribution and should be performed before any dispersion experiments of sewage, using fluorescent tracer dyes (Hale 1971), are conducted.

REFERENCES

Avčin, A., B. Vrišer & A. Vukovič 1979. Ecosystem modifications around the submarine sewage outfall from Piran sewage system. *Slovensko morje in zaledje.* 2/3:295-299 (*In Slov.*).
Faganeli J.; 1982: Nutrient dynamics in seawater column in the vicinity of Piran submarine sewage outfall (North Adriatic). *Mar. Poll. Bull.* 13(2):61-66.
Hale A.M.; 1971: The feasibility of using continuous dye injection for underwater flow visualization. *Limnol. Oceanogr.* 16(1):124-129.
Malej A.; 1980: Effects of Piran underwater sewage outfall upon surrounding coastal ecosystem (North Adriatic). *Journées Étud. Pollutions* 5:743-748. Cagliari: C.I.E.S.M.

Tracer Hydrology 97, Kranjc (ed.) © 1997 Balkema, Rotterdam, ISBN 90 5410 875 4

Separation of groundwater-flow components in a karstified aquifer using environmental tracers

R. Nativ
The Hebrew University of Jerusalem, Israel

G. Günay & L. Tezcan
Hacettepe University, UKAM, Ankara, Turkey

H. Hötzl & B. Reichert
University of Karlsruhe, Germany

K. Solomon
The University of Utah, Salt Lake City, Utah, USA

ABSTRACT: In an attempt to identify the hydraulic connections among the various outlet points in the Travertine Plateau, southern Turkey, groundwater was analyzed for stable and radioactive isotopes, CFCs and helium. The upgradient springs, belonging to the Kirkgozler system were proved to be a mixture of recent and older water on the basis of their low ^{14}C values and their measurable tritium and CFCs. Downgradient springs discharging along the Mediterranean coast contain contributions from the Kirkgozler system, but a larger proportion of the recent water component that could be contributed from direct precipitation on the travertine, and recharge from watersheds to the east and west of the Travertine Plateau. This larger portion is evidenced by the water's increased tritium content, enriched oxygen-18 and deuterium values (suggesting a lower recharge altitude), a decreased ^{14}C content, atmospheric helium and CFCs contamination.

1 SCOPE

The Antalya Travertine Plateau is located on the Mediterranean coast of Turkey (Fig. 1). The plateau covers ~615 km^2 and the average thickness of the travertine deposits is about 300 m. The stepwise plateau is comprised of three levels. The upper step of the plateau (~300 m asl) borders the foothills of the carbonate rock masses (Mesozoic) of the Beydaglari Mountains which belong to the western Taurus Range. The City of Antalya is located on the Middle (also called Varsak) Plateau (~40-150 m asl). The Lower Plateau is submerged beneath the Mediterranean Sea and its extent is unknown.

Large volumes of groundwater discharge from the intensively karstified Taurus Mountains, and flow on and through the Travertine Plateau to the Mediterranean Sea. The Kirkgozler Springs form a major discharge point (~15 L/s) along the contact zone between the mountains and the Upper Travertine Plateau (Fig. 1). Some of the discharged water is channeled for downstream irrigation and hydroelectric-power generation. Another fraction of the discharged water at the Kirkgozler Springs back-percolates into the highly porous travertine, and joins two other groundwater components: (1) the deeper groundwater from the Taurus Mountains which bypasses the Kirkgozler Springs, by laterally seeping into the travertine; and (2) the travertine

groundwater, recharged directly by precipitation and runoff on the Travertine Plateau. Some of the mixed groundwater flowing through the travertine discharges at the Dudenbasi Springs (18 L/s) located on the Upper Plateau, at the boundary between the Upper and Middle plateaus (Fig. 1). The water continues from the spring to the Mediterranean Sea along the Duden River. Another fraction is assumed to flow along a deeper flow path, discharging directly into the Mediterranean Sea via a series of unnamed submarine springs (of unknown discharges) near Antalya in the submerged Lower Plateau (Fig. 1).

A recent plan to expand the irrigated areas on the Upper Plateau near the Taurus Mountains by using more of the naturally discharged water at the Kirkgozler Springs is currently being evaluated. If the volume of the back-percolating water from the springs into the travertine is significant, the increased water use upstream will reduce the volume of water available downstream in the Middle Plateau for power generation, a clearly undesirable result. The plan may, on the other hand, result in the redemption of some of the lost water which currently discharges from the Lower Plateau, feeding the submarine springs at the Mediterranean Sea's bottom.

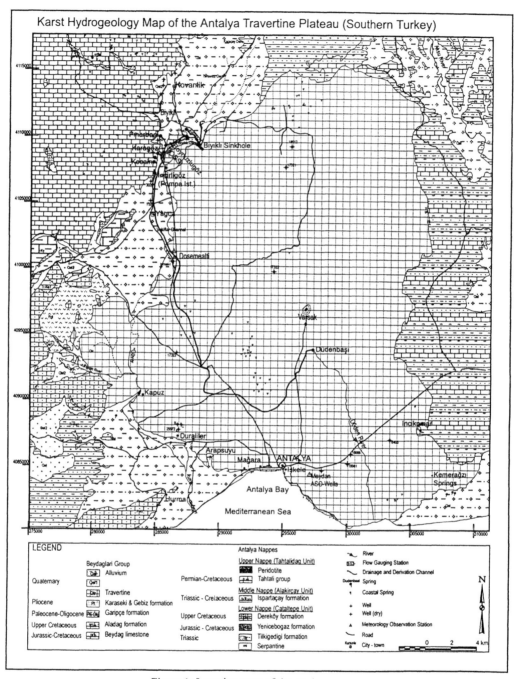

Figure 1: Location map of the study area.

2 METHODS

In an attempt to assess the relative volumes of each groundwater component in the travertine, environmental tracers that might differently tag the various components were recruited. The working hypothesis was that the high Taurus Mountains should discharge groundwater depleted in $\delta^{18}O$ and δD into the Travertine Plateau, whereas the locally

270

recharged groundwater in the travertine should be isotopically heavier, due to the lower recharge altitude. Because of the longer route, residence time of the groundwater (reflected by tritium, CFCs or carbon-14 content) recharged in the Taurus Mountains should be longer than that of the locally recharged water.

Monthly precipitation samples were collected during 1995 and 1996 at the Antalya and Dosemealti weather stations located in the Middle and the Upper plateaus, respectively (Fig. 1). Samples from 12 springs, two sinkholes and one well were collected in the travertine plateaus. Their spatial distribution was as follows: Five of the Kirkgozler Springs, discharging at the contact zone between the Taurus Mountains and the Upper Travertine Plateau; two sinkholes in the Upper Plateau, along the flow path from the Kirkgozler Springs to the Dudenbasi Spring; the Dudenbasi Spring on the Upper Plateau, at the boundary between the Upper and Middle plateaus; and six springs and one borehole, discharging in the Middle Plateau (Fig. 1).

Samples from most groundwater sampling points were collected in April, July, September and December of 1995, to reflect seasonal variations. Each sample was analyzed for oxygen-18, deuterium and tritium. At the end of September 1995, during base-flow conditions and prior to the winter rainy season, 11 groundwater samples from the 15 aforementioned sampling points were also collected for carbon-14 and chlorofluorocarbons (CFCs). Of these 11 samples taken for CFCs, five were also analyzed for helium, nitrogen, neon and argon.

3 RESULTS

The spatial distribution of the tracers' concentrations was heterogeneous, as expected from a karstified aquifer, where flow paths are complex, variable in size, and allow differential flow rates, flushing and dilution. Consequently, groundwater is a mixture of recent and older water components, as displayed by the measurable tritium and CFCs (1.9-8.4 TU and 0.84-3.27 pmoles/kg, respectively) and low percentage of modern carbon (12-64 pmc) documented in the water samples.

Springs in the Travertine Plateau discharge along the contact zone between the Taurus Mountains and the plateau and along the three steps of the plateau. Because of the >2000 m altitude difference between the mountains and the plateau, springs discharging along the contact zone were hypothesized to reflect the higher recharge altitude and possibly a longer flow path in the mountains. Groundwater in the travertine was expected to display the local conditions of a much lower recharge altitude and shorter flow path. Indeed, the Kirkozler Springs discharging along the contact zone are recharged at a higher elevation than the water in the Middle Plateau, as deduced from the more depleted oxygen and hydrogen isotopic compositions (Fig. 2, Table 1). However, the observed difference in isotopic compositions (< 1 and 5 ‰ for oxygen-18 and deuterium, respectively) was surprisingly low, suggesting that the difference in recharge altitude is not 2000 m, as hypothesized, but smaller. The direct discharge of higher-elevation groundwater in the springs of the Middle Plateau (bypassing the discharge zone of the Kirkgozler Springs) could lead to the observed depleted isotopic composition and explain also the small differences between the groundwater in the upper and middle plateaus. However, this bypass cannot explain the relatively enriched isotopic composition characterizing the Kirkgozler Springs. If indeed their recharge altitude is ~2500 m, their $\delta^{18}O$ composition should be more depleted than the observed -7.76 ‰. On the basis of our observations, a lower recharge altitude (1000-1500 m) is suggested for these springs.

Groundwater in the Upper Plateau contains less of the recent water component and more of the older component with respect to groundwater in the Middle Plateau. This is manifested by lower tritium

Table 1: Ranges and mean values of various isotopes in precipitation and groundwater from the Travertine Plateau

	Oxygen-18	Deuterium	Tritium	Carbon-14	Carbon-13	CFC-11	CFC-12	Helium-4	He3/He-4
Antalya rain	-5.07 (-8.43--3.39)	-25.88 (-51.25--9.93)	4.5 (3.8-6.2)						
Dosemealti rain	5.63 (-10.44--3.42)	-30 (-67.08--21.66)	6.9 (5.3-7.8)						
Upper Plateau groundwater	-7.76(-8.81--7.21)	-45.28 (-50.26--39.69)	4.2 (1.9-5.9)	19 (12.1--22.4)	-3.26 (-4.1--2.2)	0.84--3.27	0.48-2.58	429.88, 991.03	2.602, 1.471
Middle Plateau groundwater	-7.09 (-8.24--6.5)	-39.75 (-49.95--32.76)	5.6 (3.4-8.4)	45 (32.7--63.6)	-6.46 (-8.6--4.3)	1.21--12.5	0.68-9.54	42.69, 58.62, 81.93	1.198, 1.079, 1.006

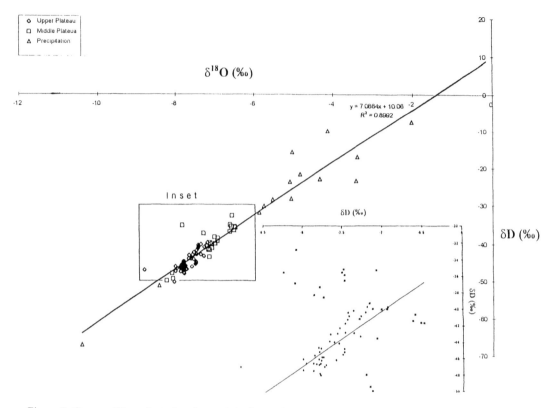

Figure 2: Oxygen-18 vs. deuterium in precipitation and groundwater from the Upper and Middle plateaus.

activity and CFCs concentration, and a lower percentage of modern carbon (Table 1). The latter could result from a long residence time, but also from more intensive dissolution of the carbon-14-devoid carbonate host rocks (suggested by the enriched values of carbon-13, the high HCO_3 and CO_2 concentrations and the frequent undersaturation with respect to calcite) and possible emissions of carbon-14-devoid CO_2 gas, possibly accompanying the observed emissions of mantle helium.

4 CONCLUSIONS
Maximum residence time of the recent water component observed in the Travertine Plateau is ~ 30 years, using tritium and CFCs concentrations as age indicators. Minimum residence time of the older water component is a few thousands years, using carbon-14 as an age indicator. Because the sampled groundwater is a mixture of older and more recent water (and most likely more than two components are involved in this mixture) the recent water could be as young as present-day, and the oldest water could be as old as 15,000 years.

ACKNOWLEDGEMENTS
This study was funded by the European Union AVICIENNE project. Special thanks go to Celal Serdar Bayri, Ugur Ozdemir and Kagan Kuyu from the Hacettepe University, UKAM, Ankara, and to Noam Weisbrod from the Hebrew University of Jerusalem who helped with the groundwater sampling.

Tracer Hydrology 97, Kranjc (ed.) © 1997 Balkema, Rotterdam, ISBN 90 5410 875 4

Drainage basin boundaries of major karst springs in Croatia determined by means of groundwater tracing in their hinterland

Ante Pavičić & Darko Ivičić
Institute of Geology, Zagreb, Croatia

ABSTRACT: The paper deals with complex hydrogeological conditions of a large karst polje, the Vrgoračko Polje, situated in Croatia, in the classical karst of the Dinarides. This polje is the lowest "hydrogeological cascade" to which groundwater flows from hypsometrically higher karst poljes: the Livanjsko, Duvanjsko, Imotsko and Rastok Poljes. The area under study is extremely rich in good-quality spring water of which only a very small part has been used for public water supply.

By pouring tracers into ponors, whose underground flow paths pass relatively close to each other, the existence of separate catchment areas of three major karst springs - Klokun, Modro Oko and (on the Adriatic Sea coast) Žrnovica - has been determined.

1. INTRODUCTION

The southern part of the Adriatic coastal karst belt of Croatia, in the wider area of the Neretva river mouth, has a large amount of still good quality spring water. Several major karst springs of this belt - Klokun, Modro Oko, Prud and Žrnovica - during low water stage, have a total water flow rate of more than 3.5 m3/s. Nowadays, public water supply uses the Klokun spring and some of the water from the Prud spring, but the Modro Oko spring water could be used as well. The coastal Žrnovica spring is under the influence of sea water and, thus, it is only occasionally used for public water supply.

The catchment areas of the major karst springs considered extend up to 60 km in their hinterlands, up to the regional watershed between the Adriatic and Black Sea drainage basins. The groundwater protection problem is related to the cartographic definition of the drainage areas of individual major springs. The remote parts of the watersheds are drained jointly and only the lowest karst poljes are drained separately, each polje to its springs. The Vrgoračko Polje is the lowest "hydrogeological cascade", as Šarin named this type of hydrogeological feature (1983), from where groundwater flows toward the mentioned karst springs. Intense agricultural activity, under the conditions of intermittent karst polje flooding and the existence of a great number of ponors, presents a real danger to water quality in the pumping sites of the Klokun and Modro Oko springs in the Neretva river valley. In order to define the catchment areas and sanitary protection zones of those springs, groundwater flow tracings were made. Within a wide area around the Vrgoračko Polje, a detailed hydrogeological exploration was made. Within it, particular attention was paid to the understanding of its structural geology framework. These detailed explorations made possible the selection of ponors in which the tracer could be injected.

2. HYDROGEOLOGICAL CONDITIONS AND GROUNDWATER FLOW

The largest part of the catchment areas of the considered springs is composed of limestones. Minor areas are built of dolomites and marly limestones. Quaternary deposits consisting of terrigenous sediments are deposited in the depressions of karst poljes.

Within the presentation of hydrogeological

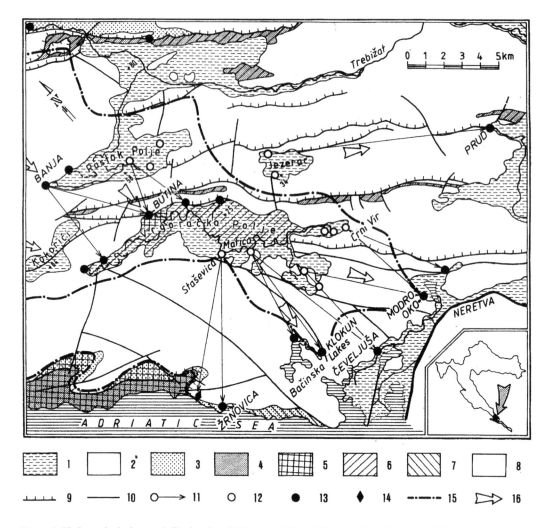

Figure 1. Hydrogeological map; 1- Rocks of variable permeability, 2- Permeable rocks, 3- Partially permeable rocks, 4- Impermeable rocks, 5- True barrier, 6- Partial (hanging) barrier, 7- Rellative barrier, mainly underground, 8- Permeable areas, 9- Reverse fault, overthurst, 10- Fault, 11- Underground hydraulic connection (ponor-spring), proved, 12- Ponor (swallow hole), 13- Karst spring, 14- Brackish spring, 15- Water divide, 16- Assumed direction of groundwater flow.

conditions and groundwater flow in the hydrogeological map (Fig.1), the rocks are separated according their hydrogeological characteristics and the terrain is classified in terms of to its hydrogeological function (after Bojanić et al., 1980).

The largest part of the studied terrain is composed of permeable deposits - limestones, calcareous breccias and alternating limestones and dolomites of Jurassic, Cretaceous and Palaeogene age. Within the tectonically fractured and deformed deposits, by later water activity, open fissures and channels were formed and large amounts of groundwater could flow through them. The thicker deposits of dolomites and/or marly limestones, due to their lesser liability to karstification processes, are shown as partly permeable rocks. Quaternary deposits in karst poljes and in surface stream valleys have alternating characteristics - as far as their permeability is concerned. Their thickness ranges between 1 m and about 10 m; in the Neretva river valley it may even reach more than 50 m. Generally, they are less permeable than their carbonate bedrock.

According to the hydrogeological role of terrains, permeable areas and barriers areas are shown. The permeable area comprises the largest part of the drainage areas concerned. Mostly all the precipitation from this terrain infiltrates into the underground where the entire runoff takes place.

Within the hydrogeological barriers inside the drainage basins considered, the following ones are shown:
- Partial or hanging barriers - formed by flysch deposits of generally small depth and Quaternary deposits in karst poljes. The groundwater flow is aggravated within those deposits resulting in the appearance of intermittent or permanent springs upstream of those barriers.
- Relative barriers - formed by dolomites and marly limestones in anticline cores directing groundwater flow parallel to the structure striking.
- True barriers - formed by Triassic dolomites and Eocene flysch in a part of the Adriatic coast (out of the watersheds of Klokun and Modro Oko springs).

A basic tectonic characteristic of this terrain is its heavy deformation, as shown in its intense folding, faulting and fracturing of deposits and in its extremely imbricated structure. Concerning the water runoff from hypsometrically higher parts of the terrain under study, starting with the Livanjsko and Duvanjsko Poljes, to the Imotsko Polje, the Tihaljina river valley and the Rastok Polje, the Vrgoračko polje is the last hydrogeological step before the final erosion base - the Neretva river valley, i.e. the Adriatic sea level. The entire terrain considerd forms the wider (combined) drainage area of the springs considered. The poljes trend longitudinally along the Dinaridic strike, NW-SE and, in general, have spring zones on their northern and northeastern sides. By groundwater flow tracing, a "cascade runoff" from one polje to the other one in accord to the strike of geological structures was confirmed. However, the model of groundwater flow simplified in this way is valid only occasionally. By tracing, the underground water flow perpendicular to geological structures has been confirmed as well as water flow and hydraulic connections under karst poljes, which mostly have the role of hanging barriers. That is to say, during low water stages, groundwater flows through the carbonate bedrock under the poljes which are filled with Quaternary deposits. This water mainly does not flow out, or does flow from the rare

major karst springs (the Opačac spring on the Imotsko Polje and Butina spring on the Vrgoračko Polje) (Bojanić et al., 1981). During high water stages, because of the low permeability of the carbonate bedrock and decreased permeability of the Quaternary deposits in the poljes, numerous intermittent springs appear, frequently reaching a discharge rate of up to several cubic metres per second.

Surface streams occur only on the karst poljes. In a smaller part of the Vrgoračko Polje, a poorly karstified tectonic block and thick Quaternary deposits form a hydrogeological barrier with which the appearance of the major Butina spring is associated. The remaining part of that polje has the role of either a hanging barrier or a permeable terrain depending on the Quaternary deposits thickness and the karstification degree of the bedrock.

Into the Vrgoračko Polje (24-26 m a.s.l.), water flows from the Rastok Polje (58-62 m a.s.l.) and the Kokorička depression from where flood water sinks through numerous ponors. The Matica river flows through the southern part of the Vrgoračko Polje and into this stream flow the water from intermittent karst springs and from the greatest permanent spring of that polje, the Butina spring. This water flows toward Jezerce and the ponor Crni Vir and much farther, through the tunnel Krotuša, is drained toward the Baćinska Lakes. Without regard for the enormous number of ponors along the stream and in the polje as well as the relatively large sizes of ponor entrances, the permeability of ponors is limited and, in the lowest part of the polje, intermittent water retentions are formed in front of those ponors. The Matica river is a "hanging river" - that is not a rare case where surface streams in karst poljes and out of them in the Dinarides are concerned (Fritz & Pavičić, 1982).

The problem of floods in the Vrgoračko Polje cannot be solved by a partial construction of reclamation channels, dams and the lower part of the surface stream and by the excavation of drainage tunnels. During the rainy season much more water flows into the polje than the ponors and the tunnel to the Baćinska Lakes can drain.

The hydrogeological conditions of the hinterland of the Baćinska Lakes and the Klokun spring are very complex. The deepest "young" karstification occurred by the end of Pleistocene when the depressions, the Vrgoračko Polje and the Baćinska Lakes, where formed. The greatest lake depth is 34 m (33 m b.s.l.). However, the depth of karstification should be greater taking into account where the sea level was in that time. By the Quaternary sea level rise, the hydrogeological conditions were also changed. The deepest "channels" were suffocated, and fine-grained poorly permeable clastic sediments were deposited in the depressions. Better hydraulic connections have been maintained locally, through a karstified carbonate bedrock.

Although a direct hydraulic connection between the Klokun spring and the Baćinska Lakes has not been confirmed, it should exist, at least in a far hinterland. It is assumed that the Klokun spring could be one of groundwater outlets in its flow toward the Adriatic Sea. The "brake" of underground flow, i.e. the flow out of water onto the ground surface, is caused by poorly permeable lacustrine clayey sediments.

3. GROUNDWATER TRACING

Within the permeable area, which comprises the largest part of the terrain considered, detailed determination of catchment areas would not be possible without groundwater tracing. During earlier explorations, the tracings were made in the catchment area of the Vrgoračko Polje and they confirmed groundwater connections between particular karst poljes.

3.1. Catchment area of the Vrgoračko Polje

This area comprises the part of terrain from where groundwater is sporadically drained (from the depression Kokorići and the poljes Rastok and Jezerac) and subsurfacely flows toward the springs on the Vrgoračko Polje and in the Neretva river valley. By tracing of ponors in the central part of the Rastok Polje (Turner, 1955, 1959), a hydraulic connection with the Butina spring and intermittent springs in the central part of the Vrgoračko Polje was determined. By tracing from the karst shaft with water, Velika Banja, which is in fact an intermittent spring used for the public water supply of the town of Vrgorac and which is situated at the northern edge of the Rastok Polje (Bojanić et al., 1980), the hydraulic connection was determined with the Butina spring and the springs in the northwestern part of the Vrgoračko Polje. The results of these tracings are in accord with the results of structural geology studies of the carbonate massif

occurring among the mentioned karst poljes. The underground water connections pass through a subsided tectonic block. The eastern boundary of the considered catchment area was determined on the basis of the hydraulic connection between the Vrcić ponor in the Jezerac Polje and the Prud spring (6.8 cm/s) which passes along a fault zone between two structures (Slišković, 1996).

3.2. Catchment areas of the springs Modro Oko and Klokun

In order to provide sanitary protection for the Klokun and Modro Oko springs, it is necessary to determine their catchment area. A model of water flow within the wide, joint catchment area is described hereinbefore. Concerning the Vrgoračko Polje altitude, earlier explorations showed that the water, which sinks between the extreme northwestern part of the polje and a ponor near the village of Staševica, flows subsurfacely toward the coastal springs Žrnovnica and another near the town of Gradac (Turner, 1959). The water, which sinks through numerous ponors in the southeastern part of the Vrgoračko Polje up to the Crni Vir ponor, flows out from the spring Modro Oko and other springs in the Neretva river valley. By that exploration work, the boundary of the lowest part of the joint catchment area of the Klokun and Modro Oko springs was defined. The tectonic blocks, determined by a structural geology analysis (Fig.2), directed researchers to a possible separation of ponors and ponor zones from which water flows toward one or another spring. Within the boundary area between the mentioned tectonic blocks, groundwater tracing was performed from the ponor Pod Spilom. The tracer was detected in the springs Modro Oko and Čeveljuša but not in the Klokun spring. However, by groundwater tracing from the ponor Crpalo, situated beside the Matica river bed, the underground hydraulic connection of that ponor with the Klokun spring and, again, with the Čeveljuša spring was determined.

In Fig. 3 and 4, where the appearance and concentrations of tracer are shown, one may see an accordance of those curves for the Klokun and Čeveljuša springs taking into account the time lag of about 100 hours. The tracer appearance and concentration curves for the springs Modro Oko and Čeveljuša are in mutual accord only in their final

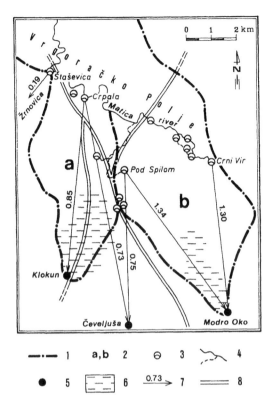

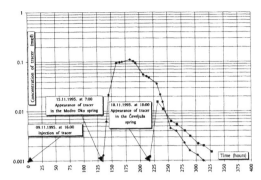

Figure 3. Discharge of tracer from the Modro Oko and Čeveljuša springs

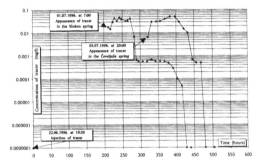

Figure 4. Discharge of tracer from the Klokun and Čeveljuša springs

Figure 2. Water catchment areas of the springs Klokun and Modro Oko; 1- Water devide, 2- Water catchment areas: a) Klokun spring, b) Modro Oko spring, 3- Ponor, 4- Surface stream, 5- Spring, 6- Underground water retention, 7- Underground hydraulic connection and flow velocity in cm/s, 8- Tectonic zone.

parts. These tracings thoroughly showed that, during the hydrological conditions characterised by medium and low water stages, it is possible to destinguish the ponors from which water flows toward the Klokun spring from those from which their water flows toward the Modro Oko spring.

A low apparent groundwater velocity from the ponor to the Klokun spring indicates the existence of underground water retention drained through the springs in which the tracer was found. The impact of tectonics in the groundwater flow is especially visible in the close spring hinterland where a regional transcurrent fault is of a particular significance. It passes along the southwestern edge of the Vrgoračko Polje and extends toward the Neretva river valley (Fig. 1). Because of a regional stress, the tectonic zone extending aside that fault is a compressed space and

has the role of a relative barrier, for the are is weakly karstified.

The northeastern tectonic block (Fig.2), due to its size and position, directs most sinking water from that part of the Vrgoračko Polje as well as groundwater from more distant hinterland toward the springs in the Neretva river valley (Modro Oko and other minor springs). A minor part of the sinking water flows through diagonal open fissures to the southeastern tectonic block, toward the Klokun and Čeveljuša springs. Hydraulic connections of those two springs, as assumed on the basis of structural geology characteristics of that part of the terrain in study, were confirmed by tracing. The results of tracing from a ponor near the village of Crpalo point out the existence of an underground (natural) water retention of a large volume and extension. It should occur in the hinterland of the Klokun spring and has a remarkable impact on the hydrological features of that spring and of the Čeveljuša spring. The authors' opinion is that the groundwater of the northeastern tectonic block flows toward the Čeveljuša spring through diagonal fissures, but more to the south from the mentioned retention.

4. CONCLUSION

Groundwater tracing is the exploration method of choice unavoidable when the catchment areas in karst terrains have to be determined. This paper deals with a case in which tracing made possible the separation of the catchment areas of individual springs occurring along the southern edge of the Vrgoračko Polje, all that within a terrain tectonically deformed and karstified where the underground hydraulic gradient between the ponors and springs is very small. Since a proper site for the injection of a tracer in a very permeable karst terrain and the amount of that tracer need to be selected, the detailed hydrogeological features of the area in study have to be understood as well as the karstification genesis and the tectonic deformations during the youngest phases of the development of the explored terrain.

The outlining of catchment area boundaries in the close hinterland of the Klokun and Modro Oko sprngs helped in a detailed determination of sanitary protection zones of public water-supply sites and a more rational use of agricultural areas. The authors finally wish to emphasize the fact that optimal tracing results in classical karst terrains may be expected if the tracing is preceded by a thorough understanding of tectonic, genetic and hydrogeological characteristics of the areas in study.

REFERENCES

Bojanić, L., Ivičić, D. & Batić, V. 1980. Hidrogeološka studija područja Aržano-Brela-Metković. Fond struč.dok. Inst. za geol. istr., Zagreb.

Bojanić, L., Ivičić, D. & Batić, V. 1981. Hidrogeologija Imotskog polja s osvrtom na značaj u regionalnom smislu. Geološki vjesnik 34, 127-135, Zagreb.

Fritz, F., Pavičić, A. 1982. Hidrogeološki viseći dijelovi rijeka Krke i Zrmanje. Zbornik, Jugosl. simp. hidrogeolo. i inž. geol., 1, 115-121, Novi Sad. Novi Sad.

Ivičić, D. & Pavičić, A. 1996. Hidrogeološki istražni radovi za zaštitne zone izvora Klokun i Modro Oko. Fond struč. dok. Inst. geol. istraž., 1-38.

Slišković, I., Kapelj, S. i Vidović, M. 1996. Hidrogeološki uvjeti sanitarne zaštite velikog krškog izvora Prud kod Metkovića. Priopćenje sa znanstveno-stručnog skupa: Zaštita prirode i okoliša i eksploatacija Mineralnih sirovina (str. 315-333), Varaždin.

Šarin, A. 1983. Hidrogeologija krškog regiona. In A. Ivković, A. Šarin & M. Komatina: Tumač za Hidrogeološku kartu SFRJ, 1:500.000, Savez. geol. zavod, 62-101, Beograd.

Turner, S. 1955. Tehnički izvještaj o bojenju ponornica na području Imotski-Neretva. Bojenje Vrgorsko jezero-more. Fond dok. RHMZ, Zagreb.

Turner, S. 1959. Bojenja Vrgorsko jezero-dolina Neretve. Bojenje br. 48 i 49. Fond dok. RHMZ, Zagreb.

Tracer Hydrology 97, Kranjc (ed.)© 1997 Balkema, Rotterdam, ISBN 90 5410 875 4

The Muschelkalk karst in Southwest Germany

T. Simon
Geologisches Landesamt Baden-Württemberg, Zweigstelle, Stuttgart, Germany

ABSTRACT: The Muschelkalk formation of the Germanic basin can be separated into three different lithlogical units. These are the Lower, Middle and Upper Muschelkalk. The upper and lower unit consist mainly of limestones which are separated by shales, whereas the Middle Muschelkalk consists of evaporite deposits.

The dissolution of evaporites causes an irregular subsidence of the overlying unit, accompanied by additional fracturing and increasing karstification. The process is apparently selfperpetuating because fresh water gaining easy access to the salt through the collapsed and fractured overlying layers. In this area we find all phenomenons of a mature karst terrain, e. g. sinkholes, areas with no surface flow and caves. Due to the alternating layers of limestones and shales in the Upper Muschelkalk the karstification is often parallel to the bedding of the layers.

We can find three different states of regional karstification. These states depend on the thickness of the overlying Keuper sediments, which retreats stepwise from the north-west to the south-east. The spatial and temporal geological and hydrogeological evolution of the karst system began in the oligocene.

More than 70 tracer tests have been carried out in the last 30 years - most of them by the Geological Survey of Baden-Württemberg. The results show a significant correlation between the lithology and the evolution of the karst system. One of the most interesting tests with two different tracers (uranine, polystyrene microspheres) is dicussed here in detail.

1 INTRODUCTION

The people living in the karst regions of the Muschelkalk in South Germany (fig. 1) always have had problems with the quantity and quality of the groundwater. Nevertheless the scientific investigations of the karst first started in 1950, at a time when more groundwater was needed for the growing population and industry. Since this time many investigations have been carried out by Zander 1973, Harress 1973, Kriele 1976, Hohberger 1977, Köhle 1980, Simon 1980 and Jungbauer 1983. Only a part of these investigations were published in international journals. This is why the most important results will be presented in this report.

2 LITHOLOGY

Above the argillaceous sediments of the Upper Buntsandstein the Lower Muschelkalk starts with dolomitic limestones, intercalated by thin marlstone beds (fig. 2). The middle part mainly consists of marl-stones with thin limestone and dolomite layers. The upper part is built up by the so called "Wellenkalk", a strongly nodular limestone (mudstone) with bioclastic marker beds. This part is the most karstified section in the Lower Muschelkalk. The basis of the aquifer system in the Muschelkalk changes from the marlstone in the middle part of the Lower Muschelkalk in the south to the argillaceous sediments of the Buntsandstein in the north. The reason is the wedging out of the marlstones to the north.

The Middle Muschelkalk is an evaporite cycle with the following setting:

    -dolomite
    -anhydrite/gypsum
    -halite
    -anhydrite/gypsum
    -dolomite.

The dissolution of the salinar sediments leads to the following effects:

1. Developement of salinar karst.

2. Reduction of the thickness to about 50 %; leading to a more intense fracturing in the Upper Muschelkalk.

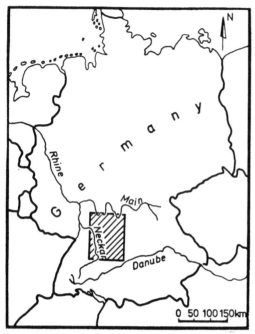

Fig. 1: Area of research in the Muschelkalk karst in South Germany (NE of Baden-Württemberg).

The fracturing due to the dissolution of the salinar sediments increases the karstification in the Upper Muschelkalk. Thus the dissolution is the force driving the karstification in the Upper Muschelkalk.

The Upper Muschelkalk consists of well bedded limestones with bioclastic marker beds and marlstone horizons. The less efficient permeability of the marlstones leads to a vertical difference in the hydraulic system and karstification. Perched aquifers are the result. Therefore the karstification is often parallel to the bedding.

3 LANDSCAPE HISTORY

The investigated area is part of the south German cuesta landscape. The limestones of the Muschelkalk and the sandstones of the Keuper form typical cuestas. These retreat stepwise to the southeast by erosion (fig. 3). Therefore the rocks of the Muschelkalk had a different position to the earth surface in the geological past. This leads to three different regions of the hydrogeological system in the Muschelkalk karst. The regions represent different states in the evolution of the karst system (fig. 4, 5).

There is no karstification in region 1. In the transition zone between 1 to 2, the karstification in the Muschelkalk started when erosion in the valleys reached the lower 30-50 m of the Keuper. At this time the dissolution of the salinar rocks in the Middle Muschelkalk starts, accompanied by an intense karstification of the Upper Muschelkalk. In region 3 the Lower Muschelkalk is also included in karstification. In the Upper Muschelkalk the karst cavities are filled up with loam.

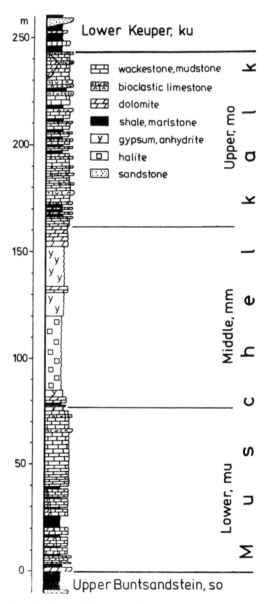

Fig. 2: General Muschelkalk lithology of Southwest Germany.

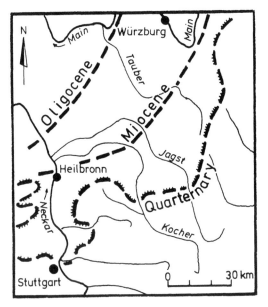

Fig. 3: The stepwise retreat of the Keuper cuesta in the tertiary and quarternary time (after Simon 1987).

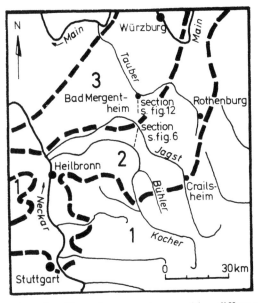

Fig. 4: The three regions with different karstification, caused by the genesis of the cuestas (s.fig. 3).

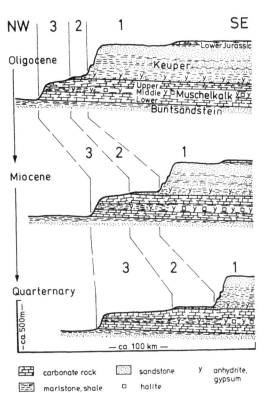

carbonate rock   sandstone   y anhydrite, gypsum

marlstone, shale   □ halite

Fig. 5: The retreat of the cuestas and the three regions in general longitudinal sections. The Sections show the different exposition of the Muschelkalk to the surface.

the Muschelkalk there is no karstification, except a small extension of the fissures by mixing corrosion. Therefore the aquifers in the Muschelkalk are joint aquifers. The transmissivities in the Upper Muschelkalk are lower than $T = 1 \times 10^{-5}$ m^2/s. The flow velocities are about 10-50 m/a. In the Middle and Lower Muschekalk only fractured dolomite and limestones are transferring water. The transmissivities are lower than $T = 1 \times 10^{-9}$ m^2/s. These results have been gained from pumping tests, isotopic analysis and groundwater-models. The region 1 represents the beginning of the evolution of the Muschelkalk aquifer system.

5 REGION 2

On the high plains between the valleys the Muschelkalk is covered by Keuper sediments, which are less than 50 m thick. The Upper Muschelkalk is in a state of intense karstification. During the quarternary era the rocks were eroded in the valleys, in some parts incised to the Buntsandstein. Therefore

4 REGION 1

In the region 1 the Muschelkalk is covered by Keuper sediments, with a total thickness of 300 m, and in the south by Jura sediments up to 700 m thick. In

281

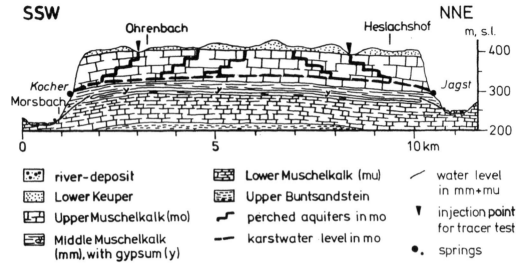

SSW   NNE

Ohrenbach   Heslachshof
                                                                    m, s.l.
                                                                    ⊢ 400

Kocher   Jagst
Morsbach   ⊢ 300

                                                                    ⊢ 200

0   5   10 km

▨ river-deposit	▨ Lower Muschelkalk (mu)	⌒ water level in mm+mu
▨ Lower Keuper	▨ Upper Buntsandstein	
▨ Upper Muschelkalk (mo)	⌒ perched aquifers in mo	▼ injection point for tracer test
▨ Middle Muschelkalk (mm), with gypsum (y)	-- karstwater level in mo	•. springs

Fig. 6: Typical section of the Muschelkalk karst between two deep incised valleys (after Simon 1980); situation s. fig. 4.

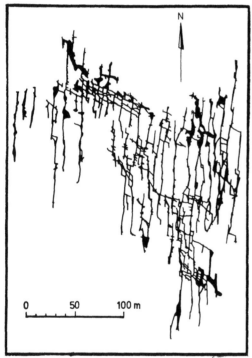

Fig. 7: Map of the "Fuchslabyrinth" cave in the karst of the Upper Muschelkalk near Rothenburg o. T. (after Rathgeber 1980).

the time was too short to allow an intense karstification in the Lower Muschelkalk and a complete dissolution of the salinar rocks in the Mid-dle Muschelkalk. A bigger part of the sulfate rocks are still there, especially below the high plains.

Due to the alternating layers of limestones and shales the karstification of the Upper Muschelkalk is parallel to the bedding. Perched aquifers beyond the shale horizons (Simon 1986) exist as well as caves parallel to the bedding (fig. 6). The longest cave is the so called "Fuchslabyrinth" (fig. 7) . More than 10 km have been explored until now. One can find sinkholes (fig. 8) and regions with no surface flow. Springs show discharge rates of up to 1,5 m³/s, but sometimes fall dry after a long time of no groundwater recharge. The transmissivities are about $T = 1 \times 10^{-3}$ m²/s. The porosities have a magnitude of 2 %.

The most important knowledge about the development of the karst system has been maintained by tracer tests. Until now there have been about 70 tests. For most of them uranine, sometimes eosin, tinopal, rhodamine B, pyranine or halite were used. One test was done with coloured polystyrene microspheres.

All of the tests show a typical breakthrough curve with a rapid increase of the tracer and a steep decrease within a few days.This can be interpreted as direct flowpaths in the karst from the injection to the observation-point. The velocities of the karst water-flow can reach up to 400 m/h. The recovery rate of the tracer is about 80 %.

One of the most interesting flow systems is the one between the two rivers Jagst and Bühler. Already 200 years ago a first test was made with chaff of corn. In 1972 the first qualitative test using the tracer uranine was made. For the 17 km long flowpath a velocity of the karstwater-flow of 40 to 60 m/h was determined. Quantitative tests were carried out in 1988 with uranine and coloured polystyrene microspheres. The results are shown in fig. 9 , 10 and 11 and tab. 1 (Simon et al. 1995).

Essential differences between the two tracers have not been found, except the recovery rate. The rate of uranine is twice as high as the one of microspheres.

Due to the time consuming investigations with the flourescens-mircroscope, the microspheres are only usable in a restricted manner. Altogether 115 hours were used by the investigation of 332 samples, but it only took 5 hours using uranine.

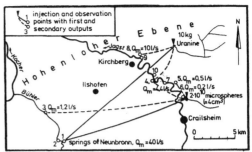

Fig. 9: Map of the tracer tests in the karst system between the rivers Jagst and Bühler (after Simon et al. 1995).

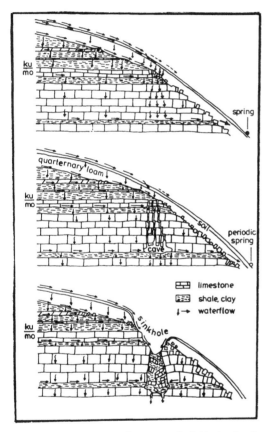

Fig. 8: Development of caves and sinkholes in the upper layers of the Upper Muschelkalk. The water from the soil, the Keuper layers and the Muschelkalk layers causes a mixing corrosion (after Simon 1982).

6 REGION 3

In this region the overlying sediments of the Keuper are only up to 10 m thick, but often the Keuper is eroded. The intense karstified Upper Muschelkalk

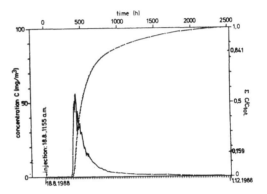

Fig. 10: Breakthrough curves (concentration-time and cumulative) of the test with uranine at the springs of Neunbronn in the karst system between the rivers Jagst and Bühler (after Simon et al. 1995).

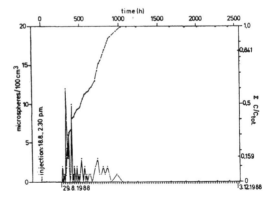

Fig. 11: Breakthrough curves (concentration-time and cumultive) of the test with coloured polystyrene microspheres at the springs of Neunbronn in the karst system between the rivers Jagst and Bühler (after Simon et al. 1995).

283

Table 1: Result of the tracer tests in the karst system between the rivers Jagst and Bühler (after Simon et al. 1995). t_{max}: time at the first appearance of the tracer since the input; t_{Cmax}: time at the maximum of tracer appearance; $t_{0,5}$: time at 50 % of the total output; v: velocity of karstwater-flow; D_L: longitudinal dispersion; α_L:longitudinal dispersivity.

	Distance (m)	Flow time (h)			Velocity (m/h)			Output (%)	D_L (m^2/s)	α_L (m)
		t_{max}	t_{Cmax}	$t_{0,5}$	v_{max}	v_{Cmax}	$v_{0,5}$			
Micro-spheres	15 978	262	312	506	61	51	32	10	16,3	1851
Uranine	17 861	347	420	509	52	43	35	20	19,1	1963

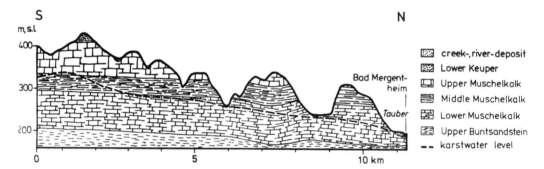

Fig. 12: Section through a Muschelkalk karst system in region 3, in the south of Bad Mergentheim (after Jungbauer 1983); situation s. fig. 4.

conducts groundwater only at its basis and only far from the recharge area. The reason for this is the permeability of the strongly weathered arcillaceous horizons and the permeable Middle Muschelkalk (s. below). So, perched aquifers are hardly existent.

The karst cavities are filled with loam up to 20 m beneath the earth surface. It is therefore difficult to find injection points for the tracer tests where the water can sink down to the aquifer. Besides that, the recovery rates in springs or wells are very low, because the loamy zone at the top of the none eroded Upper Muschelkalk retards the tracer. The breakthrough curves show a flat increase and a very flat decrease. The recovery rates reach only up to 10 %, but this lasts a month or even more.

The salinar rocks of the Middle Muschelkalk are dissolved, except for a few meters of gypsum in some small areas. The residual sediments are a little bit less permeable, but they let almost the whole groundwater pass to the Lower Muschelkalk. The average annual groundwater recharge is about 4 l/s x km².

The Lower Muschelkalk is strongly karstified, especially near the valleys. Cave-like flowpaths exist. From the tracer tests flow velocities of up to 400 m/h have been found. The breakthrough curves look like those of the Upper Muschelkalk in region 2: very steep increases and steep decreases and a small recovery rate after a few days. The recovery rate can reach up to 80 %. The transmissivities amount up to $T = 1 \times 10^{-2}$ m²/s, porosity up to 2 %.

Especially in bioclastic and very pure limestones with a thickness of 1 to 2 m a strong karstification can be noticed parallel to the bedding. The only cave, open for visitors, in the Muschelkalk karst of South-Germany is situated in and below such a bioclastic limestone (35 km west of Bad Mergentheim; Fritz 1976). In this highly karstified rocks springs can be found, which have discharge rates of up to 500 l/s after a longer rain periode or in the late winter, when the snow is melting.

The recharge areas of the karst aquifer in the Lower Muschelkalk are situated in the deep valleys (fig.12), some of them are eroded down to the Buntsandstein, the basis of the Muschelkalk aquifer system.

7 CONCLUSIONS

The differences in lithology of the German Muschelkalk lead to a complicated aquifer system. As a consequence of the development of the cuesta landscape different states of karstification exist, which represent the evolution of the karst in the Muschelkalk. This causes very different characteristics in the concerned regions. The transmissivities, the porosities, the discharge rates of springs and wells and the results of tracer tests are dependent on the evolutionary states of the Muschelkalk karst.

REFERENCES

Fritz, G. 1976. Die Eberstädter Tropfsteinhöhle-die einzige Schauhöhle im Muschelkalk Südwestdeutschlands. *Abhandlungen zur Karst- und Höhlenkunde.* A 12/2: 3-36. München.

Harress, H. M. 1973. Hydrogeologische Untersuchungen im Oberen Gäu. *Dissertation Univ. Tübingen*: 147 p. Tübingen.

Hohberger, K. 1977. Grundwasserbilanz, Chemismus und Stoffaustrag im Einzugsgebiet der Tauber oberhalb von Bad Mergentheim. *Dissertation Univ. Tübingen*: 188 p. Tübingen.

Jungbauer, H. 1983. Karsthydrogeologische Untersuchungen im Muschelkalk zwischen Hohenloher Ebene und dem Taubergrund südlich von Bad Mergentheim, Nordwürttemberg. *Dissertation Univ. Stuttgart*: 183 p. Stuttgart.

Köhle, H. 1980. Hydrogeologische Untersuchungen im Einzugsgebiet der Stuttgart-Bad Cannstatter Mineralquellen. *Dissertation Univ. Tübingen*: 326 p.; appendix: 120 p. Tübingen.

Kriele, W. 1976. Hydrogeologische Untersuchungen im Muschelkalkkarst des westlichen Unteren Gäus zwischen Pforzheim, Vaihingen/Enz und Sindelfingen. *Dissertation Univ. Tübingen*:161 p. Tübingen.

Rathgeber, T. 1980. Höhlenvermessung und Höhlenpläne. In M. Warth (ed.), *Höhlen. Stuttgarter Beiträge zur Naturkunde.* C 13: 5-10. Stuttgart.

Simon, T. 1980. Hydrogeologische Untersuchungen im Muschelkalk-Karst von Hohenlohe. *Arbeiten aus dem Institut für Geologie und Paläontologie an der Univ. Stuttgart.* N.F. 75: 68-215. Stuttgart.

Simon, T. 1982. Ursachen für die Erdfallentstehung im Muschelkalk-Karst. *Laichinger Höhlenfreund.* 17 (2): 47-60. Laichingen.

Simon, T. 1986. Schwebende Schichtgrundwasser-Stockwerke im Oberen Muschelkalk und ihre Bedeutung für die Verkarstung. *Jahreshefte des geologischen Landesamts Baden-Württemberg.* 28: 245-265. Freiburg i. Br.

Simon, T. 1987. Zur Entstehung der Schichtstufenlandschaft im nördlichen Baden-Württemberg. *Jahreshefte des geologischen Landesamts Baden -Württemberg.* 29: 145-167. Freiburg i. Br.

Simon, T., Hinkelbein, K. & Käss, W. 1995. Markierungsversuche im Bereich der Jagstversickerung bei Crailsheim (Hohenlohe). *Jahreshefte des geologischen Landesamt Baden-Württemberg.* 35: 407-432. Freiburg i. Br.

Zander, J. 1973. Hydrogeologische Untersuchungen im Muschelkalk-Karst von Nord-Württemberg (östliche Hohenloher Ebene). *Arbeiten aus dem Institut für Geologie und Paläontologie an der Univ. Stuttgart.* N.F 70: 87-182. Stuttgart.

Tracer Hydrology 97, Kranjc (ed.) © 1997 Balkema, Rotterdam, ISBN 90 5410 875 4

Development of tracer techniques for natural and artificial recharge to confined and unconfined aquifers, India

B. S. Sukhija, D. V. Reddy & P. Nagabhushanam
National Geophysical Research Institute, Hyderabad, India

ABSTRACT: Evaluation of groundwater recharge, the most important parameter for scientific management of groundwater resources is extremely vital for India because of its arid and semi-arid conditions, predominantly being underlain by hard rocks and having peaked monsoonal precipitation. This paper highlights the development of new tracer techniques as well as utilisation of isotopic and geochemical tracers for assessment of natural as well artificial groundwater recharge for both confined and unconfined aquifers. In order to circumvent the general difficulty of small time averaged recharge determined using injected tritium, the validity of chloride method for longer time averaged recharge was ascertained. Further, we have enhanced the scope of tracer methods for estimating groundwater recharge for confined aquifers by demonstrating a new methodology of conjunctive utilisation of environmental and injected tracers. Using different tracer methods, the range of natural precipitation recharge for this sub-continent is estimated to be about 20-200 mm/a (3-25% of average annual rainfall). Thus a rather limited natural recharge and increased groundwater use throughout the sub-continent calls for water conservation as well as groundwater recharge augmentation by artificial recharge structures. To assess the efficacy of artificial recharge structures like percolation tanks, we have developed a simple, inexpensive and reliable method using environmental chloride as a tracer.

1. INTRODUCTION

The Indian land mass having semi-arid to arid conditions, two thirds of its area occupied by hard rocks like basalts and granite/gneiss, consists of diverse hydrogeological character: aeolian deposits in the Thar Desert of Rajasthan, alluvial tracts in Gujarat, western Uttar Pradesh, Punjab and Haryana, and semiconsolidated sediments in the southeastern coastal belt (Fig. 1). Most of its area receives bulk of precipitation (average 1200 mm) during June to September (except for some parts in Tamil Nadu and northeast), though there is large range of variation from 11,420 mm/a in the northeast to 170 mm/a in Rajasthan. Contrarily northwestern region has the highest pan-evaporation rate of 2000-3000 mm/a while the northeast region has the lowest 1500 mm/a. Such a varied scenario calls for proper assessment and management of groundwater resources. Groundwater recharge being a key parameter for such endeavours was given high priority in India, and therefore groundwater recharge studies using environmental and artificial tritium were initiated shortly after successful development of tracer technique by Zimmermann et al., (1967) (e.g. Sukhija, 1972; Sukhija & Rama, 1973; Goel et al., 1975; Sukhija & Shah, 1976). More recently, the injected tritium technique has gained popularity in India, because of the continued reduction in environmental tritium levels and movement of bomb tritium peak to greater depths (Athavale et al., 1980 & 1983; Chandrashekaran et al., 1988; Rangarajan et al., 1989; Atha-

vale & Rangarajan, 1988). Use of environmental chloride in arid and semi-arid regions was successfully demonstrated by Allison and Hughes (1978) in Australia and Edmunds and Walton (1980) in Libya and Cyprus. In India, Sukhija et al., (1988) first utilised environmental chloride in conjunction with injected tritium to ascertain the validity of chloride method.

Tracer methods have been widely used for estimating direct recharge to phreatic aquifers. No attempt had been made for confined aquifers. We developed a methodology of combined use of environmental (isotope and geochemical) and artificial tracers to evaluate recharge for confined aquifers (Sukhija et al., 1996a). Injected tritium method, in general, provides recharge rates averaged for only one or two hydrological cycles because of the sampling problems. On the other hand, we demonstrated that if chloride method is valid, it can provide recharge rate averaged over a period of about decade and half (Sukhija et al., 1988). Thus chloride method wherever applicable circumvents the difficulty of small time averaged recharge.

Indian sub-continent, with its limited natural recharge (3-25% of average annual rainfall; averaged to ~8-10%) and increased water use, calls for conservation as well as augmentation of groundwater by artificial recharge structures. Artificial recharge structures such as percolation tanks, which store surface runoff from monsoonal streams, subsequently contribute to groundwater, are becoming increasingly important. However appropriate methodology for estimating their efficacy was not developed. We have developed a simple and reliable technique of chloride mass balance in the percolation tank to estimate the efficacy of artificial recharge structures (Sukhija et al., 1997 in press). The method should find wide application in future.

This paper summarises the results of natural recharge measurements using tracers in India, and development of new tracer techniques for assessment of recharge to confined aquifer and also recharge from artificial recharge structures.

2. TRACER METHODS FOR NATURAL RECHARGE MEASUREMENTS IN INDIA

Fig.1 shows the various geologic environs and selected basins in India for recharge measurements. For estimating natural recharge, tracer methods (geochemical and isotopic) have promise for semi-arid and arid regions than other methods (Allison, 1988). The pioneering work in the development of tracer methods was done at Heidelberg by Munnich and his co-workers (Zimmermann et al., 1967). They studied movement of soil water in the unsaturated zone and demonstrated the use of tritium tracer to evaluate groundwater recharge using piston flow model. This model was found to be valid in majority of geological provinces of India. In case of non-applicability, the method provided minimum value of recharge, where recharge occurred via preferred path ways of high permeable channels is not accounted for. Sharma (1988) showed that some times by-pass flow could be the dominant process and account for as much as 50% of recharge.

The techniques utilised are classified as: (i) environmental isotope tracers, (ii) environmental geochemical tracers, and (iii) artificial (injected) tracers. The environmental tracer techniques make use of the natural abundance of the tracers in the environment, and thus provide regional information (qualitative and quantitative) about recharge for a large time interval (~2-3 decades depending upon the hydrogeology of the formation and climatic conditions); while the artificially added tracers provide detailed information on a local scale and for relatively small time intervals (limited to duration of the experiment).

2.1 Environmental tritium method

Two methods (peak and total tritium) were employed for evaluating recharge using thermonuclear tritium. In the peak method, using the principle of piston flow, the 1963-64 bomb tritium peak was identified within the soil profile, and recharge was estimated from the moisture displaced above the peak. In the case of the total tritium method recharge was estimated comparing the total tritium present in the profile (unsaturated & saturated) with total tritium precipitated (1952 onwards, taking into account the radioactive decay) at a given site. Details of methodology are discussed in Sukhija (1972), Sukhija and Rama (1973), and Sukhija and Shah

(1976). Environmental tritium profiles (Fig.2a) were used to estimate recharge in sandstones, weathered granites and tillites (Sukhija, 1979).

2.2 Environmental chloride method

The environmental chloride method (Allison & Hughes, 1978; Edmunds & Walter, 1980) is useful under certain conditions; for example, surface runoff is negligible, precipitation (p) is the only source of groundwater recharge and chloride, water movement is predominantly by piston flow, the soil profiles are well drained, and there are no other sources or sinks of chloride. In such cases, groundwater recharge can be obtained from the mass balance of chloride in the steady state soil profile using the following relation:

$$R_g = (C_p \times P)/C_s \qquad (1)$$

where R_g is groundwater recharge, C_p is chloride concentration in precipitation, C_s is chloride concentration of soil profile, and P is precipitation. The validity of the method was tested for its application in Coastal Aquifers of Pondicherry (Sukhija et al., 1988, Fig. 2b) by using the combined approach of injected tritium and chloride. The method has provided the recharge rate for Pondicherry and Neyveli Aquifers comprising of semi-consolidated formation of Cuddalore sandstone.

2.3 Injected tritium method

The recharge rate is estimated by determining the moisture displaced in the soil profile in response to the rainfall input during the time interval between the injection of the HTO tracer and collection of soil samples. The depth to which moisture is displaced is obtained by determining the centre of gravity of tritium distribution with depth (Athavale, 1980). The injected tritium method has been extensively used for alluvial tracts of Uttar Pradesh, Punjab and Haryana (Goel et al., 1975, Fig. 2c), aeolian deposits of Thar Desert (Sharma & Gupta, 1985), weathered granites and gneisses (Athavale et al., 1983; Rangarajan et al., 1994), and semiconsolidated sediments of Pondicherry and Neyveli (Sukhija et al., 1988 & 1996a; Rangarajan et al., 1989).

3. RESULTS AND DISCUSSIONS

3.1 Natural groundwater recharge rates

Table 1 shows the measured values of recharge in different basins of India, the method used and the source of the results. As mentioned before, the northwestern Rajasthan and western Gujarat have scant rainfall (200-400 mm/a) accompanied by high evaporation (2000-3000 mm/a), and therefore very limited recharge (15-55 mm) is expected. It is believed that recharge in arid regions of Rajasthan probably takes place during those wet years when the rainfall exceeds the average value and also the field capacity of the sand is reached. The recharge rate of 8% determined by Sharma and Gupta (1985) with artificial tritium method may not be completely valid because of highly variable nature of recharge in such conditions. Much more work is needed.

In the alluvial tracts of Gujarat, estimated recharge rates (7-10% or 60 mm) are similar among three methods used (environmental tritium, injected tritium & mathematical modelling; Sukhija & Shah, 1976), but are different from those obtained by using inventory, storage, chloride and empirical formulae methods. This comparison focusses the limitations of conventional methods generally employed.

The alluvial tracts of Indo-Gangetic Plains in Uttar Pradesh, Punjab and Haryana have an estimated recharge rate of about 20% of average annual rainfall (Table 1), which is about 2-3 times greater than that of Gujarat. The principal controlling factor probably seems to be climate, though soil and topographic conditions do not greatly differ, but the average annual evaporation is much greater in Rajasthan and Gujarat (3500 mm) than that of Punjab, Haryana and Uttar Pradesh (2000 - 2500 mm).

In the semi-consolidated sediments, measured average recharge rates are 15-25%, though highly variable as observed by Rangarajan et al., (1989). Results obtained by environmental chloride, and the injected tritium methods agree within 30% (Table 2, Sukhija et al., 1988).
The basalts have a recharge rate of 55-60 mm/a, which is about 9-12% of average annual rainfall. This low recharge value is attributed to the presence of clayey soils as well as high evaporation rates. Recharge rates for consolidated formations in the Lower Maner Basin, near Hyderabad, are 7-10%. Recharge studies carried out in the granitic and gneissic complex of Peninsular India, highlight the role of fractures and joints in controlling groundwater recharge.

Sukhija and Rao (1983) have shown that an excellent correlation exists between density and intensity of fractures and recharge and groundwater age, as determined by carbon-14 and tritium. Further it was shown that the recharge rates can vary from 1.2 to 8% (Athavale et al., 1980; Athavale & Rangarajan, 1988), and are controlled by depth to water table, and sand content in the unsaturated zone.

Another important point which emerged from the study of Athavale and Rangarajan (1988) in the Auropally watershed, near Hyderabad, was that almost same amount of rainfall in 1984 (563 mm) and in 1985 (583 mm) yielded two different recharge values; 32 and 17 mm respectively. The fact that the recharge differed by a factor of two was attributed by them to the spacing of rain storms. Uniformly spaced storms, during 1984, resulted in higher recharge, whereas closely spaced storms in 1985 resulted in higher runoff and lesser recharge.

Table 1 Average annual groundwater recharge in different rock types in India using tracers (Sukhija et al., 1996b).

Rock type	Area	Recharge averaged during	Ave. Annual Rain-fall	Recharge Percent	Method used	Source
Aeolian Deposits	Thar desert Rajasthan	1982-83	328	3-10	Inj. ^{3}H & conventional	1a&1b
Aluvium	Gujarat	1952-69	688	3-11	Env. ^{3}H	2
	U.P	1971-72	990	22	Inj. ^{3}H	3
	Punjab	1972	460	18	"	4
	Haryana	1973	470	17	"	5
Semi-consolidated	Pondicherry	1970-85	1200	19	Env.Cl/	6
		1984-85	1080	28	Inj. ^{3}H	7
	Neyvely	1985-87	1200	15-27	"	8a&8b
Basalts	Kukadi basin		612	7.5	"	9
	Godavari-Purna basin		652	8.6	"	10
	Jam basin	1988-91	866	9-12	"	11
Consolidated rocks (granite/ Gneiss)	Lowere Manner (S.S)	1976	1250	8	"	12
	Marvanka basin	1979	550	19	"	13
	Manila Water shed	1986	390	6	"	14
	Vedevati Basin	1978	565	2-7	"	15
	Noyil basin	1979	715	10	"	16
	Vattamali Karai basin	1979	460	13	"	17
	Ponnani basin	1979	1320	5	"	18
	Chitravati Basin	1981	615	4	"	19
	Kunderu Basin (S.S)	1982	615	5	"	20
	Auropally Water Shed	1984-85	573	3-6	"	21

1a - Sharma & Gupta (1985); 1b - United Nations (1976);
2 - Sukhija & Shah (1976); 3,4 & 5 - Goel et al.,(1975);
6 & 7 - Sukhija et al., (1988);
8a - Rangarajan et al.,(1989); 8b - Sukhija et al.,(1996)
9&10-Athavale et al.,(1983); 11-Rangarajan et al.,(1994)
12-21 - Athavale & Rangarajan (1988).

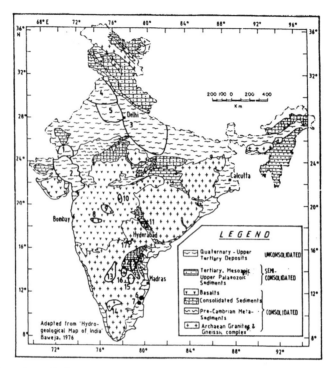

Figure 1. Hydrogeology of India, showing areas selected for recharge studies. 1) Aeolian deposits of Western Rajasthan; 2-5) Alluvial tracts of Gujarat, Western Uttar Pradesh, Punjab, Haryana; 6-7) Semiconsolidated sediments of Pondicherry and Neyveli; 8-10) Kukadi, Godavari-Purna and Jam basins in basalts; 11-12) Lower Maner basin and Kunderu basins in consolidated sediments; and 13-17) Vedavati, Noyil-Vattamalai Karai-Ponnani, Chitravati, Marvanka, Auropally basins in Archaean gneissic complexes.

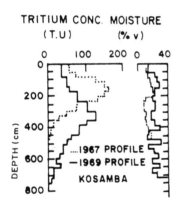

Figure 2a Typical environ-
mental-tritium profile in
alluvial tracts of Gujarat
(Sukhija, 1972).

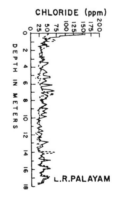

Figure 2b Depth variation
of environmental chloride
in soil moisture, Pondicherry
(Sukhija et al., 1988).

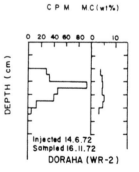

Figure 2c Typical injected
tritium profile in alluvial
tracts, Punjab (Goel et al.,1975).

290

Table 2. Average annual recharge for semiconsolidated formations, estimated by chloride and injected tritium.

Parameter	Pondichery aquifer[1]	Neyvely aquifer[2]		
	Chloride method	Injected tritium method	Injected tritium method 1985-86	86-87
Number of sites	3	3	54	26
Date of injection	-	6/1984	7/1985	7/1986
Date of sampling	-	3/1985	2/1986	2/1987
Average annual rainfall (mm/a)	1200[3]	1080	1367	746
Recharge range (mm)	160-310	180-300	0-550	0-220
Average annual recharge (mm/a)	220	260	161	112
Average annual recharge (% rainfall)	19	28	11.8	13.4

1 Sukhija et al.(1988) 2 Rangarajan et al.(1989)
3 Average of 35 years (1950-85)

3.2 Methodology for estimation of groundwater recharge for confined
aquifers using environmental and injected tracers

The novel methodology evolved comprises elucidating the intake area of the confined aquifer based on environmental carbon-14 (Fig. 3) (supported by tritium, carbon-13 & chloride data), and then utilising the recharge rates determined for the principal recharge area using artificial and geochemical tracer methodology, to estimate the amount of direct precipitation recharge.

The methodology was demonstrated in the Neyveli Groundwater Basin situated in Tamil Nadu, India. The confined aquifer mainly comprised sandstones. From a series of samples collected in different years, it was shown that the northwest belt is essentially the principal recharge area, wherein very large number of groundwater samples have 'Modern' carbon-14 ages, bomb tritium was present in number of samples. From the carbon-14 dates, the movement of recharge water from the Principal Recharge Area to the confined aquifer was evidenced, and the extent of recharge area (600 sq. km) was delineated. Using chloride and injected tritium profiles for the elucidated recharge area, 15.5% of the average annual rainfall was estimated as recharge, and thus 111 mcm/a was estimated as total input to the aquifer. It is only a lower limit since no inter-aquifer flows were considered.

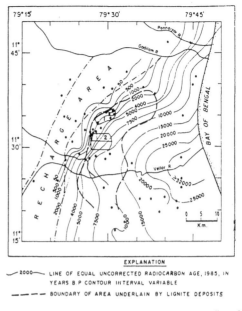

EXPLANATION
—2000— LINE OF EQUAL UNCORRECTED RADIOCARBON AGE, 1985, IN YEARS B.P. CONTOUR INTERVAL VARIABLE
— — — — BOUNDARY OF AREA UNDERLAIN BY LIGNITE DEPOSITS

Fig.3 Delineation of recharge area based on radiocarbon ages (1985 sample collection) for Neyveli confined aquifer.

3.3 Methodology for evaluation of artificial recharge through Percolation Tanks using chloride

Of various possible artificial recharge methods in Peninsular India, percolation tanks are currently the most popular. Typically, there are a number of wells on the down stream side of the percolation tank (Fig. 4) and the cost of construction of a medium size percolation tank is about half a million rupees. As large number of such structures are constructed every year, it was vital to develop a simple, reliable and inexpensive methodology to evaluate the efficacy of such percolation tanks. Sukhija and Reddy (1987), from the comparison of chloride concentration of well waters and percolation tank water, established a qualitative method in determining the effectiveness of the recharge structure. However, later, based on a series of environmental chloride measurements on tank water using chloride balance approach, a method was developed to evaluate the recharge from the tank to groundwater and evaporation (Sukhija et al., 1997).

The total chloride content of tank water at any time is estimated from the volume of tank water at that time and its chloride concentration which is measured regularly (say weekly). As there is no loss of chloride by evaporation the chloride concentration of the tank water should increase with time because of evaporation. This generally happens, as there are long dry spells of several weeks after the monsoon. The percolating water, however, takes the dissolved chloride with

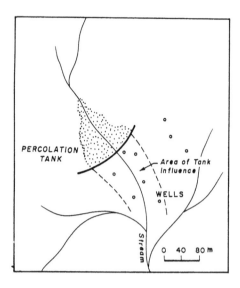

PERCOLATION
TANK

Area of Tank
Influence

WELLS

Stream

0   40   80 m

Fig.4. Schematic representation of a percolation Tank.

it. Thus, by measuring the chloride concentration and volume of water in the tank at different times (after the monsoon), a reliable estimate of tank percolation and evaporation can be made, assuming no seepage/leakage from the dam, no aditional sources of chloride other than the natural input from precipitation and runoff before impounding, and no sink of chloride other than that of percolation.

The balance is made after filling the tank in rainy season and before its drying up. The percolated fraction (recharge) is calculated by using following equation.

$$V_1C_1 = V_2C_2 + (1-f)(V_1-V_2) \ C_p \ (2)$$

where $C_p = \acute{O}C_iV_i/\acute{O}V_i$ (3)

is the time weighted average concentration of chloride in percolation water, and

$(1-f) = (V_1C_1-V_2C_2) /C_p (V_1-V_2) (4)$
where V_1 is volume of water in the tank after monsoon at time t_1;
C_1 is chloride concentration in the tank water at time t_1, V_2 is volume of water in tank at time t_2; C_2 is chloride concentration in tank water at time t_2; V_1-V_2 is water loss from the tank between time t_1 and t_2; $f(V_1-V_2)$ is loss of water by evaporation; and $(1-f) (V_1-V_2)$ is fractional loss by percolation.

The analysis of percolation data (Table 3) shows that percolation rates, as determined using chloride method for 1993-94 show a falling trend very conspicuously when compared with corresponding time period of 1992-93. Such a variation is not observed using water

balance method. The reduction in percolation rates can be expected due to siltation of the tank. Further more, unlike as in the case of chloride method, the water balance method does not indicate any significant time variation (seasonal or annual). The water balance method makes use of questionable evporation rate. The chloride method on the other hand, evaluates evaporation and recharge using directly measured quantities with some valid assumptions. Thus the latter can be considered more reliable.   Therefore, the environmental chloride method for estimating recharge to groundwater from the percolation tank is simple, inexpensive, practical and provides authentic measured values at various stages of silting and desilting.

Table 3. Seven week average percolated fractions and volumes for the Singaram percolation tank

| Time period | Percolated fraction | | | | percolated vol. Cl method m³ | |
| | 1992-93 | | 1993-94 | | 1992-93 | 1993-94 |
	(1)*	(2)*	(1)*	(2)*		
25 Nov.- 15 Jan.	0.34	0.53	0.24	0.44	534	439
3 Dec.- 22 Jan.	0.37	0.49	0.13	0.43	526	230
10 Dec.- 29 Jan.	0.40	0.47	0.37	0.55	490	843
17 Dec.- 5 Feb.	0.41	0.47	0.24	0.53	537	515
24 Dec.- 12 Feb.	0.32	0.47	0.19	0.50	422	362
1 Jan.- 19 Feb.	0.33	0.50	0.23	0.48	420	421
8 Jan.- 26 Feb.	0.27	0.49	0.12	0.50	323	270

(1)* Chloride method (2)* Water balance method

4. CONCLUSIONS

Tracer methods have been extensively utilised in India for evaluating groundwater recharge, which is found to vary from 3 to 25% of average annual precipitation, depending upon the hydrogeological and hydrometeorological conditions. New techniques and methodologies were developed and demonstrated for estimating recharge to confined aquifers, and as well as recharge from artificial recharge structures

Acknowledgements:

The authors are grateful to Dr. Harsh K. Gupta, Director, NGRI, for inspiring us, and according permission to publish the paper; Dept. of Science and Technology, Govt., of India, for sponsoring a project on "Artificial Recharge Structures: Evaluating Their Functional Efficacy".

REFERENCES
Allison,, G.B. & M.W. Hughes 1978.   The use of environmental chlor   ide and tritium to estimate   total recharge to an unconfined   aquifer. Australian J. of Soil Res., v.16, p.181-195.

Allison, G.B. 1988. A review of   some of the physical, chemical and isotopic techniques avail able for estimating groundwater recharge, in Simmers, I., ed., Estimation of natural recharge of groundwater: NATO ASI Series, v. 222, p. 49-72.

Athavale, R.N. 1980. Injected radioactive tracers in studying Indian geohydrological problems- present status and future needs: Proc. of the workshop on Nuclear techniques in Hydrology, held at Hyderabad, India, 19-21 Mar., 1980, p.59-76.

Athavale, R.N., R. Chand & R. Ran- garajan 1983. Groundwater recha- rge estimates for two basins in the Deccan Trap Basalt format ion: Hydrological Science J., v. 28, no. 4, p. 525-538.

Athavale, R.N., C.S. Murti & R. Chand 1980. Estimation of rec- harge to the phreatic aquifers of Lower Maner Basin by using the tritium injection method: J. of Hydrol., v.45, p. 185-202.

Athavale, R.N. & R. Rangarajan 1988. Natural recharge measure ments in the hard rock regions of semi-arid India using tritium injection- a review, in Simmers, I., ed., Estimation of natural recharge of groundwater: NATO ASI Series, v. 222, p.175-194.

Baweja, B.K. 1976. Hydrogeological map of India. Central Ground Water Board, India, 20 p.

Chandrasekharan, H., S.V. Navada, S.K., Jain, S.M., Rao & S.P. Singh, 1988. Studies on natural recharge to the groundwater by isotope techniques in arid west- ern Rajasthan, India,in Simmers, I., ed., Estimation of natural recharge of ground water: NATO ASI Series, v. 222, p. 205-220.

Edmunds, W.M. & N.R.G. Walton 1980. A geochemical and isotopic approach to recharge evaluation in semi-arid zone; past and pre sent: Proc. of Sym. on Arid Zone Hydrology, Investigations with Isotope Techniques, Intl. Atomic Energy Agency, Vienna, p. 47-68.

Goel, P.S., P.S. Datta, Rama, S.P. Sangal, Hans Kumar, Prakash Bah- adur, R.K. Sabherwal & B.S. Tanwar 1975. Tritium tracer stu- dies on groundwater recharge in the alluvial deposits of Indo-Gangetic plains of Western U.P., Punjab and Haryana, in R.N. Athav- ale & V.B. Srivastava, ed., Approaches and methodolo- gies for development of ground- water resources: Proc. of Indo- German workshop held at Hydera bad, India, p. 309-322.

Rangarajan, R., R.N. Athavale, D. Muralidharan, S.D. Deshmukh, & N.T.V. Prasada Rao 1994. Natural recharge measurements in Jam river basin for four

hydrolo- gical cycles using tritium tagging method: Natl. Geophy. Res. Inst., Hyderabad, India, tech. rep. no. NGRI-94-GW-152, 62 p.

Rangarajan, R., S.D. Deshmukh, D. Muralidharan & T. Gangadhara Rao 1989. Recharge estimation in Neyveli grounwater basin by tritium tagging method: Natl. Geophy. Res. Inst., Hyderabad, India, tech. rep. no. NGRI- 89-ENVIRON-65, 63 p.

Sharma, M.L. 1988. Recharge easti- mation from the depth distribu tion of environmental chloride in the unsaturated zone- Western Australian examples, in Simmers, I. ed., Estimation of natural recharge of groundwater: North Atlantic Treaty Organisation, Adv. Sci. Inst. Series, v.222, p.159-173.

Sharma, P. & S.K. Gupta 1985. Soil water movement in semi- arid climate - an isotopic investiga tion: Proc. of stable and radiocative isotopes in the study of the unsaturated soil zone: Intl. Atomic Energy Agency, Vienna, no.IAEA-TECDOC- 357, p. 55-70

Sukhija, B. S. 1972. Evaluation of groundwater recharge in semi- arid regions of India using environmental tritium: Ph.D Thesis, University of Bombay, 139 p.

Sukhija, B.S. 1979. Groundwater recharge rates in semi-arid regions of India using environ- mental tritium, in S.K. Gupta and P. Sharma, eds., Current Trends in Arid Zone Hydrology, held at Ahmedabad, India, Today and Tomorrow printers, New Delhi, India, p. 103-116.

Sukhija, B.S. & Rama 1973. Evalua- tion of groundwater recharge in the semi-arid regions of India using the environmental tritium: Proc. of Indian Aca. of Sci., v.77, no.6, p.279-292.

Sukhija, B.S. & C.R. Shah 1976. Conformity of groundwater rech arge rate by tritium method and mathematical modelling: J. of Hydrol., v.30, p. 167-178.

Sukhija, B.S. & A.A. Rao 1983. Environmental tritium and radio carbon studies in the Vedavati river basin, Karnataka and Andhra Pradesh, India: J.of Hydrol., v.60, no.1/2, p.185-196.

Sukhija, B.S. & D.V. Reddy 1987. Study of artificial recharge through percolation tank by environmental chloride method. Geophy. Res. Bull., v.25, no.1, p 27-31.

Sukhija, B.S., D.V. Reddy, P. Nagabhushanam & R. Chand 1988. Validity of the environmental chloride method for recharge evaluation of coastal aquifers, India: J. of Hydrol., v. 99, p. 349-366.

Sukhija, B.S., D.V. Reddy, P. Nagabhushanam, Syed Hussain, V.Y. Giri & D.J. Patil 1996a. Environmental and injected trac ers methodology to estimate direct precipitation recharge to a confined aquifer: J. of
  Hydrol., vol. 177, p. 77-97.

Sukhija, B.S., P. Nagabhushanam & D.V. Reddy 1996b. Groundwater recharge in semi-arid regions of India: An overview of results obtained using tracers. Hydrogeology J., v. 4 no.3.

Sukhija, B.S., D.V. Reddy, M.V. Nandakumar & Rama 1997. A method for evaluation of artificial recharge through percolation tanks using environmental chlor ide. Groundwater (in Press)

United Nations 1976. Groundwater survey in Rajasthan and Gujarat, India. United Nations Develop ment Programme, tech. rep. no. DP/UN/IND-71-614/4,24 p.

Zimmermann, U., K.O. Munnich & W. Roether 1967. Downward movement of soil moisture traced by means of hydrogen isotopes: Geophy. Monograph no.11, in Glenn E.S., ed. Isotope techniques in the Hydrologic cycle, Amer. Geophy. Union, Washington, DC, p28-36.

Tracer Hydrology 97, Kranjc (ed.)© 1997 Balkema, Rotterdam, ISBN 90 5410 875 4

Water balance investigations in the Bohinj region

Niko Trišič, Marjan Bat, Janez Polajnar & Janko Pristov
Hydrometeorological Institute of Slovenia, Ljubljana, Slovenia

ABSTRACT: Research team of the Hydrometeorological institute of Slovenia is working on water balance analysis of Slovenia for the period 1961 to 1990. In the broader area of Julian Alps there are some discrepancies in parameters constituting the balance (precipitation, evapotranspiration, runoff). To get better insight of these anomalies, the research team performed in the year 1996 a tracing test in the area of Bohinj.

1 INTRODUCTION

Bohinj with its vicinity is an alpine mountainous area between the Sava Dolinka, the Soča and the Bača rivers, located in the central part of Eastern Julian Alps. The highest mountain peak is Triglav (altitude 2863m), in its vicinity is a number of peaks higher than 2500 metres (Razor, Škrlatica, Špik, Prisojnik, Jalovec, Rjavina). South of Triglav range the mountains are of lower altitude, while at Pršivec and Vogar there is a steep drop to the Bohinj lake to the altitude of 500 metres. Bordering to Bohinj depression there is high Komna plateau to the west (altitude 1500 to 1600 m), while to the south there is mountain range of Bohinj ridge (Vogel, Rodica, Črna

prst, Kobla). East of Bohinj lake there is the Sava Bohinjka valley cut between karst plateau of Pokljuka and Bohinj ridge. The river is draining the waters of Bohinj area, and some fifteen kilometres downstream at Radovljica it is joining the Sava Dolinka to form the Sava proper. (Fig. 1)

The divide between Adriatic and Black Sea basin is still uncertain due to the intricate hydrogeology of the area. Discrepancies of water balance parameters led the team to focus its research to the meeting point of the Sava Bohinjka, the Sava Dolinka and the Soča catchments.

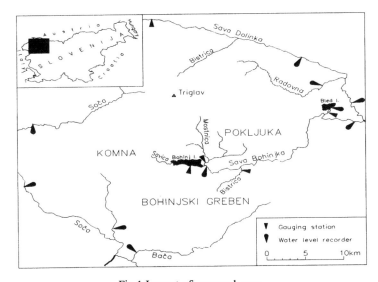

Fig. 1: Layout of reseaerch area

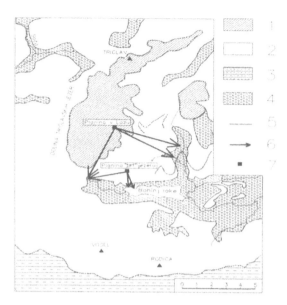

LEGEND: 1-Slatna overthrust: masive and layered limestone-well permeable; 2-Krn nappe: Dachstein limestone-well permeable, impermeable rocks at the base; 3-Inner Dinarides; 4-Quartar: well permeable and impermeable sediments; 5-Border of the overthrust; 6-Proved water connection; 7- Injection point.

Fig.2: Generalized tectonic units and their hydrogeological function

2 HYDROGEOLOGY

Geological structure of the Sava Bohinjka is the part of the overthrust of Eastern Julian Alps. It consists of several nappes characterised by the drive direction to the southwest and northwest. The overall structure is syncline sinking to the east, later cut by large number of faults. The majority of overthrusts volume are limestones with rocks of low permeability at its base. (Fig.2).

Krn nappe or overthrust of Julian Alps is the largest tectonic unit of Eastern Julian Alps. It covers the area of Bohinj, Bohinj ridge, Krn and Bovec in the Soča valley, while at the north it spreads to the Sava fault in the valley of the Sava Dolinka. Lithological constituents of this area are upper Triassic carbonate rocks, the major part being Dachstein layered karstified limestones. At overthrust base are impermeable Carnian layers of marls and mudstones.

In the area of Triglav lakes, Dachstein limestone is covered by Jurassic impermeable marls, that enabled the formation of the lakes.

Slatna overthrust is covering Krn nappe east of Triglav lakes and an area to Rjavina at the east. Lithologicaly it is of Carnian non layered or thick layered limestone. Fault plane between the two overthrusts is at least partly impermeable.

Both the Dachstein limestone of Krn nappe, as well as upper Triassic limestone of Slatna nappe are to the great depth karstified and well permeable aquifers. Big number of faults cut the area into the smaller units that are more or less hydrogeologically independent.

Catchment of Bohinj is drained by rare but water abundant karst springs flowing to the Bohinj lake (the Savica, the Govic) or to the Sava Bohinjka (sources of the Mostnica and the Bistrica). Catchment boundaries, and consequently the areas of catchments of the Sava Dolinka in the Upper Sava valley, the Soča upstream of Most na Soči and Bača are not well defined.

3 WATER BALANCE INVESTIGATION

The catchment area of the Sava Bohinjka is within the area of highest precipitation in the entire Alps. The input data to rainfall distribution map were corrected by taking into account influence of the wind and the previous wettness of the rain-gauge. The estimate of wind correction factor included also data on water content of the snow cover. The part of the catchment, upstream of water gauging station Sveti Janez has an annual precipitation 3500 mm on average, with a high reaching almost 4000 mm in the southern part. The precipitation amount is decreasing to the east, thus in the Mostnica catchment being in the range of 2600 mm (Fig.3: Rainfall distribution map 1961-1990). It was not possible to use uniform vertical rainfall gradient in precipitation analysis,

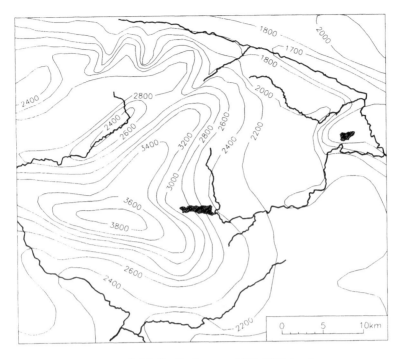

Fig.3: Rainfall distribution map 1961-1990 (mm)

since it changes locally quite a lot, depending on the position of the valleys.

In water balance of Bohinj area there is a major discrepancy between calculated runoff (as difference of precipitation and evapotranspiration) and runoff measured at gauging stations Stara Fužina at the Mostnica, Sveti Janez at the Sava Bohinjka and Bohinjska Bistrica at the Bistrica.

Unrealistic values of calculated specific runoff are caused by incorrect data on catchmnet areas, determined by taking into account surface

topography only. It is obvious that the considered area of the Mostnica is too big, while the catchment of the Bistrica is too small. The calculated specific runoff for the neighboring parts of the Soča basin is within the range of 76 to 78 l/s/km2. Thus, it seems also that the considered area of the catchment of the Sava Bohinjka is a little bit too small. But compared with calculated runoff, discharges are 10% too small, and aestimated catchment area too large.

The topography of the area is not the only governing factor, but hydrogeology too has to be

TABLE 1: WATER BALANCE FOR THE PERIOD 1961-1990

Gauging Station (g.s.)	Mean discharge as per precipitation (m³/s)	Measured mean discharge (m³/s)	Difference (m³/s)
Mostnica g.s. Stara Fužina	5.29	3.28	-2.01
Sava Bohinjka g.s.Sv. Janez	9.19	8.34	-0.85
Bistrica g.s. Boh. Bistrica	1.03	3.18	+2.15

Source to the discrepancy between the values of calculated and measured discharge could be detected at a glance by inspection of data on

catchment area at each gauging station, and calculated specific runoff:

TABLE 2: SPECIFIC RUNOFF (q)

Gauging Station (g.s.)	Catchment area (km²)	Measured discharge (m³/s)	Specific runoff (l/s/km²)
Mostnica g.s. Stara Fužina	78.11	3.28	41.9
Sava Bohinjka g.s. Sv. Janez	93.99	8.34	88.7
Bistrica g.s. Boh. Bistrica	13.44	3.18	236

taken into account to determine a real catchment boundaries. The aim of investigations of the first phase at the northern coast of Bohinj lake, was to define underground connections of the catchment of the Mostnica to the catchment of the Sava Bohinjka and Bohinj lake. There is a difference of measured outflow from the lake as compared to the measured. inflow indicating the need to increase the area of the Sava Bohinjka on account of the Mostnica catchment. The majority of this inflow is entering lake through the springs below the lake surface along the north-western coast

To define the catchment of Bohinj lake the tracer was injected at Planina pri jezeru and Planina pri Lazu, while the samples were taken at the Savica, the springs at the lake coast and tributaries of the Mostnica.

The locations of the subsurface springs at lake were defined in the winter period. The ice cover of the lake was melted at source locations due to the inflow of groundwater, warmer than the water in the lake.

Tracing experiment started at 5th July, 1996. Standard dye tracers Uranine and Rhodamine were used in tracing experiment, while the sampling was performed by the automatic samplers. Injection took place shortly after heavy storm rainfall of 2nd and 3rd July, at the time when the flow in the streams was already decreasing. The circumstances were favourable from hydrological viewpoint. The high turbidity of the water was already gone, while the high water volume in the karst aquifer enabled high relative velocity of underground flow. It is to be expected that in such circumstances all the possible flow paths are active, resulting in good estimate of groundwater flow directions.

The tracing proved the direct underground flow from the area of Planina pri jezeru to the south, feeding the Govic spring, the springs along the coast west of the Govic and the source of the Savica. The flow from Planina pri Lazu is to the source of the Savica, as well as to the Snedčica spring and the Suha brook in the Mostnica catchment. This bifurcation is most likely just one of the several in the area. Tracing experiment did not indicate any direct flow from Planina pri Lazu to the lake itself. Due to the bifurcation it is not possible, for the time being, to divide the Savica catchment from the Mostnica, and at least at the high discharges, this two catchments should be analysed together as a whole. Following this phase of tracing it is still not possible to determine catchment areas with a high degree of accuracy. Hopefully, that will enable the future systematic tracing experiments. The tracers itself, used so far, proved to be reliable in the area investigated.

In the future we plan to inject tracers to the further sinkholes at the pastures in the alpine region to the north, in the area of Triglav and Triglav lakes,

at Komna plateau to the west, and at Bohinj ridge to the south. This tracing experiments should include also sampling at the Sava Dolinka and the Soča catchments, thus being more complex, also involving the tricky transport of the equipment to the high Alps.

REFERENCES:

Buser,S. 1986a. Geological map of SFRY 1:100.000, Sheet Tolmin and Videm L33-44 L33-63. Federal Geological Institute Belgrade.

Buser,S.1986b.LegendtothesheetTolminandVidemL L33-63. Federal Geological InstituteBelgrade.

Janež, J. 1995. Groundwater pollution potential of the mountain huts. Internal report of Geologija Ltd Idrija.

Jurkovšek, B. 1986a. Geological map of SFRY 1:100.000, Sheet Beljak and Ponteba L33-52 L33-51. Federal Geological Institute Belgrade.

Jurkovšek, B. 1986b. Legend to the sheet Beljak and Ponteba L33-52 L33-51.Federal Geological Institute Belgrade.

Pristov, J. 1996. Water balance of Slovenia. Internal report (in print). Hydrometeorological Institute of Slovenia Ljubljana.

Lajovic, A. 1981. The Govic. Our Caves No. 23-24, Ljubljana.

Veselič, M. !979. Water resources of the Sava Bohinjka and the Sava Dolinka.Annual internal report, Phase II, Geological Survey of Ljubljana.

Tracer Hydrology 97, Kranjc (ed.)© 1997 Balkema, Rotterdam, ISBN 90 5410 875 4

The use of tracer tests in UK aquifers

A. T. Williams
Hydrogeology Group, British Geological Survey, Wallingford, UK

ABSTRACT: The British Geological Survey and the UK Environment Agency are currently funding a project on the various aspects of tracer testing applicable to the major UK aquifers. Flow in these aquifers, the Chalk and the Permo-Triassic sandstone, occurs predominantly in fractures and fissures, with significant contributions from the matrix in the sandstones. Tracer tests are not often performed in such aquifers.

The overall aim of the project is to encourage the use of tracer tests in UK aquifers. This will increase our knowledge of flow processes within these aquifer materials. In the past, most tracer tests that have been undertaken in the UK were designed to investigate specific problems. With this project we hope to encourage the routine use of tracer tests in a manner similar to that of pumping tests. In this way a body of information can be built up about the properties of our aquifers, which will help with the protection of the groundwater resource and aid modelling studies.

The main output of the project will be a multi-volume manual that will provide guidance in the many, different facets involved in designing and carrying out tracer tests.

1. INTRODUCTION

The Hydrogeology Group of the British Geological Survey is currently engaged in a three year project joint, funded by the Environment Agency, investigating the use of tracer tests in British aquifers. The overall aim of the project is to produce a 'Manual of Best Practice' for the design and use of tracer testing in the UK.

The Manual will cover all aspects of tracer testing including information on detection techniques, types of tracer available, their advantages and disadvantages, methods for interpreting the results and a review of tests performed in the major aquifers of the UK.

1.1 *BGS Interest*

The main emphasis of the work carried out by BGS is on the production of 'protocols' which will enable practitioners to design and perform tests with a better chance of success.

The BGS interest in the work is to develop a tracing technique which is easily applicable and which will allow a significant body of data to be gather on the main UK aquifers. These data, which will relate to transport parameters in the aquifers, will be mainly of use in numerical modelling work. To this end, the emphasis of the field work has been on producing a robust protocol for the use of radial-flow tracer tests in a variety of aquifer types.

1.2 *EA Interest*

The Environment Agency interest in the project is directed at the possible use of tracer tests during the process of source protection zone delineation. Flow in the main UK aquifers in dominated by fracture flow, at least at small scale (10-1000 m). This means that numerical modelling, based on a continuum approach will not always adequately predict a 50 day protection zone. The data provided by the BGS radial flow tests will eventually be of use in refining these continuum models, but only when a significant body of information has been collected. In the mean time, well designed and carried out tracer tests are seen as a valuable tool in defining zones in many areas.

The EA are also responsible for licensing tracer tests in the UK in as much as a 'consent to discharge' is required before anything artificial is introduced to groundwater. This means that their personnel need to have up-to-date information on the toxicity of possible tracing materials

2 REVIEW OF TRACER TESTS

A review of tracer tests reported in the UK showed that very few tests had been performed in the main aquifers of the Cretaceous Chalk and the Permo-Triassic sandstones. Most reported tests had been performed in the karst formations of the Carboniferous and Jurassic limestones. These tests, which are mostly point-to-point natural gradient tests were not considered in this project as they are, by their nature, site specific.

Out of a total of 49 non-karst tests identified so far, 41 are in the Chalk aquifer, three in the Triassic sandstone aquifer, three in the Lincolnshire limestone and two in superficial disturbed material. Most of the Chalk sites are situated in the Chalk of East Anglia, and these represent the results of research work by the University of East Anglia.

The tests carried out were divided into natural gradient and forced gradient tests. Natural gradient tests have been performed at 34 sites. These may be subdivided as follows (sometimes several flow routes were monitored per site):

Borehole dilution tests	18 sites
Borehole to spring tests	4 sites
Swallow hole to spring tests	4 sites
Swallow hole to borehole tests	3 sites
Swallow hole to stream	1 site
Soakaway to borehole test	1 site
Borehole to borehole tests	2 sites
Borehole to tunnel test	1 site
Tank to borehole test	1 sites
Sea to borehole	1 site
Saline plume monitoring	1 site
Unknown	1 site

Forced gradient tests were carried out at 17 sites. At all these sites radial flow tests from an injection borehole to a pumped borehole were undertaken, and in addition at one site flow from a soakaway to a pumped borehole was observed.

The types of tracer used may be summarised as follows:

Fluorescein	32 tests
Amino-G-Acid	13 tests
Rhodamine WT	9 tests
Photine CU	5 tests
Bacteriophage	4 tests
Bacteria	1 test
Bromide	1 test
Diphenyl brilliant flavine (DY96)	1 test
Rhodamine Ken-Acid	1 test
Salt	1 test
Saline water plume	1 test

A total of 10 tests were undertaken in order to establish groundwater connections. Nine of these were undertaken in Chalk, of which five were to establish a connection between a swallow hole and spring or borehole.

At fifteen sites tests were undertaken to obtain seepage velocities and hydraulic conductivities only (borehole dilution tests). At 23 sites the purpose of the test was to gain an understanding of the groundwater flow mechanisms; at around half of these sites an attempt was made to model the aquifer properties.

3 TESTING DURING PROJECT

During the project some tracer tests will be performed. The aim of this field work is twofold. One aim is to compare different types of tracer tests to enable the design of the 'best' test. This means the type of test that gives most useful information for a reasonable amount of effort. For instance a drift and pump-back test may give almost as much information about the flow properties of a particular aquifer as a full scale radial-flow test, but at half the cost and in a shorter time. In which case we would recommend the use of this sort of test for that aquifer. However, it is likely that this may not be true in other aquifers, and so a different test might be recommended.

The other aim of the field work is to allow us to consider the practical details of performing a tracer test. The aim of extending the use of tracer tests within the UK is being achieved by preparing a robust protocol which will enable hydrogeologists to perform a tracer test with a good chance of success. In the terms of this project a test is deemed successful if sufficient tracer is recovered to provide a discernible breakthrough curve and the aquifer is not left contaminated with a significant quantity of 'unflushed' tracer.

3.1 Radial flow tracer tests

So far a series of radial flow tests has been carried out, drawing on the experience of the University of East Anglia who have performed the majority of such tests in the UK. The tests have been performed in both chalk and sandstone and the differences in the breakthrough curves are great.

In the Chalk, fluorescein and amino-G-acid were injected into the full length of two observation boreholes at distances of 70 m and 200 m from an abstraction borehole pumping at 84 l/s. Both injection holes and the abstraction hole were 100 m deep and so penetrated the same parts of the formation. Some salt (sodium chloride) was also added with the fluorescent tracers in the injection holes and a borehole dilution test performed by logging the conductivity in these holes during the test. The pump was turned on after the tracers had been emplaced in the boreholes.

The dilution test showed that the tracer left the top of the boreholes rapidly and after a day no tracer remained. Water samples collected at the abstraction borehole showed that the tracer from the nearer borehole reached the abstraction point within the first 15 minutes and the peak had diminished to background levels within 3 hours. Tracer from the further injection borehole appeared at the abstraction well after 6 hours of pumping and persisted for over 100 hours.

A second injection was made at the nearer borehole, whilst the pump was still running. This time the tracer took longer to reach the abstraction point, peaking after 1 hour, and subsiding within 2 hours.

The sandstone test was similar in that a fluorescent tracer (amino-G-acid) and salt were added to the full length of an observation borehole. In this case the tracer was added whilst the pump was running at 44 l/s. The injection well is 20 m from the abstraction point and the tracer was first detected 48 hours after injection. Concentration reached a peak after 100 hours and levels were back to background after 300 hours.

So far the work suggests that there is a difference (as would be expected) in the breakthrough curves depending on when the pump is turned on. It is planned to do more sandstone tests

3.2 *Natural Gradient Test*

A secondary aim of the project is to develop tracer testing as a means of validating well head protection zones delineated by numerical modelling. In pursuit of this aim, a large scale tracer test has been carried out in the Chalk aquifer in Yorkshire. This aquifer is dominated by fracture flow and has a highly variable transmissivity distribution.

To date, fluorescent dyes have been used to trace flow over distances of 3 km in one valley and 1 km in another and bacteriophage tracers have been used

to trace flow to the public water supply well. The bacteriophage have not been detected at the PWS, but have been detected at springs down hydraulic gradient. This shows that the numerical modelling for this area was not able to accurately predict the area providing water to the well. This work has shown that tracer testing to define protection zones is fraught with difficulties and that the tests need to be well designed if useful results are to be produced.

4. CONCLUSIONS

Comparatively little tracer testing has been carried out within the UK. The project described here is aimed at making it easier for tests to be performed successfully. A protocol for a 'standard' tracer test will be promoted in the hope that a large number of such tests will be performed. This will provide a useful body of information about differences in behaviour within the different aquifer types prevalent in the UK.

The multi-volume manual produced by this project will provide a digest of information which will prove useful to the EA in their role as a licensing authority as well as facilitating the design of tracer tests to validate protection zones.

5. ACKNOWLEDGEMENTS

This work has been carried out with the assistance of members of the Hydrogeology Group and Fluid Processes Group of BGS. This paper is published by permission of the Director of the British Geological Survey (Natural Environment Research Council).

Contamination transport and protection

Tracer Hydrology 97, Kranjc (ed.) © 1997 Balkema, Rotterdam, ISBN 90 5410 875 4

Groundwater exploration and contaminant migration testing in a confined karst aquifer of the Swabian Jura (Germany)

H. Behrens, W. Drost & M. Wolf
GSF, National Research Center for Environment and Health, Institute of Hydrology, Neuherberg, Germany

J. P. Orth & G. Merkl
Institute for Water Quality Control and Waste Management, Technical University of Munich (TUM), Garching, Germany

ABSTRACT: For the investigation of the transport behaviour of bacteria and chemical contaminants under the conditions of a deep confined karst aquifer a test field was established. As a basis for the migration experiments the hydraulic parameters of the host rock as well as chemical and isotopic data of the karst water were characterized. These data recommended the installation of a dipol with a distance of 200 m between injection and observation well. Groundwater potential gradient and flow velocity were controlled by variation of artesian outflow. The soluble chemical contaminant was isoproturon. *Escherichia coli* K 12 were used as faecal bacterial tracer and compared to fluorescent microbeads (1 µm diameter) as an abiotic particulate tracer. Fluorescent dyes were used as conservative reference. The dissolved contaminant showed conservative transport behaviour. In contrast, only small fractions of bacteria and microspheres arrived at the observation well. Thus a mechanism of strong elimination of particulate materials in such a well conductive aquifer is suggested with increasing elimination at decreasing water flow velocity.

1 INTRODUCTION

In respect to vulnerability of groundwater quality in karst systems, the transport properties of contaminants by groundwater flow is an important factor which can easily be detected by tracer tests. So far, studies with this method have mainly been done with injection of the tracers near-to-surface, mostly using ponors, dolines, or fissures for injection of the tracers (Seiler & Behrens 1992). Under such circumstances, by observation of tracer breakthrough at springs or wells the transport characteristics in the individual parts of the flow path (e. g. unsaturated and saturated zone) are difficult to separate as the result is obtained as a combination of the different migration processes. Subject of this paper is a study in which tracers were injected via a borehole directly into the saturated flow system of a confined karst aquifer with the aim to obtain specific information on transport processes under such a condition.

Two types of contaminants (bacteria, pesticides) and a particle tracer (fluorescent microspheres) were injected together with proved conservative tracers (fluorescent dyes, bromide) as a reference for the behaviour of the transporting medium itself. The karst aquifer under investigation was beforehand subject of hydraulic studies using different methods (head observations, hydraulic testing, hydrochemical and isotope hydrological analyses, groundwater tracing) which are detailed by Drost et al. (1997).

2 DEVELOPMENT OF THE TEST SITE

The investigation area (Figure 1) is situated in the transition of the Swabian Jura carbonate plate into the Danube valley, a few kilometers north of the district town Dillingen a. d. Donau (Bavaria). The test aquifer consists of karstified massive (mostly zoogenous) Malm dolomite rocks; partial dedolomitization caused a cavernous structure. The aquifer (thickness >300 m) is overlapped by a 40 m thick layer of fine-grained Tertiary (predominantly sand and silt) and above by a 12 m thick layer of Quaternary sediments (Pleistocene Danube gravels). The site is of the deep karst type and belongs to the overburden karst zone of the Swabian Jura. The lapping causes an aquiclude effect resulting in a confined karst aquifer. The piezometric head of the confined groundwater is a few meters above terrain surface and wells show artesian overflow. In the south, a few kilometres downstream of the test area the karst aquifer discharges into the Pleistocene gravel layers.

In this area in 1979 the Bavarian state water resources authority (Bayerisches Landesamt für Wasserwirtschaft) carried out a groundwater prospection. Three wells (A, B, C), recovering the artesian confined karst aquifer, were installed and several investigations done (pumping tests, hydrochemical analyses, well monitoring and logging). A few years later the area was selected to investigate the transport of bacterial and chemical contaminants in order to obtain criterions for dimensioning groundwater protection zones in karstic areas. For that purpose in 1987/88 two more wells (1, 2) were drilled at an upstream site for injection of tracers from which a mean water flow time to wells A–C over a period of 30–100 days was expected. However, the investigations, done between 1988 and 1990, revealed that the distance between the injection and sampling wells (1,4–3,6 km) was too large for the planned transport tests. Therefore in 1991 a smaller test field was established by installation of a further injection well (C1, depth 140 m) 200 m upstream of well C. By this a small-scale dipol was obtained (Figure 2 and Table 1) where the transport of contaminants was studied in 1991/92. Before positioning well C1, horizontal electromagnetic measurements (Slingram method) were performed in order to reveal faults or fracture zones in the karst aquifer (Niedersächsisches Landesamt für Bodenforschung 1993).

Numerous measurements of the piezometric level in the standpipes showed a clear transmission of hydrostatic pressure between injection well and sampling well. Pressure at well C1 and potential gradient in the test field depend on the discharge rate at well C (Table 2). When the discharge rate at well C is changed, a new piezometric level at well C1 adapts within a few minutes.

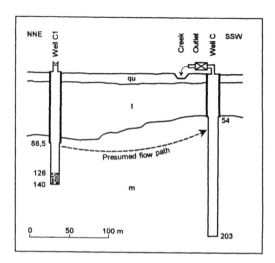

Figure 2: Vertical section of test field with injection well C1 and sampling well C (to scale, except well diam.).

qu = Quaternary sediments (Pleistocene Danube gravels)

t = Tertiary sediments (predominantly sand and silt)

m = karstified massive Malm rocks with cavernous structure

Environmental isotope techniques had been used to get information on the karst water movement in the undisturbed aquifer (for methodical/technical details see IAEA 1983). In the period September 1988 until May 1995 groundwater samples for isotope analyses were collected by pumping from different depths of each well. All samples were analysed for ^2H, ^3H, and ^{18}O contents and some for ^{14}C and ^{13}C contents (for results see Table 4).

The $\delta^{18}O$ values of the Malm karst groundwater are in the order of -10 ‰ and scatter within the range of the analytical error. Temporal variations of the stable isotope contents are not observed in the groundwater. These results suggest groundwater ages of at least several years.

The tritium content of groundwater was found to be in the range between <1 and 35 TU (tritium units – s. a. Table 4) and tends to correlate with the ^{14}C content due to the influence of "bomb" produced ^{14}C.

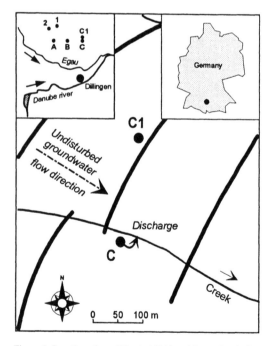

Figure 1: Location plan of the test field and its regional site. Position of injection well C1 and sampling well C. Thick lines: Faults or fracture zones, revealed by horizontal electromagnetic measurements (Slingram method).

Table 1: Characteristics of the test wells (see also Figure 2).

Well	Purpose	Depth [m]	Lap thick-ness [m]	Unlined section [m]	Artesian discharge [l/s]
C1	Injection well	140	80	88–126 [1]	10–13 [2]
C	Sampling well	203	51	54–203	27–35

[1] Lowest 14 m of the borehole filled with sloughing and/or with sealing material.
[2] Well closed during experiments.

Table 2: Characteristic values of potential difference and gradient in the test field as depending on the discharge from well C.

Outflow rate Q at well C [l/s]	Potential difference $\Delta\Phi_{C1\to C}$ [m]	Potential gradient $I_{C1\to C}$ [m/km]
0 [1]	− 0,27	− 1,35
0,50	0,06	0,30
2,70	+ 0,17	+ 0,85
≈ 35 [2]	+ 3,40	+ 17

[1] State of rest (no discharge from any well) at the start of the test campaign.
[2] 3 h after full opening of well C (start of the first tracer test).

The recharge pattern of the confined aquifer justifies the application of a combined exponential / piston flow model (EPM) for age estimates based on the tritium contents. Using the code MULTIS (Richter & Szymczak 1992) ages between 50 and 160 years were obtained. These ages correspond to tritium ages of 100 to 200 years in the shallow karst of the Franconian Jura (Seiler et al. 1996).

The measured ^{14}C contents range from 52 to 70 pMC (percent modern carbon − s. a. Table 4). The calculated initial ^{14}C contents using measured chemical data, the extended chemical mixing model (Wolf et al. 1993) and the geochemical code PHREEQE (Parkhurst et al. 1980) show values of about 57 pMC which are comparable with the measured values in the ^{3}H-free groundwaters. Therefore the ^{14}C model ages of these groundwaters are below 2 000 years. The measured δ^{13}C values of the groundwaters (− 13.9 to − 12.3 ‰) are typical for waters which have evolved in carbonate aquifers.

Table 3: Mean chemical data of karst groundwater in the test field.

Parameter	Unit	Value
Temperature	°C	11.9
Electrical conductivity	μS/cm	582
pH value	−	7.2
Oxygen	mg/l	0.15
Calcium	mg/l	96
Magnesium	mg/l	15
Sodium	mg/l	2.3
Potassium	mg/l	1.0
Hydrogen carbonate	mg/l	340
Sulphate	mg/l	21
Chloride	mg/l	11
Nitrate	mg/l	0.27
Iron	μg/l	118
Manganese	μg/l	221

By the long residence time in the confined aquifer, the karst groundwater is nearly free from oxygen (reducing environment). This correlates with high contents of iron and manganese and likewise very low concentrations of nitrate (Table 3). Anthropogenic influences on water characteristics were not found. Water temperature proves the depth position of the storey with no seasonal variation. Microbiological analyses according to the standard drinking water tests (parameters: coliforms in 100 ml and bacterial count in 1 ml) yielded no detection.

3 SELECTION OF TRACERS AND THEIR ANALYTICS

In an aquifer, the transport of water borne contaminants can be influenced by sorptive retardation, elimination or degradation of the materials in question. These effects can be detected and quantified if the reappearance of contaminants is compared with that of non-reactive (conservative) tracers which represent the movement of water as the transporting medium. Furthermore the reference tracers supply information on hydraulic parameters like the mean residence time of the traced water and dispersion processes on the flow path under investigation.

3.1 Conservative reference water tracers

Dyes selected for the water tracing were uranin (Colour Index No. 45350) and eosin (45380). Furthermore the supposed most reliable reference tracer bromide was used. All reference tracers were detected at water samples in the laboratory. Fluorescent dyes were analysed qualitatively and quantitatively by synchronous scanning spectrofluorometry (Behrens 1973) with detection limits of 0,002 ppb for uranin and 0,006 ppb for eosin (with spectrofluorometer PERKIN-ELMER LS-5B). Bromide was detected by ion chromatography (DIONEX 2010 i) with detection limit of a few ppb. While there was no background found in the investigated karst water for the dye tracers, the bromide showed a background concentration of about 13 ppb. All reference tracers were injected in form of pre-prepared aqueous solutions. Aliquot fractions of these solutions were used for preparation of calibration standards.

3.2 Contaminant tracers

As a representative for bacterial contamination a non-pathogenic *E. coli* strain (K 12 "wild type", ATCC* 23716) was selected. Sufficient amounts for tracer injection (10^{11}–10^{13} cells) were obtained by cultiva-

---

* ATCC = *American Type Culture Collection* (Rockville, Maryland, USA).

307

Table 4: Results of $\delta^{18}O$, $\delta^{13}C$, δ^2H, ^3H and ^{14}C isotope measurements together with the twofold standard deviations (2σ). The 2σ errors of $\delta^{18}O$, δ^2H and $\delta^{13}C$ are 0.15, 1.0 and 0.4 δ ‰, respectively. 1 TU = 0,118 Bq/kg H_2O; 100 pMC = 0.226 Bq/g C. The stable isotope contents are given as δ values which are the per mil deviations from the international standards V-SMOW (^2H and ^{18}O) and PDB (^{13}C).

Well	Sampling date	Depth	$\delta^{18}O$	δ^2H	^3H	^{14}C	$\delta^{13}C$
[–]	[–]	[m below surface]	[‰]	[‰]	[TU]	[pMC]	[‰]
A	13.09.88	*			3.8±1.7		
	26.01.90	*	−9.61	−67.5	4.6±0.7	53±4.7	−12.3
B	13.09.88	*			<0.8		
	28.07.89	*	−9.74	−67.7	<0.9	53±2.6	−13.9
	29.01.90	*	−9.65	−67.4	<1.3	52±4.7	−13.5
	17.10.90	55	−9.73	−70.5	<1.1		
		68	−9.85	−69.5	<1		
		85	−9.88	−69.9	<1		
		112	−9.63	−70.3	<0.7		
		130	−9.75	−69.6	<1.1		
		Mean	−9.77	−70.0	<1		
	Mean	–	−9.75	−69.3	<1		
C	13.09.88	*			12±2		
	29.01.90	*	−9.61	−67.7	14±1.1	58±5	−13.0
	18.10.90	73	−9.92	−70.8	16±1.3		
		82	−9.92	−69.2	17±1.3		
		94	−9.84	−69.7	15±1.3		
		112	−9.83	−69.8	15±1.3		
		144	−9.87	−69.3	16±1.3		
		Mean	−9.88	−69.8	16		
	Mean	–	−9.83	−69.4	15		
C 1	11.07.91	*	−10.00	−71.0	17±1.2		
	06.11.91	95	−9.94	−70.5	17±1.9		
		100	−10.06	−70.0	16±1.9		
		115	−10.02	−70.2	17±1.9		
		124	−10.07	−70.5	18±1.9		
		Mean	−10.02	−70.3	17		
	Mean	–	−10.02	−70.4	17		
1	26.01.90	*	−9.81	−67.8	14±1.2	64±5.8	−12.8
	17.10.90	40	−9.88	−67.6	15±1.3		
		49	−9.80	−69.2	16±1.3		
		55	−9.91	−68.0	16±1.3		
		66	−9.98	−69.0	14±1.3		
		Mean	−9.89	−68.5	15		
	Mean	–	−9.88	−68.3	15		
2	17.01.91	☉	−9.90	−68.7	24±1.7		
			−9.99	−70.5	32±2.5		
			−9.88	−68.4	33±2.4		
	23.05.95	*	−9.92		26±3.1	70±4.9	−13.2
	Mean	–	−9.92	−69.2	29		

* Samples taken from the artesian outflow or from the pumping discharge (well 1) – no defined depth.
☉ Samples taken from the outflow of the artesian well (no defined depth) at three different times within this day.

308

tion on nutrition media (MERCK No. 7881 and 7882). For detection of the bacteria water samples were taken under sterile conditions from the outlet of the sampling well. Bacteria counts were performed by membrane filtration method: filtration of the samples and incubation of the filters on agar disks MERCK No. 1342 and 4044, the latter showing less interference by other microorganisms.

For comparison of the bacteria transport with non-biotic fine-dispersed tracer particles, according to experiences by Käss (1992) fluorescent microspheres were selected as simulating tracer for fine-dispersed particulate matter like bacteria: yellow-green fluorescing type YG and red fluorescing type PC RED, both of 1 µm nominal diameter and originally developed for measurements of blood circulation in veins, obtained from POLYSCIENCES Inc. (Warrington, PA, USA). The microspheres were detected by filtration of water samples over membrane filters and counting the particles by fluorescence microscopy. EDTA was added to the samples to avoid precipitation of interfering particles of oxyhydrates and calcium carbonate. Detection of yellow-green fluorescing microspheres was less disturbed by unavoidable background particles than that of the red-fluorescing beads.

As a representative for a soluble chemical contaminant the herbicide isoproturon [N-(4-isopropyl-phenyl)-N',N'-dimethylurea] was chosen. The substance was obtained in form of a 500 g suspension in a volume of 1 ℓ (traded product: "®Arelon flüssig", made by HOECHST) which was diluted to 100 ℓ in water for the injection. The active substance has a solubility in water of 170 mg/l at 25 °C (Perkow 1988); after injection, its concentration fell soon below this value by further hydrodynamic dilution. For the detection water samples of 1 ℓ were taken in glass bottles. The tracer was separated from the water by solid phase extraction with ready-for-use extractors (BAKERBOND C-18), re-extracted with methanol, and quantitatively determined by HPLC. With this procedure the detection limit for isoproturon in the sampled water was about 0,005 ppb.

4 PERFORMANCE OF FIELD WORK

The tracer injections into well C 1 were done with sufficient amounts of water which had been collected from this well prior to the injections. To be able to work with a free water surface in the injection well, the head of the artesian well was extended with a 5 m long tube of 6 " diameter. Except the bacterial tracers, all tracers were pumped through a flexible ½" tube into the injection horizon at approximately 95 m below ground. Simultaneously with the tracer injection, tracer-free water was pumped into the open end of the well to flush the tracer into the aquifer surrounding the well (downward velocity in the well

tube 2–5 cm/s). After end of the tracer feeding, under continuation of the flushing from the top sufficient water was pumped through the flexible tube to guarantee full release of the tracers in the depth. In total, the water injected in each tracer test amounted to about 4000 ℓ. Simultaneous injections of several tracers were done from homogeneous mixtures of these tracers. In the case of the bacterial tracers it was provided to perform the injection with sterile equipment. However the flexible tube could not be sterilized under the given field conditions. Therefore the suspension of E. coli was released in well C 1 from a sterilized motor-driven dosing syringe under the same flushing condition as described above; during injection the dosing unit was vertically moved within 90–95 m depth.

All injections were done quasi-instantaneously (Dirac pulse). For confirmation of flow path connection between wells C 1 and C, and to obtain information on the residence time distribution (RTD) characteristics, in a first campaign the conservative reference water tracers were injected solely (five tests). In the following campaigns the conservative reference water tracers were used to accompany the contaminants (9 tests). Altogether 14 tests were performed; for this presentation tests with typical results were selected (Table 5).

5 RESULTS AND DISCUSSION

In Figure 3 the residence time distribution curves (RTD) of simultaneously injected uranin and bromide are shown together with that of eosin which was injected at a different time. Both measurements were performed at a runoff from well C of 30 l/s and indicate:

- The shape of the RTD curves is typical for dispersive flow as generally obtained as well in porous media as also in cavernous systems and does not indicate flow paths with branching.
- The simultaneously injected reference tracers uranin and bromide show identical RTDs.
- Also under same discharge condition (30 l/s from well C), the mean residence times (MRTs) and tracer yields of not simultaneously injected tracers show slight variations which may depend on variations of the wider hydraulic field.
- The longitudinal dispersion in the test field is relatively high; the mean value of the Péclet number (Pe) is $Pe \approx 4$, corresponding to a dispersivity of 50 m.

With decreasing discharge rates from well C (30 → 11 → 5 l/s) lower flow velocities between C 1 and C and lower tracer recoveries were observed (Table 5). As a plausible reason for the latter can be seen, that the axial direction of the test dipol C 1 → C does not coincide with the general groundwater flow direction (see Figure 1); thus with decreasing discharge from

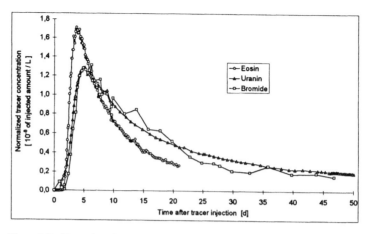

Figure 3: Residence time distributions (RTDs) of the conservative reference water tracers eosin, uranin and bromide at a discharge rate of 30 l/s from well C. Uranin and bromide were simultaneously injected, whereas eosin was injected during a different test. Tracer concentrations for comparability normalized to the injected amounts.

well C increasing fractions of the tracer clouds tend to pass the test field without reaching well C. Because the appearance of the tracers during the available observation time was not sufficient complete at the lower discharge rates, times to the tracer maximum appearance (TMT) are used for comparison of the different experiments instead of MRTs. By the same reason, for comparison, tracer recoveries are generally given for 50-day periods (with extrapolation to this period, if the observed intervalls were shorter).

The transport of both particulate tracers (bacteria and microspheres) differed strongly from simultaneously injected reference tracers (Figure 4 and 5). The breakthrough of the particles begins at the same time as the dissolved tracers, but the particle concentration decreases already very soon and disappears completely while the passage of the dissolved tracers still increases and continues. By this fact the relative yield of the particles is very low in comparison to that of the reference tracers indicating a strong hold back of the particles in the aquifer.

Table 5: Main technical data and results of the selected tracer tests.

Number of selected test	Test parameters			Results	
	Injected tracer	Injection amount	Discharge rate [l/s]	TMT[1] [d]	Recovery[2] [%]
I	Eosin	150 g	30	3,8	41
II	Bromide	3,4 kg	30	5	56
	Uranin	200 g		5	59
III	Uranin	100 g	11	9	16
IV	Microspheres	$4,55 \cdot 10^{11}$	30	2,8	1,9
	Eosin	300 g		4,1	49
V a	E. coli	$1,8 \cdot 10^{11}$	30	2,7	1,8
V b	E. coli	$3,7 \cdot 10^{12}$	11	5,2	0,04
V c	E. coli	$2,0 \cdot 10^{13}$	5	5,8	0,001
VI	Isoproturon	500 g	5	12,5	3,2
	Eosin	150 g		12	4,3

[1] Time until appearance of the tracer maximum.   [2] Tracer recovery over periods of 50 d, partly extrapolated.

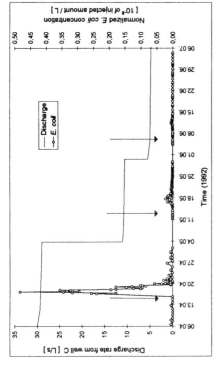

Figure 4: RTDs of microspheres and of the conservative reference water tracer eosin at a discharge rate of 30 l/s. The two tracers were injected during the same test. Tracer concentrations for comparability normalized to the injected amounts.

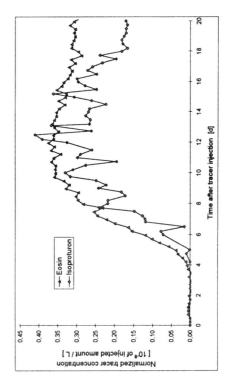

Figure 6: Reappearance (RTD) of *E. coli* in well C, injected into well C 1 at three different discharge rates from well C. Arrows marking the injection times of *E. coli*. Tracer concentrations for comparability normalized to the injected amounts.

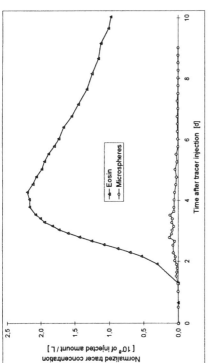

Figure 5: RTDs of *E. coli* and of the conservative reference water tracer bromide at a discharge rate of 30 l/s. The two tracers were injected during the same test. Tracer concentrations for comparability normalized to the injected amounts.

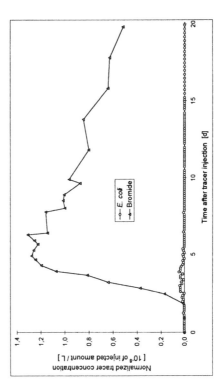

Figure 7: RTDs of the herbicide isoproturon and the conservative reference water tracer eosin at a discharge rate of 5 l/s. The two tracers were injected during the same test. Tracer concentrations normalized to the injected amounts.

Furthermore it was found that with decreasing water discharge from well C (30 → 11 → 5 l/s) the yield of particulate tracers decreases much stronger than that of dissolved reference water tracers (Figure 6). While the recovery of reference tracers decreases at a ratio of 1000:320:100, the analogous *E. coli* decrease ratio is 1000:22:1. The effect as shown in Figure 6 for *E. coli* was also observed for the fluorescent microspheres. This altogether suggests that the particles are only transported in well permeable paths of the aquifer (solution channels) in the center stream lines with the highest flow velocities; they appear to be trapped in regions with low water motion (e. g. dead-end pores) from which dissolved tracers can escape by diffusion. Extrapolated to dynamics of the low natural groundwater flow, particles of the investigated types should more or less not be transported at all. Filtration as a mechanism for elimination of the particles is less probable because retention by that should not be effected so strongly by variation of the hydraulic condition as observed. In respect to groundwater protection the results on particle transport give valuable information on a beneficial effect of low flow velocities. On the other hand they show that microspheres are well suited as reference for transport of bacteria but not as conservative water tracers under such conditions.

The transport of the herbicide isoproturon was not distinctly affected by retardation or elimination processes (Figure 7). The larger scatter of its analytical concentrations as compared to the dissolved reference tracer is probably due to larger analytical errors. Under the given conditions, if at all, less than 20 % of the herbicide was eliminated; the reason for the relatively weak losses – may it be sorption or degradation – could not be revealed by the available informations.

ACKNOWLEDGEMENTS

The authors gratefully acknowledge the financial support by the Bayerisches Landesamt für Wasserwirtschaft – BLfW, Munich (Bavarian state water resources authority). The study was initiated by BLfW section director A. Rothascher. Thanks are due to consultants for handling and analytical techniques of special contaminants and tracers: Mrs. Dr. I. Alexander, and Mrs. Dr. I. Wizigmann, both Munich (*E. coli*); Prof. W. Käss, Umkirch/Freiburg (microspheres). Isotope analyses were carried out under the supervision of W. Rauert and P. Trimborn by Mrs. A. Olfmann, Mrs. A. Schmitt and Miss P. Seibel. Technical field work and reference tracer analytics were accomplished by D. Jurrat, E. Reichlmayr, G. Teichmann and W. Weindl (GSF). Students and the technical staff of the TUM collected the samples, performed the field campaigns, and carried out analyses.

REFERENCES

Behrens, H. 1973. Eine verbesserte Nachweismethode für Fluoreszenzindikatoren und ihre Anwendung zur Feststellung von Fließwegen im Grundwasser. *Z. dt. geol. Ges.*, 124: 535–544.

Drost, W., H. Behrens, W. Rauert, J.P. Orth, R. Netter & G. Merkl 1997. Groundwater exploration in a confined karst aquifer of the Swabian Jura (Germany). In: Gültekin, G. & A. I. Johnson [eds.] 1997. Karst Waters & Environmental Impacts. *Proc. 5th Int. Symp. Field Seminar Karst waters and Environmental Impacts, Antalya / Turkey, 10–20 Sept. 1995*: 181–187; Rotterdam, Brookfield: Balkema.

IAEA International Atomic Energy Agency 1983. *Guidebook on Nuclear Techniques in Hydrology*. 1983 edition; Vienna (IAEA).

Käss, W. 1992, with contributions of H. Behrens, H. Hötzl, H. Moser, H. D. Schulz. *Lehrbuch der Hydrogeologie, Vol. 9, Geohydrologische Markierungstechnik*. Stuttgart: Borntraeger.

Niedersächsisches Landesamt für Bodenforschung 1993. Elektromagnetische Messungen bei Bergheim. – Report, Arch. No. 109 622; Hannover.

Parkhurst, D. L., L. N. Plummer & D. C. Thorstenson, 1980. PHREEQE – A computer program for geochemical calculations. – US Geol. Survey, *Water Resources Investigations*, 80–96, Washington D. C.

Perkow, W., with assist. of H. Ploss 1988. *Wirksubstanzen der Pflanzenschutz- und Schädlingsbekämpfungsmittel*. 2nd ed. Hamburg, Berlin: Parey. – [Loose-leaf edition in 2 files]

Richter, J. & P. Szymczak 1992. *MULTIS. A computer program for the interpretation of isotopehydrogeology data based on combined lumped parameter models*. Bergakademie Freiberg, Lst. f. Hydrogeologie.

Seiler, K.-P. & H. Behrens 1992. Groundwater in carbonate rocks of the Upper Jurassic in the Frankonian Alb and its susceptibility to contaminants. – In: Hötzl, H. & A. Werner [eds.] 1992. Tracer Hydrology. *Proc. 6th Int. Symposium on Water Tracing, Karlsruhe/Germany, 21–26 Sept. 1992*: 259–266; Rotterdam, Brookfield: Balkema.

Seiler, K.-P., H. Behrens & M. Wolf 1996. Use of artificial and environmental tracers to study storage and drainage of groundwater in the Franconian Alb, Germany, and the consequences for groundwater protection. *Proc. Isotopes in Water Resources Management*; IAEA, Vienna, 135-145.

Wolf, M., H. Batsche, W. Graf, W. Rauert, P. Trimborn, K. Klarr & C. von Stempel 1993. Isotopehydrological study of groundwaters from overlying rocks of the Asse salt anticline, Germany. – In: *Paleohydrological Methods and their Applications*. Proc. NEA Workshop, 9./10.11.1992, 207–218, NEA/OECD, Paris.

Tracer Hydrology 97, Kranjc (ed.) © 1997 Balkema, Rotterdam, ISBN 90 5410 875 4

Use of artificial and natural tracers for the estimation of urban groundwater contamination by chemical grout injections

M. Eiswirth, R. Ohlenbusch & K. Schnell
Department of Applied Geology, University of Karlsruhe, Germany

ABSTRACT: In many urban areas grout injections have been used to seal porous soil within the last decades. Especially the silica hydrogels have been applied in foundation engineering practice due to their various applicabilties and economical advantages. For example in the City of Berlin from 1990 to 1995 about 100.000 m^3 of silica gels have been injected within porous aquifers. Therefore the environmental authority of Berlin insisted on detailed investigations of potential groundwater contamination risks. This paper presents the use of artificial and natural tracers as well as transport modelling to quantify potential changes in groundwater chemistry and to predict both longevity and toxicity of silica grouts.

1 INTRODUCTION

Construction of horizontal grouted diaphragms, which are connected to impermeable vertical cut-offs, are required when excavations extend below the water table and groundwater lowering techniques, for one reason or another, cannot be used and an impermeable formation does not exist at a suitable depth for the vertical cut-off to be keyed into (Tausch 1985, Nonveiller 1989). Such diaphragms are usually 1-2 m in thickness. Because of the uplift pressure, they must be constructed in at a level significantly below the proposed base of the excavation in order to prevent blow-out into the excavation (Bell 1993). The design of a successful grouting programme requires the selection of a suitable grout material, and the correct drilling equipment, procedures and grout-hole pattern (Cambefort 1969, Karol 1990).

The most widely used chemical grouts are silicate grouts gel especially in medium and coarse sands. These chemical grouts are true solutions (Newtonian fluids) which contain no suspended solid particles and lack shear strength (Greenwood et al. 1984, Littlejohn 1985, Schulze 1992).

Silicate-aluminate grouts are dilute solutions of sodium aluminate as a hardener and sodium silicate. They are most viscous of the chemical grouts. Due to there low viscosity they are able to penetrate fine sands and sandy silts. On ageing the gel shrinks, becomes opalescent and cracks (Kutzner 1991).

The commonly applied silica gel has a silica /

soda ratio of 3.3 by weight. With the ratio of 2 vol. % sodium aluminate ($Na_2O_m \cdot Al_2O_3 \cdot 18.9\ H_2O$), 18 vol. % filtered silica gel ($Na_2O \cdot (SiO_2)_n \cdot 27.6 \cdot H_2O$) and 80 vol.% water the following chemical reaction occurs (Brauns et al. 1996):

$$Na_2O \cdot (SiO_2)_{3,3} + 0.16\ (Na_2O)_{1,72} \cdot Al_2O_3 + 1.11\ (H_2O)$$
$$\rightarrow 0.04\ Na_8(AlO_2)_8(SiO_2)_{82,5} + 2.23\ Na^+OH^-$$

This equation elucidates that each mol applied silica gel produces 0.04 mol hydrogel mixture with albitic mineral composition and 2.23 mol caustic soda solution. As a result the pH value as well as the contents of sodium, silicate, aluminium and organic compounds are increasing significantly in the drainage water and temporary in the adjacent groundwater (Hötzl 1996). All chemical grouts are hydrogels and, such, may redissolve on prolonged contact with groundwater. The process of dissolution governs the permanence of the grout and grout treatment as well as the long term toxicity, if any, associated with such treatment (Brandl et al. 1987, Donel 1981, Darimont et al. 1984)

In the City of Berlin from 1990 to 1995 about 100.000 m^3 of silicagels have been injected within porous aquifers because groundwater lowering by dewatering techniques cannot be used for the large excavations due to environmental hazards. The environmental authority of Berlin (SenSUT) therefore insisted in 1995 on detailed investigations of potential groundwater contamination risks due to silica grout injections (Böhme 1996).

The Department of Applied Geology (AGK) and

the Institute of Soil and Rock Mechanics (IBF), both from the University of Karlsruhe got the order to investigate the potential risks of silica grout injections. In October 1995 a report on laboratory tests and evaluation of silica grout injected excavations have been presented to SenSUT (Brauns et al. 1996). In February 1996 the Department of Applied Geology, University of Karlsruhe selected a test excavation in Berlin in order to carry out detailed field investigations (Fig. 1).

2 GEOLOGY AND HYDROGEOLOGY

The regional geology of the City of Berlin is dominated by the Pleistocene upland areas of Barnim in the North and Teltow in the South. Both areas consist of sand, gravel and till layers as well as Tertiary quartz sand with interbedded brown-coal seams. In between the glacial Warsaw-Berlin stream channel is situated, consisting mainly of Pleistocene valley sand overlain by Holocene sands, organic mud and peat. Underneath both units the Oligocene Rupelian clay forms the main regional aquiclude in around 200 m below surface (Kloos 1986).

The selected excavation site Ringcenter II is situated in the southern part of the Barnim upland area straight at the boundary to the Warsaw-Berlin stream channel (Fig. 1).

At this location glacial sands with a thickness of around 2 m are underlain by till with a maximum thickness of 12 m thinning-out to the stream channel boundary in the south. The till layer is underlain by a partial present silt followed by the main aquifer consisting of medium to coarse grained glacial sands.

To get a precise picture of the local geological conditions 40 exploration wholes with a maximum depth of 60 m have been drilled. The results show an extreme lateral variation of the thickness of the till layer. Within 10 m the thickness can vary from 0.3 to 3 m.

The hydrogeological situation in the area of the excavation Ringcenter II is characterised by a multi-layered aquifer formation. The upper storey is located in the upper glacial sands above the till with unconfined conditions. The water content is generally small and depending on precipitation the aquifer desiccates during periods of aridity.

The lower storey is located in the lower glacial sands and is bounded above by the silt and/or till at nearly. 22 m above mean sealevel (m a.m.s.l.), below by Miocene brown-coal clay at about 90 m below sealevel. The aquifer is confined with an estimated average thickness of around 110 m. In the Warsaw-Berlin stream channel the dividing till layer is eroded during the generation of the channel so that only one aquifer with unconfined conditions is present.

Fig. 1: Map of the excavation Ringcenter II in Berlin, Germany and map of 4 main construction sites of the excavation with the position of the groundwater observation wells (bgm = base grout membrane).

For the confined aquifer pumping tests, tracer tests (Chapter 3) and grainsize analysis enabled the estimation of the following hydrogeological parameters:

K = hydraulic conductivity = $4.5 \cdot 10^{-4}$ m/s
n_e = effective porosity = 0.25
S = storativity = $2.2 \cdot 10^{-3}$
v_a = transport velocity = 0.3 m/d

For the hydrological and hydrochemical investigations 28 of the exploration wells have been lined out to groundwater observation wells (Fig. 1). Two of them were supplied with online multi-parameter probes (B 16, B 17), the rest have been sampled manually.

Regional contour maps of the water table and 2D numerical models of the groundwater flow have been elaborated with the finite element simulation system FEFLOW®, WASY GmbH 1996, to show the regional and local influence of the slurry walls as well as the silica grout diaphragm on the dynamic and chemistry of the groundwater.

The numerical model elucidates a regional groundwater flow from Northeast to Southwest with a gradient I of about 0.002 (Fig. 2). Only in the neighbouring area of the excavation the groundwater flow changes direction but turns back after passing the excavation.

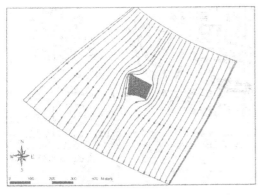

Fig. 2: Numerical modelled regional groundwater streamlines with isochrones of 500 days.

With the numerical model it was possible to elaborate the exact position for the groundwater observation wells to place them directly in the downstream of the excavation. Consequently the hydrological and hydrochemical observation of the entire area downstream of the excavation Ringcenter II was permitted.

The analysis of the groundwater for the hydrochemical expertise have identified the water of hydrocarbon, alkaline origin with a slight corrosive character. The parameters of pH-value, specific electrical conductivity (SEC), and chemical ion composition show values below the groundwater protection list of the City of Berlin. For further detailed information see chapter 3.

3 SITE INVESTIGATIONS

In 1995 the Department of Applied Geology, University of Karlsruhe, started with detailed field investigations on a selected excavation in Berlin. Besides geological and hydrogeological studies extensive chemical analysis have been carried out within 15 groundwater observation wells surrounding the excavation every week. The results of laboratory column and batch tests together with the groundwater analysis after grouting enabled the selection of groundwater parameters as natural tracers. In comparison to the transport behaviour of the natural tracers Na^+, Al^{3+}, SiO_2 the artificial tracers uranine, pyranine and sulforhodamine were injected within different tracing experiments.

3.1 Artifical tracers

On 06/05/1996 500g Uranine have been injected in a grout supply pipe in construction site I just before grouting started (Fig. 3). Samples have been taken in groundwater observation well B 24 and B 17 daily until 10/22/1996. In none of the samples the tracer uranine could be detected. However the tracer appeared at the surface in construction site I inside the excavation. Obviously it was transported to the surface with the injected silica gel.

Fig. 3: Picture of permeation grouting in excavation Ringcenter II (grout supply pipes with tube-à-manchette).

On 08/02/1996 500g of pyranine have been injected in groundwater observation well B 21. Samples have been taken in groundwater observation well B 24 (distance B 21-B 24 = 1.8 m) until 09/23/1996. Fluorescence analysis of the samples were all negative.

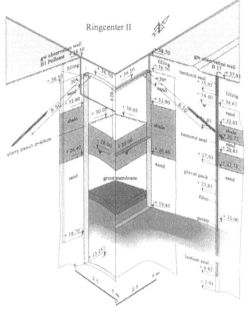

Fig. 4: Schematical map and cross-section of the Southwest corner of the excavation Ringcenter II.

To accelerate transport velocity in groundwater another tracer test combined with a pumping test in B 24 was carried out (Fig. 4). It started on 09/23/96 with the injection of 250g of sulforhodamine in B 21. This forced gradient tracer test lasted for 90 h and gave the breakthrough curves showed in Fig. 5 and 6.

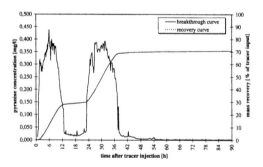

Fig. 5: Pyranine breakthrough curve of groundwater observation well B 24.

Although the breakthrough curves are significantly different both curves lead to an estimation of the transport velocity v_a of about 0,30 m/d under natural flow conditions with a hydraulic gradient I = 0,002.

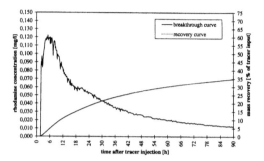

Fig. 6: Sulforhodamine breakthrough curve of groundwater observation well B 24.

These results verified the data yielded from grainsize analysis and pumping tests. The obtained aquifer parameters will be incorporated within the numeric transport model (Chapter 5).

3.2 Natural tracers

In order to estimate the influence of silica grouts on the adjacent groundwater hydrochemical analysis of 28 groundwater observation wells as well as such of the drainage water have been carried out frequently (Fig. 1).

3.2.1 Drainage water

The analysis of drainage water carried out during the dewatering period showed sodium, aluminium and silica as the chemical parameters significantly increasing in groundwater that contacted with silica gel (Fig. 7). Therefore these parameters can be used as natural tracers to estimate groundwater contamination by silica grouts.

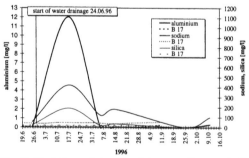

Fig. 7: Variation of the aluminium, sodium and silica content of the drainage water from excavation Ringcenter II.

Beyond this, the formula in chapter 1 elucidates the formation of caustic soda solution which causes an distinct increase of the pH-value. As a result humic substances are leaching out of the organic components of the sediment. This is expressed by the significant increase of the DOC and SEC (Fig. 8).

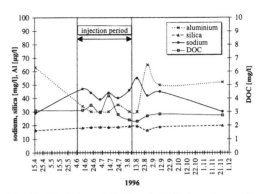

Fig. 10: Variation of SEC and pH-Value of groundwater observation well B 16.

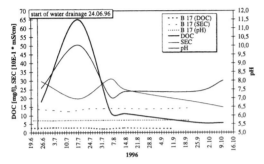

Fig. 8: Variation of the DOC, SEC and pH of the drainage water from excavation Ringcenter II.

Therefore it must be mentioned that there is strong hint that the influence of silica grouting on groundwater is reduced to a zone of only a very few meters downstream of the silica grout diaphragm. This distance varies depending on the aquifer type.

3.2.2 Groundwater

In contrast to drainage water the groundwater showed no significant hydrochemical variations caused by the grouting. For example the groundwater observation wells B 16 and B 17 in a distance of about 7,5 and 12,5 m downstream the excavation indicated any significant changes in water chemistry. All hydrochemical variations in groundwater observation well B 16 shown in Fig. 9 and Fig. 10 are not related to the silica gel grouting. They are in fact of natural origin.

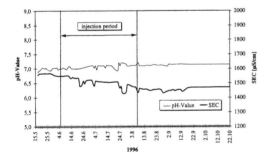

Fig. 9: Variation of aluminium, silica, sodium and DOC of groundwater observation well B 16.

The hydrochemical data of all the other observation wells underline these results. Additionally it should be considered that more than six months passed after grouting period had started.

3.3 Hydrochemical modelling and mass balancing

During the dewatering period of the excavation, slow percolation of groundwater through the gel diaphragm causes the leaching water become progressively more saturated with dissolved gel solids, and this, in turn, reduces further gel dissolution. It is for this reason that the process is slowed down (Chapter 3.2.1). In order to determine the life expectancy and toxicity of grout diaphragms whilst acknowledging the chemical complexity of silica grouts, attempts have usually been empirical rather than fundamental (Müller-Kirchenbauer et al. 1984). Some thirty years ago (1969) Cambefort studied silicate grout dissolution using permeation flow. The work indicated that the rate of dissolution of silica from silicate gels was directly proportional to the amount of water passing through the gel since the dissolution was apparently controlled by the degree of saturation by silica in the washing water.

In the period after grouting and before starting the dewatering of the excavation washing flow is very likely to occur due to the stratified or inhomogenous nature of grouted ground and true permeation flow through the gel diaphragm is unlikely. In summery, whether a silicate grout is subject to permeation or washing flow, they cannot be regarded as intrinsically permanent. This empirical approach does not indicate how the gel dissolves and in what way the dissolved material finds its way into the water passing through and thence away. This aspect of dissolution is important when estimating likely contamination and toxicity levels. Therefore a more fundamental approach is used to describe the dissolution process of silica grouts (Hutchinson 1968). This approach takes also into account, that material residing within the gel network can move through the gel either due to the passage of water resulting from an hydraulic gradient and / or as a result of concentration gradients of dissolved material causing diffusion.

Estimating toxic hazards of silicate grouts is related to those factors influencing grout permanence since materials comprising the basic gel structure and that associated with the gel left over after gelation reaction is complete may dissolve into the groundwater and represent a potential toxic threat. The maximum concentration of free leached grout component can be calculated theoretically:

$$C_{max} = \frac{k_f grout \cdot C}{k_f soil}$$

where C = concentration of leachable component [mg/l], $k_f$grout = permeability of grout [m/s].

In practice however, the free leachable material is reacting again and often building up hydroxides and complexes as well it will be supplemented by material dissolving from the gel structure which is, in turn, controlled by the dissolution constant S. Therefore mass balancing for leachates from gel diaphragms must include hydrochemical calculations for the leachable material (Schössner 1994). For the excavation Ringcenter II a total of 3967,89 m³ grouts have been injected (Tab. 1).

Tab. 1. Silica gel grouting data of excavation Ringcenter II.

construction sites	injection period	injection points	injected grout mass
site I	06/05 - 06/18/96	459	278,93 m³
site II	07/01 - 07/29/96	1480	1151,64 m³
site III	06/12 - 07/10/96	1763	1406,06 m³
site IV	07/11 - 08/07/96	1422	1131,26 m³
total	**06/05 - 08/07/96**	**5124**	**3967,89 m³**

For the injection of the silica grout diaphragm at excavation Ringcenter II following materials have been used:

sleeve grout:	112,2 tons Solidur 274 S
filtered silica gel:	1056,51 tons
sodium aluminate:	108,96 tons 200 SB
water:	3172,8 m³ (gel grout)
	449 m³ (sleeve grout)

The following calculations (Tab. 2) shows exemplarily for sodium compounds theoretically washed out of the diaphragm and pumped to the surface with the drainage water. Until October 1996 (Tab. 2) the sodium output with the drainage water is only about 7% of the input mass of the silicate grouts.

Using the equation in chapter 1, the injected 1056,51 tons of silica gel (= 1,396·10⁶ mol) produces theoretically max 3,113·10⁶ mol NaOH equivalent to a theoretically leachable sodium mass of 71,6 tons. Hydrochemical analysis of the injected silica grouts (sodium aluminate + silica gel + water) yielded a sodium concentration of 2 vol. %. With the total gel injection mass of 4338,3 tons (= 1056,51 tons + 108,96 tons + 3172,8 tons) a total of 86,77 tons sodium is put into the ground with the gel. This means at least 15,17 tons of sodium (= 17,5 %) must be fixed within the gel diaphragm and are not leachable to groundwater.

In order to investigate the reactions and reaction products within the silica gel solution and the adjacent groundwater geochemical calculations with PHREEQE (Apello et al. 1993) have been carried out. Using the chemical analysis of the silica gel solution applied at excavation Ringcenter II the theoretical gel solution composition was calculated. Following the calculated occurrence of the mean silica gel components in different species are listed as percentage of mean element species (% activity):

Ca - species:	Ca²⁺	67,471%
Na - species:	Na⁺	99,721%
K - species:	K⁺	99,831%
Fe - species:	FeOH⁺	99,584%
Al - species:	AlOH₄⁻	100,000%
Si - species:	H₃SiO₄⁻	96,308%
Cl - species:	Cl⁻	100,000%
S - species:	NaSO₄⁻	52,610%
N - species:	NO₃⁻	79,447%
B - species:	H₂BO₃⁻	99,486%
P - species:	CaPO₄⁻	57,761%

Tab. 2: Sodium concentrations in drainage water and output mass calculation.

start 06/24/96 sodium	drainage water output m³	interval d	output during interval m³	mean rate of production m³/d	sodium conc. g/m³	sodium background g/m³	sodium output (gross) g	sodium background g	sodium output (net) g	input g	output %
06/25/96	10	1	10	10,00	54	40	13500	10000	3500		
07/17/96	260	22	250	11,36	420	40	1705200	162400	1542800		
08/05/96	4320	19	4060	213,68	120	40	187200	62400	124800		
08/14/96	5880	9	1560	173,33	180	40	3708000	824000	2884000		
10/10/96	26480	57	20600	361,40	100	40	267000	106800	160200		
10/23/96	29150	13	2670	205,38							
sum		**121**					**5880900**	**1165600**	**4715300**		
	chemical analysis (sodium content = 2 Vol.%)									8,51E+07	5,54
	reaction equation (with 1056 to silica gel):									7,26E+07	6,50

The geochemical calculations with PHREEQE yielded saturation indexes (SI) of the silica gel solution. For SI = 0, there is equilibrium between the mineral and the solution; SI < 0 reflects subsaturation, and SI > 0 supersaturation which suggest precipitation. Following the calculated positive SI for minerals theoretically precipitate within the gel solution are listed:

Hydroxap (6,2); SiO_2 (a) (1,3); Chalcedony (2,0); Quartz (2,5); Gibbsite (2,5); Kaolinite (10,8); Albite (12,5); Anorthite (9,7); Microcline (16,0); Muscovite (25,2); Ca-Montmorillonite (14,2); Hematite (17,9); Goethite (8,3); $Fe(OH)_3a$ (2,4)

As shown above, in the silica gel solution applied at Ringcenter II mainly feldspars and foids will be produced and precipitated.

Beyond the above mentioned calculation the mixing model of PHREEQE have been applied for the given question formulation. Knowing the groundwater chemistry at excavation Ringcenter II it was possible to model the mixing / titration of the silica gel solution with the groundwater. We used the mean hydrochemical groundwater composition at B 15 and calculated the mixing / titration of different silica gel solution volumina to the groundwater. The results of the model calculations with PHREEQE yielded for nearly every mixing ratios following species distributions (% activity):

Ca - species:	Ca^{+2}	74%
Mg - species:	$MgHCO_3^+$	80%
Na - species:	Na^+	77%
	$NaHCO_3$	22%
K - species:	K^+	100%
Fe - species:	$FeHCO_3^-$	68%
Mn - species:	$MnCO_3$	60%
Al - species:	$AlOH_4^-$	97%
Si - species:	H_4SiO_4	99%
Cl - species:	Cl^-	100%
C - species:	HCO_3^-	80%
S - species:	$NaSO_4^-$	64%
N - species:	NH_4^+	98%
B - species:	H_3BO_3	98%
P - species:	HPO_4^{2-}	44%

As shown above mainly carbonates and aluminates will be produced. For example the aluminium compounds occurs to 97% as aluminate which is regarded as immobil. Just so $NaHCO_3$ is formed and less than 77% of the total injected sodium is theoretically mobil in the mixing solution of silica gel and groundwater. This means with the drainage water until October 1996 more than 23 % of the total injected sodium have been washed out of the gel diaphragm.

With the results of the geochemical modelling and the laboratory tests (Chapter 5) a numerical simulation of the leachates from the gel grout diaphragm will be carried out. In conclusion it could be already stated that NaOH is leached out of the sili-

cate grout diaphragm, but due to reactions within the buffering capacity of the aquifer the dissolved Na^+ and OH^- ions will be fixed mainly as immobile carbonates and aluminates.

4 CONCLUSIONS

Thirty weeks after grout injection no significant changes in groundwater chemistry are detectable. Only minor increases of sodium and silicate occur in one observation well 8.5 m downstream of the excavation. Pumping tests and forced gradient tracer tests yielded an actual groundwater velocity of $3.5 \cdot 10^{-6}$ $m \cdot s^{-1}$ within the sandy porous aquifer.

The results of our investigations shows already that the influence of silica grouting in porous aquifers is not as much alarming as it was supposed to be in the past (Schwarz 1996). Regarding these facts further effort goes into detailed investigation of hydrochemical and mineralogical processes in the zone of direct contact between groundwater and silica gel as well as the adjacent area of some few meters downstream of a silica grout diaphragm (Chapter 5). The results of laboratory analysis (column tests, permeameter tests, X-ray diffractometry) combined with the hydrochemical analysis will be used as the basic input data for the 3D transport modelling. These results will be available in April 1997. The main effort of modelling concentrates on quantifying the potential toxic threat and predicting the geochemical balance of the system for a period of about 10 years.

5 OUTLOOK

5.1 Laboratory tests

In order to carry out detailed investigation within the limited reaction zone between silica gel and groundwater special laboratory test have been designed. Permeation and column test with original sediment from the excavation Ringcenter II started in the beginning of the year. The main scientific goal of these test will be the determination of chemical reaction isotherms. With the obtained isotherms it should be possible to include complex geochemical processes within the 3D transport modelling.

5.2 Numerical simulation of leachate transport and balancing

With the help of a 3D numerical transport model of the effluence of leachate and contaminants out of the silica grout diaphragm a quantitative evaluation on the long-term behaviour of the diaphragm on the groundwater quality shall be elaborated. The required hydrological and hydrochemical parameters have been and will be produced by different scaled batch and percolation tests. With the results of the time-dependant chemical concentrations the sorption and desorption isotherms can be quantified with the software program COTAM®.

These chemical species dependant isotherms will be incorporated into a 3D numerical transport model, using the finite element simulation system FE- FLOW®, WASY Berlin.

Acknowledgements This paper is presented with the permission of the company ECE. The authors would like to express their gratitude to ECE for permission to publish the results and for funding the research work. The views expressed are those of the authors, and do not necessarily represent those of the above company.

6 REFERENCES

Appelo, C.A.J. & Postma, D. 1994. *Geochemistry, groundwater and pollution.* 536 p., Rotterdam: Balkema.

Bell, F.G. 1993. *Engineering treatment of soils.* London: Chapman & Hall.

Böhme, M. 1996. Auswirkungen von Baugruben mit Weichgelen oder Beton auf die Grundwasserqualität. *Bauen im Grundwasser*, 25. Seminar der FGU Berlin, 26.-27.02.1996: 119-131.

Brandl, H., Plankel, A. 1987. Vergleichende Untersuchungen an chemischen Bodeninjektionen. *Mitt. d. Inst. f. Bodenmechanik und Felsbau* Vol. 4. TU Wien.

Brauns, J., Kast, K., Hötzl, H. & Eiswirth, M. 1996. Anwendung von Weichgelsohlen zur horizontalen Abdichtung von Baugruben durch Injektionssohlen - Fragen zur Beeinflussung der Grundwasserqualität. *Mitt. Abt. Erddammbau und Felsmechanik Vol. 6.* Universität Karlsruhe.

Cambefort; H. 1969. *Bodeninjektionstechnik.* Wiesbaden and Berlin: Bauverlag GmbH.

Darimont, T., Bariwsky, G., Milde, G., Oetting, R. 1984. Grundwasserbeeinflussung durch chemische Bodeninjektionen auf der Basis von Wasserglas. *gwf- wasser/abwasser* 125 (12): 608-612.

Donel, M. 1981. Beeinflussung der Wassergüte durch Umströmung von Injektionskörpern. *Tiefbau, Ingenieurbau, Straßenbau* 5: 318-328.

Greenwood, D.A. & Thomson, G.H. 1984. *Ground Stabilisation: Deep compaction and grouting.* Institution of Civil Engineers. London: Thomas Telford Press.

Hötzl, H. 1996. Der Einfluß von tiefreichenden Baumaßnahmen auf Grundwasserqualität und Grundwasserströmung. *Bauen im Grundwasser*, 25. Seminar der FGU Berlin, 26.-27.02.1996: 85-101.

Karol, R.H. 1990. *Chemical Grouting.* 2nd Edition, pp. 465, New York Basel: Marcel Decker Inc.

Kloos, R. 1986. Das Grundwasser in Berlin - Bedeutung, Probleme, Sanierungskonzeptionen. in: Der Senator für Stadtentwicklung und Umweltschutz (Hrsg.): *Besondere Mitteilungen zum Gewässerkundlichen Jahresbericht des Landes Berlin.*

Kuno, G., Kutara, K & Miki, H. 1989. Chemical stabilisation of soft soils containing humin acid. Proc. *12th Int. Conf. On Soil Mechanics and Foundation Engineering.* Rio de Janeiro, Vol. 2, pp.1381-4.

Kutzner, C. 1991. *Injektionen im Baugrund.* Stuttgart: Ferdinand Enke Verlag.

Littlejohn, G.S. 1985. Chemical grouting. *Ground Engineering*, 18: Part 1, No. 2, pp. 13-18; Part 2, No. 3, pp. 23-8; Part 3, No. 4, pp. 29-34.

Müller-Kirchenbauer, H., Borchert, K.-M., Aurand, K., Milde, G., Donel, M. et al. 1984. Grundwasserbeeinflussung durch Silikatgelinjektionen. *Veröff. Grundbauinst.* 11. TU Berlin.

Nonveiller, E. 1989. *Grouting Theory and Practice.* Amsterdam: Elsevier.

Schössner, H. 1994. Injektionen in den Baugrund - Anforderungen und Prüfungen aus wasserhygienischer Sicht. *Wasser & Boden* 5: 62-64.

Schulze, B. 1992. Injektionssohlen - Theoretische und experimentelle Untersuchungen zur Erhöhung der Zuverlässigkeit. *Veröff. Inst. Bodenmechanik und Felsmechanik.* Universität Karlsruhe.

Schwarz, W. 1996. Der Aspekt der Umweltverträglichkeit bei Bauverfahren des Spezialtiefbaus. *Bauen im Grundwasser.* 25. Seminar der FGU Berlin, 26.-27.02.1996: 57-65.

Tausch, N. 1985. A special grouting method to construct horizontal membranes. *Proc. Int. Symp. On Recent Developments in Grout Improvement Techniques.* Bangkok: Balkema, pp. 351-62 (eds. A.S. Balasubramanian et al.).

Tracer Hydrology 97, Kranjc (ed.) © 1997 Balkema, Rotterdam, ISBN 90 5410 875 4

Agriculture – Potential polluter of waters in karst region in Slovenia

B. Matičič
Biotechnical Faculty, University of Ljubljana, Slovenia

A B S T R A C T: The Karst region occupies approximately 44 % of the territory Slovenia. Upland catchment areas, where agriculture takes an important part in the economic activity of peasants, supply water down in the valleys where it is as spring water used for drinking. Agriculture is a possible source of pollution of drinking water.

The objective of the study has been to examine present linkages among farming practice and nitrate content of water through identification of mineral surpluses at regional and farm level that could affect the quality of drinking water. 'Trnovsko-Banjska' plateau has been chosen as study sample area.

There are mainly shallow soil types on limestone found in this region. Soils have low water holding capacity and therefore, during the intensive rainfall, the processes of leaching of fertilisers can occur.

Surface mineral balances have been evaluated for the year 1991 for the region of Dol-Otlica, for the community of Ajdovščina as well as for farms on 'Trnovsko-Banjska' plateau for detecting possible agricultural point polluters. The methodology proposed in the EU has been used. It has been observed that Nitrogen net-balance surpluses are low (in region the N surplus is about 36 kg/ha, while average value for Slovenia is 56 kg/ha).

In the region of 'Trnovsko-Banjska' plateau the average yields and uptake by crops are low; non-point source pollution caused by mineral fertilisation in this region can not be considered a serious problem. The high Nitrogen surpluses can be caused by high animal density per ha. The stocking rate over 2,1 LU/ha can cause net-balance surplus over 100 kg/ha; in this case organic fertilisation can be considered a serious non-point pollution source.

Lack of dung yards and cesspools and/or lack of properly built dung yards and cesspools in upland catchment area can cause serious point source pollution of drinking water in the valleys.

1. INTRODUCTION

The agriculture in karstic regions of Slovenia is less intensive but still represents the main peasant's economic activity.

The objective of our study has been to determine the relationship between the soil water balance and mineral balance in the Karst region of 'Trnovsko-Banjska' plateau in western part of Slovenia above Vipava valley and to find out if the possible excessive use of fertiliser and/or high intensity of animal husbandry in upland catchment area on 'Trnovsko-Banjska' plateau could affect the quality of drinking water down in Vipava valley. 'Trnovsko-Banjska' plateau has been taken as study sample area.

The altitude of 'Trnovsko-Banjska' plateau is about 800 m. In this region mainly shallow soil types (with depth of 10-50 cm) on limestone are found with low water holding capacity (22-142 mm) and high rate of infiltration.

The amount of precipitation in 'Trnovsko-Banjska' plateau is very high. The average annual value (1951-1980) in meteorological station Otlica was 2457 mm. The extreme wet year was 1965 with 3233 mm of rain while the extreme dry year was 1973 with 1833 mm of rain (Source: Klimatografija Slovenije, Padavine, HMZ 1989).

The amount of precipitation in 1994 (the year of our evaluation of water and mineral balance) was 1822

Table 1: Structure of livestock on farms in karst region, Slovenia, 1994

Region	Livestock units	% of total livestock					UAA	kg N/ha	kg P/ha	kg K/ha
		Cattle	Pigs	Poultry	Sheep	Other				
SLOVENIA 1991	748836	58,4	16,6	22,7	0,4	1,9	1,26	89,82	50,80	92,05
SLOVENIA 1994	649916	62,6	16,0	19,6	0,3	1,6	1,10	77,51	42,34	76,47
farm 7	12,3	84,8	4,9	0,5		9,8	0,94	66,18	28,96	82,88
farm 13	11,1	93,7	5,4	0,9			1,05	74,91	33,36	96,60
farm 16	10,3	81,4	5,8	1,2		11,6	1,98	139,25	61,54	169,23
farm 14	16,3	88,2	3,7	0,7		7,4	1,42	99,49	43,48	128,70
farm 3	12,4	93,7	4,8	1,5			0,77	55,58	24,66	61,22
farm 2	7,3	87,7	12,3				1,04	76,71	35,14	91,14
farm 4	10,3	94,2	5,8				0,76	54,67	24,22	72,44
farm 9	6,3	90,5	9,5				1,05	76,33	34,50	96,33
farm 6	11,0	94,5	5,5				0,91	65,12	28,76	84,30
farm 10	3,6	83,3	16,7				1,03	76,57	36,00	96,00
farm 8	4,6	87,0	13,0				1,02	75,11	34,67	96,89
farm 15	6,0	100,0					2,01	140,94	60,40	201,34
farm 1	3,0	90,0	10,0				0,40	29,20	13,20	34,67
farm 5	6,3	90,5	9,5				0,75	54,20	24,50	68,40
farm 11	8,6	93,0	7,0				1,23	88,29	39,43	119,43
farm 12	11,0	94,5	5,5				1,10	78,80	34,80	102,00
AVERAGE*	8,8	90,4	8,0	0,9		9,6	1,01	72,33	32,16	91,63

* average for all 16 farms

Table 2: Landuse and yields on farms in karst region, Slovenia, 1994

Region	Arable and grassland ha	Arable land %	Grassland %	% of arable land								
				Cereals	Potatoes	Fodder maize	Green fodder	Cereals	Potatoes	Sugar beets	Fodder maize	Grassland
SLOVENIA 1991	569411	39,0	61,0	53,8	13,8	15,4	15,3	4,7	13,8	45,1	35,6	4,1
farm 7	13,0	0,2	99,8		66,7	33,3			18,0		15,0	2,5
farm 13	10,6	2,8	97,2		100,0				18,0			2,5
farm 16	5,0	4,0	96,0		100,0				16,0			2,7
farm 14	11,5	13,0	87,0		33,3		66,7		19,0			2,8
farm 3	10,0	19,0	81,0	10,5	10,5		89,5	3,0	18,0			2,7
farm 2	7,0	28,6	71,4		25,0		75,0		17,0			2,8
farm 4	7,5	1,3	98,7		100,0				16,0			2,8
farm 9	6,0	16,7	83,3		100,0				15,0			2,7
farm 6	12,0	0,8	99,2		50,0				15,0			2,7
farm 10	3,5	2,9	97,1		100,0				15,0			2,6
farm 8	4,5	2,2	97,8		100,0				15,0			2,7
farm 15	3,0	2,7	97,3		100,0				15,0			2,5
farm 1	7,5	1,3	98,7		100,0				15,0			2,5
farm 5	8,5	0,6	99,4		100,0				15,0			2,5
farm 11	7,0	11,4	88,6		62,5	37,5			16,0		15,0	2,7
farm 12	10,0	5,0	95,0		100,0				16,0			2,7
AVERAGE*	7,9	7,0	93,0	10,5	78,0	35,4	77,0	3,0	16,2		15,0	2,7

* average for all 16 farms

mm (from Jan.-Nov.); the evapotranspiration in this period was 628 mm (used modified Penman's equation - by Doorenbos,J., Pruitt,W.O., FAO Paper 24, 1977). The evapotranspiration according to this evaluation represents only 40 % of the amount of precipitation. During the period of intensive precipitation, therefore, the processes of leaching of fertilisers can occur.

It has been decided to evaluate regional and farm

322

mineral balance for hilly karstic region of 'Trnovsko-Banjska' plateau in order to identify vulnerability related to the nitrate problems in this less intensive agricultural region.

2. GROUNDWATER AND SURFACE WATERS

The pollution of groundwater and surface waters by nitrates, nitrites, phosphorus and ammonium was monitored for the last four years (1991-1994, data base: 'State monitoring of waters').

Average annual values for nitrate concentration (in mg/l) in groundwater in western part of Slovenia, where karst prevails, in the region of Goricia was 39,08 in 1991 (maximal value: 51,36 mg/l), 51,70 in 1992 (maximal value: 76,61 mg/l), 43,84 in 1993 (maximal value: 69,08 mg/l) and 44,06 in 1994 when the maximal value of NO3 was 67,97 mg/l. (The average ammonium concentration in mg/l in this region was 0,01-1991, 0,02-1992, 0,04-1993, 0,02-1994. The average NO2 concentration was 0,02-1991, 0,01-1992, 0,01-1993 and 0,01-1994. The average P2O5 concentration was 0,07-1991, 0,06-1992 no data for 1993 and 1994).

The average concentration of NO3 in rivers for this region in mg/l varied between 3,28 and 3,81 (maximal value was observed in 1992 being 7,53 mg/l).

3. NITROGEN BALANCE AT REGIONAL AND FARM LEVEL

For mineral balance the main agricultural crops that occupy 92 % of arable land have been taken into evaluation. Nitrogen, phosphate and potash supply was calculated by the number of livestock in the region and at each farm and the nitrogen, phosphate and potash content of liquid manure as well as the statistical data on trade of mineral fertilisers were taken into evaluation. Two nitrogen balances have been evaluated:

- GROSS-BALANCE taking into consideration nitrogen input from mineral fertiliser and
   animal wastes, minus nitrogen uptake by harvested crops (as being output).

- NET-BALANCE taking into consideration mineral fertiliser, animal wastes and deposition
   from the atmosphere as input and nitrogen uptake by harvested crops and ammonia losses
   to the atmosphere as output.

The evaluation has been done using normative approach and methodology that has been used in EU countries.

Agriculture on 'Trnovsko-Banjska' plateau is extensive, animal husbandry is prevailing. Landuse, yields, use of mineral fertilisers and livestock population on selected farms is presented in Tables 1 and 2.

Surface mineral balance was evaluated for 534 ha of arable land in region Dol-Otlica on 'Trnovsko-Banjska' plateau, for 16.145 ha of arable land in Ajdovščina community (Dol-Otlica is part of Ajdovščina community).

Nitrogen balance on farm level (for 16 farms on 'Trnovsko-Banjska' plateau) was evaluated for detecting possible point polluters in this region.

4. NITROGEN SURPLUSES AS POSSIBLE SOURCE OF WATER POLLUTION

Average net-balance nitrogen surplus for Slovenia is about 56 kg N/ha. Higher values that can be considered vulnerable for the pollution of groundwater and surface waters can be found in regions with high intensity of animal husbandry in Eastern part of Slovenia.

Average low nitrogen surpluses on regional level are found in Western part of Slovenia where less intensive agriculture prevails. Average net-balance nitrogen surplus for Ajdovščina and Dol-Otlica on regional level was 36 kg N/ha; this value can not be considered as possible non-point source of pollution of groundwater and surface waters (Figure 1). Livestock density in Dol-Otlica is 0,81 LU/ha (Livestock Unit per ha).

The high nitrogen surpluses can be caused by higher animal production. In Dol-Otlica the average yields and uptake by crops are low; stocking rate over 2,1 LU/ha can cause net-balance surplus over 100 kg/ha what can be considered vulnerable for groundwater and surface waters.

The average nitrogen net-balance surplus on selected farms has been 36 kg N/ha and is varying between 13 and 87 kg N/ha (See Figures 2 and 3). On average nitrogen input from mineral fertiliser was observed very low - 11 kg/ha, while nitrogen input from organic manure was 72 kg/ha (See Table 3). Livestock density in selected farms was between 0,4 to 2 LU/ha. The average phosphate surplus was found 27 kg/ha and the average potash surplus was 57 kg/ha.

323

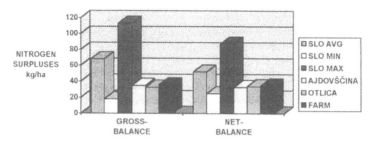

Figure 1: Gross and net balance of nitrogen surpluses

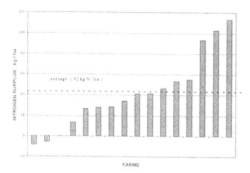

Figure 2: Gross nitrogen balance, farm level, Trnovsko Banjška planota

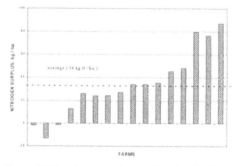

Figure 3: Net nitrogen balance, farm level, Trnovsko Banjška planota

Taking into consideration all the evaluated average data it could be concluded that the region 'Trnovsko-Banjska' plateau can not be identified as vulnerable for nitrogen leaching into the groundwater. But in these regions with limited growing conditions for agricultural crops (climate, soil depth) just small increase in livestock density can cause nitrogen surpluses over 100 kg N/ha. For this reason the restrictions regarding application of chemical fertiliser and manure on hilly and karst regions have to be more rigorous than in plains.

On the other hand it was found out that in many cases dung yards and cesspools on farms are not built and/or are poorly built. In this case liquid manure can cause serious problem
as being point polluter of groundwater.

5. NATIONAL NITRATE POLICIES

There are several regulations in force in Slovenia that are supposed to control water and food quality in connection to nitrates.

Slovenian legislation is quite strict as far as standards on drinking water, food quality or quality of agricultural products is concerned regarding nitrate (as well as other chemical elements or toxic substances). According to EU Nitrate Directive the maximum standard for nitrate in drinking water (according to adopted value by WHO) is 50 mg NO_3/l of water. (Soren, R. , Boie S. F, 1994). Slovenian legislation on the other hand has set up standard of maximum nitrate content in drinking water being 44 mg NO_3/l of water.

Bottled water is not supposed to contain any NO_2, while regular water is allowed to contain up to 0,005 mg/l of NO_2-N under regular conditions and not more than 0,05 mg/l of NO_2-N in irregular conditions. And if our Slovenian legislation, which almost entirely corresponds to Nitrate Directive and Code of Good Agricultural Practice, is followed and obeyed, there should be no fear in future to expect the agriculture to be polluter of ground water and surface waters.

The most important regulation regarding expected processes of change in agriculture is supposed to be The Regulation on animal excrement's management. This regulation gives different norms. The most important are the following:

a) The highest quantity of manure allowed to be used on agricultural land as well as limitations for the use of the manure in specific soil conditions:

Table 3: Nitrogen balance - farm level - Trnovsko Banjška planota

Region	INPUT				OUTPUT	BALANCES	
	1	2,1	2,2	2,3	3	4,1	4,2
	Nitrogen	Nitrogen from agric. production			Nitrogen	Nitrogen balances	
	from the atmosphere kg/ha	Mineral fertiliser kg/ha	Liquid manure kg/ha	Total N supply kg/ha	uptake kg/ha	Gross balance kg/ha	Net balance kg/ha
SLOVENIA 1991	15,5	47,2	89,8	137,0	70,8	66,3	56,3
SLOVENIA 1992	15,5	43,5	81,1	124,6	53,3	71,3	62,5
SLOVENIA 1993	15,5	57,3	81,9	139,2	52,1	87,1	78,1
SLOVENIA 1994	15,5	60,7	77,5	138,3	91,0	47,2	39,5
farm 7	15,5	0,0	66,2	66,2	37,5	28,6	24,3
farm 13	15,5	18,4	74,9	93,3	38,2	55,1	48,1
farm 16	15,5	15,4	139,3	154,6	41,1	113,5	87,2
farm 14	15,5	53,0	99,5	152,5	59,0	93,6	79,2
farm 3	15,5	13,1	55,6	68,7	69,0	-0,3	-1,4
farm 2	15,5	0,0	76,7	76,7	82,5	-5,8	-13,3
farm 4	15,5	1,1	54,7	55,8	42,2	13,6	12,7
farm 9	15,5	7,5	76,3	83,8	42,5	41,3	33,9
farm 6	15,5	3,7	65,1	68,8	40,6	28,2	24,2
farm 10	15,5	4,3	76,6	80,9	39,4	41,5	34,0
farm 8	15,5	0,0	75,1	75,1	40,8	34,3	27,3
farm 15	15,5	0,0	140,9	140,9	37,9	103,0	76,3
farm 1	15,5	0,0	29,2	29,2	37,7	-8,5	-1,8
farm 5	15,5	10,2	54,2	64,4	37,6	26,8	26,0
farm 11	15,5	0,0	88,3	88,3	41,8	46,5	35,5
farm 12	15,5	16,0	78,8	94,8	41,3	53,5	45,4
AVERAGE*	15,5	10,5	72,3	82,9	47,0	35,9	29,7

* average for all 16 farms
column 2.3 = 2.1 + 2.2
col. 4.1 = 2.1 + 2.2 - 3
col. 4.2 = 1 + 2.1 +2.2*0.7 -3

-The maximal allowed intensity of raising animals is 3 LU/ha for cattle or 2 LU/h
for pigs and poultry.
- Application of organic manure is not allowed during winter time on frozen soil.
-Application of organic fertiliser is not allowed on soil saturated with water.
-Application of organic fertiliser is not allowed in temporarily flooded areas.
-Application of organic fertiliser is not allowed near water streams (10 m away
from the stream) and in the depressions where there is no run-off of water.
-It is not allowed to apply liquid manure on bare soil in the period from Nov. 15 till
Feb. 15.
-Application of organic fertilisers on water aquifer protected areas has to be done in
agreement with the local regulations valid for those areas.
-In the vicinity of spring water and in underground water pumping areas waste
water can not be drained to spring water or underground water in any case.
b) The highest quantity of N, P2O5, and K2O allowed to be used per hectare is 210 kg N, 120 kg P2O5 and 300 kg K2O.
c) Animal wastes should be stored in a suitable arranged dung yards and cesspools. Dung yards and cesspools are supposed to be arranged in the way that there is no danger of leaking through and pouring over the underground water.
d) It is set up 5 years grace period needed for the adjustment of farms to these regulations
as follows:
- the adjustment of the number of animals (LU) according to the area of land available on the farm,
- possible rent of additional land according to the contract, the construction of necessary dung yards and cesspools for hard and liquid manure according to the restrictions.

The extension service is obliged to take care of the transfer of necessary knowledge to the farmers. The control over implementation of mentioned regulations is supposed to be done by agricultural inspection, belonging to Ministry of agriculture.

6. CONCLUSIONS

Nitrate leaching into ground water and surface waters influenced by agricultural production is supposed to be a problem in the karstic region of Eastern Slovenia - 'Trnovsko-Banjska' plateau under certain conditions; point source pollution due to the lack of dung yards and/or cesspools or higher concentration of animals per ha can cause the problem with nitrate pollution in the groundwater. Therefore a nitrate policy is being in the phase of preparation in order to reduce mineral surpluses in agriculture and to meet the standards of nitrate in drinking water.

Mineral balances at national, regional and farm level were calculated based on the 'corrected normative approach'. In Ajdovščina community and 'Trnovsko-Banjska' plateau
region the nitrogen net-balance surplus is less than 36 kg N/ha while average net-balance surplus for Slovenia is about 56 kg N/ha.

In 'Trnovsko-Banjska' plateau the average yields and uptake by crops are low and therefore non-point source pollution caused by mineral fertilisation in this region is not considered a serious problem. The high nitrogen surpluses can be caused by high animal density per ha. The stocking rate over 2,1 LU/ha can cause net-balance surplus over 100 kg N/ha; in this case organic fertilisation can be considered a serious pollution source

The average net nitrogen surplus in private farms in other parts of Slovenia is 46 kg/ha. It is a little bit higher than Slovenian average in 1994 (40 kg/ha). While in state farms is nearly three times higher than Slovenian average - 117 kg/ha.

In the Karst region of 'Trnovsko-Banjska' plateau with limited growing conditions for crops (climate, soil depth, shallow soil) just small increase of livestock density can cause considerable nitrogen surplus. For that reason the restrictions for the application of chemical fertiliser and manure on hilly karstic regions had to be more rigorous than in plains.

Slovenian legislation intends to level this situation with quite strict regulations which are in agreement with EC Nitrate Directive and Code of Good Agricultural Practice.

7. REFERENCES

Brower, F., M. et all.: Mineral balances of the European Union at farm level, Agricultural Economics Research Institute, Hague, 1994

The Council Directive of 12 December 1991 concerning the protection of waters against pollution caused by nitrates from agricultural sources, (91/676/EEC), Official Journal of the European Communities, 31. 12. 91 No L 375/1

Leskošek, M.:, Gnojenje, Kmečki glas 1993, Ljubljana

Matičič, B., Avbelj, L., Vrevc, S., Jarc, A.: Water pollution by nitrate in Slovenia, DELO, March 19, 1995

Matičič, B., Avbelj, L.: Water Pollution by Nitrate in Slovenia: Future Standards and Policy Instruments, ICID Proceeding: 6th Drainage Workshop - DRAINAGE AND THE ENVIRONMENT, TOPICS No. 4, Ljubljana, April 1996, 9 p.

Soren, R., Boie S. F.: National and EC Nitrate Policies, Kobenhavn, 1994

Schleef, K.-H., Kleinhanβ, W.: Mineral balances in agriculture in the EU, Part I: The regional level, Braunschweig, 1994

Statistical yearbook Republic of Slovenia, 1991, Ljubljana 1991

Statistical yearbook Republic of Slovenia, 1994, Ljubljana 1994

Zupan, M. et all.: Water quality in Slovenia year 1992, 1993, 1994, Hydrometeorological Institute of Slovenia, Ljubljana,

Water - Quality Survey, Unesco/WHO, 1978, Manual on water - Quality Monitoring, WMO, 1988.

Tracer Hydrology 97, Kranjc (ed.) © 1997 Balkema, Rotterdam, ISBN 90 5410 875 4

Tracer tests applied to nitrate transfer and denitrification studies in a shaly aquifer (Coët-dan basin, Brittany, France)

Hélène Pauwels, Wolfram Kloppmann & Jean-Claude Foucher
BRGM, Research Direction, Orléans, France

Patrick Lachassagne & Anne Martelat
BRGM, Research Direction, Montpellier, France

ABSTRACT: The simultaneous injection of a conservative tracer (Br^-) and a reactive contaminant (NO_3^-) has helped elucidate pollutant behaviour and transport mechanisms in a shaly aquifer in Brittany. Rapid nitrate removal by autotrophic denitrification due to the presence of pyrite in the solid phase has been demonstrated. Tracers are partly transported along larger fractures, and partly along interconnected "microfissures" where chemical conditions seem more favourable for denitrification.

1. INTRODUCTION

Surface waters in Brittany (western France) are highly contaminated by nitrate as a result of intensive agricultural activity, and such pollution also affects groundwater. An important factor when predicting the evolution of groundwater contamination is that of contaminant residence time in the aquifer, as one process of nitrate removal from groundwater is natural denitrification (Hiscock et al., 1991). Denitrification can be either heterotrophic or autotrophic, and special attention must be paid to reaction mechanisms and kinetics. A geochemical study has demonstrated that both mechanisms occur in the groundwater of the Coët-dan basin in Southern Brittany (Pauwels, 1994). Based on hydrochemical and hydrogeological monitoring which started in 1992, artificial tracer tests with nitrate have been carried out to study in situ denitrification and solute transport processes in a shaly, semiconfined pyrite-rich aquifer at depths as much as 100 m.

2. GEOLOGICAL AND HYDROGEOLOGICAL SETTING

The Coët-dan basin is a small drainage basin (12 km^2) underlain by Proterozoic (530 Ma) schists comprising sandstone, siltstone and claystone (Fig. 1). The test site is near the outlet of the basin, where six cased wells (4-97 m deep) were drilled along a profile of about 20 m near the Coët-dan

river. Screened intervals give access to various parts of the aquifer. Beneath alluvial loam, a zone of highly weathered bedrock (<7m thick) was encountered underlain by siltstone. Pumping tests and monitoring of piezometric levels revealed that the local hydrogeological setting consists of a superficial zone (low permeability, high storage coefficient and dominant interstitial porosity) underlain by a fissured locally highly permeable aquifer corresponding to the unweathered bedrock. Although hydraulic connection between the two zones is rather poor, the vertical flow component cannot be neglected; during the high-water period, ascending flow contributes to the recharge of the shallow aquifer and the river, whereas at low water, the flux is inverted and shallow groundwater penetrates into the deep aquifer.

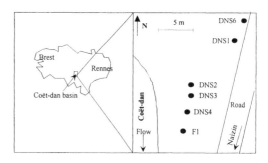

Fig. 1: Location of the test site in the Coët-Dan basin, Brittany, France

3. HYDROCHEMICAL BASELINE CONDITIONS:

High anthropogenic nitrate and chloride concentrations have been measured in the shallow aquifer (200 mg.l^{-1} of NO_3). However, nitrate concentration at depth varied with time from the detection limit to around 10-15 mg.l^{-1}. During the descending flow period, a mixing of shallow and "deep" groundwater occurs, causing nitrate and chloride contamination of the "deep" groundwater by the shallow groundwater. The relative contribution can be calculated by recording concentrations of chloride, which presents an inert behaviour. Measured NO_3 concentrations are lower than might be expected after mixing. It can be concluded that denitrification takes place very rapidly, predominantly by an autotrophic process implying iron sulphide dissolution and sulphate release (Pauwels et al., 1996):

$$5 \; FeS_2 + 14 \; NO_3^- + 4H^+ \longleftrightarrow 7N_2 + 10 \; SO_4^{2-} + 5 \; Fe^{2+} + 2 \; H_2O$$

4. EXPERIMENTAL SET-UP:

Tracer tests with the injection of nitrate and bromide were carried out between wells F1 and DNS1 over a distance of 15.2 m between the injection point (F1) and the production well (DNS1). The two wells are screened between 32 m and 82 m and between 1 m and 97 m respectively. DNS1 was pumped during the whole duration of the test with a mean constant pumping rate of 0.39 l s^{-1}. A resulting drawdown of 10 m was observed in DNS1. After injection, the water column of F1 was continuously mixed by recirculation (extraction near the water surface, reinjection at the bottom level). Sampling was performed at appropriate time steps by an operator during the day and by an automatic sampler at night for DNS1 and F1 in order to establish the real injection function, which differs from a pure Dirac-pulse.

Initially, only NaBr (5 kg) was injected into F1, which enabled the tracer distribution with depth to be studied in the production well DNS1 using a downhole sampler. Secondly, after bromide concentration in the production well fluids had decreased below the detection limit, a single pulse of 3 kg of NaBr and 5 kg of $NaNO_3$ was injected into F1. Bromide was chosen as a reference to study the behaviour of nitrates as chemically it is a very conservative tracer with low sorptive properties.

5. RESULTS AND INTERPRETATION:

The first indications of both tracers arrived virtually simultaneously only 5 h after injection (Fig. 2). The shape of the first part of the tracer curves is identical with a well defined maximum 20 h after injection. Differences then appear in the tailing of the curves where bromides decrease much slower than nitrates. Global tracer recovery is about 75% for bromides and 50% for nitrates.

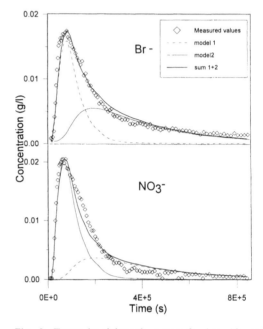

Fig. 2: Tracer breakthrough curves for bromide and nitrate in DNS1 and analytical curves modelled with CATTI

In order to model the observed breakthrough curves, we chose a two dimensional dispersion model (CATTI) designed to simulate tracer tests in a porous homogeneous 2D aquifer of infinite lateral extension (Sauty et al., 1989). Hydrochemical logging of DNS1 during the first tracer test with Br- (Fig. 3) has shown that the tracer propagates preferentially at depths below 50 m which implies transport in the unweathered fissured bedrock. As shown by pumping tests (Martelat & Lachassagne, 1995), this zone behaves approximately like a homogeneous porous aquifer, thus justifying the choice of the model. It is assumed that the tracers propagate under the predominant influence of the hydraulic gradient imposed by pumping in DNS1. Radial convergent flow conditions were assumed

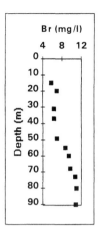

Fig. 3: Bromide concentrations in the production well DNS1 after the first injection of NaBr in F1

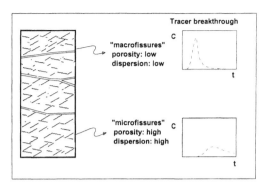

Fig. 4: Schematic representation of a double fissure porosity aquifer

during modelling. The measured variations of the input function (e.g. injected tracer mass in F1) with time were accounted for by a convolution procedure. Calibration of the transit curve of the conservative tracer Br- by an equivalent unimodal porous aquifer model is not satisfactory as the peak and tail of the tracer breakthrough cannot be correctly fitted simultaneously.

We have therefore taken into account a different conceptual model representing the fissure aquifer by an equivalent porous aquifer with a double porosity (Fig. 2). Part of tracer transport takes place in "macrofissures" or open fractures. Their volume is small compared to the total volume of the aquifer which means that equivalent porosities are low and that tracer transport relatively rapid. Transport paths are probably fairly direct and hydrodynamical dispersion low.

Another fraction of the tracer passes through a network of "microfissures" representing the major part of the total fracture volume. Equivalent porosity and kinematic dispersion is expected to be higher, and average linear velocities lower than in the "macrofissures". Figure 4 illustrates this highly schematic model.

The porosity of the equivalent porous medium reflecting the volume of fissures relative to the total rock volume is about eight times higher for the "microfissures" than for the fractures, and longitudinal dispersivity varies by a factor of two. Almost the same amount of Br^- and NO_3^- passes through the "macrofissures" (28% and 25% of the total injected mass). A striking fact is the nitrate loss

in the second medium; recovery of tracer transmitted through the "microfissure" network is only 22% for nitrates compared to 45% for Br.

6. CONCLUSIONS:

Two principal conclusions may be drawn from the observed tracer behaviour:

1. Nitrate removal in the aquifer by autotrophic denitrification is rapid and can be observed even during a tracer test of short duration (10 days).

2. Nitrate reduction seems to be mainly due to a "microfissure" type porosity - reduction can be neglected in larger fractures. Two reasons are proposed:

- Transport through larger fractures is relatively rapid not leaving the time for significant denitrification.

- Chemical conditions are favourable for denitrification in the "microfissure" network: the surface of the rock matrix available for autotrophic reactions implying disseminated pyrite is high and strongly reducing micro-environments may be encountered where nitrate decomposition is accelerated.

Acknowledgements: This is BRGM contribution N⁰. 97006. Technical equipment on site was funded by CST-BVRE. Rowena Stead is thanked for editing the English.

7. REFERENCES:

Hiscock K.M., Llyod J.W. and Lerner D.N. (1991). Review of natural and artificial denitrification of groundwater. Wat. Res., 25 (9), 1099-1111.

Martelat A. and Lachassagne P. (1995). Bassin versant représentatif expérimental du Coët Dan (Naizin Morbihan)- hydrogéologie : détermination des caractéristiques hydrodynamiques du système aquifère au lieu dit Le Stimoes. BRGM Report N° R38474DR/HYT95, pp54.

Pauwels H. (1994). Natural denitrification in groundwater in the presence of pyrite: preliminary results obtained at Naizin (Brittany, France). Mineralogical magazine, vol 58A, 696-697.

Pauwels H., Martelat A., Lachassagne P., Foucher J-C. and Legendre O. (1996). Evolution des teneurs en nitrates dans l'aquifère du bassin versant du Coët Dan (Naizin, Morbihan). Proceedings of ESRA 96-L'eau souterraine en région agricole- Poitiers 9-12 Septembre 1996. S4 81-84.

Sauty J.P., Kinzelbach, W. and Voss, A. (1989). Computer aided tracer test interpretation. BRGM Report N° 89 SGN 217 EEE, pp72.

Tracer Hydrology 97, Kranjc (ed.) © 1997 Balkema, Rotterdam, ISBN 90 5410 875 4

Tracer experiments on the site of NPP Jaslovské Bohunice – The Slovak Republic

J. Plško, M. Kostolanský, T. Kovács & J. Benko
Ekosur, Piešťany, Slovak Republic

J. Hulla
Slovak Technical University, Bratislava, Slovak Republic

ABSTRACT: At the site of Jaslovské Bohunice, in the west part of the Slovak Republic, are located three nuclear power plants (NPP). The first NPP, after an accident ceased production in 1977. The radionuclides penetrated the surrounding groundwater. Additionally an extensive system for monitoring and protection was constructed. This enables one to observe groundwater flow and its pollution. Tracer methods applied to this monitoring system included: one-borehole methods, more-borehole experiments, and analysis of the radioactive substances in the groundwater. In this paper attention is given not only to methodical problems, but also to problems involving the evaluation, interpretation and use of the measurement results.

1. INTRODUCTION

We presented information about the influence of the NPP Jaslovské Bohunice (Fig. 1) on groundwater during the International Conference About Nuclear Pollution in Prague (Plško, Kostolanský & Polák 1993), the First International Congress on Environmental Geotechnics in Edmonton (Hulla, Plško & Taha 1994) and the Second International Congress on Environmental Geotechnics in Osaka (Plško, Kostolanský, Benko, Hulla, Kovács 1996).

Fig. 1. Nuclear complex in Jaslovské Bohunice

At present a more detailed study has been made of the hydrogeological conditions and transport characteristics, about the groundwater contamination, and significant progress has been obtained for the numerical modelling of transport processes too.

2. MONITORING SYSTEM

Radioactive contaminants penetrated through the unsaturated zone (loess, variable thickness of 2 - 20 m) in a vertical direction into flowing water (gravel soils with a different content of sand particles, with a thickness of 15 - 20 m) - Fig. 2. Deeper layers are formed by clays. The groundwater table is located at a depth of about 20 m, in the territory of the NPPs.

The monitoring system of the hydrogeological boreholes is constructed in the surroundings of each NPP, both inside and outside of its territory.

There are, at present, 143 boreholes. Some of them are in Fig. 3. It is possible to use almost every borehole in cases where there is a need for drawing the contaminated water. The hydrogeological monitoring is regularly performed as radiological monitoring.

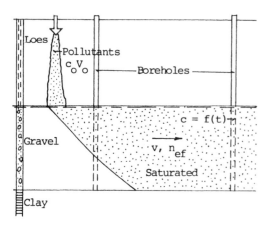

Figure 2. Geological conditions and penetration of the contaminants into the groundwater

3. GROUNDWATER FLOW

It is possible to obtain an idea of the groundwater flow direction from Fig. 3. These directions basically don't change during the year. The isolines of the water levels that are based on the results from the measurements and from the modelling by the known program Modflow constructed did. Hydraulical gradients moved in the range i = 0,0004 to 0,004.

The permeability coefficients from the pumping experiments in gravelly soils were for the territory of the NPPs in the range $k = 5.10^{-4}$ up to 2.10^{-2} m.s^{-1}, but in the surroundings lesser values were obtained (6.10^{-5} up to 9.10^{-3} m.s^{-1}). The pumping experiments were evaluated using different methods. Besides these tracer and grain methods were used too.

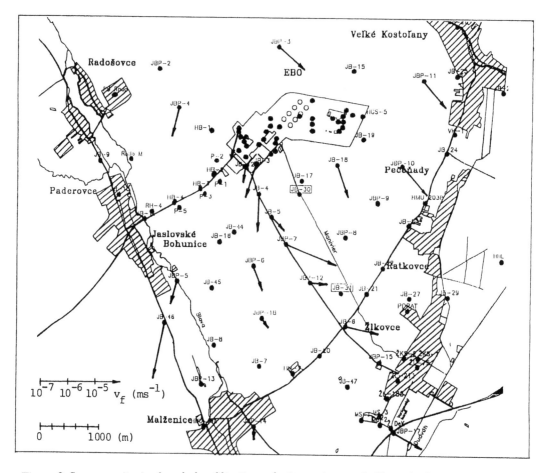

Figure 3. Some monitoring boreholes, filtration velocity vectors, and villages in the NPPs surrounding

One-borehole tracer method was used for determining the filtration velocity. This method is based on the vertical water flow in the borehole measurements. Vertical flow is determined especially by the connection of the differently pressured horizons. For the measurements, the radioactive $Na^{131}I$ and conservative NaCl solution were used as tracers. The filtration velocity is calculated after Halevy et al. (1967) according to the following equation

$$v_f = \frac{\Delta q_v}{\alpha \, d \, \Delta h} \qquad (1)$$

Δq_v is either an increase or decrease vertical discharge in the part of the borehole with the high Δh, d is the inside diameter of the perforated tube, α is the drainage influence of the borehole.

In Fig. 4 the results of the measurements in the borehole N-6, as depth dependences of the vertical discharge and filtration velocity are given. Some notes from these results are written in point 4.

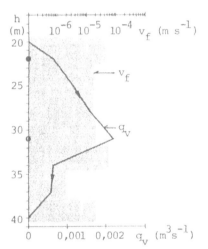

Figure 4. Results of the vertical flow measurements in the borehole N-6: h - depth, q_v - vertical discharge, v_f - filtration velocity, ●, ○ positions for water sample takings

After eq. (1) we obtain the filtration velocities for the explicit depth of the borehole. Average values for the whole borehole

$$\overline{v_f} = \frac{\sum v_f \, \Delta h}{\sum \Delta h} \qquad (2)$$

are given in Fig. 3 in vector form for the NPPs area and surrounding.

The results of the statistical evaluation of the filtration velocities (total 443 measurements values) are shown in Fig. 5. Medium value is $v_f = 7.5.10^{-7}$ m.s^{-1}, average value $\overline{v_f} = 2.10^{-6}$ m.s^{-1} can be crossed with probability $P = 25$ %.

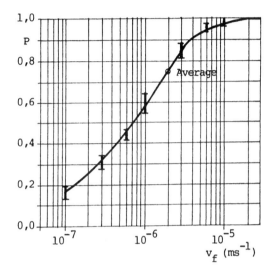

Figure 5. Empirical values and theoretical distribution of the filtration velocities (v_f), P - probability

Weibull´s theoretical function gives relatively good realism to the measured values

$$P\{v_f\} = 1 - \exp\left(-\frac{v_f}{\beta}\right)^{\vartheta} \qquad (3)$$

with the coefficients $\vartheta = 0.722$, $\beta = 1.23.10^{-6}$ m.s^{-1}.

For each borehole by means of tracer method the permeability coefficient from Darcy´s law was estimated (k_T). Except for these results, we used the permeability coefficients from the pumping tests (k_P) and from the grain analyses after Beyer-Schweiger (k_B) and Carman-Kozeny (k_C) formulas. Statistical analysis results for the tested area are shown in Fig. 6. It can be noted that, coefficients of permeability k_C are generally less than k_B. The median and average value from vertical movement tracer method is lower. It can be clearly seen that pumping tests $(k_P = 4.6.10^{-3}$ m.s$^{-1})$, vertical flow tracer measurement median $(k_T = 5.1.10^{-3}$ m.s$^{-1})$, and Beyer-Schweiger esti-

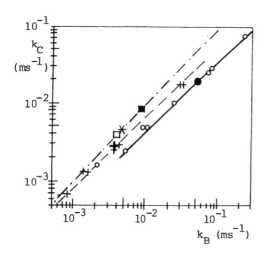

Figure 6. Correlation dependence between coefficient of permeability in the test area (k_B - Beyer-Schweiger, k_C - Carman-Kozeny) from grain size analysis (● - average, + - median), vertical flow tracer measurements (■ - average, □ - median), and from the pumping tests (*).

mated median ($k_B = 3{,}7.10^{-3}$ m.s^{-1}) give comparatively the same results.

4. POLLUTANTS TRANSPORT

In our conditions (Fig. 2), after impulse penetration of pollutants into flowing water through the unsaturated zone, three-dimensional transport takes place. In some, with not very long distances from the penetration, this transport changes to two-dimensional. The simplified mathematical model for two-dimensional transport of conservative pollutants after Bear (1972), Klotz & Moser (1974) is:

$$\frac{\partial c}{\partial t} = D_L \frac{\partial^2 c}{\partial x^2} + D_T \frac{\partial^2 c}{\partial y^2} - v \frac{\partial c}{\partial x} \qquad (4)$$

The solution of this equation according to Lenda & Zuber (1970) is

$$c(x,y,t) = \frac{c_o V_o}{2 h n_{ef} \sqrt{\pi D_L t}} \exp\left[\frac{-(x-vt)^2}{4 D_L t}\right] *$$
$$* \frac{1}{2\sqrt{\pi D_T t}} \exp\left[\frac{-y^2}{4 D_T t}\right] \qquad (5)$$

where, D_L and D_T are the longitudinal and transverse coefficients of dispersion respectively, v - the distance velocity in x direction ($v = x/t_o$), c_o - the initial concentration, V_o - the initial volume, c - concentration at time t, x and y - the coordinates, h - the depth of water, n_{ef} - effective porosity, t_o - the time for maximum concentration.

In some cases it is possible that two-dimensional transport of pollutants is substituted by one-dimensional and is characterized after Bear (1972), Klotz & Moser (1974)

$$\frac{\partial c}{\partial t} = D_L \frac{\partial^2 c}{\partial x^2} - v \frac{\partial c}{\partial x} \qquad (6)$$

From Lenda & Zuber (1970) mathematical solution of eq. (6) gives

$$c(x,t) = \frac{c_o V_o}{2 A n_{ef} \sqrt{\pi D_L t}} \exp\left[\frac{-(x-vt)^2}{4 D_L t}\right] \qquad (7)$$

Longitudinal dispersivity is

$$\alpha_L = \frac{D_L}{v} \qquad (8)$$

If we are interested in c_{max} only, and take into acount also adsorption and decay, from equations (7) and (8) can be derived

$$\frac{c_{max}}{c_o} = \frac{V_o}{2 A n_{ef} \sqrt{\pi \alpha_L x R}} \exp\left(-0{,}693 \frac{x R}{v T}\right) \qquad (9)$$

where c_{max} is the maximum concentration at distance x, R - the retardation factor, T - the half life of the radioactive tracers, or pollutants.

In the next parts of our paper we analyse some field experiments, problems of the longitudinal dispersivities, and real and predicted radioactive pollution in our conditions.

4.1 Field experiments

In the NPPs area a system of experimental boreholes is located for the tracer measurements - Fig. 7. In this system two kinds of experiments were performed.

The first experiment was done on the natural groundwater flow conditions and important re-

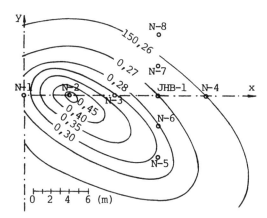

Figure 7. Experimental boreholes and groundwater level lines at the artificial infiltration of water to the N-2

sults were published in the paper Hulla, Plško and Taha (1994).

The second experiment was done in 1996 under non-natural conditions. During this experiment the water discharge was pumped $q = 0,004$ $m^3.s^{-1}$ permanently into the borehole N-2. Groundwater isolines in this conditions are given in Fig. 7.

The stationary flow was injected into borehole N-2 water solution of the NaCl (beginning 10.7.1996 at 10^{00} h., ending at 10^{30} h., $q = 0,0049$ $m^3.s^{-1}$). The changes of the conductivity, concentration NaCl and Cl in the depth and time dependences are given in Fig. 8. NaCl solution outflow from the boreholes to the gravel soils at a depth of 21 to 31 m approximately in horizontal directions, deeper was transported in the borehole in vertical directions and outflow from their down part.

Depth and time dependences of conductivity in some boreholes during the experiment were measured. In Fig. 9 these dependences are given for the borehole N-6 (distance from N-2 is 9,8 m). Influence of the vertical flow in the borehole is possible to see very clearly.

In all boreholes suitable positions for water samples taken from the vertical flow measurements were established. These positions are characterized by the most intensive inflow of water from the soils to the borehole ($\Delta q_v / \Delta h$ = max). For example in Fig. 4 this positions are fixed. Water samples with the pump from the borehole takings and chemical analyse for Cl

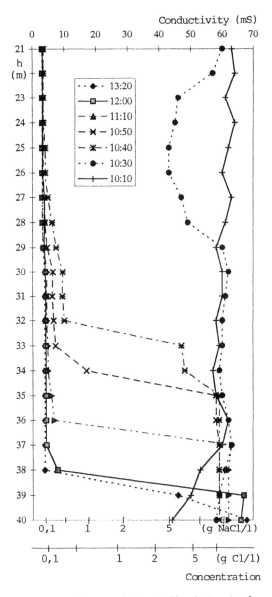

Figure 8. Dilution of the NaCl solution in the injection borehole N-2 (time of injection 10^{00} - 10^{30} h)

concentrations in labor were performed.

Distance velocities in this relatively small area of the experiments moved in a relatively great range between $v = 9,9.10^{-5}$ to $1,9.10^{-3}$ $m.s^{-1}$ and they are given in table 1.

335

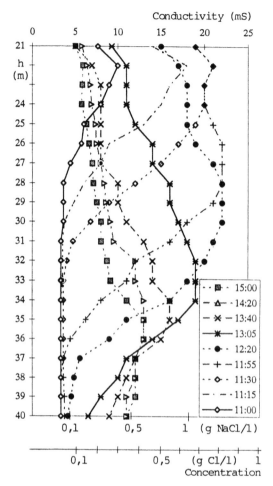

Figure 9. Transport of NaCl solution through the borehole N-6 (time of injection 10^{00} - 10^{30} h. into N-2)

4.2 *Longitudinal dispersivities*

In our experiments, performed in natural and artificial conditions help the equations for one-dimensional and two-dimensional transport, and longitudinal dispersivities were evaluated. Their dependences on the distances are given in Fig. 10 together with the data, that was obtained from literature resources.

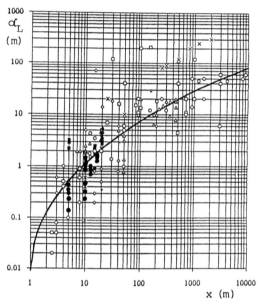

Figure 10. Dependence of the longitudinal dispersivities (α_L) on the distance (x) from the tracer experiments in Jaslovské Bohunice: ■ - natural flow, ● - artificial radial flow; ▲ - Čunovo (Danube gravel), – Beims (1982), △ - Seiler (1985), □ * - Gelhar (1986), o - Luckner & Šestakov (1986), x - Drost, Hoehn & Kováč (1991)

Table 1. The distance velocities at the tracer experiment

Transport between	Distance r (m)	Time t_o (s)	Velocity $v = r/t_o$ (m.s^{-1})
N-2 - N-1	5,1	5220	9,8 E-4
N-2 - N-3	5,0	2640	1,9 E-3
N-2-JHB-1	9,7	72000	1,3 E-4
N-2 - N-6	9,8	7200	1,4 E-3
N-2 - N-7	10,6	46800	2,3 E-4
N-2 - N-5	10,7	14400	7,4 E-4
N-2 - N-8	12,5	36000	3,5 E-4
N-2 - N-4	14,9	151200	9,9 E-5

Theoretically the longitudinal dispersivity would not be dependent on the distance. The results obtained from the field experiments are obviously different. Some parts of the conservative tracers or pollutants should be fixed inside of the dead pores. It is necessary to calculate either with the adsorption and constant value of longitudinal dispersivity, or with the variable dispersivity, according to Fig. 10.

4.3 Predicted and real pollution

Results of the experiments are possible to evaluate in dependence between relative concentrations c_{max}/c_o and distance x from the eq. (9). We can make for the stable tracers or pollutants a small correction to this equation in the form

$$\frac{c_{max}}{c_o} = \frac{K V_o}{2\,A\,n_{ef}\,\sqrt{\pi\,\alpha_L\,x\,R}} \qquad (10)$$

(K is the correction factor).

Measured values obtained from our NPP experiments and predicted dependence are given in Fig. 11.

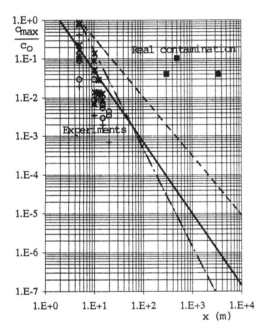

Figure 11. Measured and predicted dependence for contaminant transport in NPP conditions: c_o - initial concentration (volume activity), c_{max} - maximal concentration (volume activity) in the distance x; $V_o \cong 10$ m^3, $h = 20$ m, $n_{ef} = 0.15$; - - - - $\alpha_L = 0.06$ m, $K = 1$, $R = 1$; —— $\alpha_L = f(x)$ - Fig. 10, $K = 1$, $R = 1$, or $\alpha_L = 0.1$ m, $R = 100$, $K = 5.357\,x^{-0.307}$; –·–·– $\alpha_L = 0.1$ m, $R = 10$, $K = 18/x$.

Dependence in Fig. 11 pays for the impulse pollution. For practical use the best results are ilustrate by continued line.

Real contamination of groundwater by tritium in NPP area at present is given in Fig. 12.

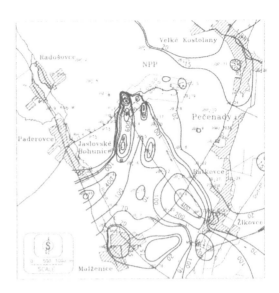

Figure 12. Isolines for tritium specific activity (Bq.dm^{-3}) in the groundwater at present.

Maximal values in the NPP area and its surroundings in Fig. 12 we marked in Fig. 11 at the definite distance. It is possible to say, that radioactive contamination had a greater volume compared to the tracer experiments, or this pollution didn´t have impulse character.

Knowledge from the tracer experiments used were for the numerically modelled transport of the tritium in the groundwater. One example from the results is given in Fig. 13. The important facts lead to the opinion that the groundwater contamination in the NPPs territory was caused not only by the accident in 1977, but other active resources in the last ten years, too.

5. CONCLUSIONS

Tracer experiments on the site of NPPs Jaslovské Bohunice gives useful results for the analyses of groundwater flow and radioactive contamination. The monitoring system enables the control of the groundwater quality, identifying the leaking contaminants and through the pumping system protection of the groundwater. The result for people, according to the radiological groundwater pollution, will not endanger health. There are not any regulatory precautions for people in the surroundings of NPPs. Despite this, other sources of clean drinking water have been secured for them.

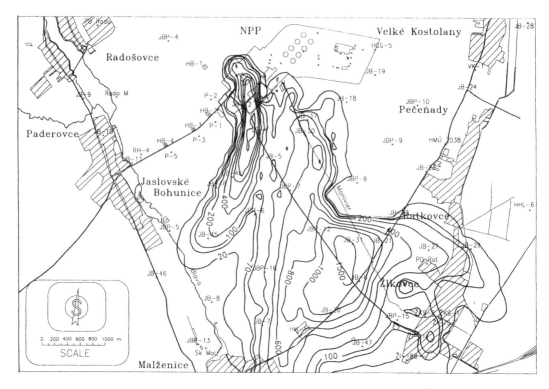

Figure 13. Isolines for the numerically modelled tritium specific activity (Bq.dm^{-3}) in the groundwater

ACKNOWLEDGEMENTS

The team of authors would like to offer their heartiest thanks to the autorities of the Nuclear Power Plants in Jaslovské Bohunice for their continuous care and financial help in solving the problems connected with groundwater protection.

REFERENCES

Bear, J. 1972. *Dynamics of fluids in porous media.* New York: American Elsevier.

Halevy, E. et al. 1967. Borehole dilution techniques: a critical review. In *Isotopes in Hydrology:* 531-564. Vienna: IAEA.

Hulla, J., Plško, J. & Taha, M. 1994. Influence of nuclear power stations on the groundwater. In W.D.Carrier III (ed.), *First ICEG:* 867-872. Richmond: BiTech.

Klotz, D. & Moser, H. 1974. Hydrodynamic dispersion as aquifer characteristic. In *Isotope techniques in groundwater hydrology:* 341-355. Vienna: IAEA.

Lenda, A. & Zuber, A. 1970. Tracer dispersion in groundwater experiments. In *Isotope hydrology:* 619-641. Vienna: IAEA.

Plško, J., Kostolanský, M. & Polák, R. 1993. Radioactive pollution and protection of groundwaters within the location of NPPs EBO. In *International conference about the nuclear pollution.* Prague.

Plško, J. et al. 1996. Radioactive pollution and protection of groundwater. In M.Kamon (ed.) *Environmental Geotechnics:* 1173-1178. Rotterdam: Balkema.

Tracer Hydrology 97, Kranjc (ed.) © 1997 Balkema, Rotterdam, ISBN 90 5410 875 4

Microbiologic activities in karst aquifers with matrix porosity and consequences for ground water protection in the Franconian Alb, Germany

Klaus-Peter Seiler
GSF-Institute of Hydrologie, Neuherberg, Oberschleißheim, Germany

Anton Hartmann
GSF-Institute of Soil Ecology, Neuherberg, Oberschleißheim, Germany

ABSTRACT: The evaluation of tracer experiments and the analyses of environmental tracers in ground waters of the Karst of the Franconian Alb demonstrates the importance of facies of limestones on tracer dilution. In bedded facies tracers propagate quickly and are detected with high recovery rates; contrary, in the reef facies tracers propagate slowly and are diluted to their detection limits (20 to 2 ng/l) within 1.5 to 2 km of distance. These differences in tracer dilution are attributed to a considerable matrix porosity in which pollutants enter by diffusion processes. Ground water out of this matrix has mean residence times of tenth of years as compared to flow velocities of hundreds of meters a day in fissures. - The existence of a matrix porosity in limestones may lead to a storage of contaminants for long times and would produce serious long-term ground water contamination problems as far as microbiological activity is absent. Recent isolation of microorganisms from karst ground waters and denitrification experiments demonstrate a considerable quick nitrate decomposition towards N_2O and N_2. With respect to pesticides there seems to exist a comparable decomposition. - From the research in the Franconian Alb a new concept of ground water protection in reef carbonates comprising two protection zones has been developed.

1 INTRODUCTION

Aquifers in consolidated rocks have commonly highly heterogeneous pore geometries and thus cover also wide ranges of hydraulic conductivities. Therefore the frequency distribution of individual flow velocities in such aquifers covers a wide range and is mostly discontinuous. Flow velocities may differ among several orders of magnitude in heterogeneous aquifers and thus create storage and drainage conditions for seepage and ground water flow. From this, many problems related to ground water exploration and to both short-term and long-term aspects of ground water protection arise, because the usual judgement of the propagation behaviour of pollutants in these aquifers is based on average hydraulic parameters from pumping and some tracer tests.

Short-term problems on ground water protection arise from high flow velocities. Long-term problems, however, are linked to low flow velocities as well as diffusive tracer/pollution exchanges between storage and drainage space.

Carbonate rocks of reef type facies are typical representatives of heterogeneous aquifers. High flow velocities and very low flow velocities linked to diffusive tracer/pollution exchanges between flow units occur in it. Artificial and environmental tracer experiments can be used effectively to gather data for short-term and long-term ground water resource assessment.

2 THE MALM CARBONATES OF THE FRANCONIAN ALB

The carbonates of the Franconian Alb, Germany, belong to two consecutive cycles of sedimentation, each of which starts with marls or marly limestones and ends with bedded limestones (Fig. 1). In the upper cycle, however, reefs replace limestones over significant stratigraphic intervals. These reefs have been transformed diagenetically into dolomites.

Bedded limestones typically lack syngenetic porosity (less than 2 vol.%), but are characterised by fissures and solution channels; fissure porosity amounts to 2 vol.%. Unlike the bedded limestones, the reef dolomites commonly have matrix porosities
- of sedimentary and early diagenetic age and
- due to weathering of dolomites

and fissure porosity
- of post-sedimentation age and
- from solution processes.

Matrix porosities in the reef facies have been measured and range mainly between 5 and 10 vol.% (Fig. 2); hydraulic conductivities of the matrix po-

rosity are lower than 10^{-7} m/s (Weiss 1987) as compared to individual flow velocities of a few hundred to some kilometres a day in fissures within the same facies. Therefore tracer experiments and an observation of environmental tracers have been conducted to obtain facies depended informations on tracer dilution.

3 RESULTS OF TRACER AND INCUBATION TESTS

The area of research covers about 1,000 km^2 out of an extensive karst area within the Upper Ju-

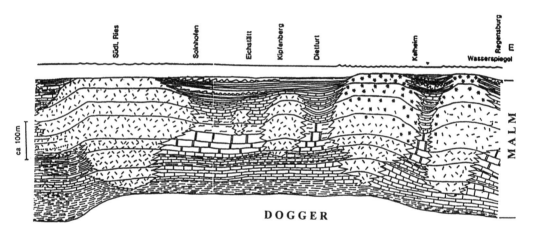

Figure 1: Generalised geologic cross section of the Upper Jurassic (Malm) of the Franconian Alb, South Germany (Meyer, Schmidt-Kaler 1989)

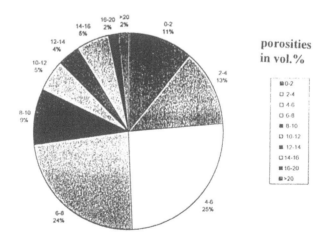

Figure 2: Matrix porosities of the reef carbonates in the Franconian Alb (results of U. Michels)

rassic. More than 150 tracer tests have been executed with fluorescent dyes (Seiler et al. 1991) that
behave conservatively with respect to flow velocities (Behrens 1971). The tracing distances ranged
from 1 to 12 km.

About half of the tracer tests have been performed in the bedded facies and both tracer recovery and flow velocities were mostly high. Tracer
tests in the reef facies, however, ended as a rule at
distances exceeding 1.5 to 2 km without recovery,
during an observation time of more than 7 years
(Seiler et al. 1991). Ground water recharge (250
mm/a) and the non-reactive tracer behaviour is the
same in both facies. Therefore differences in the
tracer dilution must be attributed to the dilution
capacity in the respective facies.

3.1 The dilution of dye tracers

The concentration-time curves for tracer tests in
the study area can be subdivided into three categories (Fig. 3).
- Curves with high concentration maximum and
  narrow geometry (curve 1 in Fig. 3) characterise
  high flow velocities (> 1.5 km/d) and low dispersivities, respectively. Such tracer experiments
  yielded recovery rates exceeding 50 % and can
  mostly be attributed to flow in solution channels
  or opened fissures.
- Curves with lower concentration maximum and
  less narrow in their width (curve 2 in Fig. 3).
  These experiments yielded recoveries between
  20 to 50 % and are attributed to flow in fissures
  with some diffusive tracer exchange between fissures with large and narrow width.
- Curves with very low concentration maximum
  over short (<1.5 to 2 km) distances (detection
  limit of fluorescent dyes 2 to 20 ng/l) and a pronounced tailing (curve 3 in Fig. 3). The recovery
  from these experiments was usually less than 1
  %. These curves have been produced by a diffusive tracer exchange between a small volume of
  water in fissures and a huge volume of water in a
  low permeable matrix. (time scale for curve 3 is
  about 100 times greater than for curves 1 and 2).

3.2 Tritium in ground waters out of both facies

Tracer tests mostly provide information about
small sectors of the ground water flow field and
preferential flow. In contrast, environmental isotopes like Tritium can be used to get areal informa-

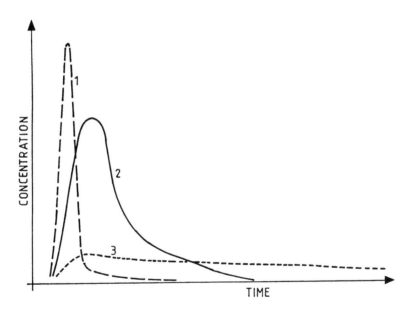

Figure 3: Types of concentration versus time curves in the Franconian Alb representing flow in. 1 = solution
channels; 2 = fissures of the bedded facies; 3 = reef facies (note that the time scale is 100 times
greater than in cases 1 and 2)

341

tion on the ground water flow field. During the dry season, for example, a clear difference exists in Tritium concentrations of ground waters out of bedded and reef facies;

- in the bedded facies Tritium concentrations in ground waters are equal to the mean for precipitation of the last few years.
- In areas with less than 1 % of tracer recovery Tritium concentrations are significantly lower than means for local precipitation.

The catchment area of Größdorf spring - as an example - is characterised of the reef facies; it has low Tritium concentrations (Fig. 4) and some of the tracer tests arrived with low flow velocities and a recovery of less than 1 %. The delineation of the recharge area on the basis of tracer tests leads to an area of about 9 km^2 (Fig. 4). Given a mean discharge of 140 l/s and a mean recharge of 250 mm/a the catchment area, however, should be about 17.5 km^2 in size. Obviously some of the tracer tests reach Größdorf spring without measurable recovery, because of dilution below detection limits over a short distance. This strong dilution results from transverse diffusion in addition to normal hydrodynamic dispersion (Seiler et al. 1989). The tracer diffusion out of fissures into the huge matrix volume in the reef facies also results in low Tritium concentrations; thus recent infiltration mixes with old waters stored in the matrix porosity.

3.3 Fate of nitrates in the reef facies

The huge storage capacity of the matrix of the reef facies for water may lead to long-term problems in groundwater quality as far as no microbiologic degradation exists. Nitrogen excesses amount in the area of research to 40 kg/(ha a) and ground water recharge ranges at 250 mm/a. From these numbers a concentration up to 16 mg N/l or 71 mg NO$_3$/l is expected in the Karst ground waters. The mean concentration of nitrate was about 40 mg/l (Fig. 5) deviating to a maximum of 80 mg/l. Contrary in the reef facies nitrates amount only to 15-20 mg/l. These differences may be attributed to either a slow charge or to biological degradation of nitrates in the matrix of the reef facies. Considering the non-reactive behaviour of chlorides in this area the relation of concentrations in both facies units is 1.5 (Fig. 5) indicate that the matrix porosity is not yet charged to the same degree with chlorides as the fissure flow in the bedded facies; the same is true for sodium and potassium. As compared to these components NO$_3^-$ and Atrazin (Fig. 5) have too low concentrations in the reef facies.

Water samples of springs and wells from areas with agriculture (A1-A2 in Fig. 6) and forestry (A3-A5 in Fig. 6) land use have been sampled. The areas A1-A4 consists of reef facies. The total bacterial numbers (DAPI-stain) and colony forming units (cfu) on R$_2$A agar (aerobic incubation, 22°C)

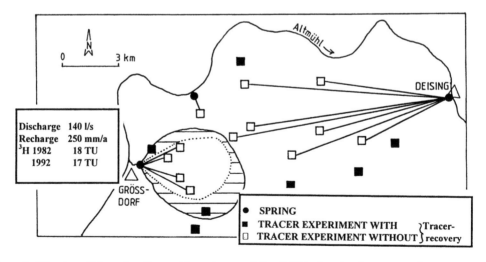

Figure 4: The size of the subsurface catchment area of the Größdorf spring by means of tracer tests (white area) and calculations (white and hatched area) on the base of mean discharge and mean recharge

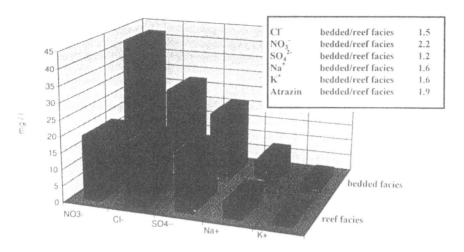

Cl⁻	bedded/reef facies	1.5
NO₃⁻	bedded/reef facies	2.2
SO₄²⁻	bedded/reef facies	1.2
Na⁺	bedded/reef facies	1.6
K⁺	bedded/reef facies	1.6
Atrazin	bedded/reef facies	1.9

Figure 5: Concentrations of agrochemicals in both bedded and reef facies in the Franconian Alb

as well as on a denitrification agar (R_2A agar + 0,5% KNO_3, anaerobic incubation, 22°C) has been determined. In addition, the denitrifying activity of some bacterial isolates was analysed.

In the samples of the agricultural areas there are higher numbers of total and viable counts as compared to samples of the forest areas. Obviously the number of microorganisms has grown according to land use because under agricultural areas better nutrition conditions for microorganisms exist than under forested areas. If this growth is linked to any incubation time is not yet well understood.

Some bacterial colonies, grown under anaerobic conditions, were isolated and their denitrifying activity was analysed. The physiological tests showed that all of these bacteria reduce nitrate to nitrite, but only a very small percentage of them produce gas in an anaerobic atmosphere and nitrate containing media.

It is, however, not only important to study microbiological activity of cultivated bacteria, but also to analyse the denitrification potential of ground waters out of the reef facies and of forested areas directly. Such ground waters have been sampled

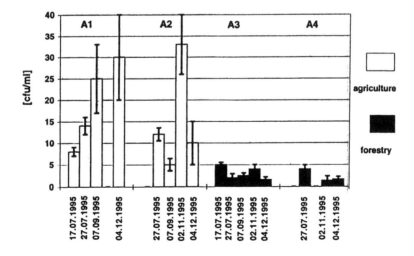

Figure 6: Colony forming units (cfu) on denitrification agar (R_2A agar + 0,5% KNO_3) anaerobic conditions in ground waters out of the reef facies

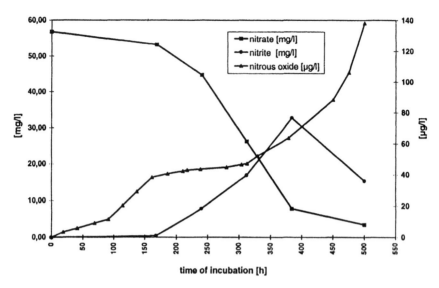

Figure 7: Development of nitrate and nitrite concentrations in anaerobically incubated water samples, 50 NO$_3^-$ mg/l and no additional carbon source was added

(300 ml), 50 mg/l NO$_3^-$ and no supplementary carbon source was added and the water samples were incubated for 3 weeks under anaerobic conditions. The DOC content of the groundwater samples of both areas was about 2 mg/l and the natural nitrate concentration was 9-12 NO$_3^-$ mg/ml. The NO$_3^-$, NO$_2^-$ and N$_2$O ratios in the water samples have been measured (Fig. 5). In sterile control samples no denitrification activity was determined. These results demonstrate, that the ground waters of these Karst areas exhibit a substantial denitrifying activity despite of very low organic carbon supply. It is hypothesised, that some autotrophic activity exists in these highly carbonated water samples (380 mg/l HCO$_3^-$).

Following these experimental results decomposition of pollutants in reef carbonates with high matrix porosities must be expected. With respect to time (Seiler et al. 1996) even disintegration of pollutants with low half time of degradation as Atrazin will be decomposed.

4   CONCLUSIONS FOR THE DELINEATION OF GROUND WATER PROTECTION AREAS

Contaminants in bedded limestones commonly cause very rapid and serious responses in the ground water. Therefore accidental spills are often difficult to be detected with the usual sampling frequency for chemical and bacteriological analyses, because of limited dispersion and very short residence times. Consequently in vases of the use of ground waters for drinking water purposes very rigid ground water protection measures had to be taken all over the catchment area.

In the reef facies, in contrast, response to both accidental and permanent pollution will only be strong in an area of 1.5 to 2 km in distance from the spring or well. In this zone tracer experiments should be executed to get informations on dilution potential and preferential flow paths. In this de lineated zone handling of potential ground water pollutants must be restricted.

At greater distances to the spring or well, dilution and disintegration potential may be sufficient to eliminate non-persistent pollutants. In this zone mean turn over times should be determined to know to which degree pollutants will be metabolised or mineralised. In this zone only handling with persistent pollutants must be restricted. Since pollutants may enrich gradually in matrix porosity and result in significant long-term risks for ground water quality monitoring programs have to supervise the development of concentrations of pollutants in this outer zones.

REFERENCES

Behrens, H. 1971. Untersuchungen zum quantitativen Nachweis von Fluoreszenzfarbstoffen bei ihrer Anwendung als hydrologische Markierungsstoffe. *Geologica Bavarica 64: 120-131; München*

Meyer R.K.F., H. Schmidt-Kaler 1989. Paläogeographischer Atlas des Süddeutschen Oberjuras. *Geol. Jahrbuch A 115: 77 p.; Stuttgart*

Seiler, K.-P., H. Behrens, H.-W. Hartmann 1992. Das Grundwasser im Malm der Südlichen Frankenalb und Aspekte seiner Gefährdung durch anthropogene Einflüsse. *Deutsche Gewässerk. Mitteilungen 35: 171-179; Koblenz*

Seiler, K.-P., P., Maloszewski, H. Behrens 1989. Hydrodynamic dispersion in karstified limestones and dolomites in the Upper Jurassic of the Frankonian Alb, F.R.B. *J. Hydrol. 108: 235-247; Amsterdam*

Seiler, K.-P., H. Behrens, H.-W. Hartmann 1991. Das Grundwasser im Malm der Südlichen Frankenalb und Aspekte seiner Gefährdung durch anthropogene Einflüsse. *Deutsche Gewässerk. Mitteilungen 35: 171-179; Koblenz*

Seiler, K.-P., H. Behrens, M. Wolf 1996. Use of artificial and environmental tracers to study storage and drainage of ground water in the Franconian Alb and the consequences for ground water protection. *Isotopes in water resources management Vol. 2: 135-146; (IAEA) Wien*

Weiss, E.G. 1987. Porositäten, Permeabilitäten und Verkarstungserscheinungen im Mittleren und Oberen Malm der Südlichen Frankenalb. *Dis. Univ. Erlangen, 211 S.*

Tracer Hydrology 97, Kranjc (ed.)© 1997 Balkema, Rotterdam, ISBN 90 5410 875 4

Simulation of pollutant-immission using geoelectric mapping of the migration of an artificially infiltrated salt tracer

R. Supper & W. F. H. Kollmann
Geological Survey of Austria, Vienna, Austria

A. Kuvaev
Moscow State University, Department of Hydrogeology, Russia

ABSTRACT

In this article a geophysical method is presented allowing the determination of the actual migration of an artificially injected salt tracer by surface geoelectric measurements. By using the information obtained with this methode it is possible to assess the effect of potential pullutant impacts on groundwater reservoirs and to clarify the location of sources of existing contaminations. Thus latent hazard potentials and reaction times required for implementation of counteractive measures can be identified. These facts play an important role in system analysis for modelling and management of groundwater quality and waste deposits.
Two case examples are presented including theoretical modelling of the infiltration process by A. Kuveav.

1 THE SALT INFILTRATION METHOD

With conventional tracer methodes only point information on groundwater velocity and flow direction can be derived involving the use of cost intensive drillings with the effect of disturbing the natural environment. The migration dynamics between drillings so far could not be mapped.
Moreover for pollutant managment it is not sufficient to know the flow direction of pure water only, but also to have the capability of prognosticating the actual contaminant migration under natural conditions (relief of aquitard, inhomagenities in porosity, ...). It is important to know if the pollutant distribution is dominated by the actual direction of groundwater flow or if the specifically heavier contaminant follows the relief of the aquitard and drifts agaist the direction (pure) of groundwater flow under the locally specific hydrogeological environment.
Therefore a methode was developed allowing the determination of the actual flow pattern by the geoelectric detection of artificially infiltrated salt tracers.
As with conventional tracer methodes a salt tracer is injected into the aquifer via a drillhole or a trench.

Once in place the tracer migrates due to pure groundwater flow directon and aquitard relief.
As the salt spreads it causes a temporally and locally variable reduction of the specific electrical resistivity of the groundwater layer which can be monitored by surface geoelectric mapping of a fixed electrode grid at distinct time sequences.
Calculating the differences to a reference measurement before the infiltration, the time dependent flow dynamics of the salt plume and the real flow paths can be determined.

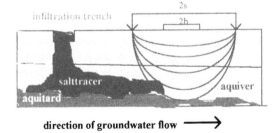

direction of groundwater flow ⟶

Fig. 1: The basic principle of the method

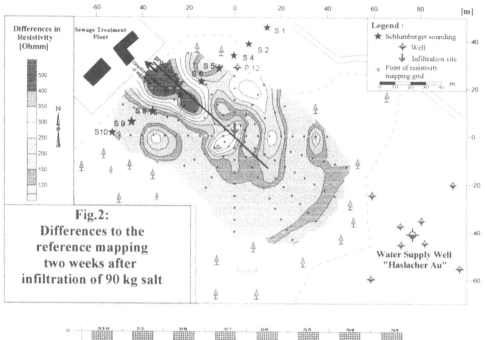

Fig.2:
Differences to the
reference mapping
two weeks after
infiltration of 90 kg salt

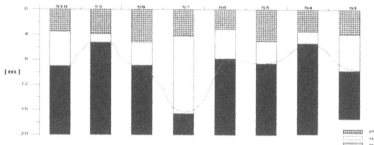

Fig.3: aquifer and aquitard profile derived from geoelectric soundings (location see fig.2)

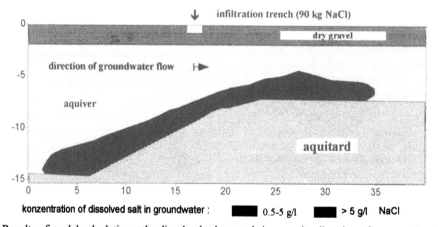

Fig.4: Results of model calculations: the dissolved salt spreads in opposite direction of   groundwater flow
according to the aquitarde relief

2 CASE EXAMPLE BACHSDORF

One tracer experiment was carried out in the area of "Haslacher Au" near Leibnitz (southern Styria, Austria). In that region a test well was situated in the area of the power plant of "Gralla" (Mur River Hydroelectric Project) by the Drinking-Water-Supply-Association "Leibnitzer Feld" (VB-H in Fig.2). This well was drilled to obtain acceptable drinking water in the future in terms of quality and quantity from bank-filtered river-water of the Auwald area. As both quality and quantity were found to be acceptable, it became necessary to define the boundaries of an additional protected area. A sewage treatment plant, located approximately 300 m upstream, presented a possible problem. Therefore an accidental release scenario caused by leakages from the collecting or settlement basins was simulated with the aid of an artificial infiltration of a salt tracer.

To plan the proposed survey properly trial measurements were performed before the start of the actual simulation. Several Schlumberger soundings were run in the test area to derive an approximate resistivity distribution. The sediments of the valley-filling consists of quaternary gravels to a depth of 8 m below surface (aquifer thickness approximately 5 m). Hydrogeologic investigations measured kf-values in the range of $2*10-2$ m/s. This data set was used to estimate the optimum value for the electrode spacing for the Wenner-mapping program by simulating the resistivity-changes according to the infiltrated salt solution with the aid of model calculations. The optimum value for the outer electrode spacing was found to be 15 meters. An input of approximately 150 kg of salt would have led to a highly significant measurable effect of about 60 Ohmm.

For the resistivity mapping program a star-like electrode spread with a radius of 20 m was used. After having received the first results the electrode-spread was then changed accordingly by enlarging the grid into the referring sector.
  A total of 130 electrodes emplaced over a rectangular grid gave us 70 mapping points.

As recent surveys showed, the most serious problem with this method is the infiltration of the salt into the aquifer. In this survey the feasibility of infiltrating through the unsaturated zone using a shallow trench was first tested. The alluvial sedimentation, covered by a humus layer of 0.3 m, underlain by coarse-grained gravel, favored the injection method.
  But in case of clay layers with thickness of several meters on top of the groundwater this method of infiltration will not be useful. Chemical reactions between salt and clay minerals might prevent the solution from passing through the aquifer. In this case wells are recommended for the injection.

On day before the salt-injection two reference-mapping programs were carried out. Then 90 kg of salt dissolved in water were injected. The saline solution had percolated within a few minutes.
At different times after the salt-injection the mapping was repeated. Differences were plotted relative to the first reference measurement. According to the fact that lower resistivity values correspond with more salt, the graphic representation (fig.2) is done in such a way that a resistivity decline after the reference measurement is shown as an increase in the difference plot.
Fig.2 shows the change in resistivity after two weeks. The salt has already drifted away from the point of injection to the NW. A minor quantity of salt still remains at the point of infiltration.
It is evident from the results obtained, that the major portion of the salt solution migrated towards the Northwest to the river Mur following the relief of the aquitard (see Fig.3) against the predicted direction of groundwater flow. The distance velocity of the main flow direction of the salt plume amounted to approximately 6.6 m/d. A pollutant having similar physical properties to the salt would therefore, if the infiltration of sewage takes place in the area between the treatment plant and at a reasonable distance from the well, present an insignificant hazard to the drinking water supply.
To estimate the role of density-convection a numerical simulation of the infiltration was done by A. Kuvaev using the modeling program MIG-2D. Fig.4 shows the results of this simulation which confirm the findings of the tracer experiment.

3 CASE EXAMPLE "HEILIGENKREUZ"

Another infiltration experiment was carried out in the area of Heiligenkreuz. Apart from method development, the experiment was undertaken to obtain answers to problems pertaining to water management. Downstream the region of the planned tracer experiment the water-supply-fountains for the regional water company were located. As quality was in danger to be polluted by accidents on the nearby main road, it became necessary to define the boundaries of an additional protected area and to find possible pollutants.

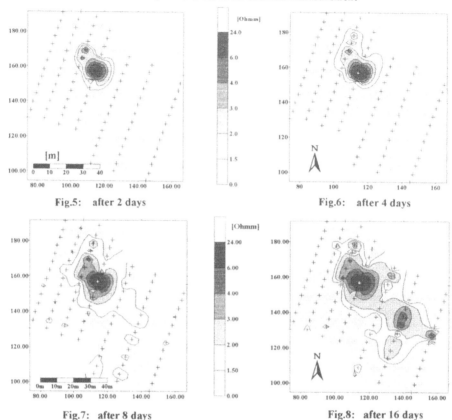

Resistivity-differences to the reference measurement

Fig.5:   after 2 days

Fig.6:   after 4 days

Fig.7:   after 8 days

Fig.8:   after 16 days

The infiltration site was chosen according to its location related to possible pollutants.

First blank measurements were carried out to assess the "as is" situation of the particular area, including any pollution that had possibly already occurred. This reference measurement was then used as the baseline for the artificial simulation of pollutant migration by tracing a leakage in the areas below the infiltration site.

A total of 60 kg of salt dissolved in 600 l water was injected upstream of the linear array of the Wenner configuration. In Fig.5-8, showing the change in resistivity after different days, a pronounced flow direction is apparent. The distance-velocity of the main flow direction of the salt amounted to 4.25+/- 1.25 m/d. Therefore a pollutant having similar physical properties to the salt would, if infiltration near the tested locations occur, present an significant and dangerous hazard to the water supply of the company.

4 CRITIQUE AND CONCLUSIONS

A critical review of the field tests carried out comes to the conclusion that, with the use of the geoelectric determination of a salt infiltration a method was developed that is simple and for certain investigations effective. Whether there is coincidence between the specifically heavier organic chemical solvents remains to be determined by columnar tests in the laboratory. Further practical experience will be gathered from the application of this method in surveys of closed-down waste repositories and the pollutant plumes emitted by these.

Flow directions as well as propagation velocity of the salt plume could be measured with the use of this method. In contrast to other tracer methods, the determination of actual pollutant flow paths forms the basis for further calculations.

For this reason real flow velocities are determined rather than distance-velocities. Furthermore, data on porosity and hydraulic conductivity can be calculated from the resistivity mapping derived therefrom.

Possible areas of application particularly pertain to aquifers at shallow to intermediate depths, that are overlain by beds of gravel. Where groundwater occurs at depths of more than 15 m, changes in resistance, occasioned by the infiltration of representative quantities of salt (50-150 kg NaCl), should not be expected to differ significantly from those caused by natural background variations. For greater depths (>12m), only an indication of flow direction can be deduced; flow velocity can no longer be determined.

Particular caution should be observed when injecting into aquifers with high clay contents (aquitard). Any change in hydro-chemical water parameters can bring about a change in clay structure and a related decrease in hydraulic conductivity. This effect can impede or even prevent the introduction of the salt solution and its dissemination and thus frustrate the evaluation of a tracer experiment on questions relating to water supply services.

This method is also an effective means of determining the originators of pollutant immission to groundwater. By defining the real flow path downstream from an inferred source of pollutants, the dispersion route of the immission can be simulated, so that either the suspicion that the source should further be considered as the originator of the pollution of a nearby well or not can be confirmed or rejected.

Finally, the enormous cost savings of the method, compared with the standard tracer methods, through savings on circular test drilling should be pointed out and which could surely be further enhanced by automatisation of the survey series. Requirements for environmental compatibility are met by using non-toxic, non-radioactive and non-carcinogenic salt tracer, which can be employed below the rooting layer of the soil and which disperses rapidly in the saturated zone by dilution after some days.

5 REFERENCES

KOLLMANN W., SUPPER R.: Geoelectric Surveys in Determining the Direction and Velocity of Groundwater Flow Using Introduced Salt Tracer, Tracer Hydrology, S.109-114, Hötzl+Werner (eds.), Balkema, Rotterdam, 1992.

KOLLMANN W., MEYER. J., SUPPER R.: "Geoelektrischer Nachweis eingebrachter Salztracer", Report of Proj. Ü34 "Schutz des Grundwassers in Tal- und Beckenlagen", year 1990-1991, Geological Survey of Austria, Vienna, 1991.

KOLLMANN W., SUPPER R.: "Geoelektrischer Nachweis eingebrachter Salztracer", Report of Proj. Ü34 "Schutz des Grundwassers in Tal- und Beckenlagen" für den Zeitraum 4.1991-2.1992, Geological Survey of Austria, Vienna, 1992.

KOLLMANN W., SUPPER R.: "Geoelektrischer Nachweis eingebrachter Salztracer", Report of Proj. Ü34 "Schutz des Grundwassers in Tal- und Beckenlagen" 3.1992-5.1993, Geological Survey of Austria, Vienna, 1993.

SUPPER R.: "Bericht über einen Salztracerversuch mit geoelektrischem Nachweis zur Simulation eines Schadstoffimmissionsereignisses im Raum Deutsch-Schützen/Eisenberg", Report of Proj. BU2, S.50-148, Geological Survey of Austria, Archiv f. Hydrogeologie, Vienna 1992, unveröffentlicht.

SUPPER R.: "Bericht über geophysikalische Untersuchungen auf der Mülldeponie Biedermannsdorf", Report of Proj.Biedermannsdorf, Geological Survey of Austria, Archiv f. Hydrogeologie, Vienna 1993.

Tracer Hydrology 97, Kranjc (ed.) © 1997 Balkema, Rotterdam, ISBN 90 5410 875 4

Environmental isotope studies and tracer testing to determine capture and protection zones in the Zechstein-Karst on the northern verge of the Thuringian Forest

C. Treskatis
Bieske + Partner Consulting Engineers, Lohmar, Germany

K. Hartsch
H + G GmbH, Freital/Dresden, Germany

ABSTRACT: On the northern downfall of the thuringian forest, groundwaters of the permo-triassic sediment series are utilized for water supply of communities and industry.

For the securing and protection of the water withdrawal installations in the thuringian Rinne valley a secure determination of the grade of protection as well as the location of the encatchment and recharge areas is essential.

The application of isotope and tracer techniques enables a distinction of the hydraulic contact between the karst, cleaved and porous aquifers of the permotriassic sediment series on top of the palaeozoic bedrocks.

According to the tritium content (exponentential modell), the mean residence time in the Zechstein springs amounts to between two and ten years.

Short-dated admittance of younger groundwater was not detectable, using either hydrochemical measurements or tritium methods alone.

A significant reduction of the δ-notation in the δ^{18}O-hydrographs of the springs and the surface water in the Rinne valley in the winter half-year 1993/94 indicates on the other hand, an abrupt and rapid admittance of younger groundwaters after strong precipitation and snow melt, originating from the southern recharge area (Zechstein outcrop).

A marking of the Zechstein and bunter sandstone waters with coloured dyes lead to the demarcation of three hydraulic systems, which each are drained within the permotriassic aquifersystem, through different springs.

The dominant apparent flow velocities amount to 2400 m/d in the Zechstein karst on the northeastern verge of the thuringian forest. The passage of groundwater from the porous aquifer of the lower bunter sandstone into the plate dolomites was confirmed by means of a multi-summited, arched breakthrough curve. The combined application of various tracer methods enabled the distinction of short-termed infiltration of waters from different encatchment areas south of the Zechstein outcrop, as well as their combination with the static and dynamic reserve in the sub-aquifers of the overlying bunter sandstone northwest of the Rinne valley.

1 INTRODUCTION

The use of groundwater for public supply demands increasing control of groundwater flow to protect encatchment areas from pollution. The most common tool to determine groundwater flow is the observation of wellwater tables and the construction of flow nets.

The determination of groundwater flow direction and flow velocities to calculate protection zones around wells and springs in semi-consolidated and karstic rocks using mathematical means is quite difficult due to the unknown flow paths.

At a study site on the northern downfall of the thuringian forest near Erfurt (s. fig. 1) different methods were used to determine groundwater flow velocities in the aquifers and the location of the capture zones due to the lack of suitable boreholes.

fig. 1: Large-scale regional map

Tracer testing and isotope measurements aimed to determine
- groundwater flow velocities and dispersivities in the uncovered and covered permian dolomites and limestones;
- the interaction between the overlying semi-consolidated bunter sandstone sediments and the karstic rocks;
- the hydraulic conditions of the captured springs;
- the influence of surface water on the springs.

The use of isotope and tracer techniques enabled to distinguish long-term and short-term water masses in the springs, which origin from different local hydraulic systems with different encatchment areas.

2   STUDY SITE

The springs and wells in the permotriassic sediments of the study site near Königsee are captured for public watersupply by the watersupply company WAZOR.

The permotriassic sediments consist of consolidated dolomites and limestones which overlay the folded palaeozoic crystalline schists of the thuringian forest (fig. 2).

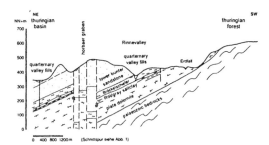

fig. 2: Geological profile NE-SW

In the study area the permian bryozoic riff-limestones and lagoon sediments arise from a littoral fazies near the middle german cristallin complex (SEIDEL 1995, JORDAN & WEDER 1995). These sedimentary rocks are covered by terrestic influenced, semi-consolidated rocks of the lower and middle bunter sandstone.

Tectonic movements resulted in dipping the sediments and applying NW-SE-striking fractures and faults due to the uplift of the crystalline bedrocks and the subsidence of the thuringian basin north of the former continent.

The thickness of the sedimentary layers discordant on top of the bedrocks increases from south to north.

Therefore the permo-triassic rocks form a multi-layered aquifersystem with alternating permeable and impervious rocks. Table 1 shows the hydrogeological characterisation of the main aquifers and impervious rocks compiled by TRESKATIS & HARTSCH (1996).

The main aquifer of the permian sediments appears as an alternating formation of plate dolomites, gypsum layers and limestones of upper and lower Zechstein age in different states of karstification and bifurcation. The groundwater circulates in fracture zones, faults and karst channels, which discharge in different springs in the Rinne valley.   The semi-consolidated sand- and siltstones of the bunter sandstone north of the Rinne valley form a porous aquifer and cover the permian carbonates. The „Bröckelschiefer"-sediments are intercalated as an impervious layer.

In tectonic fault zones, e.g. near Königsee, the claystones are fractured only slightly into blocks.

Tab. 1: Compilation of hydrogeologic characteristics of stratigraphic units in the vicintity of Königsee (acc. TRESKATIS & HARTSCH 1996)

Unit	Thickness	Hydrogeological characteristics
Quarterny sediments	up to 10 m	porous aquifer: sands, gravels with local loams and clays
lower and medium Bunter Sandstone (sm/su 3)	up to 220 m	porous aquifer: semi-consolidated sandstones and siltstones
„Bröckel-schiefer"	up to 20 m	aquitard: claystone and siltstones with local faults (hydraulic connection to „Leine"-carbonates)
upper Zechstein (permian)	up to 25 m	cleaved aquifer: plate dolomites
grey salt clay (permian)	up to 20 m	aquitard: claystones
lower Zechstein (permian)	up to 90 m	karstic aquifer: dolomites, marlstones, lagoon carbonates with bryozoic riffs
palaeozoic bedrocks	-	aquitard, cleaved zones acting as in local aquifer with connection to the permian carbonates.

Therefore hydraulic connections between the bunter sandstone and permian sediments are plausible. The generell groundwater flow direction tends towards the dipping of the permian and triassic sediments and follows the local stream Rinne.

3 MEASUREMENTS AND RESULTS

3.1 *Enviromental isotope studies*

Single-borehole hydraulic tests such as pumping tests and large-scale water table measurements which were performed to define groundwater protection zones are not able to give evidence of the local permeabilities and flow directions in the above described multilayered aquifer system.

Chemical analysis show SO_4-dominated waters in the permian carbonates and expected HCO_3-dominated groundwater in the sandstone aquifer (TRESKATIS & HARTSCH 1996). An association to the postulated capture zones was not evident. While the Köditz-springs are SO_4-dominated the Königsee-springs are classified as predominant HCO_3-influenced.

The hydrochemical pattern of the springs alone could not be used alone to determine the capture zone and recharge area of the waters.

Enviromental isotope measurements show a good correlation of $\delta^2H/\delta^{18}O$-data in the study site with the meteoric water line. In tectonic grabens isotopic compositions differ according to the $\delta^2H/\delta^{18}O$-relation in recent precipitations (fig. 3).

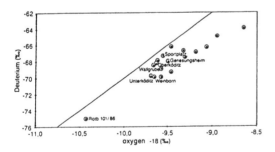

fig. 3: $\delta^2H/\delta^{18}O$-correlation of groundwater samples in the vicinity of Königsee / Rinne valley in relation to the europen meteoric waterline

Tritium data indicate an dominant influence of recent water in the springs of the Rinne valley while deep waters from the sandstone aquifer north of Königsee are free of tritium.

The Königsee-springs contain tritium amounts between 9,4 ± 1,5 T.U. and 18,0 ± 1,5 T.U. (1992-93). The surface waters contain recent amounts of 8,9 ± 1,6 T.U. The output of the springs is therefore mainly recent and lightly bomb-tritium-influenced. The mean residence times were calculated with the exponential model (completely mixed reservoir model):

Königsee-springs: 2 - 3 years
lower Köditz-spring: about 6 years
upper Köditz-spring about 10 years

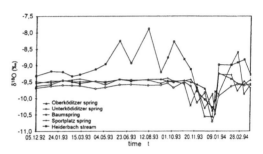

fig. 4: Oxygen-18-values of springs and streams

The exponential model was used to estimate recharge in the open, uncovered karst north of the thuringian forest. Piston flow conditions can be postulated in the covered karst north of the Rinne valley.

The annual variation of stable isotopes in the springs is low while surface water samples (e. g. Heiderbach data) show a characteristic seasonal variation (fig. 4). The discharge isotope concentrations in the springs does not correlate with the values of precipitation and surface waters.

In the Winter of 1993/94 the isotope values of the springs and the surface water tended to lower $\delta^{18}O$ notations (see fig. 4). The continuous decrease in $\delta^{18}O$ during winter is therefore attributed to the increasing proportions of waters coming from the higher outcrops of the permian carbonates south of Königsee.

Heavy autumn rains and melting water from seasonal snow falls removed the long term dynamic storage as separated pulses. A direct infiltration of surface water to the springs could not be verified because of former correlations of seasonal $\delta^{18}O$ values.

The isotope studies show, that the long- and short-term infiltrations recharging the springs must be searched not only in the uncovered karstic rocks but also in the main sandstone aquifer.

3.2 Tracerfield experiments

3.2.1 Method

Groundwater tracer experiments were performed to verify the direction and locations of short-term and long-term infiltrations in the main aquifers.

The tracer testing was performed in three locations with different hydrogeological boundary conditions:
- surface water sink of Pennewitz, near the outcrop of permian carbonates at the southern end of a large polje. At this location surface water coming from the covered karstic aquifer near the regional water divide sinks into the uncovered karst.
- well Nr. 23 near the airfield north of Pennewitz (borehole, 200 m deep in the lower bunter sandstone).
- sinkhole near Aschau at the southern outcrop of permian carbonates with a high groundwater level.

The masses of dye and salt tracers are summarized in tab. 2. During the entire tracer test groundwater samples were collected at the Königsee- and Köditz-springs and selected boreholes in the Rinne valley according to the standards layed down in KÄSS (1992).

Tab. 2: Tracer injection times and locations

tracer	date	location	input
uranine	19.01.94 (9.00 h)	sinkhole Penne-witz	3 kg dissolved in water
eosine	10.01.94 (15.00 h)	borehole 23 airfield	2,5 kg dissolved in water
salt-tracer	17.01.94 (10.55 h to 11.40 h)	sinkhole Aschau near Mühlberg	20 m³ dissolved in 10 m³ water

The continuos registration of salinity was performed by data loggers in springs, boreholes and surface waters.

3.2.2 Results

First evaluation of tracer test data compromised of plotting all observed breakthrough curves of dye and salt tracers. Mass recovery, transport velocities and

longitudinal dispersivities were calculated using analytical solutions of the 2-D transport equation (see 3.3). Figures 5 and 6 show the breakthrough curves measured in the Königsee-springs and in a deep borehole completed in the plate dolomites (Werkö-borehole). The variation of salinity in different boreholes and springs shows fig. 7. Calculated velocities and distances of proven tracer appearences are summarized in tab. 3 und 4. Discharge arrows resulting from proven dye and salt appearence are shown in fig. 8.

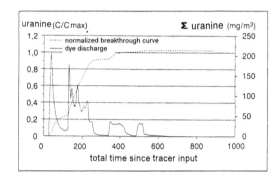

fig. 5: Breakthrough curves of Königsee springs (sporting field)

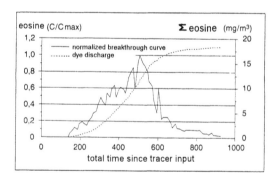

fig. 6: Breakthrough curves of Königsee springs (Werkö-borehole)

The first arrival of dye tracer uranine occured in the Königsee-springs within 26 to 32 hours while peak concentrations are attained after 38 hours. Subsequently uranine concentrations follow a „sawtooth"-pattern of gradual tailing up to 600 hours after tracer injection. During the tracer experiment flow conditions were not influenced by heavy rainfalls. Each of the periods of in- and decreasing dye concentration corresponds therefore with flow systems of different flow and storage capacities.

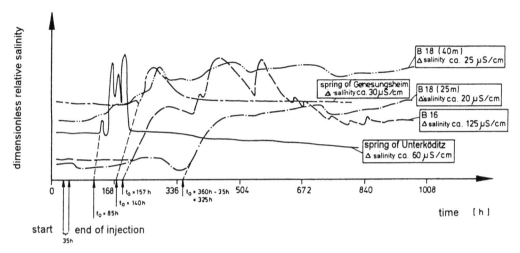

fig. 7: Breakthrough curves of salt tracer

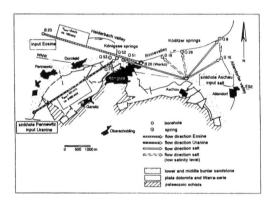

fig. 8: Geological map with tracer dispersion

Tab. 3: First arrival (t$_a$) and maximum peak concentrating time (t$_{max}$)

output loca-tion	t$_a$ [h]	t$_{max}$ [h]	c$_{max}$ [mg/ m³]	dye
Baum-spring	26 - 32	38	5,800	uranine
Schieß-haus-spring	26 - 32	38	5,480	uranine
Werkö-bore-hole	113 - 121	505	0,854	eosine

Tab. 4: Flow velocities calculated from t$_a$

input-loca-tion	output	L [m]	v$_{max}$* [m/d]	v$_{dom}$* [m/d]
sink-hole Penne-witz	Baum	3.850	3.190	2.400
	Sport-platz	3.850	3.850	2.400
bore-hole B 23	Werkö bore-hole	4.000	820	190

Flow velocities calculated in the karstic reservoir of Pennewitz polje range from 2400 m/d to 3190 m/d depending on the first detection of the dye uranine and the increase to peak concentration. The mass recovery was expected low because of unindentified uncaptured discharges in the Rinne valley.

The first arrival of eosine from the airfield borehole 23 occured within 113 to 121 hours in the plate dolomite covered by pleistozene gravels. Peak concentration was attained after 505 hours. The gradual, multi-peaked rise of dye concentration of > 500 hours as shown in fig. 6 is followed by a sharp maximum peak and a shorter „saw tooth"-tailing. This breakthrough curve measured in Werkö-borehole represents the flow path from the porous aquifer (lower bunter sandstone) to the karstic reservoir underneath with flow velocities ranging from 190 m/d to 820 m/d.

Flow velocities of the salt tracer injected in the sinkhole could be calculated from the salinity logs with peak concentration velocities from 3,7 m/h (89 m/d) to 23,5 m/h (564 m/d).

357

3.3 Hydrodynamic model for mass transport

The tracer breakthrough curves indicate water movements in reservoirs with different flow and storage conditions. Due to the low mass recovery and the long distance between the injection site and the observation points the 3-D general transport equation was simplified to 2-D. According to MALOSZEWSKI (1992) the transport equation is reduced to the form:

$$D_L \frac{\delta^2 C}{\delta x^2} + D_T \frac{\delta^2 C}{\delta y^2} \cdot v \frac{\delta C}{\delta x} = \frac{\delta c}{\delta t} + \frac{n-1}{n} \frac{\delta Cs}{\delta t} \text{ with}$$

$C_s = f(c)$ describing the mass transfer of tracer between solute and solid matrix.

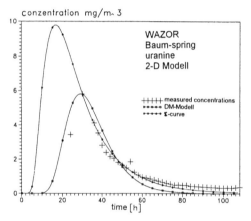

fig. 9: Modelling of dye tracer uranine

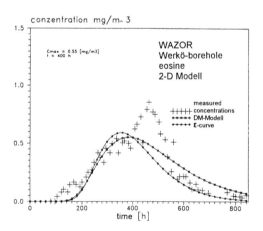

fig. 10: Modelling of dye tracer esoine

The 2-D model could be used on following assumptions, which are postulated for the karstic reservoir discharged by the Königsee-springs:
- tracer transport follows parallel layers or channels with separate flow system and constant transport parameters;
- the exchange between the layers or channels is small;
- the tracer concentration measured in the springs is the weighted mean concentration of all channels or layers;
- the tracer mass is proportional to the flow rate in each layer or channel.

The analytical solutions documented in WERNER (1991) were used to fit tracer data with the C_{max}-concentrations as shown in fig. 9 and 10. Depending on the solution the following transport parameters were calculated:

URANINE-breakthrough:
v_a: 85 - 115 m/h
= 2040 - 2070 m/d
mean transit time: 33 - 45 h
D/v_x: 0,057 - 0,190
D_L: 219 - 730 m
EOSINE-breakthrough:
v_a: 8 - 9 m/h
= 192 - 216 m/d
mean transit time: 427 - 478 h
D/v_x: 0,026 - 0,054
D_L: 105 - 214 m

4 INTERPRETATION OF ENVIROMENTAL ISOTOPE AND TRACER DATA

Using enviromental and artificial tracer techniques one obtained a synoptic hydrogeological and hydraulic model as shown in fig. 11:
- quarterny valley fills as main drainage for ascendent and descendent groundwaters from bunter sandstone and permian carbonates.
- open and covered karstic reservoir in carbonates such as plate dolomites and limestones divided by claystones with hydraulic connections to bunter sandstone aquifer in faulted zones; channel flow with different storage and flow conditions discharging in spring lines.
- discharge of bunter sandstone porous aquifer into the permian carbonates as shown by eosine appearance in plate dolomite borehole.
- permian karstic aquifer west of Königsee palaeozoic schist outcrop with direct short-term recharge from Pennewitz polje (see fig. 8).

- permian karstic aquifer east of Königsee bedrock outcrop without direct hydraulic connection to the holokarst system of Pennewitz (see fig. 11).

Within the karstic aquifer system near Königsee one could distinguish three hydraulic systems with different storage and flow characteristics:

- upper dainage system (proven by salt tracer appearance): drainage of unsaturated and saturated zone of karstic reservoir south of Köditz with direct connection to overlying bunter sandstone.

- intermediate drainage system (proven by uranine appearance): vertical superposition of unsaturated and saturated zones of Pennewitz holokarst polje zone west of Königsee bedrock outcrop.

- lower drainage system (proven by eosine appearance): permian plate dolomites with contacts to lower bunter sandstone northwest of Königsee. The karstic reservoir is not drained by the springs in the Rinne valley. Discharge zone has still to be proven.

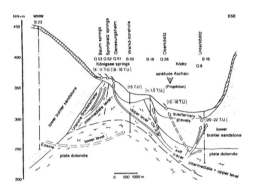

fig. 11: Synoptic profile with hydraulic levels

5   CONCLUSIONS

The determination of capture zones for karstic springs in multi-layered aquifer systems in areas with few boreholes must be supported by enviromental and artificial tracer techniques.

The result of the studies indicates the inhomogenity of water flow in the different karstic and porous aquifers as well as the hydraulic interaction of subaquifers through faulted or missing impervious layers.

Channels with high velocities and low mean residence times exist near zones containing water with high mean residence times. The protection of groundwater in such aquifers demands the identification of short-term drainage channels with high flow velocities and low cleaning abilities as

well as the influence of extrem seasonal events on the mean residence times of the discharged and captured water.

The identification and determination of the recharge and protection zones of karstic spring must be proven by calculating up water balances which are in preparation at time.

REFERENCES

Jordan, P. & Weder, H. (1995): Hydrogeologie - Grundlagen und Methoden, Regionale Hydrogeologie.- 603 p.; Stuttgart.

Käss, W. (1992): Geohydrologische Markierungstechnik.- Lehrbuch der Hydrogeologie Band 9, 519 p.; Berlin-Stuttgart.

Maloszewski, P. (1992): Mathematical modelling of tracer transport in different aquifers: Results from ATH test fields.- in: Hötzl, H. & Werner, A.: (ed.) (1992): Tracerhydrology, p. 25 - 30.

Seidel, G. (1995): Geologie von Thüringen, 556 p.; Stuttgart.

Treskatis, C. & Hartsch, K. (1996): Isotopenhydrologische Untersuchungen und Tracerversuch im Zechsteinkarst am Nordrand des Thüringer Waldes.- in: Grundwasser 1/2, p. 69 - 78; Berlin-Heidelberg.

Werner, A. (1991): Auswertung von Tracerversuchen im porösen Medium mit analytischen Lösungen der Dispersionsgleichung, 93 p.; Dipl. Arbeit AGK, Karlsruhe (not published).

Tracer Hydrology 97, Kranjc (ed.) © 1997 Balkema, Rotterdam, ISBN 90 5410 875 4

Hydrogeological investigations at the mineral springs of Stuttgart (Muschelkalk karst, South West Germany) – New results

W. Ufrecht
Amt für Umweltschutz Landeshauptstadt Stuttgart, Germany

ABSTRACT: Since the early Pleistocene, artesian springs located in the valley of the river Neckar near Stuttgart have been discharged mineralized water enriched with carbonic acid. From the 19th century the confined mineral water, rising up along faults, has been drawn from 19 wells which were drilled into the Lower Keuper and karstified Upper Muschelkalk (Trias). Alltogether they discharge 275 l/s. Due to geothermic and hydrochemical investigations in the river Neckar and the quaternary gravel deposits, additional natural outflows of mineral water could be located so that the total discharge increased to approximately 500 l/s. Whereas the karstwater in the Upper Muschelkalk generally shows low concentrations in the upstream of the mineral wells, different states of mineralization have been found in the discharge area. The geochemical evolution of the mineral water can be explained sufficiently by means of ^{34}S-isotope determinations of evaporite sulfate and dissolved sulfate in water. It can be shown that the hydrochemical character of the mineral water must be understood as the result of mixing processes between low mineralized karstwater, tempered brines from the Buntsandstein and Middle Muschelkalk and water from the Gipskeuper. The δ^{18}O and δ^{2}H-signature of the karstwater is determined by the altitude effect. In the discharge area the signature is superimposed by rising brines which are strongly depleted in ^{18}O.

1 INTRODUCTION

In central Württemberg (South West Germany) well-known and widely used mineralized water, enriched with carbonic acid, is discharged from the karstified Upper Muschelkalk (Trias). One of the most important and best known are the artesian mineral springs of Stuttgart, located in the valley of the river Neckar. After those of Budapest they are the second largest mineral water system in Europe. Due to the appearance of travertine deposits containing fossils, spring discharge has been proved since the early Pleistocene. The mineral water was already used in Roman times and again in the Middle Ages. Spa activities developed last century, flourishing between 1835 and 1870. Nowadays they are used for cures as well as in three leisure pools.

From last century the confined mineral water, rising along faults and being discharged into the valley of the river Neckar, has been drawn from 19 wells, which were drilled into the Lower Keuper and karstified Upper Muschelkalk. The wells reach depths between 30 and 60 m. Two further wells with depths of 135 m and 477 m respectively open up the transition to Middle Muschelkalk (Gottlieb Daimler-Quelle) and the strata from Buntsandstein to cristalline basement (Hofrat Seyffer-Quelle).

Initiated and coordinated by the city of Stuttgart (Local Water Authority) since 1989, intensive hydrogeological, isotopic and hydrochemical investigations have been carried out at the mineral springs of Stuttgart and the adjacent aquifer system of karstified Upper Muschelkalk in order to determine their origin and genesis. The results are quoted for a transient flow and transport model. This will be used in developing an improved concept for the qualitative and quantitative protection of the mineral water system (UFRECHT 1996).

2 GEOLOGICAL CONDITIONS

The outcrop and distribution of strata are extensively determined by faulting tectonics. The controlling tectonic element is the Fildergraben, together with its accompanying faults striking NW-SE. On the western upper block of the Fildergraben, Upper Muschelkalk outcrops in the area of the socalled Upper Gäu in a narrow strip, extending in N-S direction. Otherwise it is covered with Keuper and quaternary loess. Within the Fildergra-

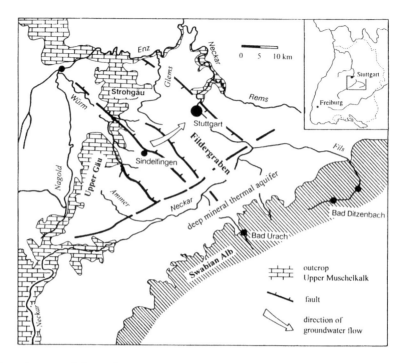

Fig. 1a: Schematic map of the study area showing the outcrop of the Upper Muschelkalk west of Stuttgart and the main faults.

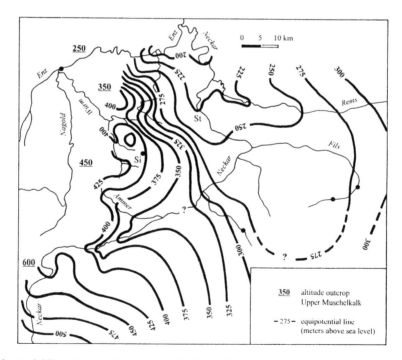

Fig. 1b: Equipotential lines (meters above sea level) for the Upper Muschelkalk
(Si: Sindelfingen, St: Stuttgart).

ben, the stratigraphic sequence ranges to Upper Keuper respectively in some areas to Lower Jurassic. Nearby the eastern Fildergraben faults, the river Neckar and its tributaries eroded the basin of Stuttgart-Bad Cannstatt. It forms the recharge area of the mineral waters of Stuttgart.

Within the Germanic Triassic, Upper Muschelkalk is built up by dolomitic limestone (Trigonodusdolomit), micritic and bioclastic limestone, interposed with beds of marlstones and shales. In the area of Stuttgart, Upper Muschelkalk is about 80 m thick. To the south thickness decreases and reaches 45 to 50 m in the foreland of the Swabian Alb. The jointed limestone and dolomitic limestone are well karstified. In the Middle Muschelkalk, dolomites, sulfates and evaporites dominate. Keuper sediments, overlying Upper Muschelkalk, consist mainly of terrestric sediments including gypsum. The latter appears predominantely in Gipskeuper.

3 HYDROGEOLOGY

Groundwater flow extends from the outcrop of karstified Upper Muschelkalk in the western upper block of the Fildergraben to the southeast, east and northeast, in most cases perpendicular to the faults. Here the main rivers (Neckar, Ammer, Enz and Glems) are coupled hydraulically with the karst aquifer. The difference of piezometric heads between the recharge and discharge area is about 225 m. The hydraulic gradient varies between 0.5 to 1 %. Southeast of this regional flow system with in the Fildergraben the piezometric heads rise in significantly straight to the foreland of the Swabian Alb. Here Upper Muschelkalk represents a productive mineral thermal aquifer (total dissolved solids about 4 to 5 g/l and 45 to 55°C) at a depth of 400 to 600 m. Following previous concepts this system is supposed to discharge about 45 l/s into the mineral springs of Stuttgart (Villinger 1982). The existence of mineral waters in Stuttgart is thus explained by a lateral mixing process of this highly mineralized groundwater component from the south with low mineralized karstwaters recharged in the Muschelkalk outcrop west of Stuttgart. The investigations of recent years however do not confirm this concept (GRAF, TRIMBORN & UFRECHT 1994; PLÜMACHER 1993).

The discharge area for the mineral springs of Stuttgart belongs to a tectonically highly disrupted fracture zone at the eastern border of the Fildergraben. Today 19 wells with a discharge rate of 275 l/s exist. Due to geothermic investigations in the river Neckar and the quaternary gravel deposits, some more „wild" outcrops of mineral water could be located. This increased total discharge to approximately 500 l/s (Armbruster 1994; Kappel-

meyer et al. 1994). In the valley of the river Neckar tectonic disruption and rising piezometric heads corresponding to depth (Upper Muschelkalk: 223 to 225 meters above sea level, Buntsandstein/cristalline basement 242 meters above sea level) enable cross formation flow along faults.

4 HYDROCHEMISTRY

Whereas in the upstream of the mineral springs karstwater in Upper Muschelkalk is generally of low concentration, you will find in the discharge area some different states of mineralization ranging from highly concentrated $Ca-Na-SO_4-Cl-HCO_3$-acidulous mineral water (total dissolved solids 3 to 6 g/l) to lower concentrated $Ca-(Mg)-SO_4-HCO_3$-mineral water (total dissolved solids 1 to 1.6 g/l). Besides sulfate in the highly mineralized waters, sodium chloride with a content of about 2 g/l is the characteristic component. It originates from the dissolution of halite. The content of sodium chloride rises with depth to 5 g/l at the transition to the Middle Muschelkalk (Gottlieb Daimler Quelle: Na-Ca-Cl-brine, total dissolved solids 12 g/l) and to 23 g/l in the Buntsandstein/cristalline basement (Hofrat Seyffer-Quelle: Na-Cl-brine, total dissolved solids 35 g/l). Apart from the Gottlieb Daimler-Quelle, in all waters the sodium chloride relation is about 1. The mineral waters are saturated with regard to calcite and gypsum. With regard to halite they are significantly unsaturated.

The hydrochemical character of the highly mineralized waters of Stuttgart confirm a strong relationship to the mineral thermal waters from the foreland of the Swabian Alb. However no hydraulic correspondence can be deduced.

The record of mantle helium reveals, that the carbonic acid, as an essential part of the mineral waters, originates from the degassing earth mantle (Schuhbeck et al 1994).

5. ISOTOPE HYDROLOGY

5.1 Oxygen-18 and deuterium

The $\delta^{18}O$ and δ^2H signature of the karstwaters are determined by the altitude effect. The altitude of the recharge area rises from north (about 300 meters above sea level) to south (about 600 meters above sea level). Over a distance of about 50 km depletion is more than 1 ‰. In the open karst between Stuttgart and the river Enz, $\delta^{18}O$-values are in the magnitude of -8.3 to -8.8 ‰. In the Upper Gäu (near the town of Sindelfingen) notations of -9.1 to -9.2 ‰ correspond with those of the karstwater in Stuttgart which is directly upstream the mineral water. This confirms the hydraulic concept that the catchment area of the Stuttgart mineral water system is extended 20 km to WSW. To the south, in

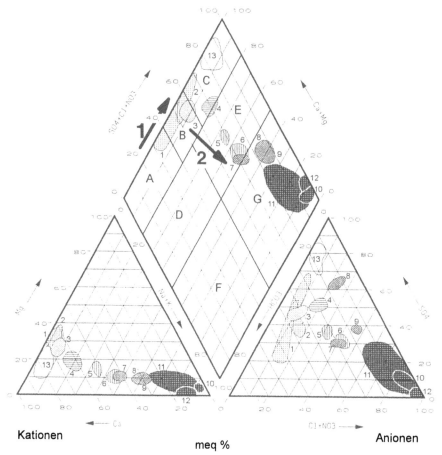

Kationen   meq %   Anionen

Fig 2: Hydrochemical characterization of groundwater in the Buntsandstein, Middle and Upper Muschelkalk and Gipskeuper by means of the PIPER-diagram. Arrow 1: mixing of karstwater in Upper Muschelkalk with groundwater of Gipskeuper. Arrow 2: mixing of karstwater (enriched with groundwater of Gipskeuper) with brines. 1: karstwater recharge area, 2: karstwater north of Stuttgart, 3: lower concentrated mineral water Stuttgart, 4-6: highly concentrated acidulous mineral water (different states of mineralization), 7-9: deep thermal mineral water (foreland of the Swabian Alb, western and eastern part), 10: brines Middle Muschelkalk, 11: brines Buntsandstein, 12: formation water Upper Muschelkalk (South German pre-alpine region).

the area of the river Ammer, $\delta^{18}O$ notations of about -9.3 to -9.4 ‰ are determined. Values between -9.6 to -10.6 ‰ are known for the mineral thermal aquifer in the foreland of the Swabian Alb. In Stuttgart the oxygen isotope ratio in the mineral water ($\delta^{18}O$ - 9.4 to -9.7 ‰) is depleted in contrast to the karstwater. The shift concurs with a rising content of total dissolved solids determined by sodium chloride and sulfate. This strong correlation can be understood as a mixing process by which hydrochemistry and the isotope signature of karstwater are superimposed by rising brines. These are rich in sodium chloride and strongly deple -

ted in ^{18}O (Hofrat Seyffer- Quelle: -10 to -11 ‰). None of the observed wells exhibit seasonal variations in $\delta^{18}O$.

Comparing $\delta^{18}O$ with $\delta^{2}H$ of the karstwaters, the different states of mineral waters in Stuttgart and the brine of the Hofrat Seyffer-Quelle fall along the meteoric water line $\delta^{2}H = 8\delta^{18}O + 5$, whereas the mineral thermal waters of the foreland of the Swabian Alb are situated left of it along the line $\delta^{2}H = 8\delta^{18}O + 10$. Variations in water temperature or in the content of carbonic acid do not influence the isotope ratios.

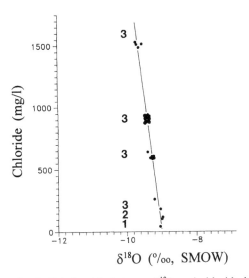

Fig. 3: Relationship between $\delta^{18}O$ and chloride in the karstwater and the mineral water of Stuttgart . 1: low concentrated karstwater , 2: lower concentrated mineral water, 3: highly concentrated acidulous mineral water (with different states of mineralization).

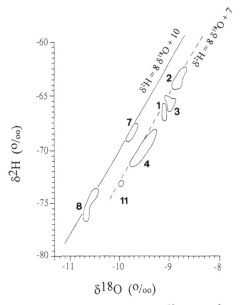

Fig. 4: Relationship between $\delta^{18}O$ and δ^2H. 1: karstwater recharge area and Stuttgart, 2: karstwater north of Stuttgart, 3: lower concentrated mineral water Stuttgart, 4: higher concentrated acidulous mineral water, 7. 8 deep thermal mineral water (foreland of the Swabian Alb), 11: brine Hofrat Seyffer Quelle Stuttgart.

5.2 Sulfur-34

The geochemical evolution of the highly concentrated mineral water can be visualized by $\delta^{34}S$ determinations of dissolved sulfate in groundwater, compared with those of gypsum sediments above and underneath the Upper Muschelkalk. The specific isotope signature from triassic formations is transferred to dissolved sulfate without any fractionation. Thus in the Upper Muschelkalk the $\delta^{34}S$ notations of dissolved sulfate are useful to find out the hydraulic correspondence with other aquifers. $\delta^{34}S$ notations of gypsum of the Middle Muschelkalk and Upper Buntsandstein are in the range of +20 ‰ CDT, whereas on an average those of the Gipskeuper are depleted by about 5 ‰ (GRAF & UFRECHT 1995).

In the open karst west of Stuttgart, $\delta^{34}S$ values between 3 to 7 ‰ and dissolved sulfate of about 100 mg/l are ascertained. The isotope signature is dominantely marked by secondary sulfate, derived from acid rain (Upper Gäu $\delta^{34}S$ 2 to 7 ‰). Where the Upper Muschelkalk is covered with Keuper and the gypsum layers are in contact with groundwater, an enrichment in dissolved sulfate (400 to 800 mg/l) and rising $\delta^{34}S$ nearly up to the isotope signature of Gipskeuper (13 to 15 ‰) is recognizable in Upper Muschelkalk. This is the case on a large scale in the city of Stuttgart, directly before the discharge area of the mineral water system. Only in the highly concentrated mineral water $\delta^{34}S$ exceeds the isotope signature of the Gipskeuper and approaches that of gypsum in the Middle Muschelkalk and Upper Buntsandstein (Röt). The mineral water of Stuttgart therefore consists of three main components:

- karstwater, recharged in the Upper Muschelkalk outcrop west of Stuttgart
- groundwater from the Gipskeuper
- brines from the Middle Muschelkalk and/or Buntsandstein

First quotations show that the amount of brines in the highly concentrated mineral water is less than 10 %.

6 ANTHROPOGENIC IMPAIRMENT OF THE STUTTGART MINERAL WATER

As the mineral springs of Stuttgart discharge directly in the city their protection in the qualitative as well as in the quantitative way is an important task. The quantitative protection becomes relevant in civil engineering, which is realized underneath the piezometric head of the confined Upper Muschelkalk. A qualitative impairment of the system results from numerous abandoned landfills and contaminated landsites. So shallow and deeper

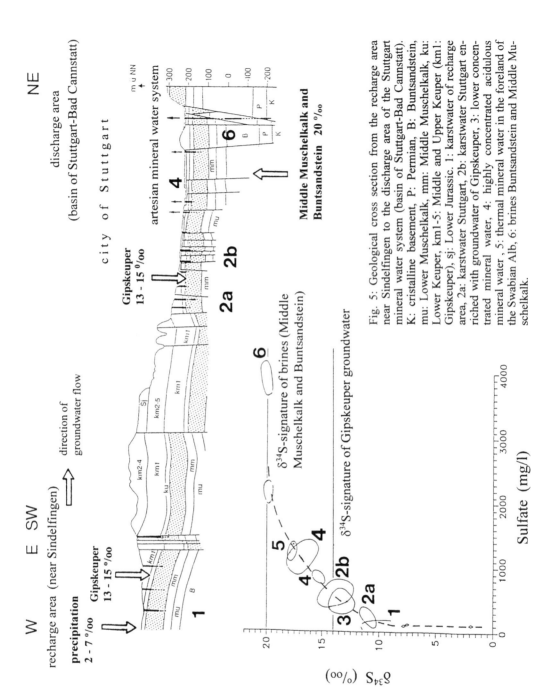

Fig. 5: Geological cross section from the recharge area near Sindelfingen to the discharge area of the Stuttgart mineral water system (basin of Stuttgart-Bad Cannstatt). K: cristalline basement, P: Permian, B: Buntsandstein, mu: Lower Muschelkalk, mm: Middle Muschelkalk, ku: Lower Keuper, km1-5: Middle and Upper Keuper (km1: Gipskeuper), sj: Lower Jurassic. 1: karstwater of recharge area, 2a: karstwater Stuttgart, 2b: karstwater Stuttgart enriched with groundwater of Gipskeuper, 3: lower concentrated acidulous mineral water, 4: highly concentrated acidulous mineral water, 5: thermal mineral water in the foreland of the Swabian Alb, 6: brines Buntsandstein and Middle Muschelkalk.

groundwater in Gipskeuper is widely contaminated, especially with chlorinated solvents. When infiltrating in the Upper Muschelkalk, the contaminants are displaced into the mineral water system. A contamination with trichloroethene of up to 20 µg/l has been proved in the low concentrated mineral water since 1984. In some wells with highly concentrated mineral water trichloroethene varies between 1 and 3 µg/l.

To obtain a qualified protection of the mineral springs the demarcation of a protection area is necessary. This is currently being realized.

7   CONCLUSIONS

On the basis of the presented data the following conclusions can be drawn:

The Stuttgart mineral water system is part of a complex regional aquifer system in karstified Upper Muschelkalk.

Due to geothermic and hydrochemical investigations total discharge of the system increased to approximately 500 l/s.

The catchment area is extended 20 km WSW of Stuttgart. The main recharge area is the outcrop of Upper Muschelkalk on the Upper block of the Fildergraben (Upper Gäu, near the town of Sindelfingen).

The hydrochemical character of the mineral water can be understood as the result of mixing processes between low mineralized karstwater, groundwater from the Gipskeuper and tempered brines from the Buntsandstein and Middle Muschelkalk.

The shallow and deeper groundwater of Gipskeuper is contaminated with chlorinated solvents. When infiltrating in the Upper Muschelkalk, the contaminants are displaced into the mineral water system.

ACKNOWLEDGEMENTS

The author is much indebted to the GSF Institute of Hydrology (Munic-Neuherberg) and to the Institut für Umweltphysik (University of Heidelberg) for isotope determinations.

REFERENCES

Armbruster, H. (1994): Ergänzende Infrarotmessungen im Neckar zum Nachweis von Mineralwasseraustritten. - In Ufrecht, W. & Einsele, G. (Hrsg.): Das Mineral- und Heilwasser von Stuttgart. - Schriftenreihe des Amtes für Umweltschutz, **2/1994**: 157-166, 7 Abb.; Stuttgart.

Graf, W., Trimborn, P. & Ufrecht, W. (1994) Isotopengeochemische Charakterisierung des Karstgrundwassers und Mineralwassers im Oberen Muschelkalk im Großraum Stuttgart unter besonderer Berücksichtigung von Sauerstoff-18 und Schwefel-34. - In Ufrecht, W. & Einsele, G. (Hrsg.): Das Mineral- und Heilwasser von Stuttgart. - Schriftenreihe des Amtes für Umweltschutz, **2/1994**: 75-115,12 Abb.; 9 Tab., Stuttgart.

Graf; W. & Ufrecht, W. (1995): Über den Einsatz von Schwefelisotopen zur hydrogeologischen Erkundung des Stuttgarter Mineralwassersystems. - Exkurs.f. u. Veröfftl. GGW, **195**: 32-34; Berlin.

Kappelmeyer, O., Smolka, K., Pinkau, G. & Dornstädter, J. (1994): Temperaturmessungen im Nekkar zum Nachweis von Mineralwasseraustritten. - In Ufrecht, W. & Einsele, G. (Hrsg.): Das Mineral- und Heilwasser von Stuttgart. - Schriftenreihe des Amtes für Umweltschutz, **2/1994**: 141-156, 7Abb., 2 Tab.; Stuttgart.

Plümacher, J. (1993):   Erkundung der regionalen Grundwasserströmung im Festgesteinsaquifer des Oberen Muschelkalks (mittlerer Neckarraum) mittels numerischer Modelle. - Dipl. Arb. Univ. Aachen, 152 S., 93 Abb., 13 Tab., 10 Anl.; Aachen.

Schuhbeck, S., Weise, S., Wolf, M. & Ufrecht, W. (1994): Isotopenhydrologische Studie zur altersmäßigen Klassifizierung ausgewählter Karst- und Mineralwässer aus dem Raum Stuttgart-Bad Cannstatt. - In Ufrecht, W. & Einsele, G. (Hrsg.): Das Mineral- und Heilwasser von Stuttgart-Bad Cannstatt und Berg. - Schriftenreihe des Amtes für Umweltschutz, **2/1994**: 117-132, 8 Abb., 1 Tab.; Stuttgart.

Ufrecht, W. (1996): Hydrogeologische Untersuchungen am Mineral- und Heilwassersystem von Stuttgart. - Proc. 11. Nat. Symp. Felsmechanik 29./30.11.1994, Aachen, 97-104, 4 Abb., 1 Tab.; Essen.

Villinger, E. (1982): Hydrogeologische Aspekte zur geothermalen Anomalie im Gebiet Urach-Boll am Nordrand der Schwäbischen Alb (SW-Deutschland). - Geol. Jb., **C 32**: 3-41, 9 Abb. 9 Tab.; Hannover.

Tracer Hydrology 97, Kranjc (ed.) © 1997 Balkema, Rotterdam, ISBN 90 5410 875 4

The use of groundwater tracers for assessment of protection zones around water supply boreholes – A case study

R.S.Ward
Fluid Processes Group, British Geological Survey, Nottingham, UK

A.T.Williams
Hydrogeology Group, British Geological Survey, Wallingford, UK

D.S.Chadha
The Environment Agency, York, UK

ABSTRACT: A series of tracer tests have been performed to identify and evaluate the hydrogeological conditions controlling groundwater flow in the vicinity of a public water supply (PWS) well at Kilham, in the Chalk aquifer of North Yorkshire, England. The unconfined area around the well has been designated a nitrate sensitive area (NSA), the boundary of which has been applied using local knowledge and professional judgement. In contrast, Groundwater Protection Zone (GPZ) modelling, which uses numerical methods to delineate capture zones around wells, produced a capture zone for the PWS well which differed markedly from the NSA boundary. In both cases, modelling does not take into account the complex fractured nature of the Chalk aquifer and so the validity of the results is uncertain. Preliminary tracer tests (using fluorescent dye tracers) focused on establishing connection and groundwater velocities between observation boreholes along valley floors up hydraulic gradient of the PWS. Results were used to direct further drilling and more focused tracer testing (using bacteriophage tracers) with multiple injections and sampling points including observation boreholes, the PWS well and springs. The results of the tracer test were used to extend the understanding of local groundwater flow in the Chalk aquifer and provide a realistic assessment and validation of the GPZ and NSA models.

1 INTRODUCTION

The Environment Agency of the United Kingdom has adopted a groundwater protection policy (GPP) of which one of the three key elements is the definition of groundwater protection zones (Environment Agency, 1992). The purpose of the policy is to enable the management and protection of groundwater on a sustainable basis by preventing its pollution. The GPP provides a policy which, although not having statutory status, enables the Agency to respond to statutory and non-statutory consultations in a consistent and uniform manner. The outcome is exemplified in the implementation of two European Union initiatives (EC Directive 91/676 and EC Agrienvironmental Regulation 2078/92) which aim to reduce aquifer pollution from agricultural sources of nitrate. The UK Government's adopted approach has been to identify Nitrate Sensitive Areas (NSA) and Nitrate Vulnerable Zones (NVZ) which have statutory measures associated with them. Both of these schemes have utilised the Agency's protection zone work to identify catchments with a (potential) problem.

Groundwater Protection Zones (GPZ) are one part of a dual strategy approach to groundwater protection by the Agency which seeks both to protect aquifers as a whole and also safeguard specific sources of water supply (e.g. PWS wells). A programme of GPZ delineation is continuing in the

UK but already attention has been paid to over 150 priority sources which were candidates for NVZ or NSA designation. One such place is Kilham in North Yorkshire where rising nitrate levels in abstracted groundwater resulted in the designation of the area around the source as a NSA. In addition, GPZ modelling has been applied in order to define the protection zone(s) around the well.

GPZ modelling utilises proven techniques (Environment Agency, 1995). However because of the complex hydrogeological conditions which prevail in many UK aquifers, results of models which are currently available for zone delineation are in many cases inaccurate because of their over-simplistic approach. Model results therefore only represent a prediction based on 'best available current technology'. Capture zone definition is an area of continuing research. The methods which can be adopted range from empirically based models to, more recently, two or three dimensional steady-state/time-variant flow and particle tracking models.

Of particular concern is the applicability of model results in the Chalk aquifer. This aquifer is commonly a multi-porosity geologic medium which contains fissures (some enlarged by solution). These fissures intersect a rock mass characterised by high porosity (up to 40%) and low hydraulic conductivity (down to 0.001 m/day) (Price, 1987). Groundwater flow is dominated by fissure flow and the heterogeneity and anisotropy of the fissuring introduces the complexities which most protection zone models are unable to account for.

'Ground truth' validation of protection zones is rarely performed and so in many cases the uncertainties in model results will prevail. Groundwater tracer testing offers a potential method for assisting in zone delineation and validation but has thus far been rarely employed. This paper describes a study being carried at Kilham to investigate the validity of the GPZ and NSA models applied to the Kilham PWS well using tracer testing. A programme of testing has been developed and partially completed to investigate the hydrogeological controls on groundwater flow in the area, their significance with respect to the protection zones and to attempt validation of the models. The programme has involved a staged approach including natural gradient observation borehole to observation borehole tracer testing to determine groundwater travel times, single borehole dilution tests to identify major flow horizons and a large scale multi-point injection and sampling prevailing gradient tracer test to test the validity of the protection zones. In addition, geophysical logging of boreholes has been performed to supplement the data.

2 GEOLOGICAL AND HYDROGEOLOGICAL CONDITIONS

The Chalk of the Wolds in North Yorkshire, England is used extensively for groundwater abstraction. The study area in this investigation (Figure 1) comprises an upland plain between 100 - 120 m (OD - elevation relative to Ordnance Datum) dissected by steep sided dry valleys. The catchment area of the Kilham PWS well is unconfined, but the Chalk is confined by glacial till down gradient of the PWS well. The PWS well is located on the northern edge of the Langtoft valley close to the confluence of the two valleys.

In the Kilham area the Chalk is 300 - 350 m thick and the dip is variable with measurements ranging from 1 - 40° (British Geological Survey, 1993). Correlation of recent geophysical logging results obtained from new and existing boreholes as part of this project have indicated apparent dips of between 1 - 2° towards the south-east in the area to the west and north-west of the PWS well.

The Langtoft and Broachdale valleys are dry valleys which converge to the south-east of Kilham village forming the Lowthorpe Beck valley. Springs in the lower reaches of both the Langtoft and Broachdale valleys form the ephemeral headwaters of the Lowthorpe Beck and additional spring augmentation occurs at Bellguy and Bracey Bridge springs further down stream.

Previous hydrogeological investigations (Environment Agency, unpublished report) have revealed the area to be very complicated with large variations in measured values of transmissivity found over small distances (Table 1). In some cases a difference of at least two orders of magnitude was

measured over a distance of 40 m. Groundwater flow direction inferred by water level data is parallel to the two dry valleys towards the south-east as far as Kilham before becoming more southerly as the Chalk becomes confined. The water level information also indicates that there is a marked difference in hydraulic gradient between the two valleys. In the Langtoft valley at high water levels a gradient of 7×10^{-3} was measured and at low water levels 2×10^{-3}. In Broachdale valley hydraulic gradients were 2×10^{-3} and 2×10^{-4} respectively. This has been suggested to indicate higher transmissivities in the Broachdale valley (Environment Agency, unpublished report) although this is not substantiated by the currently available data (Table 1). The measured hydraulic properties demonstrate the variable nature of the Chalk and it can be seen that high transmissivities are not restricted only to the valleys, e.g. West End Farm (WEF) which was drilled and tested as part of this study. Tracer testing (described later) revealed that this transmissivity can be attributed to one particular fissure horizon within the borehole at 100 m below ground level close to 0 m OD.

The degree of connectivity and spatial extent of fissures is however uncertain although the variability in properties observed may indicate that flow may be occurring along discrete bedding plane fissures which are not laterally extensive but display channelling (Foster and Milton, 1977). Faulting, for which there is some evidence in the area, may also have resulted in discontinuity of bedding plane fissures (Kirby and Swallow, 1987).

Table1. Aquifer properties around Kilham.

Location		Borehole 1	Borehole 2	
Name	Code	T (m^2/d)	T (m^2/d)	S (%)
Henpit Hole	HPH	11028	355	1.8
Middledale	MD	6318	-	-
Little Kham Fm	LKF	7698	6588	2.2
Broachdale	BRD	10298	-	-
Bartondale	BTD	-	17	0.3
Tancred Pit	TPB	4272	10257	0.7
West End Fm	WEF	-	909	0.8
K'ham PWS	KPWS	T=850-47000	S = 0.3-0.7	

3 DEFINITION OF PROTECTION ZONES

At Kilham, definition by the Environment Agency of the NSA and GPZ have employed different techniques. For the NSA, the method adopted for defining the area was based on an empirical approach. An area was calculated by dividing the annual licence abstraction of the PWS well by the average annual recharge. The area was then increased by ten percent to allow for a degree of over protection. Apportionment of the area was then based on professional judgement and local knowledge, the limits following field boundaries. In line with the accepted view, at that time, that flow in the Chalk is preferentially along dry valleys rather than the interfluve areas and the dry valley areas are the most vulnerable, the NSA extended primarily along the two dry valleys (Figure 2). The subsequent GPZ delineation modelling employed numerical modelling which utilised measured or estimated aquifer properties and their spatial variability to determine the protection zone for the Kilham PWS well. Sensitivity analysis, to take into account the uncertainties in aquifer parameters, was also performed to produce zones of confidence for the numerical model results. The model, which is based on a porous medium approach, indicates that the capture zone is relatively narrow but extends beneath the interfluve region towards the north-west (Figure 2). This result has been produced despite high transmissivity being assigned to the valleys and low transmissivity to the interfluve (the West End Farm result was not available at the time of modelling). Because the model does not take into account the fissured nature of the Chalk aquifer and there is disagreement with the NSA model, the validity of the model results are uncertain. It was proposed therefore that a practical approach should be adopted and a tracer test programme implemented to provide information of flow direction, connectivity and produce realistic travel times for groundwater in the area in order that the modelling results could be assessed and if necessary improved.

4 TRACER TEST PROGRAMME

Because of the complex nature of the Chalk aquifer within the area and the geographic scale of the problem, it was decided to adopt a staged experimental approach which considered all of the available hydrogeological information and the practical constraints of tracer testing.

4.1 Constraints

Tracer testing can provide an ideal method for solving certain specific problems, however a number of constraints exist which must be considered for each individual application of tracer testing to improve the likelihood of success and minimise uncertainty. These include non-detection as a result of inappropriate monitoring, selection of unsuitable injection points, poor understanding of hydrogeological conditions, environmental impact of tracers, incorrect selection and quantities of tracer. A major concern at Kilham was tracing to a Public Water Supply (PWS) well. At the scale of experiment proposed here, it is extremely difficult, if not impossible, to calculate accurately the quantity of tracer which will enable detection but not result in a degradation of abstracted water quality. To ensure acceptance by regulatory authorities a demonstration of a suitable responsible approach was necessary.

4.2 Development of tracer test programme

A desk study was initially performed to review available geological, hydrogeological and geophysical information, identify and evaluate suitable tracer injection and detection points and assess and select suitable tracing agents. A test programme was therefore developed which was aimed at maximising the success of the tracer testing. This programme included the following elements:

1. Preliminary tracer testing. Observation borehole to observation borehole tracer tests over the small to intermediate distances (40 - 3100 m). Interpretation of results, recommendations for further testing and design of future experimental programme (if appropriate).

2. Implementation of large scale tracer test to PWS well and other discharge points (over a distance of up to 5500 m) for validation of protection zones and enhanced understanding of local hydrogeology.

4.3 Preliminary tracer testing

The purpose of this preliminary testing was to investigate connections along the valley floor over minimum distances using available boreholes to improve understanding of the prevailing hydrogeological conditions and determine travel times where connection was proven. These data were then used to design the strategy and procedures for further testing which would be focused on assessment and validation of the protection zone modelling.

Observation boreholes along each of the dry valleys were selected for these tracer tests. The aim was to trace between the boreholes, which in each valley were located in a line parallel to groundwater flow. Tracer injection was at those boreholes nearest the top of the valleys (Broachdale borehole (BRD) and Hen Pit Hole (HPH)) and sampling was at the closest observation borehole(s) down gradient (Middledale (MD), Little Kilham Farm (LKF), Bartondale (BTD) and Tancred Pit (TPB)) (Figure 1). Due to operational difficulties and borehole instability, the tracer injections were performed at a single horizon within each of the boreholes. This horizon was associated with major flow (based on geophysical information).

Photine CU and fluorescein were selected as the tracers because of their ability to be detected by passive detector (cotton wool and charcoal for each tracer respectively). Passive detectors were selected to minimise costs and enable continuous sampling. The detectors which adsorb the tracers from the groundwater were positioned at levels within the monitoring boreholes which were believed to have significant flow associated with them. The detectors were routinely collected (initially on a daily basis but with gradually decreasing frequency) for analysis and replaced. Analysis was performed by inspection of the cotton wool detectors under UV light and the

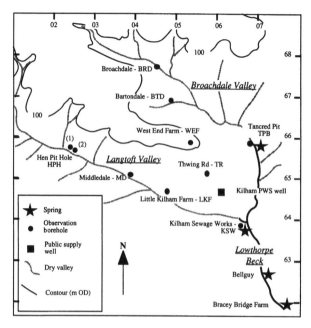

Figure 1. Location of study area showing boreholes and springs.

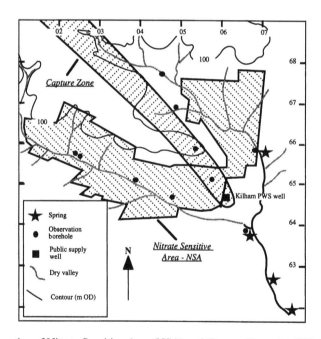

Figure 2. Location of Nitrate Sensitive Area (NSA) and Capture Zones for Kilham PWS well.

charcoal samplers by fluorimetry after desorption of tracer (Smart, 1976). The nature of the sampling enables primarily a qualitative analysis with an positive or negative result for a given sampling period. However, an attempt was made to produce semi-quantitative results for the charcoal samplers by using a known mass of charcoal and volume of eluent for desorption. This provided a crude breakthrough curve at Tancred Pit which indicated two peaks (Figure 3).

4.3.1 *Preliminary test results*

The injection boreholes were monitored to ensure that the tracer had entered the aquifer. In each case, over 99.5% of the tracer had migrated from the borehole after 18 hours. Sampling at the down gradient boreholes was continued for over three months. During this time tracer from Broachdale was detected at Tancred Pit but no tracer was detected at Bartondale, Middledale and Little Kilham Farm. To test the connection between Middledale and Little Kilham Farm, a further test was performed with injection of Photine CU at Middledale. Tracer arrival at Little Kilham Farm proved a connection between the two.

Table 2. Approximate groundwater velocities based on first arrivals.

Langtoft Valley			
HPH BH1	->HPH BH2	:	>50 m/day
MD	->LKF	:	480 m/day
Broachdale			
BRD	->TPH	:	440 m/day

The approximate groundwater velocities determined from the first arrival of tracer are shown in Table 2. The lowest value was measured at Hen Pit Hole, although this may reflect the sampling frequency. In reality the velocity would probably be much higher. Even if a velocity of 50 m/day was accurate, tracer arrival at Middledale would be expected in less than 40 days if a connection existed and so adequate time was allowed for the tracer to arrive.

Interestingly, a similarity exists between the velocities for the two valleys where connections were proven with movement of groundwater along the two valleys being extremely rapid (Table 2).

4.4.*Large scale tracer test*

The results of the preliminary testing proved that connections existed between boreholes spaced up to 3000 m apart. These connections were despite the apparent dip of the Chalk. With a dip of 2°, a vertical displacement in strata of approximately 105 m would be expected over a horizontal distance of 3000 m. This would mean that the stratigraphic horizon in which the tracer was injected at Broachdale would pass below Tancred Pit borehole. There must therefore be an element of upward flow for tracer to reach the sampling borehole. Between Hen Pit hole and Middledale borehole the same conditions occurred but in this case no tracer was detected implying no upward movement of tracer if the bedding plane transmitting tracer passed below Middledale borehole. Between Middledale and Little Kilham Farm, taking into account a similar dip, the horizon of tracer injection intersects the Little Kilham Farm borehole.

The data from the preliminary tests were used to design the large scale test. In addition, a new borehole was drilled to supplement the existing borehole array. This new borehole, at West End Farm (WEF), was positioned in between the two dry valleys on the ridge of the interfluve running between them. This borehole was drilled to a depth of approximately 110 m to penetrate a similar depth and sequence of Chalk as the boreholes along the floor of the valleys.

The new borehole plus Tancred Pit borehole (TPB) and Little Kilham Farm (LKF) were selected as tracer injection points and Kilham PWS well (KPWS), Kilham Sewage Works borehole (KSW), Bellguy springs (BGS) and Bracey Bridge Farm spring (BBS) were chosen as sampling points. Due to the nature of the monitoring points and the need for quantitative analysis, automatic water samplers were used. Tracer injection location within each of the injection boreholes was again chosen to correspond to the zone of maximum flow. To ascertain this with confidence, single borehole dilution tests were performed in each of the injection boreholes prior to the large scale test. In each borehole, except at Tancred Pit, a single discrete horizon was identified. At Tancred Pit tracer dilution within the whole borehole was so rapid that no single location could be identified.

Table 3. Tracer injection details for large scale tracer test.

Injection borehole	Injection depth (m OD)	Tracer	Quantity (pfu)
LKF	0.0	SM	5×10^{15}
TPB	-7.0	EC	1×10^{15}
WEF	0.0	MS2	3×10^{15}

SM - Serratia Marcesens, EC - Enterobacter clocae
MS2 - E-coli - MS2

Because of the remaining uncertainties about tracing to a PWS well and the problems of water quality, it was decided to use bacteriophage tracers. Although non-toxic, any phage present in the water would be killed by routine chlorination. The tracers

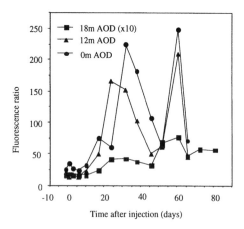

Figure 3. Tracer breakthrough curves at
Tancred Pit Hole.

also have the advantage of being colourless, tasteless
(at low concentrations) and of being detectable at
even lower concentrations than most other tracers
(Skilton and Wheeler, 1988). Detection of these
tracers was achieved by microbiological plate
counting. Details of tracer injection are shown in
Table 3.

4.4.1 *Results of large scale test*

Sampling was continued for over 100 days at each of
the sampling locations. The only tracer to be detected
was the E-coli - MS2 injected at West End Farm.
This was detected at Kilham Sewage Works borehole
and at Bellguy spring. The distance from the
injection point to each of these sampling points is
2400 and 3800 m respectively. Their breakthrough
curves for the period between injection and the last
sample in which tracer was detected (after 30 days)
are shown in Figure 4. Although the breakthrough is
not continuous for any extended period, two separate
'packets' of tracer breakthrough occur at Bellguy
spring, the firsts after 9 days (1) and the second after
27 days (2). At Kilham Sewage Works only three
samples were found to contain tracer (between 16
and 19 days after injection). Based on the arrival
times of each of the 'packets' of tracer the
approximate groundwater velocities are: at Kilham
Sewage Works, 130 - 150 m/day, and at Bellguy
spring, (1) 300 - 475 m/day and (2) 125 - 140
m/day.

4.4.2 *Discussion of results*

The results are extremely interesting in that the tracer
injected into the interfluve borehole was detected
nearly 4 km away. Not only that, but the tracer <u>was</u>

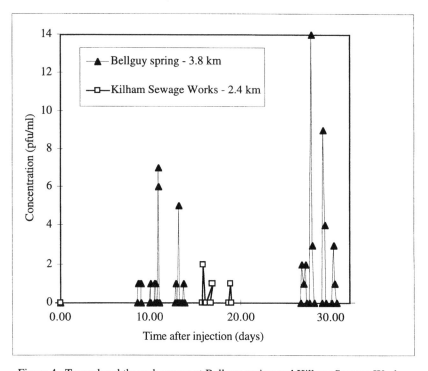

Figure 4. Tracer breakthrough curves at Bellguy spring and Kilham Sewage Works.

detected at one intermediate borehole (2.4 km) but not at another, Kilham PWS well (1.5 km). Both of these boreholes lie on a straight line, geographically, between West End Farm and Bellguy spring (Figure 1). It is apparent therefore that the tracer did not enter the 'true' capture zone of Kilham PWS well but passed either around or below it. Correlation of geophysical logs had revealed that although the strata were dipping, the saturated horizons within the West End Farm borehole intersected Kilham PWS well. Therefore preliminary conclusions indicated that the movement of tracer did not follow a linear path but passed around the PWS well before passing the Kilham Sewage Works borehole and entering the capture zone of Bellguy spring. The occurrence of two 'packets' of tracer at Bellguy spring may also be indicative of two distinct flow paths. Tracer from Tancred Pit and Little Kilham Farm boreholes was not observed at any of the sampling points suggesting that flow lines passing through these boreholes do not intersect the PWS well, Bellguy spring or Bracey Bridge Farm spring capture zones.

It was decided that because of the unusual nature of the test results, another borehole should be drilled between West End Farm and Kilham PWS well to enable an additional tracer test to be performed whilst sampling for the existing test was continuing. A location 450 m from the PWS and to the north-west was selected (Figure 1). The borehole (Thwing Road (TR)) was drilled to a depth which ensured penetration to the same relative depth as the PWS well. Subsequent geophysical logging of this borehole provided important and valuable information and enabled a considerable refinement and re-interpretation of the structure of the Chalk. Correlation of the logs with the other boreholes confirmed that the Chalk dip was greater than had been assumed from earlier interpretations. This meant that the horizon into which the tracer had been injected at West End Farm did not in fact intersect the PWS well, as had been thought previously, but pass in excess of 25 m below. A tracer was injected at the Thwing Road borehole as a column and a repeat injection also performed at West End Farm to confirm the results already described. This phase of the testing is continuing and the results are not yet available but it is hoped that the further work will continue to improve the understanding of the local hydrogeology.

5 CONCLUSIONS

The tracer testing and associated geophysical testing has confirmed the complexity of the hydrogeological conditions in the Kilham area. Groundwater flow is generally very rapid and is controlled by fissuring within the aquifer. Velocities of up to 475 m/day have been measured and connections over several kilometers have been proven. The results of testing have shown that groundwater flow in some cases is restricted to bedding plane fissures and no cross-bed

(vertical) flow occurs (e.g. tracing to Kilham PWS well) and in other cases it does (e.g. tracing to Tancred Pit hole). A possible explanation is that at the locations where vertical movement of water is implied, relative to the Chalk structure, the sampling points are, or are very close to, springs. The development of the Chalk is such that at these foci of groundwater flow, enhanced fissuring may have developed to enable vertical flow or they may be related to faulting.

The results to date have also shown that the capture zones for both the NSA and GPZ modelling are not accurate because of the complexity of the Chalk and the effect of the dipping strata. The problem of delineation of capture zones becomes extremely difficult and a truly three-dimensional approach which is able to consider the complexities of the system is required. From the current work, it would appear that the capture zone may extend only a limited distance upgradient with it getting progressively shallower. The further testing at the new borehole should help to establish the likely extent of the capture zone up gradient of the PWS well.

Overall, tracer testing has proved invaluable in assessing the applicability of existing protection zone models. Whilst it is accepted that drilling new boreholes will not always be possible, tracer testing does offer an extremely promising method although financial costs can be high.

6 ACKNOWLEDGEMENTS

We are grateful to John Watkins of Alcontrol UK and our colleagues, Linda Brewerton, John Davis (British Geological Survey) and John Aldrick (Environment Agency) who assisted in this work, which was jointly funded by the Environment Agency and the British Geological Survey. This paper is published by permission of the Director of the British Geological Survey (Natural Environment Research Council) and the Environment Agency.

REFERENCES

British Geological Survey 1993. Sheet 64 - Great Driffield. Solid and Drift Geological Map. Provisional Series.

Environment Agency (formerly National Rivers Authority) 1983. *Unpublished Report for Yorkshire Water Authority , Rivers Division.* Groundwater Resources of the Chalk of East Yorkshire: Kilham Pumping Test 1982. Authors; I.C. Barker, D.S. Chadha, R Courchee & A.S. Robertson. 59 pp.

Environment Agency (formerly National Rivers Authority) 1992. *Policy and Practice for the Protection of Groundwater.*

Environment Agency (formerly National Rivers Authority) 1995. *Guide to Groundwater Protection Zones in England and Wales.*

Foster , S.S.D & V.A. Milton. 1974. The permeability and storage of an unconfined chalk aquifer. *Hydrol. Sci. Bull.*; 19:485-500.

Kirby, G.A. & P.W. Swallow 1987. Tectonism and sedimentation in the Flamborough Head region of north-east England. *Proc. Yorks. Geol. Soc.* 64(4): 301-309.

Price, M. 1987. Fluid Flow in the Chalk of England. *From* Goff, J.C. & B.P.J Williams (eds). 1987. *Fluid Flow in Sedimentary Basins and Aquifers,* Geological Society Special Publication No. 34: 141-156.

Skilton, H.E. & D. Wheeler 1988. Bacteriophage tracer experiments in groundwater. *Journal of App. Bacteriology*, 65: 387-395.

Smart, P.L. 1976. The Use of Optical Brighteners for Water Tracing. *Trans. British Cave Research Assoc.* 3(2): 62-76.

Tracer Hydrology 97, Kranjc (ed.) © 1997 Balkema, Rotterdam, ISBN 90 5410 875 4

Development of a tracer test in a flooded uranium mine using *Lycopodium clavatum*

Christian Wolkersdorfer
IFG, Ingenieurbüro für Geotechnik, Niederkaina, Germany

Irena Trebušak
IGGG, Institute of Geology, Geotechnics and Geophysics, Ljubljana, Slovenia

Nicole Feldtner
Technische Universität Clausthal, Clausthal-Zellerfeld, Germany

ABSTRACT: The polymetallic Niederschlema/Alberoda uranium deposit in the Saxonian Erzgebirge (Ore Mountains) has been flooded since 1991. The objectives of the tests were to investigate the quality and rate of flow within a large part of the flooded mine to predict the mass flow of the pollutants. Based on the results of a first tracer test with *Lycopodium clavatum* in mid 1992 a second one was conducted at the end of 1995.

Four insertion and two sampling points were chosen and at each sampling point up to 800 g of coloured spores were inserted by using a newly developed insertion apparatus: LydiA (*Lycopodium* Apparatus). Beginning one day after insertion, at each sampling point two samples per weekday were taken. Out of the 15 samples an aliquot amount of material was counted and resulted in a reasonable good recovery rate of 2 %.

It could be shown, that the mean speed of the mine water within the investigated part of the mine ranges between 3 and 8 m min^{-1} and that the different parts of the mine are hydraulically well connected with each other. Therefore it may be that the pollutants within the flooded mine are transported by convective flow resulting in an exchange from deeper parts of the mine into higher ones.

1 INTRODUCTION

1.1 The Niederschlema/Alberoda Mine

The Niederschlema/Alberoda uranium mine is located 28 km south-west of Chemnitz in a densely populated area of Saxony. It has a depth of nearly 2000 meters and consists of 50 main levels with more than 50 shafts connecting the different levels of the mine. The length of the mine works sums up to 4150 km (Büder & Schuppan 1992) and the total volume to $36 \cdot 10^6$ m³ (Meyer, pers. comm.). For 45 years the mine has been operated by the SDAG Wismut, a Sowjet-East German Company. Due to economic, environmental and political problems of uranium mining and milling, in 1990 the Government of East Germany decided the mine's closure (Bundesminister für Wirtschaft 1993).

In January 1991 the controlled flooding of the mine, which will be completed as early as 2003, began. During the mine water's circulation through the open mine works it will be enriched in various ele-ments, some of them being toxic to both humans and the environment. Therefore the up to date remediation plans for the Niederschlema/Alberoda mine include treatment of the discharged water as soon as the water table will reach the water-adit Markus-Semmler (Gatzweiler & Mager 1993).

At the beginning of the flooding only little was known about the hydrogeochemical, thermal, and hydrodynamic processes during the flooding. Hitherto numerous consultant works and a Ph.D. thesis have been undertaken on these subjects, but the hydrodynamic conditions during the flooding and within the water body have only been outlined in general (Wolkersdorfer 1996, Merkel & Helling 1995).

Already in 1991 a tracer test was considered and the Wismut GmbH entrusted one of the authors (C.W.) with establishing a program for a tracer test, which will be suitable for the special conditions in the mine water. Based on the results of a first tracer test in the mid of 1992 (Wolkersdorfer 1993, 1996) a second one was conducted at the end of 1995. Some results of these tests, especially of the latter one, will be described here.

At the beginning of the tracer test on November 16th 1995 the mine was flooded as far as the level -726 (approx. -390 mNN). Due to a rapid raise of the water table to level -720 (approx. -386 mNN) the test had to be stopped on November 27th 1995, one week earlier than scheduled.

In the deepest parts of the mine rock temperatures of 70 °C have been measured. It is not known if the water there currently has the same temperature, but in January 1991 a water temperature of 52 °C was observed at level -1800 (G. Fröhlich, pers. comm.) and between June 1992 and December 1994 a maximal water temperature of 41 °C with an average of 36 °C was measured (Wolkersdorfer 1996). The total volume of the mine water within the flooded part of the mine had reached $13.9 \cdot 10^6$ m³ at the time of the 1995 tracer test.

1.2 Objectives of the Tracer Test

Main objective of the tracer test was to investigate the quality and rate of flow within a large part of the flooded mine. In that way the mass flow of the pollutants might be predicted. The working model for the development of the tracer test was that there is convective flow within the shafts and that they are connected hydraulically through the adits.

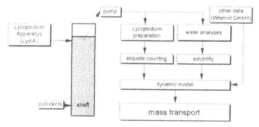

Fig. 1: Schema of the tracer test with *Lycopodium clavatum* and the questions connected to the test.

There are four questions which should be solved by the tracer test in the Niederschlema/Alberoda uranium mine:

- How high is the mine water's speed in the flooded mine works?

- Which paths through the mine works are used by the water?

- Are the dams (bulkheads) at the onsetting stations still effective?

- Is the similar chemical composition of the mine water in different shafts caused by mixing of the water?

In conjunction with chemical analyses and physicochemical data of the mine water it will be possible to adopt the working model. Furthermore it will be possible to estimate the flow rate of the water. All this data brought together will at a later stage be used for modelling the hydrodynamic regime within the flooded mine (Fig. 1).

1.3 Using Lycopodium clavatum as a Tracer

Lycopodium clavatum (clubmoss, Käß 1992) is widely utilised for the tracing of underground waterways in karstic regions (Atkinson 1973, Maurin 1976, Smart & Smith 1976, Plumb 1985, Gascoine 1985, Dechant & Hacker 1986, Smart et al. 1986, Käß 1992), one of the most extensive having been conducted in the NW Dinaric Karst (Hötzl et al. 1976). Hacker et al. (1983) conducted tracer tests with spores in basalts.

First hints for the possible use of *Lycopodium clavatum* as tracer are given by Timeus in 1910 (cited in Käß 1992) but not until 1953 the method and its positive application was described in the literature (Mayr 1953). Initially only natural spores served as tracers, but since Zötl's (cited after Käß 1992) and Dechant's (1959) work also coloured spores could be used. The latest improvement was the introduction of fluorescent spores by Käß (1982), speeding up the time consuming counting.

To our knowledge, no tracer test with drifting material has ever been conducted in a flooded mine. Over and above no method is known to release drifting materials into a water body in a certain depth. Extensive studying of the literature yielded no similar tracer test at all. Tracer studies in the Stripa mine (Sweden) or Konrad mine (Germany) investigated the fracture network between the galleries by the use of radioactive tracers and dyes (Birgersson et al. 1992, Kull 1987). Horn et al. (1995) used natural U-Th-ratios as tracers in a flooding experiment at the Königstein mine (Germany).

2 METHODS

2.1 Preparation of Lycopodium and LydiA

The colouring of spores has been described by several authors and is summarised by Käß (1992). Both, non-fluorescent and fluorescent spores are frequently used for tracer tests in karst aquifers and their benefits compared to non-coloured spores or dyes are well known (Käß 1982). Besides the advantage of fluorescent spores, to be easily countable, their disadvantage is that, due to spectral overlaps, only three different fluorescent dyes can be used in one single test (Smart & Smith 1976).

This restriction caused the conclusion to use non fluorescent spores for the tracer test in the Niederschlema/Alberoda uranium mine. Based on the results of Dechant (1960), Eissele (1961), and Drews & Smith (1969) six dyes were chosen for dying the spores according to a slightly modified method introduced by Dechant (1960): bismarck brown, fuchsin, safranine (saffron), nile blue, malachite green, and crystal violet, the first two for the 1992, the others for the 1995 test.

One of the main problems while conducting a tracer test with *Lycopodium clavatum* is to avoid contamination, not only in the laboratory, but at the insertion points. Dechant & Hacker (1986) describe one possible method for contamination free insertion. For the tracer test in the Niederschlema/Alberoda mine another problem had to be solved:

Release of spores has to be accomplished in a specified depth at a certain time without contamination of the water body above the final injection position.

During the 1992 test the abandoned piping system for the water drainage could be used for insertion. By opening and closing different slide valves at pumping station 296 II (level -996) the spores were inserted at two depths at distinct localities. To prevent contamination between the two insertions the piping system had to be flushed for 19 hours with 2000 m³ of infiltration water. Although this method solved the above mentioned problem, it is unreasonable for a tracer test in a larger scale. Therefore LydiA (*Ly*copo*dium A*pparatus), a sonde for inserting spores at certain depths, was constructed and will be applied for a patent. A chemical lock and

holders to be used with a cable winch assured the solution of the above mentioned problem.

As could be shown, the lock dissolves within 6 to 10 hours after insertion into the warm mine water (Wolkersdorfer 1996). In the laboratory the most part of the spores injected within the first 30 minutes and without jerky movements of LydiA it takes up to 24 hours to complete injection. For this reason, to be on the safe side and to guarantee injection of all the spores into the mine water, all LydiAs were moved jerkily after 25...26 hours of insertion.

2.2 *Spore budget*

Käß (1992) summarises several formulas for calculating the amount of spores needed for a tracer test. All of them are for an open water system with one or more swallets or springs (Brown & Ford 1971, Atkinson et al. 1973). In contrast to this situation a flooded mine with no connection to a drainage gallery might be described as a quasi closed water system (during the tracer test's duration the total volume of the water was increased by a mere of 0,5 % by infiltration water). Therefore none of these formulas could be used for the test.

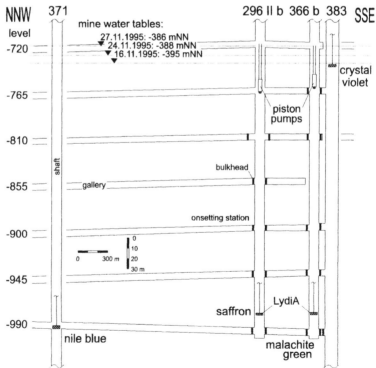

Fig. 2: Geometric details of the 1995 tracer test in the Niederschlema/Alberoda mine.

In view of the fact that 90…99 % of the spores might be lost (Maurin & Zötl 1960, Wolkersdorfer 1996) and that at least four spores per pumping hour should be detectable, the undermentioned formula was deduced. It is based on the hypothesis that the water in the mine is well mixed by convective flow and the spores are even distributed in the water.

$$m = \frac{1}{55000} \frac{V}{r \cdot q}$$

m: total mass of spores, g
V: volume of enclosed fluid, m³
r: recovery rate of spores, %
q: pumping capacity, L min⁻¹

Assumed that $13.9 \cdot 10^6$ m³ of water would be in the mine, the recovery rate to be 1 %, and the pumping capacity be between 0.3 and 0.4 L min⁻¹, the amount of spores needed for the tracer test was calculated to range at least between 630 g and 840 g per colour. As has been shown by the 1992 tracer test, where the recovery rate was assumed to be 0.5 %, this is enough to obtain reasonable results.

For the tracer test at the Niederschlema/Alberoda mine, four LydiAs were loosely filled with the coloured spores and carefully assembled in the laboratory to prevent contamination. Tab. 1 lists the used masses of spores and some of the test's details.

2.3 Insertion and Sampling

Insertion, sampling and preparation of samples were done on different places and levels to minimise sampling errors. While the sonds were lowered from level -550, where also the sample preparation took place, sampling was conducted at level -720.

In the mine each LydiA was fixed to a mount and slowly lowered to the appropriate depth using cable winches which were hold at the final depth during the whole duration of the tracer test. This was to prevent contamination of the water body while pulling out the opened LydiAs.

Tab. 1: Mass of spores, used dyes and depths of insertion and sampling points (shafts) in mNN.

Insertion/ sampling point	mass of spores g	colour of dye	depth mNN
371	804.2	nile blue A	-652
296 II b	805.3	safranine T	-639
366 b	722.6	malachite green	-639
383	764.0	crystal violet	-397
296 II b	–	–	-423
366 b	–	–	-419

In shafts 371, 296 II b and 366 b the LydiAs were lowered up to 260 m below the mine water's surface, close to the onsetting stations at the main level -990 (see Fig. 2 for details). Unfortunately, the glueing of LydiA failed in shaft 383 after 20 meters of lowering and nearly half of the crystal violet coloured spores were spread over into the return air. Contrary to the original planning this LydiA was then lowered only 10 m below the mine water's surface, according to 20 m below the onsetting station of level -720.

Because of the mine's operational sequence, insertion of LydiA in shafts 366 b and 383 had to be carried out before, in shafts 296 II b and 371 after starting of sampling. As the opening time of the LydiAs was to be scheduled 6…10 hours after insertion this was not to be seen as a disadvantage for the test.

Two mini piston pumps (Pleuger Mini-Unterwasserpumpe, Pleuger Worthington GmbH/Hamburg), one in shaft 296 II b, 28 m below the water level, the other one in shaft 366 b, 25 m below the water level, were installed. All of the discharged water was then fed into a filter system on level -720 to recover the spores. The use of mini pumps with pumping capacities well under the net water inflow, which was assumed to be 3000 L min⁻¹, avoids the establishment of a significant flow field to the pumps. Their capacity depends on both, the delivery head and the water level. In shaft 366 b the pumping capacity ranged from 0.39…0.41 L min⁻¹ and in shaft 296 II b from 0.31…0.55 L min⁻¹.

For the 1992 tracer test a vertically installed plankton net was used (Maurin & Zötl 1960, Atkinson 1968, Gardener & Gray 1976, Käß 1992). Unfortunately the wooden frame was quickly destroyed by either bacteria or mould and the pores of the nylon nets were blocked by too much sediment, bacteria and mould. Therefore a new filter system with coarse as well as fine nylon nets had to be established to stand the warm and humid environment within the mine. Both, a cheap price and easy use, even for unskilled persons, had to be guaranteed.

Based on these experiences a filter system was constructed, which consists of three plastic parts, normally being used in sewage disposal (Marley GmbH/Wunstorf; DN 150: DIN 19534), and two nylon nets fixed between the plastic parts. The upper net has a mesh size of 335 μm (NY 335 HC, Hydro-Bios/Kiel), the lower lycopodium net one of 30 μm (NY 30 HC), both nets being 270*270 mm in size. Although Käß (1992) recommends a 25 μm mesh size for the nets, a 30 μm net was installed under the special circumstances with the mine water having 61 mg L^{-1} filter residue in average (Wolkersdorfer 1995). According to the investigations of Batsche et al. (1967) this mesh size should yield at least 80 % of the total spore number. In the case of the tracer test described here, it will be definitely more, as the net's pores will soon be partly blocked by sediment.

At the time of sampling the whole filter system on level -720 was changed for a new system with cleaned plastic parts and unused nets. The used system was dismounted on level -550, the 335 μm net and the plastic parts carefully rinsed, and the rinsing water as well as the 30 μm net collected in a plastic bag for later evaluation. Details of the sampling results are listed in Tab. 2 and are discussed in chapter 2.5.

2.4 Sample Preparation and Counting

Most of the 15 samples contained noticeable amounts of Fe-oxides. Therefore up to 0.6 g of edetic acid (Idranal 2) were added, according to the suggestions of Käß (1992). This method removes both, Fe-oxides and carbonates to reduce the amount of suspended particles in the samples.

After one day of reaction the samples were filtered through 8 μm cellulose nitrate filters (Sartorius/Göttingen) with 50 mm diameter, using Nalgene plastic filters for membrane filtering with a vacuum pump. This is an adapted method introduced by Käß (1982) for the counting of fluorescent spores. After each filtration the Nalgene filters, the filter unit, and the working tables in the laboratory were cleaned to exclude any kind of contamination during sample preparation.

Every filter with the filter residue was then stored in a Petri dish bottom, covered with a Petri dish top and air dried in the laboratory. Further details for sample preparation can be found in Eissele (1961) or Käß & Reichert (1986). After drying, all filters were mounted between two glass plates, 56 * 63 mm in dimension. Prior to mounting each glass plate was cleaned with ethyl alcohol and acetone in a separate room. Finally the glass plates were fixed with tape and labelled with the sample number.

For the investigations a Leitz Ortholux 2 Pol-BK microscope with wide angle oculars (10*) and two different objectives (25*, 40*) was used, resulting in a magnification of 250 and 400 respectively.

Nearly all the samples contained high amounts of sediment. Nevertheless the samples could be used because the spores normally are on the top of the sediment as has been stated by Eissele (1961). Unfortunately, after drying some of the membrane filters were covered with a thin film of small, transparent crystals as a result of the adding of edetic acid. In spite of this film it was possible to identify the spores because the crystal's thickness is always less than the spore's diameter.

Due to this high amount of suspended material the printed net on the membrane filters, which should have been used for aliquot counting, could not be seen. By dividing every sample into four quarts, another aliquot method to reduce the necessary time for counting was used. Interestingly, the dyes nile blue and malachite green, which were easily distinguishable prior to the tracer test, were hard to distinguish after having been in the mine water. Therefore the use of two persons for counting had two reasons: to improve the statistical significance of the aliquot counting and to avoid wrong interpretations as a result of distinct colour sights. It could be shown that the differences in the numbers of spores per colour and the total numbers of spores per quart are not high enough to prove a significant statistical difference between colours and quarts.

Tab. 2: Results of the aliquot counting of the 1995 tracer test with *Lycopodium clavatum* in the Niederschlema/Alberoda mine. time: pumping time in hours and minutes. distance between sampling and insertion points in m.

Samp-ling point	sample	safra-nine T 296 II b	green 366 b	nile blue A 371	crystal violet 383	pump-ing time
296 II b	296-17	1	26	24	72	72:54
	96-20	174	160	188	824	22:50
	96-21a	96	244	82	564	3:32
	96-21b	60	80	40	106	42:58
	96-23a	354	598	358	1112	4:30
	96-23b	322	254	522	752	19:16
	96-24a	172	452	180	784	4:20
	96-24b	38	12	16	38	67:45
sum		1217	1826	1781	4635	
distance. m		216	776	2159	736	
366 b	6-17	30	30	132	192	91:59
	66-20	260	670	352	852	23:47
	66-21a	70	100	164	518	3:59
	66-21b	28	58	34	82	44:54
	66-23a	254	526	206	1008	2:52
	66-23b	266	454	230	602	20:44
	66-24a	242	292	94	202	4:12
sum		1150	2130	1212	3456	
distance. m		780	220	2723	172	
total	–	2367	3956	2622	7708	–

During the twelve days of the tracer test 15 samples were collected, thereunder two blind samples. The latter were taken to proof if there were still spores from the 1992 test in the water or not. As could be shown, no spores of this previous test were found.

Tab. 2 lists the results of the counting as well as the pumping times and the names of the coloured spores inserted in each of the four shafts. The pump was running for 238 hours at sampling point 296 II b and 192 hours at point 366 b, having pumped 4460 L and 4470 L of mine water, respectively. On the basis of the aliquot counting, the total number of spores found are 2367 for safranine T, 3956 for malachite green, 2622 for nile blue A and 7708 for crystal violet. The total loss was 98 %.

All the spores in samples 296-17 and 6-17 arise from contamination either during sample preparation in the laboratory or, more probable, during filter installation in the mine. Since it was not possible to have two teams in the mine, one for installation of filter systems and LydiA, the other one for changing the filter systems, contamination could not fully be omitted on the first day of the tracer test. Therefore it must be considered that approximately 30 spores of each sample are due to contamination. The high number of crystal violet spores in both blind samples is a result of the destroyed LydiA in shaft 383. Return air from shaft 383 moves on level -540 to shaft 366 b and from there to shaft 296 II b, finally reach-

ing the day shaft 371. In view of the fact that the filter system for sample 6-17 was filled with mud at the time of sample collection, it was not possible to have it sealed before insertion of the crystal violet spores, resulting in the relatively high number of violet spores in this sample. However, sample 296-17 was taken after the incident in shaft 383. Therefore, the spores transported through the air respond for the 72 crystal violet spores in the sample. Nothing can be said about the reason for the 132 nile blue spores in sample 6-17. It must be considered that they are the result of contamination.

Already in the first samples (96-20 and 66-20), spores from each insertion place are existing (Tab. 2, Fig. 3). Because the numbers are significantly higher than in both blind samples no contamination will be taken into account. This result is interesting in so far as the shortest distances between the sampling and insertion points are within an interval of 200 m and 2700 m (Tab. 2). Therefore it could not be expected to find nile blue spores already in samples 96-20 and 66-20.

Based on the shortest flow distances, mean opening time for LydiA, as well as 16 %, 50 % and 84 % quartiles of travelling time, the maximal mean speed of the spores can be calculated to be 8 m min^{-1} (Fig. 4). The mean speed for all spores in the first set of samples is approximately 3 m min^{-1}, fitting well with the results of flow meter measurements in shaft 371 II b (Wolkersdorfer 1996).

As can be seen in Fig. 3 the number of spores arriving at the sampling points is not evenly distributed over the length of the experiment. A first maximum is found 17 hours, a second one 65 hours after opening of the chemical lock. Between these two maximums only a minimum of 10…12 spores per hour was found. The second maximum is again followed by a period of less spores arriving at the sampling points and finally there is a third maximum after 89 hours. In the case of sampling point 296 II b it is followed by a period of 68 hours with less than 1 spore per hour arriving. The sampling at 366 b had to be stopped due to the fast rising water table.

This is an interesting result, as the number of spores per hour was

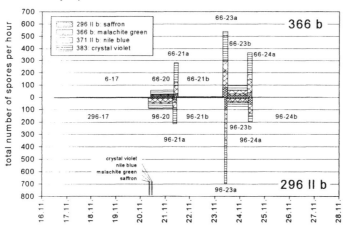

Fig. 3: Results of the tracer test in the Niederschlema/Alberoda mine given in spores per hour for the hole duration of the test. The width of each box represents the pumping time as shown in Tab. 1. Also shown, between the 20th and the 21st, are the insertion times of the four LydiAs.

expected to raise to a certain level and staying constant by the end of the test. Not until 5 to 6 weeks after injection, the numbers were expected to fall to 10 spores per hour, as has been revealed by the first tracer test in 1992 (Wolkersdorfer 1996). Contrary to the 1995 test, sampling of the spores in the 1992 test started 42 days after insertion, due to high Radon concentrations at the expected sampling point. Therefore no data was available to estimate the distribution of spores within time and space of the mine or to compare both tests.

Four explanations for the uneven temporal distribution are possible: sampling errors, breakdown of power supply, convection in the shafts, and varying flow paths.

Sampling errors can be excluded because it seems highly unlikely that in *each* case the samples of point 366 b and 296 II b should be faulty.

Breakdown of power supply, too, must not be taken into account, as this would have been noticed by the dispatcher's office.

Varying flow paths would cause separate arrival times for different coloured spores. As this is not the case, varying flow paths as a reason for the uneven distribution will not be taken into account, at the very least for the duration of the test.

As a result of the short time needed for the majority of the spores to flow out from LydiA (approx. 30 minutes), the spore cloud will be relatively small at the beginning of the test. Supposed, the water in a single shaft is dominated by convective transport, the spore cloud transported thereby will pass the pumps

several times. This would result in concentration maximums and minimums as they can be seen in Fig. 3.

Comparing the total numbers of safranine and malachite green spores at sampling points 296 II b (ratio 1:1.5) and 366 b (ratio 1:1.9) it is evident that safranine spores in both cases are fewer than the malachite green ones. Expecting the flow regime in both shafts to be similar, the number of safranine spores at sampling point 296 II b should have been higher than observed. Moreover, the result is interesting as the mass of inserted spores (1.1:1) is indirect proportional to the number of observed ones. Obviously the upward movement of the water in shaft 366 b is more developed than in shaft 296 II b.

3 CONCLUSIONS

The tracer test with *Lycopodium clavatum* in the mine water of the Niederschlema/Alberoda mine has produced useful results. Insertion of the drifting material by the use of specially constructed sonds (LydiA) with a chemical lock and inserting them at a certain depth proved to be possible.

It could be shown that the mine water's speed averages 1 m min^{-1} in the vertical shafts and 6 m min^{-1} in the horizontal galleries. The mean speed of the water was about 3 m min^{-1}. Within less than 24 hours spores from all four insertion localities reached the two sampling points, thus showing a good hydrodynamic connection of all the investigated shafts and galleries. The similar chemical composition of the mine water in different, even far apart shafts must therefore be considered as a result of the water's convective flow. Furthermore, the tracer test implicates, that the chemical similarities of water analyses in different depths of a single shaft are also due to a good mixing of the water.

The dams (brick walls) on the onsetting stations seem to have no noticeable influence on the large scale mixing of the water, although the results of the tracer test gives no clarity about small scale movements in the vicinity of two dams. Nevertheless, it seems that the dams do not reduce the mixing significantly at all. Unfortunately the tracer test gave no satisfying answer to the question of the exact water paths. The first sampling

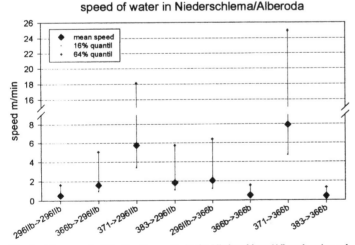

Fig. 4: Mean speed of the flooding water in the Niederschlema/Alberoda mine calculated for each injection and sampling location.

interval is too long compared to the water's speed, therefore only general estimates of paths might be done. It is obvious that the former main level -990, and after having been flooded, also level -720 act as good horizontal paths for the water. They seem to distribute the water everywhere over the investigated part of the mine. Consequently it must be considered that also water from deeper parts of the mine, and pollutants, will reach the upper parts within a relatively short time, most certainly in less than a day. It is important for further tracer tests to have a sampling interval of 6...12 hours, especially for the first samples.

Important transport paths for water and pollutants in the mine are therefore the main levels and the shafts connecting them. Even places far apart from each other and off the main galleries are bind up into the water stream fair well. Based on the results of the tracer test it must be expected that water which is in contact with highly mineralised parts of the mine is included in the general convection cell.

As the motor for the fast convection the high temperatures (up to 70°C) in the deeper parts of the mine must be supposed. In all probability, highly contaminated water will therefore always be within the portion of the mine taking part in the convective mixing. This is usually the whole water body under the last level to be flooded. Consequently, a proposed treatment plant would have to treat contaminated water for a relatively long time period.

As the dams on the onsetting station do not have the desired effect of stopping the interchange of water, another method must be considered instead. Plugging the vertical connections at a very high level in the mine would stop the transport of contaminants from deeper parts into higher parts. It can therefore be assumed that vertical sealing of the mine makes a treatment plant superfluously.

4 ACKNOWLEDGEMENTS

The authors would like to thank the Wismut GmbH/ Chemnitz, especially J. Meyer, J. Schreyer, G. Lein and G. Fröhlich for the helpful support during the whole tracer test. Thanks to R. Schecke for his counting of spores. One of us (I.T.) received a grant from the Slovenian Ministry of Education and the German Academic Exchange Service/Bonn, the other one (C.W.) from the Hanns-Seidel-Foundation/ Munich. Support has also been given by the German Science Foundation/Bonn under contract Re 920/1-2 (Univ.-Prof. G. Reik, PhD).

5 REFERENCES

Atkinson, T.C. 1968. Tracing Swallet Waters using Lycopodium Spores. *Trans. Cave Research Group of Great Britain.* 10(2): 99—105.

Atkinson, T.C., Smith, D.I., Lavis, J.J. & Whitaker, R.J. 1973. Experiments in tracing underground waters in limestones. *Journal of Hydrology.* 19: 323—349.

Batsche, H., Bauer, F., Behrens, H., Buchtela, K., Hribar, F., Käß, W., Knutsson, G., Mairhofer, J., Maurin, V., Moser, H., Neumaier, F., Ostanek, L., Rajner, V., Rauchert, W., Sagl, H., Schnitzer, W.A. & Zötl, J. 1967. Vergleichende Markierungsversuche im Steirischen Karst. *Steirische Beiträge zur Hydrogeologie.* 1966/67: 331—404.

Birgersson, L., Widen, H., Aagren, T. & Neretnieks, I. 1992. *Tracer migration experiments in the Stripa mine 1980—1991,* Vol. STRIPA-TR–92-25. Stockholm: Swedish Nuclear Fuel and Waste Management Co.

Brown, M.C. & Ford, D.C. 1971. Quantitative tracer methods for investigation of karst hydrological systems. *Trans Cave Research Group of Great Britain.* 13(1): 37—51.

Büder, W. & Schuppan, W. 1992. Zum Uranerzbergbau im Lagerstättenfeld Schneeberg-Schlema-Alberoda im Westerzgebirge. *Schriftenreihe Gesellschaft Deutscher Metallhütten- und Bergleute.* 64: 203—221.

Bundesminister für Wirtschaft 1993. Wismut – Stand der Stillegung und Sanierung. *BMWI Dokumentation.* 335: 1—35.

Dechant, M. & Hacker, P. 1986. Neue Entwicklungen in der Methode des Sporennachweises. *International Symposium of Underground Water Tracing.* 5: 149—155.

Dechant, M. 1960. Das Anfärben von Lycopodiumsporen. In Maurin, V. & Zötl, J. (eds), *Untersuchung der Zusammenhänge unterirdischer Wässer mit besonderer Berücksichtigung der Karstverhältnisse,* Vol. 12: 145—149. Graz: Beiträge zur alpinen Karstforschung.

DIN Deutsches Institut für Normung e.V. 1979. *DIN 19 534. Teil 1: Rohre und Formstücke aus weichmacherfreiem Polyvinylchlorid (PVC hart) mit Steckmuffe für Abwasserkanäle und -leitungen.* Berlin: Beuth Verlag.

Drew, D.P. & Smith, D.I. 1969. Techniques for the Tracing of Subterranean Drainage. *British Geomorphological Research Group Technical Bulletin.* 2: 1—36.

Eissele, K. 1961. Erfahrungen mit der Sporentrift-methode. *Jahrheft geologische Landesanstalt Baden-Württemberg* (5): 345—350.

Gardener, G.D. & Gray, R.E. 1976. Tracing Subsurface Flow in Karst Regions Using Artificially Colored Spores. *Bulletin Association Engineering Geology* 13: 177—197.

Gascoine, W. 1985. Black Mountain West – The Catchment for Llygad Llwchwr. *Caves and Caving*. 27: 30—31.

Gatzweiler, R. & Mager, D. 1993. Altlasten des Uranbergbaus – Der Sanierungsfall Wismut. *Die Geowissenschaften*. 11(5-6): 164—172.

Horn, P., Hölzl, S. & Nindel, K. 1995. Uranogenic and Thorogenic Lead Isotops (^{206}Pb, ^{207}Pb, ^{208}Pb) as Tracers for Mixing of Waters in the Königstein Uranium-Mine, Germany. *Proceedings Uranium-Mining and Hydrogeology, Freiberg, Germany; GeoCongress*. 1: 281—289.

Hötzl, H., Maurin, V. & Zötl, J. 1976. Results of the Injection of Lycopodium Spores. *International Symposium of Underground Water Tracing*. 3: 167—181.

Käß, W. & Reichert, B. 1986. Tracing of Karst Water with Fluorescent Spores. *International Symposium of Underground Water Tracing*. 5: 157—165.

Käß, W. 1982. Floureszierende Sporen als Markierungsmittel. *Beiträge zur Geologie der Schweiz – Hydrologie* (28 I): 131—134.

Käß, W. 1992. *Geohydrologische Markierungstechnik, Lehrbuch der Hydrogeologie*, Vol. 9. Berlin Stuttgart: Bornträger.

Kull, H. 1987. Geochemical and retention investigations in the framework of a tracer test in Konrad mine/Salzgitter-Bleckenstedt. In Technische Universität München, Commission of the European Communities, Gesellschaft für Strahlen- und Umweltforschung m.b.H. München (eds), *Chemistry and migration behaviour of actinides and fission products in the geosphere*, Vol. 9201069: 160—161. München: Gesellschaft für Strahlen- und Umweltforschung m.b.H.

Maurin, V. & Zötl, J. 1960. Untersuchung der Zusammenhänge unterirdischer Wässer mit besonderer Berücksichtigung der Karstverhältnisse. *Beiträge zur alpinen Karstforschung* (12): 1—179.

Maurin, V. 1976. Drifting Materials. *International Symposium of Underground Water Tracing*. 3: 228—229.

Mayr, A. 1953. Blütenpollen und pflanzliche Sporen als Mittel zur Untersuchung von Quellen und Karstwässern. *Anzeiger der Österreichischen Akadademie der Wissenschaften Mathematisch naturwissenschaftliche Klasse* (1953): 94—98.

Merkel, B. & Helling, C. 1995. Proceedings Uranium-Mining and Hydrogeology 1995, Freiberg, Germany. *GeoCongress*. 1: 1-583.

Plumb, K. 1985. Littondale Hydrological Study Weekend. *Caves and Caving*. 27: 30.

Smart, P.L. & Smith, D.I. 1976. Water Tracing in Tropical Regions, the Use of Fluorometric Techniques in Jamaica. *Journal of Hydrology*. 30: 179—195.

Smart, P.L., Atkinson, T.C., Laidlaw, I.M.S., Newson, M.D. & Trudgill, S.T. 1986. Comparison of the Results of Quantitative and Non-Quantitative Tracer Tests for Determination of Karst Conduit Networks: An Example from the Traligill Basin, Scotland. *Earth Surface Processes and Landforms*. 11: 249—261.

Wolkersdorfer, C. 1993. *1. Bericht zum Kooperationsvertrag zwischen der WISMUT GmbH und der Abteilung Ingenieurgeologie*. Clausthal: unpublished report.

Wolkersdorfer, C. 1995. Die Flutung des ehemaligen Uranbergwerks Niederschlema/Alberoda der SDAG Wismut. *Zeitschrift für geologische Wissenschaften* 23(5/6): 795-808.

Wolkersdorfer, C. 1996. Hydrogeochemische Verhältnisse im Flutungswasser eines Uranbergwerks – Die Lagerstätte Niederschlema/Alberoda. *Clausthaler Geowissenschaftliche Dissertationen*. 50: 1—216.

385

Aquifer parameters and modelling

·

Tracer Hydrology 97, Kranjc (ed.) © 1997 Balkema, Rotterdam, ISBN 90 5410 875 4

A deuterium-calibrated compartment model of transient flow in a regional aquifer system

Michael E. Campana
Department of Earth and Planetary Sciences, University of New Mexico, Albuquerque, N. Mex., USA

William R. Sadler, Neil L. Ingraham & Roger L. Jacobson
Water Resources Center, Desert Research Institute, Las Vegas, Nev., USA

ABSTRACT: We use the spatial distribution of deuterium to calibrate a 30-cell compartment model of a 19,000-km^2 ground-water basin underlying the Nevada Test Site and environs in southern Nevada-California, USA. Basin throughflow is 58.9 x 10^6 m^3/year. Approximately 40% of this flow originates as subsurface inflow into the basin; 60% is recharged within the basin. Major recharge areas are the Spring Mountains and the Fortymile Canyon/Wash-Stockade Wash area, contributing 9.6 x 10^6 m^3/year and 7.6 x 10^6 m^3/year, respectively. Our model calculates mean/median ground-water residence times and cumulative residence time distributions. The lowest mean (6,700 years) and median (1,400 years) residence times are found beneath Fortymile Canyon/Wash; the highest mean (26,300 years) and median (21,600 years) residence times occur beneath Yucca Flat in the central part of the flow system. Residence times are not necessarily correlated with position in the flow system; some waters show a decrease in mean/median residence times along their flow paths, indicating dilution of old water with relatively young recharge water. This is corroborated by δD values: -90 to -102 permil for local recharge and -98 to -117 permil for regional ground water. Older ground water, depleted in deuterium and presumably recharged during a colder, wetter climate, is enriched along its flow path by mixing with younger, more enriched ground water.

1 INTRODUCTION

Four decades of nuclear testing have served as an impetus for numerous studies of the ground-water flow system beneath the Nevada Test Site (NTS) and vicinity, southwestern USA. A more recent impetus is the possible location of a high-level nuclear waste disposal site at Yucca Mountain, adjacent to the western boundary of the NTS. Possible radionuclide migration to the accessible environment is a concern; therefore, knowledge of the nature and extent of the NTS ground-water flow system, hereafter referred to as the NTSFS, is of paramount importance. We use the environmental stable isotope deuterium (D) to calibrate a simple compartment model of the ground-water system beneath the NTS and vicinity. Deuterium has the advantage of being stable and essentially conservative once in the saturated zone. Our model is based upon that of Feeney et al. (1987) but encompasses a larger area, provides better information on ground-water residence times, and, more importantly, attempts a transient simulation by treating each compartment as a linear reservoir.

2 HYDROGEOLOGY

The study area lies between 36 and 38 degrees north latitude and 115 and 117 degrees west longitude and extends 19,000 km^2 (Figure 1). The area is in the southern Great Basin section of the Basin and Range physiographic province, with a topography of north-trending, block-faulted mountain ranges separated by alluvial basins. Elevations in the study area range from about 3,500 meters above mean sea level in the Spring Mountains to below sea level in Death Valley.

Precipitation, temperature, and plant communities in the area are generally a function of elevation. The average annual precipitation increases as a function of elevation, from fewer than 8 cm in Death Valley and the Amargosa Desert to greater than 70 cm in the upper reaches of the Spring Mountains. Therefore, the climate can be arid on the valley floors while subhumid at higher elevations. Most of the precipitation occurs during winter as a result of Pacific Ocean fronts, but some occurs during summer as high intensity thunderstorms. Winters are short and mild, while summers are long and hot except at the higher altitudes. Because of the primarily arid conditions, no

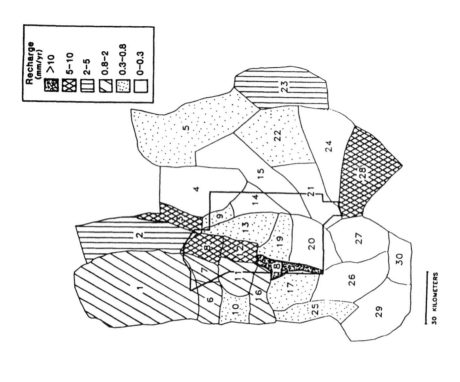

Fig 2. Recharge rates (mm/yr).

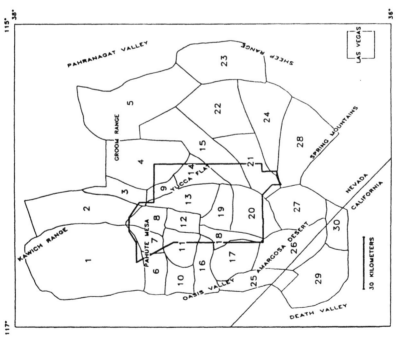

Fig 1. Study area and model cell network.
Nevada Test Site outlined in heavy line.

390

major perennial streams exist in the study area except local drainage from major springs.

Ground water is recharged by precipitation in the higher elevations in the north, east, and southeast and by stream-channel infiltration during the infrequent flow events. It also enters the system as subsurface inflow from the north and east. Generally, ground water flows to the south toward areas of discharge in Oasis Valley, Ash Meadows, Death Valley and Franklin Lake playa (southern Amargosa Desert).

The geology of the region is complex and has been well-described by Winograd and Thordarson (1975). They identified eleven hydrogeologic units (see their Table 1, pp. C10-C11) in the region ranging in age from Precambrian to Quaternary; five of these are aquitards, six are aquifers. Of these units, two of the most important are the lower clastic aquitard, comprised of Precambrian to Cambrian quartzites and shales, and the lower carbonate aquifer, comprised of Cambrian through Devonian limestones and dolomites. The former's great thickness, an aggregate of about 3000 meters (m), and areal extent make it a major element in controlling regional ground-water flow; the latter (aggregate thickness of about 4500 m) underlies much of the study area and is the major conduit for regional ground-water flow in the area. The upper clastic aquitard and upper carbonate aquifer are similar to the lower clastic aquitard and lower carbonate aquifer respectively but are less important because of their lower thicknesses and more limited areal extents. Other hydrogeologic units worth noting are the basin-fill aquifer, important in the Amargosa Desert, and the various volcanic-rock aquifers and aquitards (Tertiary tuffs and associated lithologies), which are up to 4000 m thick in the western portion of the study area.

The area's structural geology has played a major role in shaping the hydrogeology of the region. Mesozoic and Tertiary folding and thrust faulting significantly deformed the Precambrian and Paleozoic rocks; Tertiary normal block faulting produced classic basin and range topography. These same forces were responsible for fracturing the aforementioned carbonate rocks, producing a very transmissive regional aquifer. The intermontane basins filled with sediments derived from the surrounding mountain ranges, in many cases producing a two-tiered flow system characteristic of the region: a shallow system developed in the basin fill overlying a deeper regional system in carbonate and other rocks.

3 THE MODEL

We use a numerical compartment model developed by Simpson and Duckstein (1976) to simulate flow in the NTSFS. The code was developed by Campana (1975) and has been applied to a variety of subsurface flow systems, including the one to the east of the NTSFS, the White River Flow System (Kirk and Campana 1990). The basic equation, applied to each cell, or compartment, is (Simpson and Duckstein 1976):

$$S(N) = S(N-1) + [BRV(N)*BRC(N)] - [BDV(N)*BDC(N)] \qquad (1)$$

where: $S(N)$ = cell state at iteration N, the mass of tracer within the cell; $BRV(N)$ = boundary recharge volume, the input volume of water; $BRC(N)$ = boundary recharge concentration, the input tracer concentration; $BDV(N)$ = boundary discharge volume, the output volume of water; and $BDC(N)$ = boundary discharge concentration, the output tracer concentration.

Tracer concentrations and water volumes crossing system boundaries and entering/leaving a cell on the boundary of the system are given the prefix "system" or "S". Thus, recharge entering a cell from outside the system has a characteristic tracer concentration SBRC (system boundary recharge concentration) and volume SBRV (system boundary recharge volume). The similar holds for discharge from the system (SBDC and SBDV).

Equation (1), a mass balance equation, is applied successively to each cell during a given iteration; discharge (BDV and BDC) from an "upstream" cell becomes recharge (BDV and BDC) to a "downstream" cell. The BDC(N) term on the right-hand side of equation (1) is the only unknown and can be determined from one of two mixing rules, the simple mixing cell (SMC), which simulates perfect mixing, or the modified mixing cell (MMC), which simulates some regime between perfect mixing and piston flow. For the SMC:

$$BDC(N) = \frac{S(N-1) + BRV(N)*BRC(N)}{VOL + BRV(N)} \qquad (2)$$

For the MMC:

$$BDC(N) = S(N-1)/VOL \qquad (3)$$

where: VOL = volume of water in the cell.

Note that the MMC simulates pure piston flow as $BRV \rightarrow VOL$ and perfect mixing as $BRV \rightarrow$ zero.

Transient flow can be treated by assuming that the outflow from a ground-water reservoir is proportional to the storage in the reservoir (Dooge, 1960):

$$S = KQ \qquad (4)$$

where: S = storage above a threshold, below which the outflow is zero; K = storage delay time of the element; and Q = volume rate of outflow from the element. Equation (4) defines a linear reservoir.

In the context of the compartment model Equation (4) for a single compartment is (Campana 1975):

$$VOL(N) = K*BDV(N) \qquad (5)$$

Equation (5) can be modified to account for the presence of thresholds in the system. If K is held constant for all N, then the system described by either of the above equations is a linear, time-invariant system; if K is a function of time or iteration number, then the system is a linear, time-variant system (Mandeville and O'Donnell, 1973).

If Equation (5) is rewritten for iteration N+1 and substituted into Equation(6), a volume conservation equation for a given compartment or cell:

$$VOL(N+1) = VOL(N) + BRV(N+1) - BDV(N+1) \qquad (6)$$

the result is

$$VOL(N+1) = VOL(N) + BRV(N+1) - \frac{VOL(N+1)}{K} \qquad (7)$$

Equation (7) simplifies to

$$VOL(N+1) = \frac{K}{K+1} * [VOL(N) + BRV(N+1)] \qquad (8)$$

At iteration N+1, all quantities on the right-hand side of Equation (8) are known, so VOL(N+1) can be calculated. Once this has been accomplished, then BDV(N+1) can be calculated from Equation (5).

Each compartment depicts a region of the hydrogeologic system, and is delineated based upon hydrogeologic uniformity and data availability.

The compartment model essentially performs a mass balance on the flow system to determine flow rates and ground-water residence times; it needs a tracer to do this. An ideal tracer is one that moves with the velocity of the water, is easy to sample for and detect, does not react chemically once in the saturated zone, and displays spatial variability. The stable isotope deuterium, which occurs as part of the water molecule HDO, comes about as close to an ideal tracer as there is. For that reason, we chose to calibrate the model with deuterium. Deuterium contents are expressed as permil deviations (δD) from a standard; we used Standard Mean Ocean Water (SMOW) as the standard.

4 NTSFS MODEL DEVELOPMENT AND CALIBRATION

Two sets of deuterium values were used: "local" values from high-altitude springs and shallow wells, with signatures of -90 to -102 permil, representing recharge water; and "regional" values from large, low-altitude springs and deep wells, with signatures of -98 to -117 permil, representing the regional flow system ground water. The latter were used in model calibration and, along with the hydrogeology (e.g., hydrostratigraphy, structure), used to subdivide the flow system into 30 compartments or cells (Figure 1). The difference between the regional and local δD values can be explained by relatively depleted subsurface inflow from higher latitudes and possibly past climatic regimes entering the study area and becoming gradually enriched along flow paths by water recharged within the study area. The overall trend of the regional δD values is gradual enrichment from north to south with the northwest area being the most depleted and the southwest being the most enriched. Some recharge δD values were estimated from the δD values in precipitation. Complete data can be found in Sadler (1990).

Initial estimates of the SBRV (both as recharge and subsurface inflow) were based on Rush (1970); Walker and Eakin (1963); and Malmberg and Eakin (1962). The initial recharge estimates were used as a starting point and as a reference during calibration.

Flow routing inputs to the model are expressed as the percentage of total discharge from each cell to each of its receiving cells or out of the model boundaries. Initial flow routing values were based on the aforementioned publications on the hydrogeology of the area. Total discharge from a cell is calculated by Equation (5).

The parameter VOL equals the volume of active water in a cell. Cell areas were measured with a planimeter from a 1:250,000 scale map. An effective porosity of 2 percent was used for the carbonate aquifer, which is the mean value of 25 samples presented by Winograd and Thordarson (1975). The 5 percent effective porosity measured for the welded tuff aquifer was chosen for the volcanic strata based on the observation that most of the flow is transmitted through the welded tuffs (Winograd and Thordarson, 1975); the higher porosities of the non-welded tuffs were not used. The upper clastic aquitard (cells 13 and 19) was assigned an effective porosity of 4 percent, the mean values of 22 samples presented in Winograd and Thordarson (1975). The basin-fill aquifer is characterized as being generally poorly sorted by Winograd and Thordarson (1975) and was assigned an effective porosity of 15 percent.

The model utilized a 100-year iteration interval and the MMC (modified mixing cell) option.

To simulate transience we increased recharge to the model and decreased its δD by 5 permil during the period 23,000 to 10,000 years before present (White and Chuma 1987).

Model calibration was accomplished by adjusting SBRV and intercellular flow routing values until the difference between the observed and simulated δD values was within ±1 permil, the analytical error for deuterium.

5 RESULTS AND DISCUSSIONS

Space does not permit the inclusion of the complete suite of transient results. The results presented below represent average values of the various quantitites (recharge, flow, etc.).

The areal distribution of recharge is shown in Figure 2. In general, higher recharge rates are present in the northern region of the model with lower rates in the southern; however, quasi-isolated areas which do not follow this trend have the highest recharge rates within the model area. The high recharge areas in Figure 2 correspond to areas of relatively enriched δD values: eastern Pahute Mesa (cell 8); Stockade Wash (cell 12); Fortymile Canyon/Wash (cell 18); Spring Mountains (cell 28); and Sheep Range (cell 23).

The total flow rate through the system averages $58.9 \times 10^6 m^3/yr$. Broad divisions of flow rates are shown in Figure 3. The lowest flow rates correspond to cells which are dominated by the presence of an aquitard (cells 13 and 19), cells immediately downgradient from an aquitard (cells 9, 14, 20, and 25), and a cell which is a moderate recharge area (cell 3). Cells 3, 13, 19, and 20 are thought to divide the Alkali Flat-Furnace Creek and

Ash Meadows subbasins. The highest flow rates correspond to a major potentiometric trough in the carbonate aquifer immediately upgradient (cell 21) from the Ash Meadows area (Winograd and Thordarson, 1975), the terminus of the Ash Meadows subbasin (cell 27), and the constriction and termination of the Alkali Flat-Furnace Creek subbasin (cells 26 and 30).

Other regional models of the area (Waddell, 1982; Rice, 1984) suggested the possibility of, but did not simulate, subsurface inflow from northern and northwestern areas. Therefore, all flow through these previous models is comprised of locally recharged water and not a combination of locally recharged and subsurface inflow water as in the present model. Our model indicates that a substantial amount (40%, or $23.6 \times 10^6 m^3/yr$) of the average total system throughflow is derived from subsurface inflow.

Mean residence times are shown in Figure 4. The youngest values are found in the cells with high recharge versus subsurface flow from upgradient cells (cell 3, 18, and 28). Cells 8 and 12 have relatively young waters due to their high recharge rates, while cell 24 receives its relatively young water from cells 28 and 23. The oldest mean residence times are found in the upper clastic aquitard cells (cells 13 and 19), downgradient from aquitards (cells 9, 14, and 15) and in areas where most of the flow originates directly or indirectly as subsurface inflow (cells 5, 15, 21, 22, 27, 1, and 6). A decrease in mean residence times along flow paths occurs in many areas and is caused by relatively large amounts of recharge downgradient. The means represent all of the water in a given cell and may include a mixture of very young water recharged locally and very old water received from upgradient cells.

Residence time distributions (RTD) provide more information on the cells' waters than simply mean or median values. Cumulative RTDs for six regions are shown in Figures 5 and 6. Figure 5 shows the RTDs (F(N)) for cells 16 (Oasis Valley/Beatty Wash), 17 (Crater Flat) and 18 (Fortymile Canyon/Wash); Figure 6 shows F(N) for cells 28 (northwestern Spring Mountains), 29 (Furnace Creek Ranch region of Death Valley) and 30 (Franklin Lake playa and vicinity).

Fortymile Canyon/Wash has the highest areally distributed recharge rate (29.4 mm) and a volumetric rate second only to the Spring Mountains. Most of the ground water beneath this region is very young - - 60% of the water is fewer than a few thousand

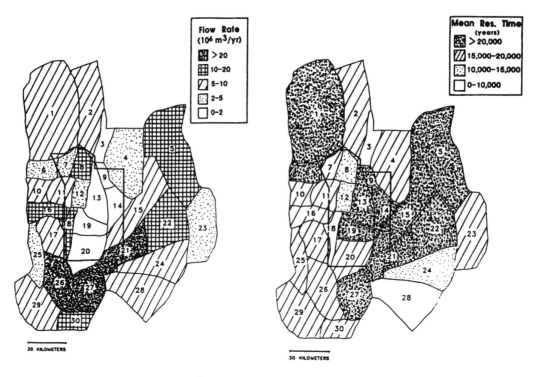

Fig 3. Volumetric flow rates (10^6 m³/yr).   Fig 4. Mean ground-water residence times.

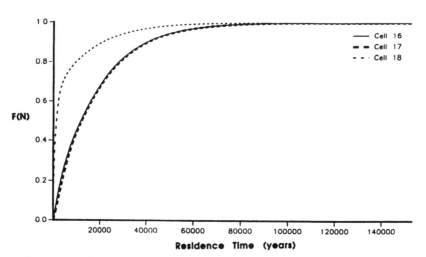

Fig 5. Cumulative ground-water residence time distribution F(N) for cells 16 (Oasis Valley/Beatty Wash), 17 (Crater Flat) and 18 (Fortymile Canyon/Wash).

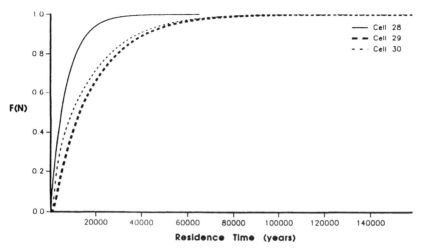

Fig 6. Cumulative ground-water residence time distribution F(N) for cells 28 (northwestern Spring Mountains), 29 (Furnace Creek Ranch area) and 30 (Franklin Lake playa).

years old. Contrast this fact with the other two regions on Figure 5 -- Oasis Valley and Crater Flat -- where 60% of the waters are at least 15,000 years old. Both of these areas are minor recharge areas; indeed, Oasis Valley is more important as a discharge area.

Figure 6 contains the cumulative RTDs of the area's major recharge area (Spring Mountains -- cell 28) and two major discharge areas (Furnace Creek Ranch -- cell 29) and Franklin Lake playa -- cell 30). These latter two cells are major discharge regions and have much older ground waters. Note that even though cell 30 is farther downgradient than cell 17 (Crater Flat) its F(N) is shifted slightly to the left relative to Crater Flat's, indicating slightly younger waters. This apparent discrepancy is easily explained by noting that Franklin Lake playa receives relatively young water (via other cells) from the Spring Mountains and Fortymile Canyon/Wash.

6 SUMMARY AND CONCLUSIONS

We used a deuterium-calibrated mixing-cell model to simulate regional ground-water flow beneath an area of approximately 19,000 km^2 in southern Nevada-California, USA. This model consists of a network of 30 compartments delineated through the integrated interpretation of general hydrogeologic characteristics of the area and deuterium data from approximately 300 sites.

Our model shows the significant contribution of subsurface inflow -- 40% of the average total system throughflow -- to the NTS regional ground-water system. This flow enters from the north and east. The eastern subsurface inflow is undoubtedly from the White River regional ground-water flow system. Winograd and Friedman (1972) estimated that about 7.4 x 10^6 m^3/yr flowed from Pahranagat Valley (part of the White River flow system and just east of cell 5) to the NTS system: Kirk and Campana (1990), using a mixing-cell model, showed as much as 5.4 x 10^6 m^3/yr discharged as underflow from Pahranagat Valley. We estimate the underflow to be 11.1x10^{6}m^3/yr.

High recharge areas within the flow system boundaries are the Fortymile Canyon/Wash-Stockade Wash area, the Spring Mountains, the Sheep Range, and Pahute Mesa. Recharge accounts for 60% of the average system throughflow.

The model provides detailed information on ground-water residence times. The position of a region in the flow path does not necessarily correlate with mean residence time as recharge can mask the effects of old subsurface inflow to a region.

7 ACKNOWLEDGEMENTS

We thank Patti Halcli for her word-processing skills. We are grateful for the financial support provided by the Water Resources Center of the Desert Research

Institute and the U.S. Department of Energy (Contract #DE-AC08-90-NV10845).

REFERENCES

Campana, M.E., 1975. Finite-state models of transport phenomena in hydrologic systems. Ph.D. dissertation, University of Arizona, Tucson, 252p.

Kirk, S.T. and M.E. Campana, 1990. A deuterium-calibrated groundwater flow model of a regional carbonate-alluvial system. *J. of Hydrology*, 119: 357-388.

Dooge, J.C.I., 1960. The routing of groundwater recharge through typical elements of linear storage. Pub. 52, General Assembly of Helsinki, Int'l. Assn. of Sci. Hydrology, 2:286-300.

Feeney, T.A., M.E. Campana and R.L. Jacobson, 1987. A deuterium-calibrated groundwater flow model of the western Nevada Test Site and vicinity. Pub. 45057, Water Resources Center, Desert Research Institute, Reno, Nevada.

Mandeville, A.N. and T.O'Donnell, 1973. Introduction of time variance to linear conceptual catchment models. *Water Resources Research*, 9(2): 298-310.

Malmberg, G.T. and T.E. Eakin, 1962. Groundwater appraisal of Sarcobatus Flat and Oasis Valley, Nye County, Nevada. Ground-Water Resources--Reconnaissance Series Report 10, Nevada Department of Conservation and Natural Resources.

Rice, W.A., 1984. Preliminary two-dimensional regional hydrologic model of the Nevada Test Site and vicinity. Report SAND837466, Sandia National Laboratories, Albuquerque, New Mexico.

Rush, F.E., 1970. Regional groundwater systems in the Nevada Test Site area, Nye, Lincoln, and Clark Counties, Nevada. Ground-Water Resources--Reconnaissance Series Report 54, Nevada Department of Conservation and Natural Resources.

Sadler, W.R., 1990. A deuterium-calibrated discrete-state compartment model of regional groundwater flow, Nevada Test site and vicinity. M.S. thesis, University of Nevada, Reno, 249p.

Simpson, E.S. and L. Duckstein, 1976. Finite-state mixing-cell models. In *Karst Hydrology and Water Resources*, Vol. 2, V. Yevjevich (ed.), Water Resources Publications, Ft. Collins, CO, 489-512.

Waddell, R.K., 1982. Two-dimensional, steady-state model of ground-water flow, Nevada Test Site and vicinity, Nevada-California. U.S. Geological Survey Water-Resources Investigations Report 81-4085.

Walker, G.E. and T.E. Eakin, 1963. Geology and ground water of Amargosa Desert, Nevada-California. Ground-Water Resources Reconnaissance Series Report 14, Nevada Department of Conservation and Natural Resources.

White, A.F. and N.J. Chuma, 1987. Carbon and isotopic mass balance models of Oasis Valley Fortymile Canyon groundwater basin, southern Nevada. *Water Resources Research*, 23(4): 571-582.

Winograd, I.J. and I. Friedman, 1972. Deuterium as a tracer of regional groundwater flow, southern Great Basin, Nevada-California. *Geological Society of America Bulletin*, 83(12): 3691-3708.

Winograd, I.J. and W. Thordarson, 1975. Hydrogeologic and hydrochemical framework, south-central Great Basin, Nevada-California, with special reference to the Nevada Test Site. U.S. Geological Survey Professional Paper 712-C.

Tracer Hydrology 97, Kranjc (ed.) © 1997 Balkema, Rotterdam, ISBN 90 5410 875 4

2D and 3D groundwater simulations to interpret tracer test results in heterogeneous geological contexts

A. Dassargues & G. Carabin
Laboratoires de Géologie de l'Ingénieur, d'Hydrogéologie, et de Prospection Géophysique (L.G.I.H.), University of Liège, Belgium

S. Brouyère
Laboratoires de Géologie de l'Ingénieur, d'Hydrogéologie, et de Prospection Géophysique (L.G.I.H.), University of Liège & National Fund for Scientific Research of Belgium

ABSTRACT: It is well known that the interpretation of tracer tests breakthrough curves can be multiple. It should be difficult to describe all the possible situations of wrong and inconsistent interpretations induced by 'automatic' calibrations solving the inverse problem without introducing any 'hard' or 'soft' geological data. When a 3D flow and transport model is applied to simulate tracer tests performed in 2D (depth averaged) conditions, a quasi infinite number of suitable parameters combinations allow to reach the calibration of the transport model. Geological data must provide informations on the different geological layers to be distinguished in the model. Inside these layers, the observed facies changes can motivate the choice of different values of the flow and transport parameters. Morphostructural analyse provides also informations on more fissured zones (in hardened formations) or on sedimentological features (in loose sediments), and shallow geophysical surveys provide new data or confirmation of the previous informations. In fact, one can not choose freely the parameter spatial distributions. Only few of the combinations are fully geologically consistent with all the collected 'hard data' and 'soft data'.

Due to differentiated values of hydraulic conductivity, effective porosity and dispersivities distinguished in the different layers of a 3D model, computed breakthrough curves can represent any complex shape of the measured curves including double peaks, strong delays, etc.

Comparisons are made between the computed breakthrough curves obtained by calibration of a 2D model and by calibration of a 3D model for a same situation corresponding to tracer tests, performed in depth-averaged conditions, in the alluvial sediments of the Meuse River (Belgium). Conclusions in terms of under- and over-estimation of the interpreted parameters values are drawn.

1 INTRODUCTION

When delineating protection zones which are based on contaminant travel times in the saturated part of aquifers, different methodologies are proposed to take the aquifer heterogeneity into account:
(1) geological and hydrogeological characterisation based on geology, hydrology, morphostructural geology and shallow geophysical prospecting,
(2) multi-tracer tests with artificial tracers, and
(3) numerical modelling of groundwater flow and transport.

During this third step, an accurate calibration of the model is required, using measured piezometric heads for groundwater flow and measured breakthrough curves for transport of solute contaminant.

The spatial distribution and values of the calibrated permeability coefficients must be consistent with pumping test results, and with geological, geophysical and hydrological data sets. The shape and the characteristics of each computed breakthrough curve is fitted on the corresponding experimental curve, so that the spatial distribution of the transport parameters can be assessed.

Usually, the detected spatial variability (laterally and/or vertically) can fully justify that heterogeneity should be invoked to explain late arrivals of tracers.

For one set of tracer-test data, comparisons can be made between computed breakthrough curves obtained by calibration of a 2D model on one hand, and by calibration of a 3D model on the other hand. Suitable combinations of the groundwater flow and transport parameters which are deduced from both calibrations can be strongly different. Advantage and drawback of each approach can be compared on a same situation. For illustration, in this paper, the discussion will concern the interpretation of tracer tests performed in depth-averaged conditions in the alluvial sediments of the River Meuse. These tracer tests, with iodide and LiCl injected in the same piezometer, have been modelled using 2D and 3D numerical codes.

The alluvial plain of the River Meuse is characterized by a fluvial sedimentation composed of gravels mixed in a sandy, silty or clayey matrix. A complete hydrogeological study has been realized at the pumping station of Vivegnis near Liège in Belgium (figure 1). The aim of the study consists in the accurate delineation of protection zones based on contaminant travel times.

The general methodology includes the determination of the experimental values and the spatial layout for hydrodynamic and hydrodispersive parameters and the use of these parameters in a groundwater flow and contaminant transport model (Biver and Meus, 1992). This model is supposed to take all detected local heterogeneities into account. After calibrations, it can be used to compute the transport travel times from different injection points to the pumping well. A good assessment of the protection zones can then be deduced (Dassargues, 1994).

The average drinkwater production of the site is about 8000 m³/day, using four pumping wells. More than ten piezometers have been drilled, and are available for measurements and tracer injections. The lithological information provided by those boreholes, added to data from many penetration tests and from geophysical survey (electric and seismic sounding methods), leads to the definition of a main gravel layer with an averaged thickness of 7 meters, overlaid by a 2 meters thick silt layer. The Primary shale and sandstone bed-rock (Namurian and Westphalian ages of Carboniferous formation), is characteristic of the substratum of the River Meuse valley and can be considered as an impervious bottom for the alluvial aquifer (figure 2). At the regional scale, the unconfined aquifer presents a 0.075 percent average gradient towards the North (Dassargues and Lox, 1991).

A first interpretation of pumping test results (following the classical analytical solutions of Theis in transient conditions, and Dupuit in steady-state), have given transmissivity values which range from 1.10^{-4}m²/s to 2.10^{-1}m²/s. Transmissivity values lower than 5.10^{-3} m²/s are found corresponding to zones where the clay content is higher. Transmissivity values higher than 8.10^{-2} m²/s correspond to clean and well sorted gravels. An averaged storage coefficient of 0.10 was estimated using the Theis solution. This value can be used as a first approximation for the effective porosity of the porous medium. In steady-state conditions, an averaged radius of influence for the production wells

was estimated to range from 230 to 810 meters. In transient conditions, a value of 440 meters was calculated. One should keep in mind that the hydrodynamic parameters obtained from those pumping test interpretations are only first approximations, given the strong assumptions under which the Theis and Dupuit expressions are valid. Moreover, these values can only be considered as 'mean values' around the wells.

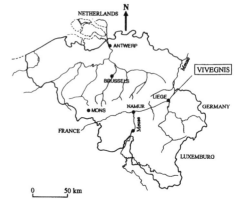

Fig.1 Location map

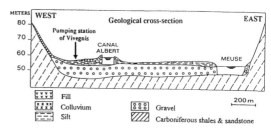

Fig.2 Transversal cross-section in the alluvial deposits of the River Meuse valley.

2 2D AND 3D GROUNDWATER FLOW MODELLING

2.1 Groundwater flow conditions and boundary conditions

A 2D regional FEM model using the code AQUA2D (Vatnaskil Consulting Engineers) was previously calibrated in steady state conditions on a measured piezometric map (figure 3) (Derouane, 1994).

In the 3D local model, it is also assumed that stabilized pumping conditions are reached, so that steady state groundwater flow is considered for modelling the alluvial water table aquifer. In the gravels, a vertical sequence was distinguished with, at the bottom, coarse and clean pebbles in a 2 meters thick layer, and finer pebbles with increasing silt and loam content in the 6 meters thick upper part. The silt layer is not taken into account as it lies definitively above the water table (i.e. in the unsaturated zone) in the modelled zone and as only saturated zone of the aquifer is considered here. Moreover, in a first approach and to simplify reinterpretation of the results, homogeneous distributions of the parameters are chosen in each horizontal layer.

The finite-element program SUFT3D (Dassargues, 1994) is used. Dirichlet prescribed piezometric heads are chosen as flow boundary conditions of the domain around the pumping well P6 (figure 3). These values are deduced from the calibrated regional piezometric map (Derouane, 1994) obtained by the 2D regional model. The lower horizontal boundary which corresponds to the bottom of the alluvial aquifer is characterized by no flow. A uniform infiltration flow rate is prescribed on the top boundary of the model.

A three layers discretization is chosen. The lower layer, representing the clean and coarse gravels (2 m thick), is overlaid by two layers, each of them having a thickness of three meters (figure 4). The

horizontal discretization was made using two meshing networks (1) a regular grid with 1292 nodes and 864 blocks, (2) an irregular and refined grid with 2604 nodes and 1848 elements (figure 5).

2.2 Heterogeneity and permeability values

Due to the different lithological nature of the alluvial sediments, it was arbitrarily decided to maintain a factor ten between the permeability values of layers 2 & 3 and layer 1 (bottom layer). Taking a mean value of the saturated thickness at 7 m, and in order to be consistent with the equivalent value of the transmissivity which was used in this zone of the 2D model, the contrasted permeability coefficients can be found as following:

$$T = 7.K_{eq} = 2K_1 + 5K_2 = 2K_1 + 0.5\ K_1 = 2.5\ K_1 \quad (1)$$

where K_1 is the permeability coefficient in layer 1 and K_2 is the permeability coefficient in the layers 2 & 3 ($K_2 = 0.1\ K_1$).

In the 2D model, an equivalent transmissivity value was calibrated at $6.8\ 10^{-2}\ m^2/s$, so that permeability values of $2.7\ 10^{-2}\ m/s$ and $2.7\ 10^{-3}\ m/s$ are deduced respectively for K_1 and K_2 (figure 6).

After calibration of the 3D local model, it is

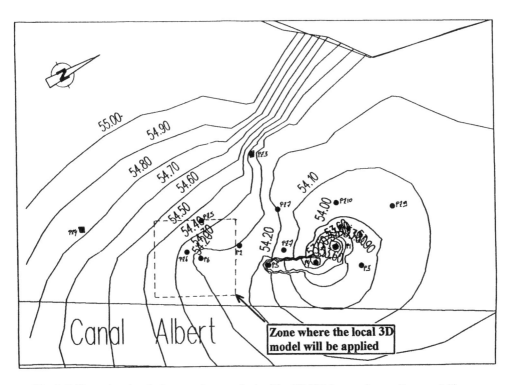

Fig. 3 Calibrated regional piezometric map obtained by 2D FEM groundwater flow modelling.

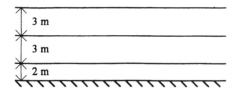

Fig.4 Vertical discretization of the local 3D model.

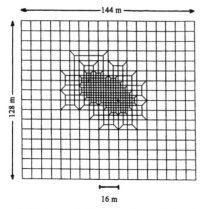

Fig. 5 Horizontal discretization of the local 3D model: refined mesh.

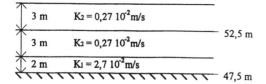

Fig.6 Vertical distribution of the permeability values.

evident that a strong contrast in computed Darcy's specific fluxes is found (figure 7). This contrasted situation will have a overwhelming influence on the transport conditions in the modelled zone.

3 MODELLING TRANSPORT IN 3D

Results from 2D transport simulations are often questionable when the porous medium is vertically heterogeneous. The choice between 2D and 3D modelling must be guided by the aims of the study, the spatial variability of the parameters describing the aquifer, and the available data set. If an accurate representation of the aquifer heterogeneity with a realistic determination of the hydrodynamic and hydrodispersive properties is needed, then a detailed

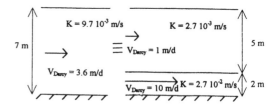

Fig.7 Effect of the chosen heterogeneity on the Darcy's specific fluxes.

3D is more adequate. However, the actual spatial distribution of the parameters to be introduced in the model is only badly known so that the uncertainty is still important.

A 2D transport simulation is often considered as sufficiently accurate and detailed when interpreting 'depth averaged' tracer and pumping tests. This approach can be accepted when the awaited values for the properties in the different layers are not too contrasted, so that equivalent values can be easily considered.

For both 2D an 3D approaches, the concept of Representative Elementary Volume (REV) is applicable so that the link between actual values and adjusted parameters in the elements of the models is difficult to assess, and in any case, depends strongly on the chosen scale (Gelhar et al., 1992). The effective velocity or advection velocity in the porous medium is given by:

$$\underline{v}_e = -\frac{K}{n_e}\underline{gradh} \qquad (2)$$

where n_e is the effective porosity of the aquifer and $\underline{gradh}$ is the piezometric gradient. When 'depth averaged' conditions are chosen, the transmissivity is used in the groundwater flow computations so that the effective velocity becomes:

$$\underline{v}_e = -\frac{T}{n_e . e}\underline{gradh} \qquad (3)$$

where e is the saturated thickness of the aquifer. For a given value of transmissivity (deduced generally from calibration of the groundwater flow model), the effective porosity as well as the saturated thickness influence the effective velocity. Consequently, a 2D calibration on the measured 'depth-averaged' breakthrough curves, cannot provide clear information on the effective porosity. In studies concerning protection zones, the first arrival of contaminant is of big importance. If we consider 2 layers with contrasted permeability values, as in figure 7, a depth averaged 2D approach should compute a global effective porosity value which is three times lower than the value found for the lower layer.

In a 3D transport model, the geometry of the different layers is supposed to be known and equation (2) is used. For given values of permeability in each layer, effective porosity values are found to calculate the effective velocity and to calibrate the computed breakthrough curve on the measured one.

Different breakthrough curves were found as a result of two tracer tests conducted in a same piezometer (figure 8). Injection of iodide and LiCl were performed in different conditions, and at different times. Many scenarios can be invoked to explain the different breakthrough curves which were obtained. One of them consists in invoking the influence of the injection mode on the tracer test results (Brouyère & Rentier, this issue). Another explanation can be given assuming that the largest part of each tracer was actually injected into the aquifer at different levels of the screened piezometer (Carabin, 1995). This last scenario imposes a 3D analysis of the results. A combination of both causes should probably be invoked.

Anyway, having two different measured breakthrough curves, the situation is ideal for testing how a 3D transport model can always be calibrated on measurements by introduction of vertical (and/or lateral) heterogeneities.

For the local 3D simulation of the transport in the radial convergent flow conditions induced by the pumping at the well P6, transport Neumann boundary conditions or 'natural condition of zero dispersive flux' are chosen allowing only an advective flux through boundaries. By this way the injected tracer (in Pz6) can go out of the modelled zone through the lateral boundaries.

The injection simulation is realized considering a constant injection flow rate during 5 time steps of 1 hour. As awaited, the computed breakthrough curve is largely influenced by the choice of the injection nodes in the 3D mesh.

In a first step, the iodide injection (in Pz6) is simulated according to the scheme of figure 8. The computed breakthrough curves at each of the pumping nodes (from bottom to the top: nodes 1 to 3) are given in figure 9. They show the strong influence of the vertical heterogeneity on the results. The less pervious layers (layers 2 & 3) have a transport behaviour which can be qualified as characteristic of 'buffer zones': the tracer reached them mainly by transversal dispersion. The transport parameters which were used, are provided in table1.

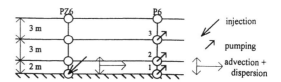

Fig.8 Injection and pumping conditions for the local 3D transport model simulating the tracer test n°1.

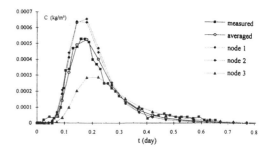

Fig.9 Measured and computed breakthrough curves for tracer test n°1.

Table 1 Transport parameters used to simulate the tracer test n°1 (a_L and a_T are the longitudinal and transversal dispersivities).

Tracer test n°1	n_e	a_L (m)	a_T (m)
layers 2 & 3	0.05	1.5	0.3
layer 1	0.07	1.5	0.3

For calibration on the results of the second tracer test (with injection of LiCl) some changes in the way to inject the tracer were needed. As observed on

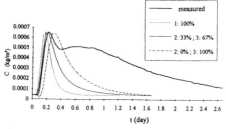

Fig.10 Measured and computed breakthrough curves for tracer test n°2 with different injection schemes (percentages of the total injected mass are mentioned).

figure 10, to simulate the first peak of the measured breakthrough curve, a good fitting can be found by changing only the distribution of the tracer injections at the different nodes of the vertical discretization.

When changing the transport parameters in layers 2 & 3, different breakthrough curves can be computed showing two peaks (figure 11). The injection distribution and the transport parameters which were used for each curve, are given in table 2.

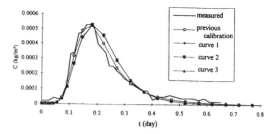

Fig. 12 Measured and computed breakthrough curves for tracer test n°1 with the different transport parameters in layers 2 & 3 .

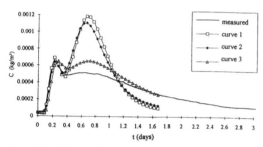

Fig. 11 Measured and computed breakthrough curves for the tracer test n°2 with different transport parameters in layers 2 & 3.

The same changes of transport parameters do not affect strongly the computed breakthrough curves for the tracer test n°1 (figure 12).

In studies concerning protection zones, it is essential to obtain an adequate calibration of the first peak. Having in mind this aspect, the first peak is better simulated with the transport parameters corresponding to curves 1 & 2. Keeping these different changes in the model to simulate again the tracer test n°1, no important discrepancy is found. The greatest changes affected the layer 3, which does not play an important role in the transport results of the tracer test n°1.

Table 2 Injection distribution and transport parameters which are used to compute the curves of figure 11.

Tracer test n°2		n_e	a_L (m)	a_T (m)
layer 3:	curve 1	0.08	0.5	0.05
	curve 2	0.075	1.5	0.05
	curve 3	0.075	5.0	1.0
layer 2:	curve 1	0.07	0.5	0.05
	curve 2	0.07	0.5	0.05
	curve 3	0.07	0.8	0.05
layer 1		0.07	1.5	0.3
Injection distr.		node 1	node 2	node 3
		0 %	3 %	97 %

4 COMPARISONS AND CONCLUSIONS

On basis of the last results, the transport parameters corresponding to curve 2, are considered (table 3) for the best calibration on the first peaks of both tracer tests. Injection distributions on the three nodes of the vertical discretization are not the same: 100 % of the iodide injection in node 1 for the tracer test n°1 and 97 % of the LiCl injection in node 3 for the tracer test n°2.

Table 3 Final transport parameters

Tracer tests n°1 & 2	n_e	a_L (m)	a_T (m)
layer 3	0.075	1.5	0.05
layer 2	0.07	0.5	0.05
layer 1	0.07	1.5	0.15

The comparison of the flow and transport parameters of this 3D model with those obtained previously by calibration of the 2D model (Derouane, 1994), can be made (table 4). Computed breakthrough curves are given in figure 13.

In table 4, the dispersivity values of the 2D model are not consistent. This is due to an important numerical dispersion. Consequently, the 2D computed breakthrough curves (figure 13) show more dispersion than awaited with the values of table 4. It is difficult to establish clear comparisons

Table 4 Flow and transport parameters obtained by 2D and 3D calibrations.

	K (m/s)	n_e	a_L (m)	a_T (m)
2D	0.0215	0.048	0.01	0.003
layer 3	0.0027	0.075	1.5	0.05
layer 2	0.0027	0.07	0.5	0.05
layer 1	0.0270	0.07	1.5	0.15

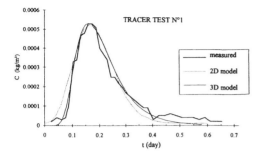

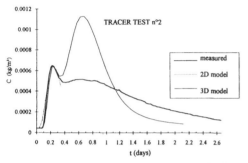

Fig. 13 Comparison of the results obtained with 2D and 3D models.

between the transport parameters obtained with the 2D and the 3D models. The main differences in values of the effective porosity are due to the contrasted value of permeability distinguished in the layers of the 3D model. As mentioned previously, the 'depth averaged' effective porosity can be largely underestimated. This can be very important when considering the delineation of protection zones.

The first arrival of tracer and the way to reach the first peak are certainly better simulated in 3D than in 2D. However, the 3D model allows to distribute the injection and the pumping on different nodes. These distributions can have a significant influence on the results. The vertical discretization allows to gain more flexibility in the introduction of the spatial variability. To calibrate the model, lateral and vertical heterogeneities can be introduced.

Advances must be considered to obtain better controlled experimental conditions and better numerical representations of the actual conditions. Tracer tests in 3D conditions with adequate use of packers, and a better control of the actual input function into the aquifer should be addressed. For modelling aspects, an explicit representation of the injection piezometer and of the pumping well (with line elements), should help to simulate more accurately the injection and pumping conditions (Lacombe et al., 1995, Sudicky et al., 1995).

REFERENCES

Biver, P. and P.Meus. 1992. The use of tracer tests to identify and quantify the processes in an heterogeneous aquifer. Tracer Hydrology. Hötzl & Werner (eds). Balkema. Rotterdam. pp. 415-421.

Brouyère, S. and C. Rentier. 1997, About the influence of the injection mode on tracer test results. This issue.

Carabin, G. 1995. Simulation 3D du transport de polluant en milieu aquifère en vue de l'interprétation d'essais de traçage. Travail de fin d'études en vue de l'obtention du grade d'Ingénieur en Hydrologie. Université de Liège. Faculté des Sciences Appliquées. Belgium. 147p.

Dassargues, A. 1994. Applied methodology to delineate protection zones around pumping wells. *Journal of Environmental Hydrology*. IAEH, vol.2, n°2, pp.3-10.

Dassargues, A. 1994. Validation of a Finite Element code to simulate the coupled problem of salt transport in groundwater. Computer Techniques in Environmental Studies V. Proc. of ENVIROSOFT'94 San Francisco, vol.1, pp173-180, CMP.

Dassargues, A. and A.Lox. 1991 Modélisation mathématique de la nappe alluviale de la Meuse en aval de Liège (Belgique). In "Le système hydrologique de la région frontalière Liège-Maasbracht, résultats des recherches 1985-1990". *Rapport et notes n° 26, CHO-TNO*, Delft, pp. 27-54.

Derouane J. 1994. Etude hydrogéologique du site de captage de Vivegnis (Plaine alluviale de la Meuse). Détermination des zones de protection. Travail de fin d'études en vue de l'obtention du grade d'ingénieur géologue. Université de Liège. Faculté des Sciences Appliquées. Belgium. 172p.

Gelhar, L.W., Welty, C. and Rehfeldt, K.R. 1992. A critical review of data on field-scale dispersion in aquifers. *Water Resources Research*.28(7): 1955-1974.

Lacombe, S., Sudicky, E.A., Frape, S.K. and Unger, A.J.A. 1995. Influence of leaky boreholes on cross-formational groundwater flow and contaminant transport. *Water Resources Research*. 31(8): 1871-1882.

Sudicky, E.A., Unger, A.J.A. and Lacombe, S. 1995. A noniterative technique for the direct implementation of well bore boundary conditions in three-dimensional heterogeneous formations. *Water Resources Research*. 31(2): 411-415.

Tracer Hydrology 97, Kranjc (ed.) © 1997 Balkema, Rotterdam, ISBN 90 5410 875 4

Interpretation of tracer tests in a granitic formation (El Berrocal site, Spain)

J. Guimerà, J. Carrera, I. Benet, M. Saaltink & L. Vives
Universitat Politècnica de Catalunya, Barcelona, Spain

M. García-Gutiérrez
Centro de Investigaciones Energéticas Medioambientales y Tecnológicas, Madrid, Spain

M. d'Alessandro
Joint Research Center, Ispra, Italy

ABSTRACT: Tracer tests at El Berrocal site were aimed to hydrological characterization, instrumentation development and data base generation for modeling purposes. They were placed at two borehole arrays in fractured granite. Tracers were exclusively non-sorbing although great differences were observed among some of them during the recovery periods. Performance of the tests was carried out with newly developed instrumentation which proved to be robust and reliable working with downhole electronics. Interpretation of the tests included several conceptual models. Those that resulted in more realistic prediction capabilities took into account the heterogeneity of the system, matrix diffusion and the hydrogeological context at local scale. Simple analytical models proved to be helpful but with limited prediction capabilities. Including in the model features and processes which are often neglected in normal permeability media, such as the bore presence or the transient flow effects during the tracer injection, was of utmost importance.

1. INTRODUCTION

El Berrocal project was an international effort to study radionuclide migration under natural conditions in a granitic environment. El Berrocal site was not considered as a potential repository site. The particular interest of tracer tests at El Berrocal was, on one hand, to enhance hydrogeological characterization and, on the other, to build know-how on performance, instrumentation development and modeling. Regional and local geological, geochemical and hydrogeological settings are out of the scope of this paper. The reader is referred to Rivas et al. (1996) for general interest of the project. Carrera et al. (1996) and Guimerà et al. (1996a) describe the hydrogeology of the site, and Guimerà et al. (1996b) report tracer tests performance and interpretation.

Purposes of the tests were purely hydrodynamical. That is, they were aimed to refine the hydrogeology of the site, thus confirming/rejecting early conceptual models. Tests suffered from common limitations of this type of media. Such limitations are usually imposed by the low K ad the heterogeneous distribution of hydrodynamical parameters which implies low available flowrates and few monitoring points among others. As a result, conventional interpretations may lead to uncertain parameter values of doubtful meaning (Malozewski and Zuber, 1992). Recently, some tests have aimed to overcome such limitations (Olsson ad Winberg, 1996; Meigs et al. 1996). They are designed, performed and modeled taking into account heterogeneity of the flow system and physical and chemical mechanisms affecting tracer's transport. They rely on previous experiences by other authors (Frick et al., 1992; Abelin et al., 1991; Gustafsson and Andersson, 1991 among others), thus gaining insight in understanding relevant processes affecting the transport of solutes.

Tracer tests at El Berrocal site were performed at the most conductive fractures between boreholes. Therefore, results were biased towards relatively high permeability zones. That would cast serious doubts on the validity of the results, in terms of reducing the uncertainty due to spatial variability. To overcome this limitation, interpretation accounted explicitly for either conductive fractures and rock matrix. Besides, rock mass double porosity models were used to simulate processes likely to occur such as matrix diffusion. Heterogeneity of the flow system enhances mixing during transport, which leads to anomalous values of hydrodynamical dispersivity (Carrera, 1993). On the other hand, matrix diffusion is known to cause retardation of solutes transport at any scale and regardless the type of medium. A review of the topic is out of the scope of the paper yet relevant information is found in Neretnieks (19839, Feenstra et a. (1984), Malozewski and Zuber (1985, 1993), Wood et al. (1990) and Motyka and Zuber (1994) among others. The effects of heterogeneity and matrix diffusion are comparable in the breakthrough curve. Although some field tests are designed to characterize matrix diffusion alone (Tsang, 1995), we contend that both can not be separated because neglecting one of them decreases the reliability of the other. In fact, in spite of conservative tracers were used, some of them never behaved the same to each other. They displayed similar first arrival times; but peak arrival times were different, corresponding the tracers of biggest molecule sizes to the latest times. Differences in tailing were even more significant. Therefore, we conclude that results are conditioned by matrix diffusion and heterogeneity and that long term predictions of contaminant transport cannot neglect them.

The objectives of this paper are to reproduce analytically and numerically the breakthrough curves of field tests, first, and second, to check the predictive skills of the different models, taking advantage of different tests performed at the same location. We show that calibration of simple models may produce better fittings than more sophisticated 3D heterogeneous but display lower predictive capabilities.

2. FIELD SET-UP

The tests involved two arrays of boreholes (Figure 1): S11-S12 (vertical) and S2-S13-S15 (inclined). Tests methodology allowed to identify preferential flow paths and to evaluate how would they be affected by the convergent flow test. Dilution tests, performed both with and without pumping, were of great help in this context.

Tracers were injected at two isolated sections of borehole S11 and recovered at one pumping section of borehole S12 at the S11-S12 test. The S2-S13-S15 tests consisted first of one flushed injection at one section of borehole 13 and second, of three decay injections at borehole S13 (two) and S15 (one). Table 1 summarizes the performances. d'Alessandro et al. (1996) and García-Gutiérrez et al. (1996) describe further details of the design and performance of the tests.

3. RESULTS AND INTERPRETATION OF TESTS AT S11-S12

3.1 Analytical interpretation

The flow field was assumed radial in spite of the geometry of the test and the fracture distribution. The pumping flowrate was distributed according to measured drawdowns and distances between intervals (Table 2). Since some stops occurred during the experiment, data most influenced by such failures were not used during the calibration. Instead, they are represented by different symbols in Figure 2.

Results of advection-dispersion model

Calibration of eosine resulted in a reasonably well fitted curve, especially the rising part of the curve (Figure 2). The model returns relatively high dispersivity values as well as thickness porosity when reproducing the tail of the curve ($\alpha_L = 4.4$ m and $\phi b =$

0.02 m). Data belonging to the first two days are not properly fitted as a counterpart of maintaining the tail of the curve well reproduced. Iodide and uranine calibration returns similar parameter. The rising part of the uranine breakthrough is specially well fitted, compared to eosine, as well as iodide. However, the model fails when reproducing the uranine tail and fits iodide rather succesfully. Differences in dispersivity are of 0.6 m, the one for uranine being smaller. Therefore, in spite of the noisy nature of the data, we can infer that processes affect selectively iodide and uranine.

Results of advection-dispersion and matrix diffusion model

Eosine results largely benefits from considering matrix diffusion, especially tail fitting. However, improving the fitting of the tail results in a worse reproduction of the rising part of the curve. Uranine tail fitting is also improved with the inclusion of matrix diffusion. Thickness porosity is reduced by a factor of about 3, while dispersivity decreases by a factor of 8. The iodide breakthrough calibration, does not lead to a significant improvement in the fitting, which was acceptable with advection-dispersion. Matrix diffusion fitting provides lower computed values than the measured data of the tail - as for uranine -, while those for advection-dispersion were consistently higher. Parameters returned for eosine -matrix diffusion-showed thickness porosity values similar to iodide and uranine, but longitudinal dispersivity is two-fold. Molecular diffusion also displays higher values than uranine and lower than iodide. In summary, accounting for matrix diffusion leads to qualitative improvements on the fit and, consistently, to more reasonable parameters than advection-dispersion model.

The fact that the parameters are so consistent lends credibility to results. It should be kept in mind, however, that tailing can be attributed to factors other than matrix diffusion such as spatial heterogeneity and by residual mass at the injection source.

3.2 Numerical interpretation of the results

The three breakthroughs were modeled numerically by using the computer code TRANSIN-III (Galarza et al., 1996). The followed steps can be summarized as conceptual model definition, numerical model construction and automatic parameter calibration.

Conceptual model

We consider the flow field reached steady-state water velocities after two days of pumping. Yet, while the flow field is steady, the transport problem is transient. Groundwater flow takes place through preferential flow paths, which were explicitly modeled. The most

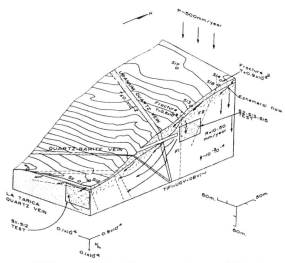

Figure 1. Block diagram of El Berrocal site involving major elements of the hydrogeological conceptual model (after Guimerà et al. 1996). Boreholes involved in tracer tests were 11-12 (vertical) and 2-13-15 (inclined).

Table 1. Summary of the performances and results of tracer tests at El Berrocal.

Interval	Action	Depth (m)	Distance to recovery (m)	Dilution volume (l)	Flushing volume (l)	Tracers	Rec (%)	10% peak arrival (d)	Peak arrival (d)	Peak C
TEST S11-S12										
S12	pumping 2 l/min	49.0-27.0	-	129.0	-	-	-	-	-	-
S11-1	Injection	64.0-48.0	22.0	118.0	385	Uranine 5 g	62.5	3.0	7.5	60 ppb
						Iodide 3.5 g	61.3	3.0	7.5	120 ppb
S11-2	Injection	47.0-25.0	14.5	140.0	300	Eosine 5 g	41.2	1.7	7.5	60 ppb
TEST S2-S13-S15										
S2	Pumping 0.09 l/min	20.79-30.10	-	53.5	-	-	-	-	-	-
S13-4	Injection	31.09-37.08	19.6	37.5	300	D$_2$O 12 l	56.3	0.5	2.2	rel. unit
					300	Uranine 15 g	40.8	0.5	4.2	255 ppb
					300	Gd-DTPA 1.1 g	43.5	0.5	2.0	435 ppb
					0	Phloxine 12.4 g	30.1	3.5	14.1	0.90 ppm
					0	Re 0.7 g	34.5	3.5	14.0	0.06 ppm
S13-2	Injection	66.66-73.70	22.2	49.0	0	B.Sulpha. 13.1 g	0.0			
					0	Iodide 11.4	0.0			
S15-2	Injection	35.64-42.63	24.6	49.4	0	Eosine-Y 13.5 g	0.0			

Table 2. Parameter values returned by different conceptual models for the test at S11-S12 (analytical model).

Tracer	Model	L(m)	Q(l/min)	Rec (%)	ϕb(m)	α_L(m)	ϕ'_m(-)	D$_m$(-)
Uranine	AD	22.0	1.375	62.54	0.0187	3.400	-	-
Iodide	AD	22.0	1.375	61.30	0.0224	4.000	-	-
Eosine	AD	14.5	0.625	41.16	0.0201	4.400	-	-
Uranine	MD	-		-	0.0088	0.565	2.000	0.5479
Iodide	MD	-		-	0.0081	0.570	2.000	0.7200
Eosine	MD	-		-	0.0078	0.904	2.000	0.6039

AD= Advection-dispersion; MD= Matrix-diffusion ϕb= thickness porosity α_L = longitudinal dispersivity; ϕ'_m = dimensionless matrix porosity

$\frac{(1-\phi c)}{\phi c} \phi_m$; D$_m$ = dimensionless molecular diffusion (.D $\frac{\pi L^2 b\phi_c}{d^2 Q}$)

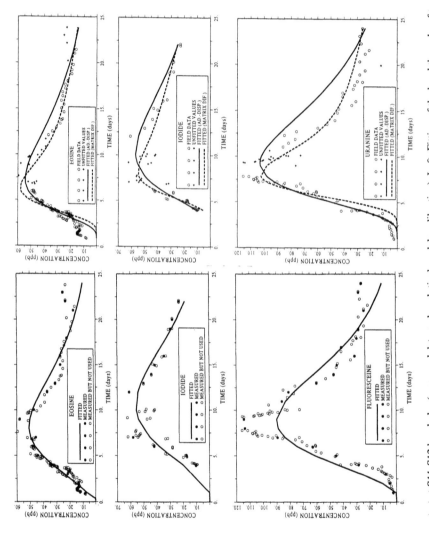

Figure 2. Right: Results from test at S11-S12 boreholes (measured data and analytical models calibration). Left: Fitting of the breakthroughs of test at boreholes S11-S12 by automatic calibration of the 3D model.

important flow paths were La Tarica vein (see Figure 1), the phreatic surface and two major fractures. Other features included in the analysis were prescribed recharge, La Tarica stream, regional discharge flow (upwards) and pumping at S12. Water injection during tracer flushing periods was neglected due to its small volume (885 l) compared to the extraction flowrate (2.88 m³/day). We considered that transport was affected only by advection and dispersion through fractures and rock matrix.

Numerical model

Further details on the flow model are reported by Guimerà et al. (1996a). Following we concentrate on those features relevant to transport. Molecular diffusivity is assumed to be constant throughout the whole domain, while one zone of α_L and α_T exists for each hydraulic conductivity-transmissivity zone. Dispersivity values are modified according to element sizes in order to ensure numerical stability throughout the whole domain. Rock formation, La Tarica vein and the phreatic surface layer are assumed to have the same porosity while fractures F11.1 and F11.7 have the same aperture. Injection, pumping and observation borehole intervals are modeled separately to simulate the behavior of the tracer.

Calibration

Table 3 shows the parameters returned by the flow model. Calibration of the three breakthroughs, which was accomplished separately led to estimated dispersivities and porosities from both rock matrix and fractures. Table 4 shows the results returned by the transport models and Figure 3 shows the corresponding fits. Estimated values of fracture aperture are constant for all tests. However, some of the dispersivity values obtained present high uncertainty. We feel that, even though more sophisticated models could improve the fitting, the resulting models would be less robust (too many parameters). The main purpose of this 3D heterogeneous model was to explain tail effects caused by heterogeneity, which in terms of fitting, were comparable to those displayed by matrix diffusion in 2D models (Figure 2).

Regarding the comparison with respect to the parameter values obtained by the analytical models, one must say that conceptual models (flow and transport) are too different to make any resembling meaningful. However, given that fittings of both models are comparable, it would appear that radial models do not realistically apply for this case, unless we make use of processes, such as matrix diffusion. Therefore, the role of matrix diffusion may not be as important as stated by the homogeneous analysis and that part of the effects attributed to it, can be explained by the heterogeneity of the media (that is, fracture flow

Table 3. Flow parameters returned by the numerical model of test at 11-12.

Parameter	Estimated value
K in N-S direction (m/d)	0.25e-3
K in vertical direction (m/d)	0.94e-4
K in E-W direction (m/d)	0.097
T of the phreatic surface (m²/d)	0.609
T of La Tarica vein (m²/d)	0.022
T of fracture 11.7 (m²/d)	0.033
T of fracture 11.4 (m²/d)	0.428
Recharge to the phreatic surface (mm)	222.3
Constant level at La Tarica stream (m)	1.01
Constant level at bottom (m)	0.9903
Constant level at the interception of the stream with fracture 11.7 (m)	0.9833
Constant level at the interception of the stream with fracture 11.7(m)	0.8455
Leakage coefficient (1/m)	100.
Pumping rate (m³/d)	-2.88

Table 4. Transport parameters returned by the numerical model of test at S11-S12.

Parameter	Eosine	Iodide	Uranine
Longitudinal dispersivity of the rock (m)	5.24	1.75	3.32
Transversal dispersivity of the rock (m)	0.26	0.20	0.20
Porosity of the rock (-)	0.23e-3	0.50e-3	0.35e-3
Longitudinal dispersivity of the fractures (m)	0.70	0.70	0.70
Transversal dispersivity of the fractures (m)	0.40	0.40	0.40
Thickness porosity of the fractures (m)	0.005	0.005	0.005
Injected mass (g)	5.02	3.35	5.05
Measured recovery (%)	35	53	55
Calculated recovery (%)	35	65	65

embedded into less pervious formation). Following it is shown that both contributions cannot be separated.

4. NUMERICAL INTERPRETATION OF TESTS AT S2-S13-S15

We present heterogeneous models with and without matrix diffusion. Less sophisticated interpretations that were carried out are reported by Guimerà et al. (1996b).

4.1 Conceptual Model

Likewise the previous case, the models used for the interpretation simulate the altered granite, the uranium quartz vein and the fractures F1 and F2 by means of two-dimensional finite elements embedded within an anisotropic 3D domain (fully 3D elements) that represents the granite rock mass. Regional flow is almost vertical.

Local flow is affected by discharge into the mine and by the presence of high permeability-preferential flows. This approach causes the tracers to travel with higher velocity through the major fractures (F1 and F2) and with lower velocity through the minor fractures of the granite rock mass. The hydraulic conductivity of rock mass is considered to be anisotropic, following major directions of joints and minor fractures. For some models a double porosity approach was used to simulate the diffusion of solutes into the crystalline rock. The double porosity approach assumes the medium to consist of a porosity with immobile water, where tracers are only transported by diffusion, and a porosity with mobile water, where they are also transported by advection and dispersion.

4.2 Numerical Interpretation

We interpreted numerically all breakthrough curves of the tracers recovered at borehole S2 (see Table 1). The interpretation was performed as follows: Flow parameters were estimated by a steady state flow model (with pumping in borehole S2), from pressure heads measured after three weeks of pumping. Parameters returned by the flow model (Table 5) were used to calibrate the transport models.

Table 5. Flow parameters returned by the numerical model of test at S2-S13-S15.

Parameter	value
K in N-S direction (m/d)	0.16e-5
K in vertical direction (m/d)	0.86e-4
K in E-W direction (m/d)	0.15e-4
T of the phreatic surface (m^2/d)	0.26e1
T of UQ vein (m^2/d)	0.12
T of fracture F1 (m^2/d)	0.48e-2
T of fracture F2 (m^2/d)	0.90e-3
Recharge to the phreatic surface (mm)	222.3
Constant level at La Tarica stream (m)	1.01
Constant level at bottom (m)	piezometry
Pumping rate (m^3/d)	-0.112

The full modeling process of tracer breakthroughs is not divided into first and second injection given that some feedback was produced between both tests. As a consequence, we ended up with different conceptual models that explain with certain success the observed breakthrough curves. The models for all the tracers were calibrated by estimating the longitudinal dispersivity α_L of the granite rock mass and of the fractures, the product of porosity times retardation (ϕR) of the granite rock mass, the product of the porosity times aperture times retardation ($\phi b R$) of the fracture and, in case of double porosity, the immobile porosity (ϕ_{im}) and the diffusion coefficient divided by the square thickness of the immobile porosity (D_{im}/b_{im}^2). As the transversal dispersivities α_T had little influence, they were kept at a fixed value.

Measured concentrations in borehole S13 were used to calculate the mass of tracer leaving the borehole as a function of time. For uranine and gadolinium this led to significant differences between the total amount of

tracer, calculated from the measured concentration, and the amount actually used. Possible explanations for this difference include: loss of tracer from the borehole before the start of the flushing, adsorption of the tracer to the equipment walls and/or an insufficient mixing within the borehole, leading to measured concentrations that are not representative. The sudden drop in rhenium and phloxine after 30 days is due to a change in flowrate. Since the flow model is steady state, concentrations after this change are not taken into account.

In order to obtain good fits for the tests without flushing by means of models with single porosity, we were forced to estimate also the injected mass (phloxine and rhenium, Figure 3 left). The difference in estimated and actual injected mass are discussed above. However, by means of models with double porosity, these fits are not as good as the fits of the single porosity model. Nevertheless, we feel that the double porosity model is better, yet it reflects in a better way the processes that we think are occurring in reality, which leads to parameter values with more physical meaning and to higher predictive capabilities of the models.

Tracers injected by flushing are modeled by means of two steady state flow models: one for the period of flushing (approximately one day) and one for the period after flushing. The problem of this approach is that we could not use the algorithm of automatic calibration, because the parameters are distributed over two models. Therefore, we first calculated the breakthroughs with the transport parameters obtained from the test of phloxine, both for the single porosity and double porosity model. Second, we use the parameters returned by the natural decay injection test, to predict the breakthroughs of the flushed injections. As it can be seen from Figure 4, the results of the double porosity model resemble better the observed data than those of the single porosity model. Then, a manual calibration (i.e., trial and error) was carried out. Taking into account that a manual calibration generally does not give as good fits as automatic calibration, rather acceptable fits could be obtained (Figure 3 right) by changing slightly some parameters (Table 6). It is expected that we can obtain better fits by means of an automatic calibration. For the test of uranine we multiplied the porosities of both the rock mass and the fractures by 1.5 and for the test of gadolinium by 0.6. Note, that these differences are much smaller than those obtained by the models that do not simulate the flushing of water. For the test of deuterium we had to reduce the injected mass. A possible explanation for this reduction is diffusion into the crystalline rock, which may be more important for this tracer than for the others.

410

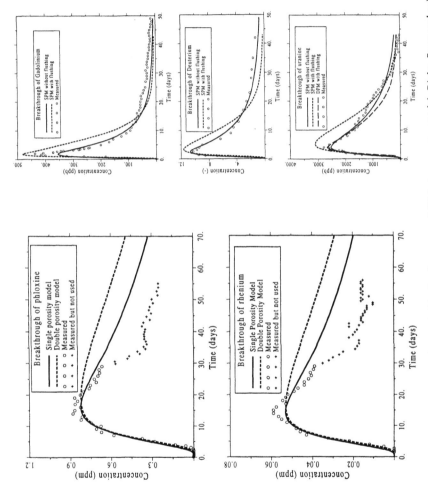

Figure 3 Left: Calibrated breakthrough curves of rhenium and phloxine by means of 3D heterogeneous model. Right: Separated calibrations of the tracers injected by flushing. Note that although fits are superior than those at Figure 4, returned parameters are less consistent. Parameters are different for all them, since chemical behavior of the tracers is not considered.

411

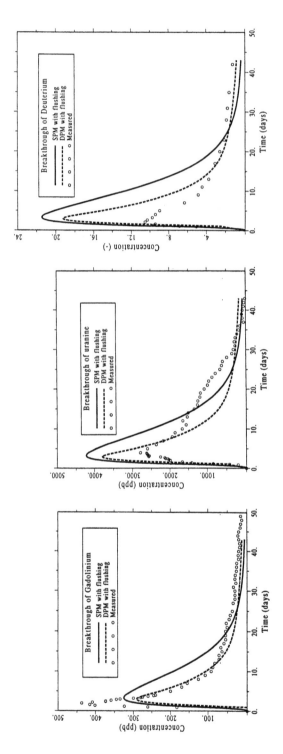

Figure 4. Breakthrough curves of flushed injected tracers. Computed values are simulations by using parameters calibrated from rhenium test. Flushing period is accounted by coupling two steady-state flow models.

5. CONCLUSIONS

Good fits could be obtained by means of relatively simple models, but led to parameters of doubtful validity (little physical meaning) for prediction purposes. Therefore they have little predictive capabilities. More complicated models (3D heterogeneous) can overcome these problems, but may encounter difficulties with respect to software and CPU time and memory, for not mentioning uncertainty of the parameters when measurements are scarce or poorly distributed.

Table 6. Transport parameters returned by the numerical model of test at 2-13-15.

Parameter	Phloxine	Rhenium	Gadolinium	Uranine	Deuterium
Single porosity model without flushing					
α_L rock (m)	0.16e1	0.13e1	0.15e2	0.13e2	0.21e2
α_L fract (m)	0.27e1	0.26e1	0.15e2	0.15e2	0.23e2
ϕR rock (-)	0.88e-3	0.72e-3	0.12e-3	0.15e-4	0.33e-3
ϕbR fract (m)	0.14e-3	0.13e-3	0.10e-3	0.13e-2	0.75e-3
Mass	63%	65%			
Single porosity model with flushing					
α_L rock (m)			0.16e1	0.16e1	0.16e1
α_L fract (m)			0.27e1	0.27e1	0.27e1
ϕR rock (-)			0.53e-3	0.13e-2	0.88e-3
ϕbR fract (m)			0.81e-4	0.20e-3	0.14e-3
Mass					54%
Double porosity model without flushing					
α_L rock (m)	0.10e1	0.10e1			
α_L fract (m)	0.10e1	0.10e1			
$D_{im}/b_{im}{}^2$	0.40e-4	0.40e-4			
ϕR rock (-)	0.32e-4	0.27e-4			
ϕbR fract (m)	0.33e-4	0.37e-4			
ϕ_{im}	0.45e-1	0.40e-1			
Double porosity model with flushing					
α_L rock (m)				0.10e1	
α_L fract (m)				0.10e1	
$D_{im}/b_{im}{}^2$				0.14e-4	
ϕR rock (-)				0.32e-4	
ϕbR fract (m)				0.33e-4	
ϕ_{im}				0.75e-1	

α_L = Longitudinal dispersivity ϕ= porosity R = retardation b = fracture aperture ϕ_{im} = immobile porosity b_{im} = thickness the immobile porosity; Mass: reducing factor applied to input mass to achieve good fits. Double porosity model with flushing was only applied to uranine

Predictive capabilities of the 3D models were verified by simulating tests under flow conditions different from those used for calibration. It is shown that both single and double porosity models led to fair predictions, but that the latter was superior. In any case, the interpretations have undoubtedly shown that models can be used to detect what kinds of processes are important for the behavior of the tracers in groundwater. For the tests performed in boreholes S2, S13 and S15, diffusion into the crystalline rock is an important process and the influence on the groundwater flow of the flushing of water cannot be neglected (in case of the test with flushing). In addition, heterogeneous flow caused by the presence of the fractures also causes tailing. Therefore, both contributions can not be separated. As seen in the S11-S12 test interpretation, tailing effects early attributed to physical-chemical processes, may appear explained by the heterogeneity. When accounting properly for different hydraulic properties of both fracture and rock mass -instead of fracture solely-, part of the tails are reproduced thanks to the late arrivals of mass traveling through less pervious medium. In fact, if only heterogeneity would be the cause of tailings, they would appear exactly the same for all tracers, which is not the case. The apparent dependence of the tail and peak retardation with respect to molecule size, seems to point out that matrix diffusion is more important than adsorption. Besides, the injection procedure also affects the type of breakthrough; this effect, plays an important role in controlling mass dilution in the rock mass and therefore, the extent to which retardation processes may appear. Finally, it results that reasonable parameters and model fits could only be obtained when features often neglected in normal permeability media (natural flow field, flushing, bore presence, etc.) were included in the interpretation.

ACKNOWLEDGEMENTS

Tracer tests and the whole project of El Berrocal, were funded by ENRESA and the CEC. Authors thanks all those individuals that made possible the design, performance and interpretation of these tests. Special thanks are devoted to Abel Ylllera, Antonio Hernández, Pedro Rivas and Benigno Ruiz (CIEMAT, Spain), José Bueno (CEDEX, Spain) and Francis Mousty (JRC, Italy). El Berrocal project is reported in topical report series by ENRESA and in task-group reports by the CEC. Limited free copies are available upon request.

REFERENCES

d'Alessandro, M.; Mousty, F.; Bidoglio, G.; Guimerà, J.; Benet, I.; Sánchez, X.; García, M. and Yllera, A. (1996) Field tracer experiments in low permeability-fractured medium: results from El Berrocal site. *J. Cont. Hydrol.* (in press)

Benet, I. and Carrera, J. (1992) Desarrollo de un programa de ordenador para la interpretación de ensayos de trazadores. *Hidrogeología y Recursos Hidráulicos,* XVII, pp. 335-349.

Carrera J. et al. (1996) Hydrogeology task group report. El Berrocal project, TGR 2, Task group reports, volume I, ENRESA, (Madrid) and CEC (Brussels).

Feenstra, S., Cherry, J.A., Sudicky, E. and Haq, Z. (1984) Matrix diffusion effects on contaminant migration from an injection well in fractured sandstone. *Groundwater,* 22(3):307-316

Frick, U. et al. (1991) The radionuclide migration experiment - overview of investigations 1985-1990 NAGRA Tech. rep. 91-04

Galarza G.; Medina A. and Carrera J. (1996) TRANSIN-III. Applications to 3D media and non-linear problems.. El Berrocal project, Vol. IV. Topical Report 17, ENRESA. Madrid.

García, M.; Guimerà, J.; Yllera, A.; Hernández, A.; Humm, J. and Saaltink, M. (1996) Tracer tests at El Berrocal site. *J. Cont. Hydrol.* (in press)

Guimerà, J.; Vives, L.; Saaltink, M; Tume, P.; Ruiz, B.; Carrera, J. and Meier, P. (1996a) Numerical modeling of pumping tests in a fractured low permeability medium. El Berrocal project, vol IV, Topical Report 14, ENRESA, Madrid.

Guimerà, J. et al. (1996b). Tracer test task group report, El Berrocal project TGR6, Task group reports, volume I, ENRESA (Madrid) and CEC (Brussels).

Malozewski, P. and Zuber, A. (1985) On the theory of tracer experiments in fissured rocks with porous matrix. *J. Hydrol.,* 79:333-358

Malozewski, P. and Zuber, A. (1992) On the calibration and validation of mathematical models for the interpretation of tracer experiments in groundwater. *Adv. in Water Resour.,* 15:47-62

Malozewski, P. and Zuber, A. (1993) Tracer experiments in fractured rocks: matrix diffusion and the validity of models. *Water Resour. Res.* 29(8):2723-2735

Meigs, L.C., Beauheim, R.L. and McCord, J.T. (1996) Design, modeling and current interpretations of the H-19 and H-11 Tracer Tests at the Wipp site. GEOTRAP Summaries. 42-45

Neretnieks, I. (1983) Diffusion in the rock matrix: An important factor for radionuclide retardation? *J. Geophys. Res.,* 19:364-370

Rivas, P. and 17 more contributors (1996) El Berrocal project. Final summary report. European Comission, Nuclear Science and Technology (in press)

Winberg, A. and Olsson, O. (1997) The Äspö TRUE experiments. GEOTRAP summaries. 38-41

Wood, W.W., Kraemer, T.F. and Hearn, P.P. (1990) Intragranular diffusion: an important mechanism influencing solute transport in clastic aquifers? *Science,* 247:1569-1572

Zuber, A. and Motyka, J. (1994) Matrix porosity as the most important parameter of fissured rocks for solute transport at large scales. *J. Hydrol.,* 158:19-46

Tracer Hydrology 97, Kranjc (ed.) © 1997 Balkema, Rotterdam, ISBN 90 5410 875 4

Geohydraulic parameters in hard rocks of SW-Germany determined by tracer tests

A.E.Jakowski
JUNG Consulting Engineers, Kleinostheim, Germany

G.Ebhardt
Institute of Geology/Palaeontology, Technical University of Darmstadt, Germany

ABSTRACT: The results of more than 700 tracer tests are shown carried out in the aquifers of the Triassic and Jurassic in Baden Wuerttemberg, SW-Germany. Based on 115 data sets the geohydraulic parameters flow velocity, dispersivity, Peclet number and apparent coefficient of the hydraulic conductivity have been evaluated. It is discussed how the dipersion can be used for assigning groundwater protection areas.

1 INTRODUCTION

Tracer tests belong to the frequently used research methods in the hydrogeological practice due to their multifarious applicabilities. Through the last four decades more than 700 tracer tests have been organised in karst and fissured aquifers by the Geological Survey of Baden-Wuerttemberg, SW-Germany (Crystalline Basement: 6, Buntsandstein: 96, Middle Triassic: 257, Keuper: 76, Lower and Middle Jurassic: 10, Upper Jurassic: 271, Tertiary: 5).

This large amount of data sets made it possible to compare the geohydraulic parameters such as flow velocity, dispersivity, Peclet number, and apparent coefficient of the hydraulic conductivity determined from tracer tests of the geological epochs of the Triassic and Jurassic.

1.1 *Regional geology*

This chapter gives a short summary of those geogical epochs in which the below evaluated tracer tests have been carried out. More information about the regional geology is given by Geyer & Gwinner (1986).

The continental influenced Buntsandstein (Lower Triassic) of SW-Germany is built up by sandstone sequences interrupted by clay interlayers. Stratigraphically follows the marine Muschelkalk (Middle Triassic) with limestones, dolomites and evaporites intercalated by marls and mudstones. The Keuper (Upper Triassic) consists of fluvial and terrestrial clays, marls and sandstones interbedded

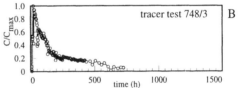

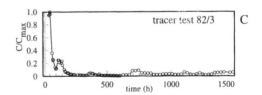

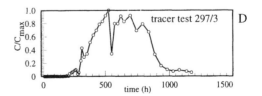

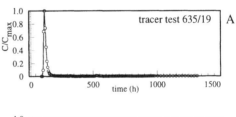

Fig. 1. Typical breakthrough-curves of the Triassic and Jurassic

by marine carbonates, dolomites, and gypsum. Most tracer tests in Baden-Wuerttemberg have been carried out in the marine Upper Jurassic, which is built up by limestones and marly beds. In the Upper Jurassic there exist two facies, the reef facies and the bedded facies.

Table 1. Recovery rates of the tracers in aquifers of different geological epochs

recovery rate %	> 10	1-10	< 1	none	total
Upper Jurassic	60	26	17	45	148
Keuper	9	11	17	17	54
Middle Triassic	32	14	57	38	141
Buntsandstein	4	5	22	24	55

1.2 Description of the field observations

The following tracers were used: the fluorescence tracers uranine (488 tests) and eosine (118), other fluorescence tracers (40), salts (46), spores etc. (27).

Fig. 1 shows some typical breakthrough curves appearing in aquifers of all geological epochs. Most frequently encountered was type A (single peak without tailing) followed by type B (single peak with tailing). Two or more peaks (C) and broad curves (D) are rarely found.

Recovery rates of > 10 % appeared especially in the Upper Jurassic karst, rarely in the Buntsandstein (see table 1). Recovery rates < 1 % were recognised mostly in the Triassic. For 30 to 40 % of the tracer tests in all geological epochs there was no recovery.

In the karstified aquifers of the Muschelkalk and the Upper Jurassic the tracer was mainly injected into natural openings, i.e. sinkholes, vugs etc. which were supposed to have good contact with the water table. In the fissured aquifers of the Buntsandstein and the Keuper the injection into open cuts and into creeks was preferred.

The above illustrated data material is quite homogeneous, so the problems mentioned by Gelhar et al. (1992) as typical for studies comparing tracer tests published by different authors are minimised: mainly fluorescence tracers are used, so the problem of nonconservative effects of the tracers is reduced; the injection and analysis of the tracers was carried out by the Geological Survey, so the mass input history is known; expert studies and publications of the Geological Survey Baden-Wuerttemberg guarantee good documented geological informations for the data analysis. Furthermore it has been attempted to match the data analysis to the flow configuration as well as to match the dimensionality of the analysis to the dimensionality of the monitoring.

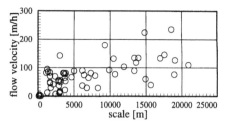

O   Upper Jurassic

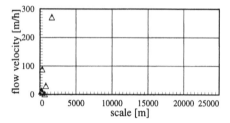

△   Keuper

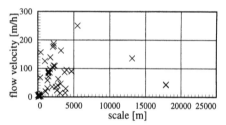

✕   Muschelkalk

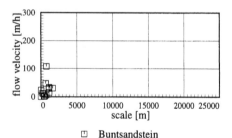

◻   Buntsandstein

Fig. 2. Flow velocity versus distance

416

2 GEOHYDRAULIC PARAMETERS IN DIFFERENT MESOZOIC AQUIFERS

2.1 *Flow velocity*

The flow velocity is the most important hydraulic parameter determined by tracer tests. Fig. 2 shows the relation of the distance between injection and detection points, i.e. the scale of observation, and the flow velocity for the karst aquifers of the Muschelkalk and the Upper Jurassic and for the fissured aquifers of the Lower and the Keuper.

In the karst aquifers the distances between injection and detection points differ between 5 and 21,000 m, the flow velocities between 0.2 m/h and 250 m/h. The mean flow velocity is 72 m/h.

In the Keuper one tracer test should be noticed with a flow velocity of 272 m/h. In fact for this test the tracer was injected into a gypsum swallow hole. Omitting this tracer test in the fissured aquifers of the Lower and the Keuper the distances (21 to 1630 m) and the flow velocities (mean value: 25 m/h) were smaller than in the karst aquifers. There was only a bad correlation between distance and flow velocity.

2.2 *Longitudinal Dispersivity*

Fig. 3 shows the relation of the distance between injection and detection points and the longitudinal dispersivity utilizing a one dimensional dispersion model.

The longitudinal dispersivities increased with the distance between the injection and the detection point. In the karstified aquifers the values spread more than in the fissured aquifers. The correlation coefficients were small, only the data of the Buntsandstein has a correlation coefficient of 0.9. In the karst aquifers the Peclet numbers were up to one scale larger than in the fissured aquifers.

According to our data the dispersion does not reach a plateau value as reported for porous aquifers by several authors. We suppose that in fissured and karstified aquifers the plateau value is reached only at distances larger than 20 km not existing in SW-Germany.

2.3 *Interpretation*

The results fit the idea of two karst water systems, the short-term and the long-term karst water. Small dispersivities at long distances between injection and detection points and high flow velocities point to short term karst water, running quickly and straight in channels without extensive spreading to the outlets.

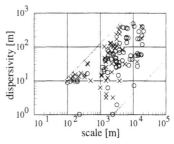

○   Upper Jurassic (65 tracer tests)
×   Muschelkalk (50 tracer tests)

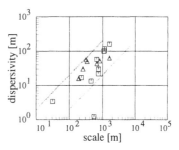

△   Keuper (7 tracer tests)
▢   Buntsandstein (11 tracer tests)

Fig. 3. Longitudinal dispersivity versus distance for karstified (above) and fissured aquifers (below)

In other cases the karst behaves like fissured aquifers with small flow velocity and Peclet numbers. The tracer is transported by long-term karst water which penetrates the matrix and is spread there.

In contrast to the karst aquifers tracer tests in fissured aquifers display small distances, hence smaller dispersivities, small flow velocities and small Peclet numbers.

3 GEOHYDRAULIC PARAMETERS IN THE UPPER JURASSIC KARST

As 40 % of the tracer tests were carried out in the carbonate karst of the Swabian Alb *(Late Jurassic)*, those were studied in detail.

3.1 *Type of injection point, flow velocity and recovery rate*

If the tracer was injected directly into the ground water via wells or boreholes, the recovery rates were small (<10%). The same concerns the mean distance and the mean flow velocity. Corresponding to the small distances also the dispersivities of the tracers were small. The opposite was observed when the tracer was injected into the unsaturated zone, in sinkholes etc. (fig. 4).

This shows that wells and boreholes usually have only poor contact to the strongly karstified and highly permeable parts of the karstified aquifers but are fed by long term karst water. If the tracer was injected into the unsaturated zone, there exists a direct connection between the injection point (sinkhole, gaping fissure etc.) and the detection points (sources etc.). The tracer is transported by short term karst water and may be detected over long distances.

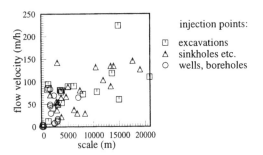

Fig. 4   Dependence of flow velocity and scale on the type of the injection points, Upper Jurassic

3.2 Peclet numbers

Due to tectonic and stratigraphic features a regional differentiation of the dispersion is observed. The eastern Swabian Alb shows Peclet numbers < 100, which means strong spreading of the tracer. This agrees with the results of pumping tests, which revealed a homogeneous and isotropic karst aquifer (Schloz 1993).

In the strongly karstified aquifer of the middle Swabian Alb (catchment areas of the "Blautopf" spring and the "Große Lauter" river) the Peclet numbers were larger than 100. In the western part Hercynic and Rhenian systems of faults cross and break up the limestones. This causes a larger dispersion of the tracers with Peclet numbers smaller than 100.

3.3 Hydraulic parameters and discharge of the karst springs

The relation between the discharge of the karst springs and the flow velocities was studied for the following springs (table 2).

Table 2. Data of the investigated springs

spring	mean discharge (l/s)	period
Aachtopf	8,170	1923-1989
Blautopf	2,300	1952-1989
Kreuzbrunnen	76.1	1978-1989
Gassenbrunnen	35.3	1978-1989
Gallusquelle	520	1956-1969

In fig. 5 the flow velocity is drawn versus the relative discharge (quotient of the discharge and the mean discharge of the given period (table 2)). The flow velocities increase with the relative discharge. Within our data no tests with different injection times but the same injection and detection points existed, so tracer tests with different injection points and perhaps different geological conditions had to be compared. This causes the irregular values in fig. 5.

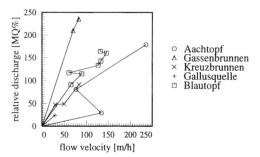

Fig. 5. Tracer flow velocity versus relative discharge of springs (Upper Jurassic)

3.4 Hydraulic parameters and facies

As mentioned in chapter 1.1 there exist two different facies in the Upper Jurassic of the Swabian Alb, the less karstified bedded facies and the more karstified reef facies. 271 tracer tests have been carried out in the Upper Jurassic, 21 of them definitely in the reef facies, 10 in the bedded facies. In the reef facies the mean distance, the recovery rate and the mean longitudinal dispersivity are larger than in the bedded facies (fig. 6, 7). The flow velocity is in the mean the same in both facies. Different results of Seiler & Behrens (1992) in the eastern Swabian Alb may be due to the intensive dolomitisation in that region.

3.5 Hydraulic conductivity

An apparent coefficient of the hydraulic conductivity

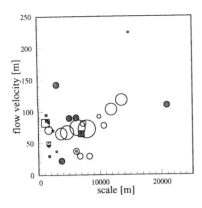

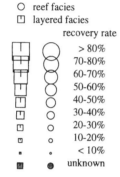

○ reef facies
▱ layered facies

recovery rate

> 80%
70-80%
60-70%
50-60%
40-50%
30-40%
20-30%
10-20%
< 10%
unknown

Fig. 6. Dependence of the flow distance, velocity and recovery rate on the carbonate facies of the Upper Jurassic

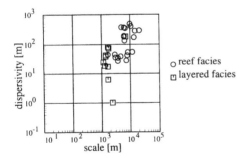

o reef facies
▱ layered facies

Fig 7. Dependence of the flow distance and the dispersivity on the carbonate facies of the Upper Jurassic

was determined from 57 tracer tests of the Swabian Alb. The mean coefficients of the hydraulic conductivity are from east to west [m/s]:

East Swabian Alb	1.4 E-1
Blautopf spring	6.4 E-2
west the Große Lauter river	3.1 E-2
around the town Geislingen	3.1 E-2
the Lauchert river	9.2 E-2
around the town Burladingen	4.6 E-2
around the town Tuttlingen	4.0 E-3
for the Aachtopf spring	9.1 E-2

These data imply a decrease of the hydraulic conductivity from East (1.4 E-1 m/s) to West (4.0 E-3 m/s). Intensive tectonics in the areas of the Lauchert river and the Aachtopf spring cause high hydraulic conductivities (fig. 8).

The hydraulic conductivity obtained from tracer tests is almost one order of magnitude larger than the intrinsic one determined by Stober & Villinger (1995) using pumping tests.

4 CONSEQUENCES REGARDING GROUNDWATER PROTECTION ZONES

Aside from regional investigations applications to the groundwater protection are of actual interest. Protection areas are assigned to wells and sources to prevent biological, chemical and radioactive pollution. Based on the assumption of diminishing risk of pollution with rising distance to the capture in Germany the three zones are assigned to the water catchments. The boundary of the protection zone II (against virological and bacteriological pollution) usually is given by the distance x_{50}, from which the groundwater needs fifty days to reach the capture (see equation 1).

$$x_{50} = v_{Cmax} \cdot 50 \text{ days} \quad (1)$$

v_{Cmax} is the flow velocity determined from the maximum tracer concentration.

Presently the dispersion has not been regarded for the determination of protection zones though by reason of this phenomenon a contaminant may reach a spring or well much faster than in fifty days.

The spatial distribution of the tracer concentration is described by an analytical solution of the transport equation which corresponds with the Gauss distribution.

The standard deviation σ of the concentration curve is expressed by:

$$\sigma = 2\alpha x = x\sqrt{2/Pe} \quad (2)$$

with the distance x [m], the longitudinal dispersivity α [m] and the Peclet number [-].

Adding one, two or three times the standard

419

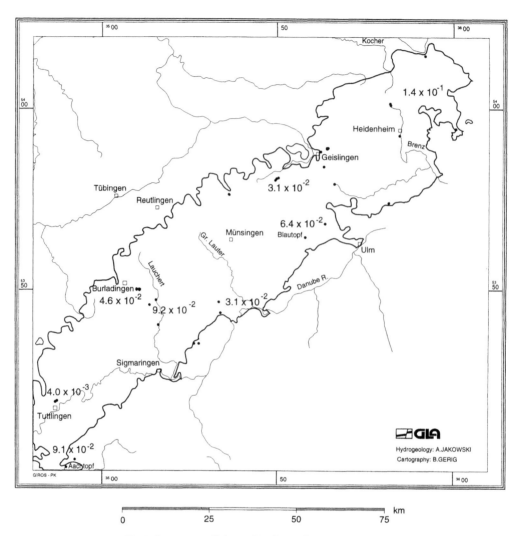

km

0   25   50   75

Fig. 8: Apparent coefficients of the hydraulic conductivity
obtained from tracer tests, Swabian Alb, SW-Germany

deviation to the boundary of the fifty-days zone means with a probability of 84.1 %, 97.7 % or 99.9 % that the pollutant needs fifty days to reach a source or well (see equation 1, 2 or 3). The boundary of fifty days calculated by the flow velocity and the dispersion x_{50_Disp} [m] can be described by the following equations:

$$x_{50Disp} = x_{50} + \sigma = x_{50} + x_{50}\sqrt{2/Pe} \qquad (3)$$

$$x_{50Disp} = x_{50} + 2\sigma = x_{50} + 2x_{50}\sqrt{2/Pe} \qquad (4)$$

$$x_{50Disp} = x_{50} + 3\sigma = x_{50} + 3x_{50}\sqrt{2/Pe} \qquad (5)$$

Spitz et al. (1980) already proposed to take a dispersivity factor into consideration when determining the boundary of fifty days which based on three times the standard deviation:

$$\gamma = x_{50\_Disp}/x_{50} = (1 + 3\sqrt{2\alpha_L/vt_{50}}) \qquad (6)$$

with the flow velocity v and the wanted residence time t_{50} (50 days). If there are no data on the longitudinal dispersivity are available, they should be estimated.

But if there is a tracer test for determining the dispersivity, it is better to take into account the quotient of the distance and the Peclet number (see equations 3 - 5) instead of the dispersivity. The last one is only valid for the distance of the test, from which it was determined, while the quotient of the boundary of fifty days and the Peclet number refers to the questioned distance.

Taking the Peclet number into account for assigning protection areas will be a safe measure. This method may be used, if there is a tracer test to be carried out or if there exist older tracer tests. (Unfortunately those aren't often disposable.) It will be senseless if the flow velocities are so large that the complete groundwater catchment would lie within the boundary of fifty days.

If a tracer test is available first the flow velocity and the Peclet number should be determined. Then it should be considered, which probability of a pollution may be acceptable and weather equation 2, 3 or 4 should be used.

SUMMARY

More than 700 tracer tests in the karstic aquifers of the Muschelkalk and the Upper Jurassic and the fissured aquifers of the Lower and Keuper have been studied. For 115 tests the hydraulic parameters of the flow velocity, the dispersivity and the Peclet number have been determined.

Especially the karst systems of the Swabian Alb in SW-Germany (Upper Jurassic) have been investigated. The dispersivities and hydraulic conductivities of several parts of the Swabian Alb have been correlated with tectonic features as well as with different facies and the influence of the individual test conditions (injection points, discharge).

It has been showed how the Peclet number can be used for assigning groundwater protection areas.

For special information see Jakowski (1995).

ACKNOWLEDGEMENT

The Geological Survey of Baden-Württemberg supported this research project. This is gratefully acknowledged.

REFERENCES

Geyer, O.F. & M.P. Gwinner 1986. Geologie von Baden-Württemberg. Stuttgart: Nägele und Obermiller.

Gelhar, L.W., C. Welty & K.R. Rehfeldt 1992. A Critical review of data on field-scale dispersion in aquifers. Water Resources Research 28 (7): 1955-1974.

Jakowski, A.E. 1995. Ermittlung der Dispersion und anderer geohydraulischer Parameter aus Markierungsversuchen in Karst- und Kluft-grundwasserleitern Baden-Wuerttembergs. Thesis TH Darmstadt.

Schloz, W. 1993: Zur Karsthydrogeologie der Ostalb. Karst und Höhle:119-134.

Seiler, K.-P. & H. Behrens 1992. Groundwater in Carbonate Rocks of the Upper Jurassic in the Franconian Alb and its Susceptibility to Contaminants. In: Hötzl, H. & A. Werner (eds), Tracer Hydrology - Proc. 6th Int. Symp. on Water Tracing, Karlsruhe, 21-26 September 1992: 259-266. Rotterdam: Balkema.

Spitz, K. & H. Mehlhorn & H. Kobus 1980. Ein Beitrag zur Bemessung der engeren Schutz-zone in Porengrundwasserleitern. Wasser-wirtschaft, 70: 365-369.

Stober, I. & E. Villinger. Hydraulisches Potential und Durchlässigkeit des höheren Malms und des Oberen Muschelkalks im baden-württembergischen Molassebecken. Jh. geol. Landesamt Baden-Württemberg (in print).

Tracer Hydrology 97, Kranjc (ed.) © 1997 Balkema, Rotterdam, ISBN 90 5410 875 4

Dual-tracer transport experiments in a heterogeneous porous aquifer: Retardation measurements at different scales and non-parametric numerical stochastic transport modelling

T. Ptak
Applied Geology, University of Tübingen, Germany

ABSTRACT: For the investigation of reactive transport processes within a physically and chemically heterogeneous porous aquifer, forced gradient tracer tests were conducted at the 'Horkheimer Insel' test site in Germany. In each of the experiments, two tracers, one practically non-sorbing and the other one sorbing, were injected simultaneously across the entire aquifer thickness at the same location. In order to examine the scale dependence of the effective retardation factor, different transport distances of up to about 30 m were chosen for the experiments. The results show, that at the scales investigated the effective retardation factor increases with the transport distance. The measurements suggest that a complex numerical flow and transport modelling approach within a stochastic framework is needed to adequately describe the transport behaviour observed. Therefore, a three-dimensional Monte Carlo type non-parametric numerical stochastic transport simulation model, accounting for parameter uncertainty and variability, is applied for the evaluation of the experiments. Based on laboratory investigations on core sample aquifer material from the field site, the sorption process is treated spatially variable in the model, not needing a predefined correlation law of hydraulic conductivity and sorption parameters. According to the first results, the transport simulation technique proved to be suitable for highly heterogeneous aquifer conditions such as at the 'Horkheimer Insel' test site.

1 INTRODUCTION

Many industrial sites causing groundwater contaminations are located in highly heterogeneous alluvial sedimentary environments. In such environments, very often reliable model predictions of non-reactive and reactive solute transport in groundwater at small to intermediate transport distances are needed for groundwater contamination risk assessments and/or for the planning of aquifer remediation activities. Within a highly heterogeneous aquifer such predictions generally require a detailed knowledge of the aquifer parameters controlling the transport processes.

At present, none of the available investigation techniques is able to provide a description of the subsurface structure and its properties in a resolution needed for a deterministic transport model in case of a heterogeneous aquifer such as at the 'Horkheimer Insel' test site, located in South Germany. Therefore, owing to the remaining parameter uncertainty, transport modelling must be performed within a stochastic framework.

Accompanying groundwater research activities comprise both experimental and theoretical work. Transport experiments are performed in order to investigate non-reactive and reactive transport, and new subsurface investigation techniques are developed to estimate the spatial distribution of parameters controlling the transport processes. The experimental data also serve for the validation of analytical or numerical stochastic solutions of the transport problem.

From stochastic groundwater flow and transport theory, closed-form (first-order) analytical solutions or combined numerical-analytical methods are available for both nonreactive and reactive solutes (e.g. Dagan, 1989; Cvetkovic & Shapiro, 1990; Selroos & Cvetkovic, 1994; Bellin & Rinaldo, 1995; Rajaram & Gelhar, 1995; Bosma et al., 1996; Miralles-Wilhelm & Gelhar, 1996). In order to obtain the closed-form solutions, a small variance of hydraulic conductivity (σ^2_{lnK}) has to be assumed in general. Other limits for a practical application may be that the aquifer is assumed to be infinite, the groundwater flow field is uniform and stationary, a difficult to obtain relationship of hydraulic conductivities and distribution coefficients (or retardation factors) is needed, and that the analytical solutions generally only provide ensemble mean results. Other authors employed numerical approaches for modelling of nonreactive and reactive solute transport. For example, Teutsch et al. (1991) and Ptak (1993) used a flexible Monte Carlo method for the analysis of nonreactive tracer transport. Tompson (1993) performed particle tracking simulations of reactive solute transport in physically and chemically heterogeneous porous media with a correlation of hydraulic conductivity and sorption properties. Burr et al. (1994) assumed that hydraulic conductivity and distribution coefficient fields are inversely correlated and that they have the same spatial correlation structure. Using a particle tracking technique, Tompson et al. (1996) investigated the impacts of physical and chemical heterogeneity on cocontaminant transport in a sandy porous medium.

In this paper, tracer experiments performed at the 'Horkheimer Insel' test site are presented, where two fluorescent tracers with different sorption properties, Fluoresceine and Rhodamine WT, representing the classes of practically non-sorbing and sorbing solutes respectively, were injected simultaneously across the entire aquifer thickness within the same groundwater monitoring well. Different distances between tracer injection and observation wells were used for each of the experiments. It was an objective of the field experiments to obtain an insight into the sorption process, to provide a data base for model validation purposes and to investigate the transport scale (or travel time) dependence of the effective field scale retardation factor. This macrokinetic sorption behaviour may be a consequence of physical and chemical aquifer heterogeneities (e.g. Miralles-Wilhelm & Gelhar, 1996).

The experimental results are compared with sorption measurements at the laboratory scale. The focus is on surface sorption and its scale dependence. Surface sorption is relevant for the transport of many polar/-ionizable groundwater contaminants, e.g. chlorinated phenols, some pesticides etc.

If sorption macrokinetics are present, the modelling approach should be based on a formulation of the sorption process that is able to account for a travel time dependent effective retardation factor, without the need to define some difficult to obtain travel time dependent (upscaled) sorption parameters a priori. Therefore, in this paper the application of an alternative modelling approach (Ptak, 1996), accounting for parameter uncertainty and variability, is suggested, that can be used for the modelling of the simultaneous transport of the nonreactive tracer (Fluoresceine) and the reactive, sorptive tracer (Rhodamine WT). In this approach, grain size distribution curves are first generated stochastically. Then hydraulic conductivity values are derived from the grain size curves using an empirical relationship, and spatially variable local scale retardation factors are computed based on grain size related sorption data from laboratory batch experiments on aquifer material samples. All data are combined in a Monte Carlo type non-parametric stochastic transport model to evaluate the tracer test measurements. Compared to other modelling approaches, there is no need for an a priori relationship of hydraulic conductivities and sorption parameters (distribution coefficients or retardation factors), or for assumptions concerning the spatial structure of the sorption parameters.

2 THE FIELD SITE

The 'Horkheimer Insel' field site is located in the Neckar valley, about 70 km north of Stuttgart. The test site (Figure 1) is presently equipped with 45 sampling and monitoring wells. The aquifer is formed by a 2.5 - 4 m thick sequence of poorly sorted alluvial sand and gravel deposits of holocene age. From pumping tests, the geometric mean hydraulic conductivity was determined to 0.012 m s^{-1}. The underlying Triassic clay- and limestone bedrock formation is several orders of magnitude less permeable and can be considered hydraulically tight.

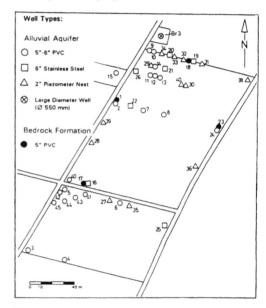

Figure 1. Site map and well location.

3 INVESTIGATION OF TRANSPORT PROCESSES AT DIFFERENT SCALES

The transport properties of the two tracers (Fluoresceine and Rhodamine WT) were investigated at the laboratory scale in batch experiments, as well as within tracer experiments at field scale. A small transport distance of 8.9 m respective 12.8 m, and an intermediate distance of 32.7 m were chosen in order to test if there is some macrokinetic sorption behaviour.

3.1 *Investigations at the laboratory scale*

Fluoresceine and Rhodamine WT are both polar and ionizable tracers. It could be shown in laboratory experiments, that for the Horkheimer Insel aquifer, where the organic carbon content (C_{org}) is low (as low as 0.2 mg/g; Strobel, 1996), it is admissible to accept

Fluoresceine as a quasi ideal, non-sorbing tracer. Rhodamine WT, however, is known to be significantly sorptive. Adsorption on mineral surfaces (clay, carbonate, quartz) may dominate the sorption if C_{org} of the aquifer material is low (e.g. Sabatini & Austin, 1991; Shiau et al., 1993). Laboratory investigations of the sorption of Rhodamine WT using the Horkheimer Insel aquifer material were therefore performed (Strobel, 1996). Drilling core samples (100 mm diameter) were subdivided into sections of about 5 cm thickness, and used for batch experiments. No significant kinetic behaviour could be observed at the laboratory scale, which is typical for surface sorption. Figure 2 shows an example of resulting vertical profiles of retardation factors R [-] computed with:

$$R = 1 + \frac{\rho_b}{\theta} \cdot K_D \qquad (1)$$

where ρ_b [kg m^{-3}] is the bulk density of the aquifer material and θ [-] is the porosity. K_D [l kg^{-1}] is the distribution coefficient.

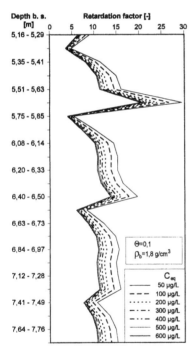

Figure 2. Vertical profiles of retardation factors R from laboratory batch experiments at one of the monitoring wells (Strobel, 1996).

K_D depends on the equilibrium concentration due to a Freundlich type partitioning of Rhodamine WT. Therefore, Figure 2 shows profiles of R for several concentrations, typically encountered within the tracer experiments performed. The values of R are in the

range between about 4 and 30 for the equilibrium concentrations shown. Due to the Freundlich partitioning, R increases with decreasing concentration.

3.2 Dual-tracer transport experiments at small scale

A forced gradient tracer test method was applied within two experiments (TT9 and TT10) with short transport distances of 8.9 m (TT9) respective 12.8 m (TT10). In the two experiments performed successively, groundwater was pumped out of a well (P11, Fig. 1) at a constant rate of 3.02 l s^{-1}. After a quasi stationary radially convergent flow field was established, the two tracers (10 g Fluoresceine and 8 g Rhodamine WT for TT9, respective 20 g Fluoresceine and 16 g Rhodamine WT for TT10) were injected instantaneously across the entire saturated aquifer thickness into a neighbouring monitoring well (P14 for TT9 respective P21 for TT10, Fig. 1). The advantage of the radially convergent flow experiments is that the test duration is reduced and effects of variation of the natural gradient are minimized.

Within each of the two experiments, seven multilevel breakthrough curves with a temporal resolution of 1 min, each representative of a vertical section of 0.3 m thickness, were measured within the pumping well for both tracers directly in-situ using fibre-optic fluorimeters and a flow separation technique. The multilevel approach was chosen in order to investigate the spatial variability of the transport parameters. A detailed description of the experimental setup, the instrumentation, the results and the interpretation is given in Ptak & Schmid (1996). Figure 3 shows as an example the multilevel breakthrough curves for both tracers, obtained from the experiment TT9.

To estimate the Rhodamine WT retardation factors, effective at the small scale of the two tracer tests, from the multilevel breakthrough curves, temporal moments equations can be applied. The n-th temporal moment is defined as (e.g. Kreft & Zuber, 1978):

$$M_{n,t} = \int_0^\infty t^n c(r,t)\,dt \qquad (2)$$

where t [s] is the time, c [kg m^{-3}] the concentration and r [m] is the transport distance. Then the arrival time of the center of mass is given by the first normalized moment:

$$m_{1,t} = \frac{M_{1,t}}{M_{0,t}} \qquad (3)$$

From the first normalized moment a mean effective transport velocity can be computed:

$$\overline{v}_{eff} = \frac{r}{m_{1,t}} \qquad (4)$$

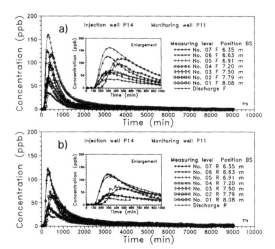

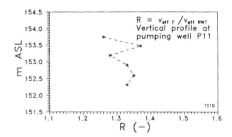

Figure 5. Effective retardation factors TT10 (Ptak & Schmid, 1996).

3.3 Dual-tracer transport experiment at an intermediate scale

Within one new experiment (TT11), again a forced gradient tracer test method was applied with an intermediate transport distances of 32.7 m (Kleiner, 1997). In this experiment, groundwater was pumped out of the same well as in the former experiments (P11, Fig. 1) at a constant rate of 3.3 l s^{-1}. After a quasi stationary radially convergent flow field was established, the same two tracers (102 g Fluoresceine and 102 g Rhodamine WT) were injected instantaneously across the entire saturated aquifer thickness into a more distant monitoring well (P22, Fig. 1).

Integral (depth averaged) breakthrough curves with a temporal resolution of 5 min were measured within the groundwater discharge line at the pumping well for both tracers directly using fibre-optic fluorimeters and a portable field fluorimeter. Details are given in Kleiner (1997).

The application of the method of temporal moments equations, as described above, yields at the intermediate transport scale investigated an effective retardation factor R of 2.17.

3.4 Comparison of the retardation factors measured at different scales

The measurements at the different investigation scales clearly indicate a scale dependence of the effective retardation factor.

At the laboratory scale, significantly larger retardation factors are obtained compared to the values from the field scale experiments. Within the tracer experiments, the field scale effective retardation factor increases with the travel distance, i.e. a macrokinetic behaviour of the sorption process is observed.

Laboratory batch experiments are known to yield higher sorption capacities, expressing in higher retardation factors, since due to the experimental procedure more sorption sites are made accessible for the reactive compound compared to a natural aquifer situation.

Within the tracer experiments at small scale (TT9

Figure 3. Multilevel breakthrough curves from experiment TT9: a) Fluoresceine, b) Rhodamine WT (Ptak & Schmid, 1996).

Assuming that Fluoresceine is practically an ideal tracer, an effective field scale retardation factor R [-] can be defined for each set of two corresponding multilevel breakthrough curves:

$$R = \frac{\overline{v}_{\text{eff Fluoresceine}}}{\overline{v}_{\text{eff Rhodamine WT}}} = \frac{m_{1,t \text{ Rhodamine WT}}}{m_{1,t \text{ Fluoresceine}}} \qquad (5)$$

Figure 4 and Figure 5 show the resulting vertical profiles of the effective retardation factor at the pumping and observation well.

The values of R are in the range between 1.27 and 1.40 for TT9, respectively 1.26 and 1.37 for TT10, when individual multilevel breakthrough curves are compared. The variability indicates the heterogeneity of the aquifer. The arithmetically averaged retardation factors (Cvetkovic & Shapiro, 1990) are 1.35 for TT9 and 1.32 for TT10. It is seen, that Rhodamine WT is retarded compared to Fluoresceine even though the organic carbon content of the Horkheimer Insel aquifer material is very low.

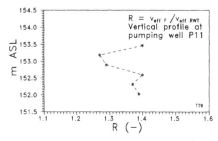

Figure 4. Effective retardation factors TT9 (Ptak & Schmid, 1996).

and TT10), in the heterogeneous aquifer most of the tracer mass is transported in highly conductive preferential pathways, since the horizontal correlation length of the hydraulic conductivity field at the test site is in the order of 10 m (Ptak & Teutsch, 1994), which is about the transport distance of the two experiments. Therefore, the probability for connected preferential pathways between the pumping and injection wells is relatively high for the small transport distances.

Within the intermediate transport distance tracer experiment (TT11), the probability for discontinuities in the high conductivity preferential transport zones increases, since the horizontal correlation length of about 10 m is smaller compared to the transport distance of 32.7 m. The tracer has therefore to pass low conductivity aquifer zones also. Low conductivity zones are generally composed of finer grains compared to the high conductivity zones. Therefore, the specific surface is higher within the lower conductivity aquifer zones, which yields higher sorption capacities in these zones, if surface sorption is controlling the reactive transport process. Since the tracer is hydrodynamically mixed into lower conductivity zones, as the transport distance becomes larger compared to the hydraulic conductivity correlation length, the effective retardation factor increases with travel distance.

This suggests that the field scale macrokinetic sorption behaviour is caused by the physical and chemical aquifer heterogeneities. This is also supported by the observations from the laboratory batch experiments, where equilibration occurs almost immediately, without any significant kinetic behaviour.

It is obvious, that transport simulations based on the laboratory scale values would overpredict the retardation of a reactive compound undergoing surface sorption. Also, the use of only one constant effective field scale retardation factor for transport predictions is not justified, due to the sorption macrokinetics observed at the investigated transport distances.

The modelling approach needed for reactive transport predictions in such a situation should account for the macrokinetic sorption behaviour. It should not be based on bulk retardation factors from batch experiments in order to avoid overprediction of retardation, and it should not require a priori equations for the scale dependence of the effective retardation factor. The formulation of such equations is generally difficult in practice, requireing costly measurements, and the equation parameters are mostly only valid for the specific site and the transport scales investigated. Furthermore, the modelling approach should account for remaining parameter uncertainty, not having restrictions like some solutions from stochastic transport theory as described above.

To overcome the difficulties, the flexible three-dimensional Monte Carlo type numerical stochastic transport model described in the following can be applied.

4 NUMERICAL STOCHASTIC TRANSPORT SIMULATION

Since aquifer sediments are a heterogeneous mixture of different lithocomponents with different grain size fractions, sorption isotherms were measured in batch sorption experiments for the different lithocomponents as well as for the individual grain size fractions of aquifer material samples of about 5 cm thickness. The analysis revealed, that it was not necessary to regard the lithological composition (quartz, Jurassic and Triassic limestones, Keuper and Bunter sandstones, calcite) as an aquifer property controlling the surface sorption process of Rhodamine WT.

Then, from the experiments (Strobel, 1996), a relationship between sorption capacity and grain size, based on a Freundlich isotherm model, was established, finally overcoming the difficulties in defining transport distance dependent retardation factors:

$$K_{Di} = \frac{1}{n_i} \cdot K_{Fri} \cdot c_{eq}^{\frac{1}{n_i}-1}$$
$$= 0.85 \cdot 0.132 \cdot d_i^{-1.0} \cdot c_{eq}^{0.85-1} \qquad (6)$$

where K_{Di} [l kg^{-1}] is the distribution coefficient at sorption equilibrium, K_{Fri} [l kg^{-1}] is the Freundlich coefficient, c_{eq} [μg l^{-1}] is the equilibrium concentration, n_i^{-1} [-] is the Freundlich exponent of a grain size fraction i, and d_i is the representative diameter [mm] of i.

The three-dimensional Monte Carlo type transport simulation technique (Ptak, 1996) itself comprises three main steps.

4.1 Geostatistical analysis of grain size distribution curves

Since equation (6) is based on grain size, information about the spatial distribution of grain size fractions has to be provided as a model input in the first step.

The structural analysis yielding parameters of the spatial structure of grain size distributions was performed using categorical variables which were defined based on measured grain size distribution curves of the 5 cm aquifer material samples (Schad, 1993). Out of 243 measured grain size distribution curves, those with similar properties (shapes) were clustered into six groups (six clusters) with the multivariate statistical method of KMEANS clustering (McQueen, 1967). Each grain size distribution curve was discretized into seven classes, represented by class variables, which describe the mass fractional contributions of individual grain size fractions to the total sample mass.

After clustering, six binary categorical variables (cluster1 - cluster6) were defined for each grain size distribution curve. Since categorical variables are exclusive, one of these six variables was assigned the value 1 and the remaining five were assigned the value

0, thus yielding six binary data sets, one for each cluster. Then, the statistics (mean and variance) of the seven class variables were computed for each of the six clusters.

Subsequently, Schad (1993) investigated the spatial correlation of the six categorical variables by computing experimental variograms and fitting exponential variogram models in vertical and two horizontal directions. For the clusters correlation lengths between 0.08 m and 0.12 m in vertical direction, and between 2 m and 10 m for the two principal horizontal directions were found, indicating a structural anisotropy of the heterogeneous aquifer.

4.2 *Generation of hydraulic conductivity and distribution coefficient fields*

In the second step, the experimental histogram of the six clusters and the cluster variogram models are used to generate conditioned equiprobable three-dimensional realizations of the categorical variable field. For the non-parametric approach used here, the three-dimensional conditional sequential indicator simulation method for categorical variables (Deutsch & Journel, 1992) is applied. This method is able to honour extreme values and allows for consideration of more than one spatial structure of the investigated data.

Each simulated cluster from the categorical variable field represents a mean grain size distribution curve, consisting of the means of the seven class variables. Subsequently, the means of these class variables are perturbed with a normally distributed random component following the individual class variable statistics estimated from the measurements. The resulting simulated grain size distribution curves are then evaluated at each nodal position of the simulation domain in terms of local hydraulic conductivity values K and local effective distribution coefficients K_D. K [m s^{-1}] is obtained using Beyer's (1964) empirical relationship between hydraulic conductivity and grain size distribution:

$$K = c(u) \cdot d_{10}^2 \qquad (7)$$

where c(u) is an empirical constant, u is defined as d_{60}/d_{10} and d_{10} and d_{60} are the diameters [mm] of the grains where 90% respective 40% of the sample mass are retained in a sieve analysis. The local effective K_D value is obtained from:

$$K_D = \sum_{i=1}^{m} x_i \cdot K_{Di} \qquad (8)$$

where m is the number of grain size fractions (here 7), x_i [-] is the mass fractional contribution of each grain size fraction, obtained from the geostatistical simulation of the grain size distribution curves, and K_{Di} is the distribution coefficient following equation (6).

4.3 *Simulations of flow and transport*

Third, flow and transport simulations are performed for the two tracers within each generated hydraulic conductivity field, with initial and boundary conditions and geometry according to the tracer experiment. The simulation domain was 2086 m x 2086 m x 2.4 m, and 29 realizations were generated.

The three-dimensional flow field was computed for each stochastic aquifer realization using the finite-difference code MODFLOW (McDonald & Harbaugh, 1984). Then, for each realization tracer breakthrough curves were simulated at the pumping well for Fluoresceine without sorption and with sorption for Rhodamine WT, using an extended version of the program MT3D (Zheng, 1991). Due to the almost immediate equilibration of the Rhodamine WT concentrations and the resulting high Damkoehler number, local instantaneous equilibrium was assumed. Simulation routines were added to the original MT3D code to allow for the grain size based description of the sorption process, which is finally introduced by a retardation factor R, computed with equation (1). In this application of equation (1), K_D is the concentration dependent distribution coefficient, computed with equation (8) and updated after each transport step. A variable effective porosity θ, following a correlation function of porosity (estimated from permeameter measurements) with hydraulic conductivity, was used for the transport calculations. No model calibration was performed using the tracer experiment results, as it was an aim to test the model predictions based only on measured input parameters. From the simulations an ensemble of realizations of the nonreactive and reactive transport processes is obtained, which can be compared with the field measurements.

5 COMPARISON OF FIELD MEASUREMENTS AND SIMULATIONS

The comparison of simulated and measured tracer transport is performed using breakthrough curves, since concentration measurements are available only at the pumping well. In this comparison, each measured and numerically simulated breakthrough curve is evaluated individually, and effective transport parameters (e.g. peak concentration, peak concentration arrival time, second temporal moment, mean transport velocity, etc.) derived from one measured breakthrough curve are compared with a numerically simulated stochastic ensemble. The ensemble variance of a transport parameter is a measure of prediction uncertainty.

As an example, in Figure 6 out of the compared effective transport parameters (Ptak, 1996), the measured and numerically simulated vertical profiles of effective retardation factors (eq. 5) are shown for the experiment TT9.

The numerical stochastic transport model yields at

428

present slightly higher retardation factors compared to the measurements. However, the differences between measurements and numerical simulation seem acceptable since the model was not calibrated with respect to transport. Calibration procedures could of course easily improve the fit.

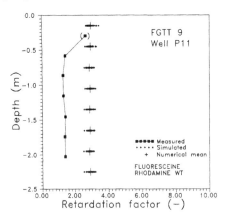

Figure 6. Comparison of measured and simulated effective retardation factors at the pumping well (P11), experiment TT9.

The stochastically simulated retardation factors are significantly smaller compared to the values from the laboratory experiments, even though equation (8), describing the local effective K_D value, is based on batch experiments. A possible reason for the slightly higher simulated retardation factors might be attributed to the disturbance of the core samples during drilling activities, where the core is drained rapidly under gravity when the core barrel is lifted to the surface. The induced hydrodynamic forces most likely distribute fine grains within the core. On the other hand, as discussed above, within a heterogeneous aquifer such as at the 'Horkheimer Insel', near the source most of the tracer mass is transported in highly conductive preferential paths, where almost no fine grains are present (open framework gravels). Since the sorption process described by equation (6) yields high K_D values for fine grains even if their mass fractional contribution is relatively small, the distribution of fine grains into the preferential transport zones during core sampling may express in higher simulated retardation factors based on laboratory measurements.

6 CONCLUSIONS

Laboratory batch experiments as well as dual-tracer forced gradient transport experiments with transport distances between 8.9 m and 32.7 m were performed in order to investigate reactive transport properties within a highly heterogeneous porous aquifer. The results demonstrate the scale dependence of the sorption parameters and a macrokinetic field scale sorption behaviour. Therefore, sorption parameters measured at one scale cannot be simply used at an another scale. To overcome upscaling problems, a flexible numerical stochastic transport simulation technique, accounting for parameter variability and uncertainty, was used to simulate both the non-reactive and the reactive transport within one of the experiments. Following the simulation results, it is believed that the simulation technique can be successfully applied in highly heterogeneous porous aquifers such as at the 'Horkheimer Insel' test site. The results from measurements and simulations are in a comparable range, even though the model was not calibrated with respect to transport.

The geostatistical approach using sequential indicator simulation of categorical variables and the grain size based formulation of the surface sorption process allow the generation of spatially variable hydraulic conductivity and retardation factor fields, without the need for an a-priori correlation function of hydraulic conductivities and distribution coefficients, or for assumptions about the spatial structure of the distribution coefficient. Since the model is based on local scale retardation factors, assuming instantaneous local equilibrium, it is suitable for the simulation of sorptions macrokinetics, caused by physical and chemical aquifer heterogeneities. Future work will comprise more numerical stochastic simulations to test the model prediction capability also at the another transport distances investigated.

REFERENCES

Beyer, W. (1964) Zur Beschreibung der Wasserdurchlässigkeit von Kiesen und Sanden. Zeitschr. f. Wasserwirtschaft-Wassertechnik, 14, 165-168.

Bellin, A. & Rinaldo, A. (1995) Analytical solutions for transport of linearly adsorbing solutes in heterogeneous formations. Water Resour. Res., 31(6), 1505-1511.

Bosma, W.J.P., van der Zee, S.E.A.T.M. & van Duijn, C.J. (1996) Plume development of nonlinearly adsorbing solute in heterogeneous porous formations. Water Resour. Res., 32(6), 1569-1584.

Burr, D.T., Sudicky, E.A. & Naff, R.L. (1994) Nonreactive and reactive solute transport in three-dimensional heterogeneous porous media: Mean displacement, plume spreading, and uncertainty. Water Resour. Res., 30(3), 791-815.

Cvetkovic, V.D. & Shapiro, A.M. (1990) Mass arrival of sorptive solute in heterogeneous porous media. Water Resour. Res., 26(9), 2057-2067.

Dagan, G. (1989) Flow and transport in porous formations. Springer, Berlin, F.R.G., 465 pp.

Deutsch, C.V. and Journel, A.G. (1992) GSLIB Geostatistical software library and user's guide. Oxford University Press, New York, 340 pp.

Kleiner, K. (1997) Erprobung von Multitracer-Analyseverfahren und Einsatz ausgewählter Tracer im Feldversuch. Diplomarbeit, Geologisches Institut, Universität Tübingen, F.R.G.

Kreft, A. & Zuber, A. (1978) On the physical meaning of the dispersion equation and its solutions for different initial and boundary conditions. *Chemical Engineering Science,* Vol. 33, 1471-1480.

McDonald, M.G. & Harbaugh, A.W. (1984) *A modular three-dimensional finite-difference groundwater flow model.* USGS Open-File Report 83-875, USA.

McQueen, J. (1967) Some methods for classification and analysis of multivariate observations. *5th Berkeley Symposium on Mathematics, Statistics and Probability,* 1, 281-298.

Miralles-Wilhelm, F. & Gelhar, L.W. (1996) Stochastic analysis of sorption macrokinetics in heterogeneous aquifers. *Water Resour. Res.,* 32(6), 1541-1549.

Ptak, T. (1993) *Stofftransport in heterogenen Porenaquiferen: Felduntersuchungen und stochastische Modellierung.* Dissertation, Heft 80, Institut für Wasserbau, Universität Stuttgart, F.R.G.

Ptak, T. & Schmid, G. (1996) Dual-tracer transport experiments in a physically and chemically heterogeneous porous aquifer: Effective transport parameters and spatial variability. *J. Hydrol.,* 183(1-2), 117-138.

Ptak, T. & Teutsch, G. (1994) A comparison of investigation methods for the prediction of flow and transport in highly heterogeneous formations. *Proc. Int. Symposium on Transport and Reactive Processes in Aquifers, April 11 -15, 1994, ETH Zürich, Switzerland,* Balkema, Rotterdam, 157-164.

Ptak, T. (1996) *Evaluation of reactive transport processes in a heterogeneous porous aquifer within a non-parametric numerical stochastic transport modelling framework based on sequential indicator simulation of categorical variables.* Proceedings of the First European Conference on Geostatistics for Environmental Applications, geoENV 96, 20.-22. November 1996, Lisbon, Portugal, Kluwer Academic Publishers.

Rajaram, H. & Gelhar, L.W. (1995) Plume-scale dependent dispersion in aquifers with a wide range of scales of heterogeneity. *Water Resour. Res.,* 31(10), 2468-2482.

Sabatini, D.A. & Austin, T.A. (1991) Characteristics of Rhodamine WT and Fluorescein as adsorbing ground-water tracers. *Ground Water,* 29(3), 341-349.

Schad, H. (1993) Geostatistical analysis of hydraulic conductivity related data based on core samples from a heterogeneous fluvial aquifer. *International Workshop on Statistics of Spatial Processes,* Bari, Italy, 27.-30. Sept. 1993.

Selroos, J.-O. & Cvetkovic, V. (1994) Mass flux statistics of kinetically sorbing solute in heterogeneous aquifers: Analytical solution and comparison with simulations. *Water Resour. Res.,* 30(1), 63-69.

Shiau, B.J., Sabatini, D.A. &Harwell, J.H. (1993) Influence of Rhodamine WT properties on sorption and transport in subsurface media. *Ground Water,* 31(6), 913-920.

Strobel, H. (1996) *Sorption eines reaktiven Tracers (Rhodamin WT) in heterogenem Aquifermaterial vom Testfeld "Horkheimer Neckarinsel.* Diplomarbeit, Geologisches Institut, Universität Tübingen, F.R.G.

Teutsch, G., Hofmann, B. & Ptak, T. (1991) Non-parametric stochastic simulation of groundwater transport processes in highly heterogeneous formations. *Proc. Int. Conference and Workshop on Transport and Mass Exchange Processes in Sand and Gravel Aquifers, Oct. 1-4, 1990, Ottawa, Canada, AECL-10308,* Vol. 1, 224-241.

Tompson, A.F.B. (1993) Numerical simulation of chemical migration in physically and chemically heterogeneous porous

media. *Water Resour. Res.,* 29(11), 3709-3726.

Tompson, A.F.B., Schafer, A.L. & Smith, R.W. (1996) Impacts of physical and chemical heterogeneity on cocontaminant transport in a sandy porous medium. *Water Resour. Res.,* 32(4), 801-818.

Zheng, Ch. (1991) *MT3D: A modular three-dimensional transport model.* S.S. Papadopulos and Associates, Inc., Rockville, Maryland, USA.

Tracer Hydrology 97, Kranjc (ed.) © 1997 Balkema, Rotterdam, ISBN 90 5410 875 4

Some aspects of the functioning of *careos* determined by tracer experiments: Example of La Alpujarra (Spain)

A. Pulido-Bosch, Y. Ben Sbih & A. Vallejos
Grupo de Investigación 'Recursos Hídricos y Geología Ambiental', University of Granada, Spain

ABSTRACT: Since Moorish times (from the eighth to the fifteenth century), and possibly much earlier (the Roman Empire), there has existed, on the southern edge of Sierra Nevada (southern Spain), a noteworthy technique of artificial recharge, known as careos". This consists of the diverting of snowmelt to a series of sites with great infiltration capacity, where the water percolates to a greater or lesser extent. A tracer test was performed at a *careo* site and control points at the known springs located downstream. The calculated transient time varied from one spring to another. As expected, it was not possible to recover all of the tracer, and there were discovered to be various flow paths through the aquiferous material, as shown by the form of the tracer recovery curve, which presented more than one peak.

INTRODUCTION

In the Alpujarra, a region situated on the southern edge of the Sierra Nevada, an ancestral practice known as *careos* continues, consisting of the use of snowmelt, during the months of April, May and June, to infiltrate into the terrain in highly permeable areas, thus guaranteeing a flow of water at the springs located downstream during the dry summer months. Furthermore this practice produces an increase in water temperature and in salt content –which is normally very low in snowmelt– and also retains ground moisture, thus permitting the development of an abundance of vegetation, which would not otherwise exist (Pulido-Bosch & Ben Sbih, 1993 and 1995).

The Alpujarra (Figure 1) receives an annual precipitation of less than 600 mm, beginning in the autumn and continuing until the spring (Alwani, 1991). Summer temperatures reach 25°C, which in winter can fall to -10°C, when precipitations occur in the form of snow. The area is mainly agricultural, irrigated following techniques inherited from the Moors, or possibly earlier (Sabovik, 1973), with a complex system of water distribution by means of an extensive network of irrigation canals, normally unlined, which also serve to recharge the aquifers.

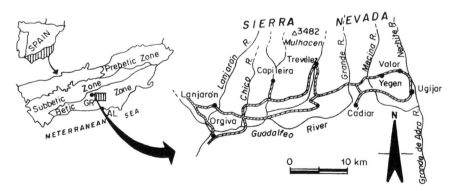

Fig. 1. Location of the study area.

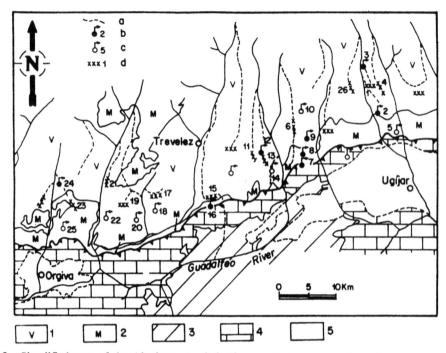

Figure 2.- Simplified map of the Alpujarra. a.- irrigation canal; b.- *remanente* or temporary spring; c.- perennial spring; d.- *careo* ; 1 and 2.- Nevado-Filabride Complex (1: Veleta unit; 2.- Mulhacen unit); 3 and 4.- Alpujarride Complex (3: phyllites and quartzites; 4: limestones and dolomites; 5.- Neogene-Quaternary materials.

Outcropping materials in the area mainly correspond to the Nevado-Filabride Complex, composed of the Veleta and Mulhacen Units, the latter being uppermost (Aldaya et al., 1979). In the southern part (Figure 2) there are outcrops of the Alpujarride Complex, lying over the above-mentioned complex. There also exist localised examples of postorogenic Neogene and Quaternary outcrops filling small basins.

From a tectonic point of view, the area is characterized by overthrust structures, composed of materials affected by multiphased metamorphism – prealpine, alpine and postalpine.

HYDROGEOLOGY FRAMEWORK

Three types of materials can be distinguished: hard rocks, carbonated rocks and, finally, plioquaternary deposits. In the first two cases, hydrogeological behaviour is determined by fractures, fissures and other discontinuities, while in the latter case the intergranular porosity is the most significant factor.

The hard rocks of the Nevado-Filabride Complex are mainly composed of micaschists, which occupy most of the surface of the study area. These micaschists form an anisotropic, discontinuous medium, through which water flows along a network of open and interconnected discontinuities. This network of discontinuities comprises the transmissive element, while the capacitative function is ensured by the smaller fissures. In addition, the wathered zone plays a relevant hydrogeological role, as does the accumulation of colluvions; these are heterometric materials that range from angular blocks to clayey silt. Consequently, the average permeability of this area is not very high.

Recharge proceeds basically from snowmelt and is characterized by the slow release of water, which favours infiltration. This effect is reinforced by the existence of a layer of vegetation over the rock, which retains surface runoff. The weathered zone is a waterbearing material, but its scant thickness and the presence of clays in the matrix mean that it has only slight regulatory influence. This is borne out by the fact that, for those springs that are principally supplied through this area, the flow falls sharply after the recharge *(careo)* period.

The other Nevado-Filabride materials, the quartzites, quartzoschists and marbles, present a certain degree of permeability due to fissuring, to dissolution and to karstification within the marbles. The transmissive elements are very localized, through particular fractures. The hard rocks of the Alpujarride Complex are composed of metapelites (mainly phyllites); these are very slightly permeable materials. When they intercalate with other rocks such as quartzites and quartzoschists, the permeability is increased due to fissuring.

The permeability of the Alpujarride carbonated rocks (mainly dolomites) in this sector is a product of fracturing and karstification. Beside the contributions of precipitation and surface runoff, they are rechargeg by the infiltration from the micaschists of the Nevado-Filabride Complex, while the impervious substratum is made up of basal phyllites. The springs associated with these materials are characterized by their high discharge, which can reach 100 l/s (Ben Sbih, 1995). These springs, like those associated with the flow through the bed rock of the Nevado-Filabride Complex, are locally known as *fuentes*; they are characterized by deeper flow and are related to fracturing.

THE *CAREOS* AND TYPES OF GROUNDWATER FLOW

We have recently identified 46 *careo* points, affecting 22 irrigation canals of different widths and depths, between Orgiva and Ugíjar. There is a close relation between the flow from the springs and the functioning of the *careos* and/or the irrigation canals. It is necessary to distinguish the temporary springs or *remanentes*, whose flow ceases a few days after interruption of that from the *careos*, or when no water is transported by the irrigation canals, from the perennial springs or *fuentes* (figure 3).

To determine the relation between the *careo* points and the springs (the qualitative aspect) and the types of flow (quantitative aspect), a tracer test was performed. We injected 6 kg of LiCl, equivalent to 1 kg of Li, into the *careo* point known as "Prado Nogal", at an altitude of 1500 m. This point was then receiving 40 l/s of the 140 l/s carried by the Cástaras irrigation canal (the longest of all those in the Alpujarra region). To determine the response of the system, we performed an instantaneous, point injection, after dissolving the tracer in water and letting it cool, in order to avoid precipitation of the product on injection, as the reaction is exothermic.

The monitoring points were fixed at the springs numbered 1, 2, 3 and 4, located at altitudes of 1390, 1350, 1330 and 940 m, respectively, and at distances of 400, 800, 1200 and 2000 m from the *careo* point (figure 3). The analytical determinations were carried out with an atomic emission spectrophotometer (ICP).

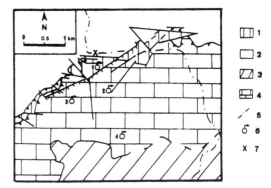

Figure 3.- Hydrogeological sketch of the tracer study area. 1 and 2: Nevado-Filabride Complex (1: micaschists from the Mulhacen Unit; 2: micaschists from the Veleta Unit); 3 and 4: Alpujarride Complex (3: quartzoschists from the Murtas Unit, 4: carbonated rock from the Cástaras Unit); 5: Cástaras irrigation canal; 6: monitored springs; 7: tracer injection point.

The springs were monitored for two weeks. The study of the type of flow of the tracer system was analyzed for springs 1 and 4. After two weeks, the tracer had not been completely recovered, implying the existence of complex flowpaths within the system. Nevertheless, from the breakthrough curves, we were able to calculate the time and average transit speed of the fastest fraction of the tracer, from the injection time (t_o) to the start of breakthrough.

The first traces were obtained 62 h after the injection, both at number 1 and number 2, while for 3 and 4, this occurred 24 h later. The maximum speed of the fastest particles, for the distinct springs under study, was proportional to the distance separating these from the *careo* point, with measured values of 6.45, 12.9, 13.9 and 20.0 m/h, respectively (Pulido-Bosch & Ben Sbih, 1995). From this, it may be deduced that the preferential flow, in the experimental conditions, is the deep one. The increase in the average speed is explained by the lesser permeability of the weathered zone, due to the existence of fine fraction, materials which retain and

433

slow the flow. The presence of Lithium at spring number 4, located over Alpujarride limestones, confirms the recharge to the latter from the micaschists of the Nevado-Filabride Complex.

The calculated speeds indicate the existence of different flow components; a quick flow through a system of open fissures, where flow speed is high (as in the case of 4); an intermediate flow corresponding to a system of fissures that are only slightly open or that are semi-obstructed (the case of 2 and 3); and, finally, a slow one, with a low, subsurface, flow speed through the weathered zone (the case of 1).

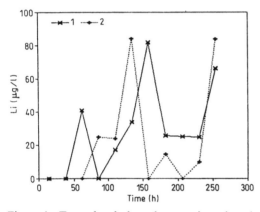

Figure 4.- Tracer break through curves in springs 1 and 4.

In no case did we obtain uni-modal breakthrough curves (figure 4), which is interpreted as being due to the absence of direct flowpaths between the injection and output points. We believe flow occurs through distinct, interconnected and ramified conduits; these ramifications communicate those conduits bearing tracer-containing water with others with no tracer, as has been described in other cases (Gaspar, 1987). Part of this water is carried to the deep flow system and is discharged outside the study area, which would explain the fact that the tracer was not completely recovered.

CONCLUSIONS

Fissuring and fracturing, together with the existence of a weathered zone, have been used for the practice of artificial recharge. The *careos* not only further the recharge at the springs, particularly in the summer, but also produce saline enrichment of the water for human consumption and favour the existence of vegetation that reduces the effects of erosion and maintains the humidity of the area.

From the results of the tracer test, it may be deduced that flow through the weathering is shallow and slow, while that related to the system of discontinuities of the bed rock is faster and deeper. Thus, the waters of the Alpujarra have been classified into three groups: surface water; the temporary springs, or *remanentes*, with low salt contents due to the short residence time; and the perennial springs (*fuentes*), with higher mineralization, due to the greater water-rock contact time.

REFERENCES

Aldaya, F. 1979. *Memoria y Mapa Geológico de la Hoja de Lanjarón*. 1:50.000 (2ª serie). IGME. 65 p. Madrid.

Alwani, G. 1991. La distribución pluviométrica en la cuenca del río Guadalfeo y su influencia en la evolución espacial y temporal de sus recursos hídricos. *III SIAGA*, 1: 43-54. Córdoba.

Gaspar, E. 1987. *Modern trends in tracer hydrology*. Vol. 2. CRC Press, Florida.

Pulido-Bosch, A. & Ben Sbih, Y. 1993. The *careos*, a traditional system for artificial recharge in the Southern Sierra Nevada (Granada, Spain). *XXIV IAH Congress*, I: 301-310. Norway.

Pulido-Bosch, A. & Ben Sbih, Y. 1995. Centuries of artificial recharge on the southern edge of the Sierra Nevada (Granada, Spain). *Environ. Geol.*, 26: 57-63. USA.

Sabovik, P. 1973. *Spanish irrigation and its control*. Thesis Univ. Yale, 564 p.

Tracer Hydrology 97, Kranjc (ed.) © 1997 Balkema, Rotterdam, ISBN 90 5410 875 4

Differentiation of flow components in a karst aquifer using the $\delta^{18}O$ signature

M. Sauter
Applied Geology, University of Tübingen, Germany

ABSTRACT: In the past two decades, the relative abundance of oxygen isotopes had frequently and successfully been used to separate groundwater discharge components in surface water, to determine mean transit times and the topographic height of precipitation and to characterise different types of karst water. The above analyses had been mainly conducted on a longterm basis. In the study presented $\delta^{18}O$ had been used to analyse single storm events in the Gallusquelle catchment of the Swabian Alb, SW. Germany, which allowed to distinguish between a fast and a slow spring discharge component. The signal of the relative abundance in spring water after storm events is characterised by a bimodal breakthrough, a short sharp peak, showing the arrival of the fast conduit flow component and a broad peak indicative of the discharge of the slow karst water. Assuming simple mixing, the fast groundwater component can be quantified to range between 5% and 10%, depending on the characteristics of the recharge event. This information is a prerequisite to quantify groundwater recharge for a karst-groundwater flow model. The model takes into account the duality of the flow system, fast conduit flow and slow flow in small fissures by employing a double-continuum approach. The calibrated flow model had been successfully used to model the breakthrough of the isotope signal.

1 INTRODUCTION

The hydrological response of a karst aquifer is strongly controlled by the geological structure of the aquifer and the mode and time variation of the recharge input. With water flowing at different rates through a karst aquifer, slowly through small fissures and very rapidly through large fractures and solution enlarged channels, the simulation of the flow and transport requires information on both systems and their interaction as well as quantitative estimates of quantity and distribution of groundwater recharge.

On the scale from conduit to diffuse flow, the aquifer investigated can be placed on an intermediate position (mixed-flow). Although the water quality parameters and the discharge react "flashy", and the response to storms is still very rapid, the flow controlling part of the aquifer is in fact the matrix blocks (fissured/diffuse system), that store and release the water over several months. Nevertheless, it is understood that the drainage is still via a tributary network of conduits. As can be shown

below, following storms, approximately 5-10% of groundwater recharge reaches the spring directly via the conduit network within one to two days after the event, causing the frequently observed flashy response in hydraulic and physico-chemical parameters.

Shuster and White (1971) could distinguish between fast and slow groundwater based on variations in Ca/Mg ratio temperature and hardness. Frequently, especially in connection with contaminant transport analysis, short term fluctuations of water quality and spring flow have been analyzed by Quinlan and Alexander (1987), with the aim of designing appropriate groundwater monitoring strategies.

Hess and White (1974, 1988) interpreted high frequency fluctuations in spring water chemistry by attributing them to different arrival times of water from tributaries at the main conduit. Dreiss (1989) explains rapid changes in water quality by surface water entry to the groundwater system close to the spring.

Dreiss (1989) comments on the lack of detailed

examination of storm responses on both, short and longterm basis. Many karst springs display both, a fast initial response and also a longterm memory effect after recharge events. A proper understanding of the aquifer system therefore asks for measurements on both time scales. Vervier (1990), Dreiss (1989), Behringer (1988), Meiman et al. (1988) and the present study attempted to fill this gap. The studies of Vervier (1990) and Meiman et al. (1988) display very complex output signals of flood pulses, difficult to interpret, caused by the additional influence of sinking streams and maturely karstified aquifers. In the present study, the dominant factors determining the output are evaluated. The breakthrough of the isotope signal had been modelled and the sensitivity of the isotope breakthrough to various hydraulic parameters was examined.

2 AREA OF INVESTIGATION

The spring catchment of the Gallusquelle was selected for the investigations. It is situated in southwest Germany on the Swabian Alb, a small mountain range (cuesta landscape), that stretches in an approximate south-west north-east direction for ca. 200 km (Fig. 1). Morphologically, the project area dips gently from an escarpment (1000 m a.s.l.) in the north-west down to about 600 m in the region of the spring. Geologically, the area is composed exclusively of carbonate rocks of the Upper Jurassic.

As shown in Fig. 2, the aquifer is formed by three geological units, the massive limestone (ki 2/3) in the south-east, the marly limestone (ki 1) in

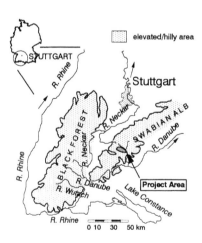

Abb. 1 Area of Investigation

the centre and the Oxford 2 in the north-west. Unconfined conditions prevail in the entire catchment. The unsaturated zone is highly karstified and reaches thicknesses between 90 and 120 m.

3 METHODS

Various time variant parameters were measured at a continuous basis. Therefore, automatic digital dataloggers (PHYTEC, PRODATA) were installed to measure spring discharge (pressure transducer), groundwater levels (pressure transducer), electrical conductivity (WTW LA 1/T-) and temperature (PT100) at half hourly intervals. Turbidity values were provided by the Zweckverband Zollernalb, representative measurements could however only be obtained during pumping hours, because the readings were taken at the waterworks and not directly at the spring. Turbidity data can be regarded as a

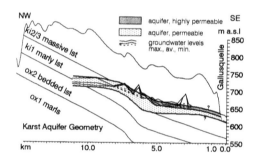

Abb. 2 Geological Cross-Section and Aquifer Geometry

measure for the quantity of suspended solids in spring water. Rainfall was also measured continuously in order to be able to exactly determine the highest intensity of rainfall at an hourly resolution (LAMBRECHT, automatic raingauge).

Groundwater levels can be measured at a resolution of 0.01 m, electrical conductivity at 0.1 μS/cm and the temperature at 0.01°C. The electrical conductivity values are temperature corrected for 25°C.

Due to the fact that electrical conductivity cannot be considered as a conservative parameter, regarding possible chemical changes on the passage from the surface to the water table and also within the phreatic zone, a suitable conservative tracer was important for the analysis of hydrograph data. Therefore, based on the experiences of Zeidler (1987) in shallow and deep karst aquifers in adjacent areas, the relative abundance of [18]O in rain and spring water was measured as well, in order to be able to identify new and old groundwater components in spring water. $\delta^{18}O$ has been frequently used as a tracer for the separation of the groundwater component in surface discharge (Fritz et al., 1976; Sklash and Farvolden, 1979; Herrmann and Stichler, 1980). In karst groundwater systems, the relative abundance of oxygen isotopes has been employed to determine mean transit times (Bertleff, 1986; Geyh and Groschopf, 1978) and to analyse single storm events (Bakalowicz and Mangin, 1980). Samples were collected at daily and half-daily intervals using an automatic sampler. The samples were analyzed by Dr. Trimborn at the Institut für Hydrologie, Gesellschaft für Strahlen- und Umweltforschung, Neuherberg.

Although $\delta^{18}O$ can usually be measured at an accuracy of $\pm\ 0.15\%$, it can be assumed that, if whole time series are measured, the accuracy is improved substantially (oral communication, Prof. Seiler, GSF). This point is critical, because frequently it has been attempted to interpret $\delta^{18}O$ fluctuations in spring water as significant, although they ranged within the error bars. However, in combination with the information from the other parameters, such assumptions appear to be justified.

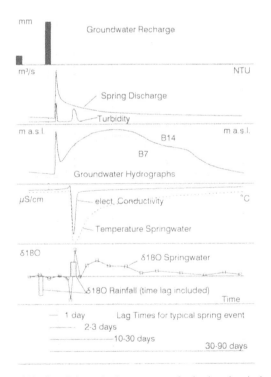

Abb. 3 Schematised response of physico-chemical and hydraulic parameters after a recharge event

4 INTERPRETATION OF TYPICAL EVENTS

The analysis of natural pulses have a major advantage compared to local borehole tests, because they deliver results, that are representative for the whole catchment and these pulses also activate parts of the groundwater system, that are very difficult to test with local field methods, such as the dominant conduit network.

The analysis however requires clean, sharp inputs, that do not show any spatial and temporal variation. The major drawback, however, is the overlapping and superposition of the signals of most events, which makes detailed examination and interpretation often difficult and ambiguous.

The variations in parameters and different effects, described below, can be summarised in an idealised rainfall event and its idealised responses (Fig. 3). The various stages are described in the following in chronological order.

The fast water component of a sharp input pulse of recharge water arrives at the water table, which leads to an increase in discharge induced by the pressure pulse in the conduit system. This increase in the potential is indicated by the peak in the water levels in B7 and B14, and the high flow velocities in the conduit network by the elevated turbidity. The arrival of the fast water components at the spring is schematised in Fig. 4. Between the time of discharge increase and the increase in electrical conductivity, groundwater, displaced from the conduits

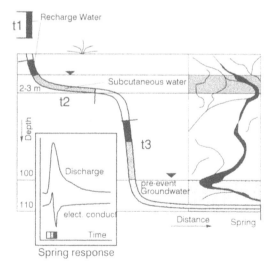

Abb. 4 Conceptual model of flow of fast water component through the karst system

arrives at the spring. After that, the discharge of fast subcutaneous water (Williams, 1983) is sometimes signalled by a slight increase in electrical conductivity and the arrival of the new fast recharge water commences with the rapid drop in electrical conductivity. The negative offset in $\delta^{18}O$ represents subcutaneous water, that stems from a previous, less important isotopically light rainfall event. Isotopically light water is displaced from the fast subcutaneous zone. The slow event water component, delayed by the "fissured system" of the subcutaneous zone, leads to an increase in the regional water levels and a second maximum in spring water $\delta^{18}O$, once it is discharged at the spring.

With the above conceptual model in mind, the actual field data can be understood better. In Figure 5 "scw" stands for subcutaneous water (Williams, 1983), "fw." and "sw." for fast and slow event water, with the figure indicating the chronological order, the rainstorm occurred in.

The presentation of all the parameters is self-explanatory. The plotting of the ^{18}O data require some further clarification. The ^{18}O graph contains two sets of information, the relative abundance of ^{18}O in the spring water (left ordinate), represented as squares and crosses and the $\delta^{18}O$ in rainfall input as bars (right ordinate). The length of the bars indicates the deviation of the isotope ratio in the rainfall from the baseline (see below) and the shading of the bar the intensity of the respective recharge. The $^{18}O/^{16}O$ ratio in precipitation was plotted relative to the deviation of the $\delta^{18}O$ (rainfall) from a value of -10.55‰. This value was evaluated as a base value, the $\delta^{18}O$ appears to tend to after rainfall events. The relative depth of recharge is indicated as the intensity of shading of the respective bar and distinguishes between < 5 mm and > 20 mm in categories of 5 mm. The lag time between rainfall and recharge appearing at the spring has already been taken into account, by shifting the bars by the respective lag.

$^{18}O/^{16}O$ ratios were analyzed in order to circumvent problems, associated with the changes, the new event water is subjected to, i.e. changes in its dissolved mineral content and temperature during its passage through unsaturated and saturated zone. However, the measured values display variations, that are frequently difficult to analyse due to the superposition of several events and small, high frequency variations in $\delta^{18}O$, which are often within the range of the analytical error. However, the Apr89 event also reveals the distinction between the two flow components, the fast and the slow. The broken line indicates an interpretation of the expected bimodal breakthrough of $\delta^{18}O$ as a response to the event of 1st April, the first peak indicating the arrival of the fast component, the second broader maximum the delayed slow event water (further details see below). Before the actual arrival of the fast water, frequently, an offset in the $\delta^{18}O$ is observed, which occurs simultaneously with the arrival of the subcutaneous water at the spring.

Next to the qualitative analysis of spring flow and water quality variations to understand how the system responds and to derive from the response information on aquifer characteristics, quantitative estimates of the different proportions of fast and slow flow components were derived using different volumetric methods (Sauter, 1992). Depending on the recharge event, the fast flow component varied between 5 and 10%.

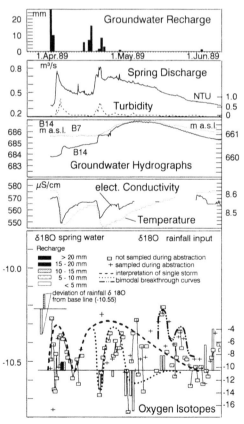

Abb. 5 Physico-chemical and hydraulic parameters measured after a recharge event (April 1989)

5 MODELLING OF NON-POINT SOURCE TRANSPORT

5.1 Transport modelling

The typical breakthrough signal for non-point source input are the time series of hydraulic and physico-chemical parameters, measured at the spring, discussed extensively above.

The relative abundance of oxygen isotopes can be employed as well as a regional tracer and displays a bimodal breakthrough after recharge events. The measured time series however reveal very complex patterns due to the superposition of several events, the influence of the subcutaneous zone and the variations in input $\delta^{18}O$.

Fig. 7 displays the model simulation of the variation in the relative abundance of the oxygen isotopes in the spring water of the Gallusquelle over a two month period, starting with the beginning of April 1989. The model is based on the calibrated double continuum flow model (Fig. 6; Sauter, 1992), without further changes using dispersivities of 10 m for the conduit and 70 m for the fissured system. The dispersivities were evaluated from tracer tests (Merkel, 1991), that tested the respective system. The tracer test in Stober (1991), conducted in the fissured system, yielded a dispersivity of 90 m.

The model curve is a result of the superposition of 10 model runs, one per recharge event, the first one from the beginning of March 1989. The "input concentration" was determined based on recharge depth and the $\delta^{18}O$ relative to the base level of -10.55‰.

The comparison between model and field data demonstrates that the model is capable of reproducing the range of fluctuations, the negative or positive (relative to base level) peaks and also the general trend of the areal input breakthrough curve. The model also reproduces the return to base level at the beginning of June 1989. Deviations of the field data from the model can be explained as follows. The initial fast drop of $\delta^{18}O$ at the 2 April 1989 is probably the result of relatively light water (in $\delta^{18}O$), caused by the early March events, which is displaced from the subcutaneous zone. The model does not yet include the simulation of the epikarst with respect to its $\delta^{18}O$ variation.

The high fluctuations of the field data in May 1989 are most likely the result of incomplete mixing within the aquifer. The model assumes complete mixing and therefore shows an average $\delta^{18}O$ value.

As apparent from Fig. 5, the recharge calculation did not produce any groundwater recharge for the middle of May 1989, although an increase in the $\delta^{18}O$ was registered and rainfall isotope data were obtained for that particular event. Due to the averaging over a large area, no recharge could be calculated and therefore no increase in $\delta^{18}O$ was produced by the recharge model.

5.2 Sensitivity Analysis

The non-point source breakthrough curve is a result of the time variation in the recharge input (subcutaneous zone, although with constant concentration), the pre-event flow and storage in both systems and the relative event flow from conduit and fissured system.

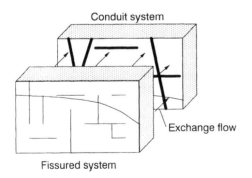

Abb. 6 Conceptual approach of double-continuum modelling

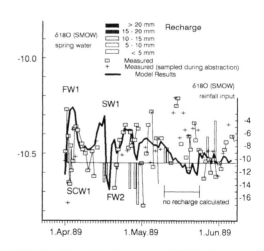

Abb. 7 Comparison between modelled and measured breakthrough

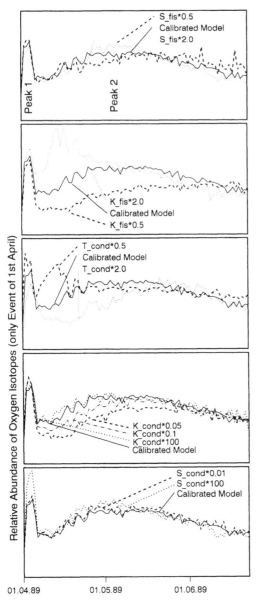

Peak 1
Peak 2

S_fis*0.5
Calibrated Model
S_fis*2.0

K_fis*2.0
Calibrated Model
K_fis*0.5

T_cond*0.5
Calibrated Model
T_cond*2.0

K_cond*0.05
K_cond*0.1
K_cond*100
Calibrated Model

S_cond*0.01
S_cond*100
Calibrated Model

Relative Abundance of Oxygen Isotopes (only Event of 1st April)

01.04.89   01.05.89   01.06.89

Abb. 8   Sensitivity of breakthrough to a change in hydraulic parameters

An exhaustive testing of all relevant parameters with and without time variation is beyond the scope of this study, the main effects are however discussed. Figs. 8 a to e display the change of the calibrated breakthrough curve with the change in various parameters as a result of tracer input due to one single rainfall event (1/2 April 1989) and dilution effects of later recharge. Especially because of

this dilution effect, separate sensitivity runs were calculated, so that the various mechanisms, controlling the changes could be identified.

Fig. 8 a shows the variation of the curve as a result of a change in the storage coefficient of the fissured system (S_{fis}). By doubling S_{fis}, the maximum concentration of peak one (conduit system) is decreased and that of peak two (fissured system) increased. The maximum concentration of the peak is also shifted towards earlier times as compared with the calibrated model. Due to the higher, zero concentration pre-event discharge, the concentration of peak one is lower, and due to the larger quantity of higher concentrated water in the fissured system (compared with the calibrated model), the event of 20 April 1989 is unable to substantially dilute the groundwater stored in the fissured system.

Doubling the hydraulic conductivity of the matrix/fracture exchange coefficient (K_fis), (Fig. 8 b), reverses the effects on the breakthrough curve, observed above. The maximum concentration of peak 1 is slightly increased, because of the increased contribution of the fissured system and almost merges with the peak two (fissured system). The quicker release of water from the slow system storage causes a substantial increase in the peak concentration of the second peak. An increase of the gradient in the conduits by the flow from the slow system causes higher flow velocities and therefore a substantial shift in the second peak.

A similar response of the breakthrough signal can be seen in Fig. 8 c. In this case however, a similar change in concentration can be observed as a result of halving the crossectional area of the conduit system (T_cond; i.e the transmissivity) at the spring. This leads to an increased gradient in the conduits further upgradient and to increased velocities. Because the head within the conduit system, close to the spring affects the head within the fissured zone, the general head increase within the fissured system augments the flow from the slow to the fast system. The general increase in the width of the peak with increasing transmissivity can probably be attributed to the larger volume of water available in the conduits, which recharge water is diluted in. Isotope breakthrough has been shown to be very insensitive to a change in the hydraulic conductivity of the conduit system (K_cond) and the storage of the conduit system (S_cond).

6 CONCLUSIONS

The main hydraulic parameters dominating the transport through a karst aquifer have been identified as the hydraulic conductivity of the fissured system (K_fis), controlling the release of water from the slow into the fast system and the transmissivity of the conduit system. Because of the complex interaction between both systems, only a numerical transport model allows the interpretation of the relative importance of every single parameter for non-point source transport. This statement applies in particular, if the breakthrough curves of several events are superimposed. This study also demonstrates that a number of variations in the isotope ratios, although frequently within the analytical error bars, can be interpreted as significant.

REFERENCES

Bakalowicz, M. & A. Mangin, 1980, L'aquifère karsti que. Sa définition, ses charactèristiques et son identification. Mm. L. sr. Soc. geol. France, 11, 71-79.

Behringer, J., 1988, Hydrochemische Kurz- und Langzeitstudien im Malmaquifer der Mittleren Schwäbischen Alb. Ph. D. Thesis, Universität Tübingen, 222p.

Bertleff, B.W., 1986, Das Strömungssystem der Grundwässer im Malmkarst des westlichen Teils des süddeutschen Molassebeckens. Abh. Geol. Landesamt, Baden-Württemberg, 12, 1-271.

Dreiss, S.J., 1989a, Regional scale transport in a karst aquifer, 1. Component separation of spring flow hydrographs, Water Res. Res., 25, 117-125.

Fritz, P., J.A. Cherry, K.V. Weyer and M.G. Sklash, 1976, Runoff analyses using environmental isotopes and major ions. In: IAEA, Interpretation of environmental isotope and hydrochemical data in groundwater hydrology. Vienna, 111-130.

Geyh, M.A. & P. Groschopf, 1978, Isotopen-physikalische Studie zur Karsthydrologie der Schwäbischen Alb. Abh. Geol. Landesamt, Baden-Württemberg, 8, 7-58.

Herrmann, A. and Stichler, W., 1980, Groundwater - runoff relationships. Catena, 7, 251-263.

Hess, J.W. & W.B. White, 1974, Hydrograph analysis of carbonate aquifers, Res. Pub. 83, Inst. Res. Land and Water Resourc., Pa State University, 63p.

Hess, J.W. & W.B. White, 1988, Storm response of the karstic carbonate aquifer of south-central Kentucky. J. Hydrol., 99, 235-252.

Meiman, J., Ewers, R.O. & J.F. Quinlan, 1988, Investigation of flood pulse movement through a maturely karstified aquifer at Mammoth Cave National Park: a new approach. Proc. Second Conference on Environmental Problems in Karst Terranes and their Solutions, Nashville, Nov. 1988, 227-263.

Merkel, P., 1991, Karsthydrologische Untersuchungen im Lauchertgebiet (westl. Schwäbische Alb). Diplom thesis, Geologisches Institut, Universität Tübingen, 108p.

Quinlan, J.F. & E.C. Alexander, 1987, How often should samples be taken at relevant locations for reliable monitoring of pollutants from an agricultural, waste disposal, or spill site in a karst terrane?. Proc. 2nd Multidisciplinary Conf. on Sinkholes and Env. Impacts of Karst, Florida Sinkhole Res. Inst., Orlando, Florida, 277-286.

Sauter, M., 1992, Quantification and forecasting of regional groundwater flow and transport in a karst aquifer (Gallusquelle, Malm, SW. Germany). Tübinger Geowissenschaftliche Arbeiten, C13, 150p.

Shuster, E.T. & W.B. White, 1971, Seasonal fluctuations in the chemistry of limestone springs; A possible means for characterising carbonate aquifers. J. Hydrol., 14, 93-128.

Stober, I., 1991, Strömungsvorgänge und Durchlässigkeitsverteilung innerhalb des Weißjura - Aquifers im baden-württembergischen Anteil des Molassebeckens. Laichinger Höhlenfreund, 26, 29-42.

Vervier, P., 1990, Hydrochemical characterisation of the water dynamics of a karstic system. J. Hydrol. 121, 103-117

Williams, P.W., 1983, The role of the subcutaneous zone in karst hydrology. J. o. Hyd., 61, 45-67.

Zeidler, N., 1987, Hydraulisch - hydrochemische Untersuchungen und Modellrechnungen im Malmkarst der Schwäbischen Alb. Unpubl. Report, Geologisches Inst. Universität Tübingen, 74p.

Tracer Hydrology 97, Kranjc (ed.)© 1997 Balkema, Rotterdam, ISBN 90 5410 875 4

Simulation of cave hydrology using a conventional computer spreadsheet

J. D. Wilcock
School of Computing, Staffordshire University, UK

ABSTRACT: Simulation of the characteristics of a karst conduit system can provide insights into possible passage configurations based on input and output waveforms. This has been well known since Ashton's work in the mid 1960s, the most common parameters of interest being flow, total hardness and pH, observed at the sinks and resurgences, and within caves. Modern software, e.g. the MS Excel spreadsheet, now allows simulation of streams and cave characteristics to be performed using personal computers, without the need for extensive programming. The simple simulation described is capable of extension using more complex time-dependent coefficients to produce a model which must both represent best practice in hydrological modelling and predict parameters which conform with observed parameters to a high degree of statistical confidence. The incorporation of artificial intelligence to predict passage configurations will be discussed.

1 INTRODUCTION

The principles of simulation of stream flow are well-known, and modelling computations were undertaken before the general availability of electronic computers (e.g. Grover & Harrington 1943). Ashton (1965; 1966; 1967) described how flow, total hardness and pH, at sinks, in the cave, and at resurgences could be used to provide insights into possible passage configurations. In the 1960s, however, it was necessary to use a main-frame computer and considerable programming skills (Wilcock 1968) to make a reliable computer model of a hydrological system. With more frequent access to computers it has become possible to employ complex techniques of simulation such as the kriging and Monte Carlo methods (de Marsily 1986). Kriging is a method for optimising the estimation of values for parameters which are distributed in space and measured at a network of points, for example the hydrogeological parameters transmissivity of an aquifer, piezometric head, and concentrations of solutes. The method is not limited to simple point estimations, but can also be used to obtain an estimation of the variance, i.e. the confidence interval of the estimation. The Monte Carlo simulation method for a stochastic process generates a large number of probabilistic realisations of a variable, and then statistically analyses the ensemble in terms of expected value, variance, histogram and distribution function. Various other numerical solutions have been employed such as finite differences, finite elements, large linear systems of equations, and the transport equation. The use of a numerical model involves data collection, choice of parameters such as size of mesh and time step, calibration of the model, and prediction using the model. Common computer programs for groundwater modelling include well characteristics, solute analytical models, and formation modelling in two and three dimensions (Oko 1994).

It is now possible to use personal computers in an interactive fashion to simulate streams and cave characteristics using standard software, such as the Excel spreadsheet package of Microsoft Office. Far from being just an accounting aid, a modern spreadsheet package allows the use of colour graphics, a wide range of charts, the representation of time series, formulae for evaluation at run time, logical decisions, buttons for the execution of complex macros written in a language such as Visual Basic, and facilities for importing hydrological data from other packages, such as word processors and databases. This paper describes a numerically-based stream and cave hydrograph model, which although simple in its original configuration, with constant coefficients and a fixed time granularity, is capable of extension to use variable coefficients and time granularities, all without significant effort in programming.

2 PREVIOUS MODELLING OF KARST FEATURES

The waveforms of input and output for cave hydrological systems have proved fascinating to cave scientists since at least the beginning of the 20th Century. The first successful recorded use of artificial flood pulses to give information on a cave system was by the Yorkshire Geological Society (Howarth 1902; Dwerryhouse 1905). These investigators raised the level of Malham Tarn, and then flooded the Tarn Sinks by releasing the dam boards. The resulting flood pulses proved normal flow connection to Aire Head Springs, and a flood connection to Malham Cove. A summary of early cave hydrological work is given by Myers (1953). However, the use of the flood pulse technique as a serious method for the investigation of the dynamic characteristics of a cave system only received attention in the early 1960s. Ashton first propounded his classic theory for the pulse hydrology of caves in 1965 (Ashton 1965; 1966) and went on to develop the theory, which was used to analyse the results of the University of Leeds Hydrological Survey Expedition to Jamaica 1963 (Ashton 1967). A notable success of the method was the prediction of the form of the Kingsdale Master Cave, later confirmed by exploration.

Some more recent examples of modelling applied to karst features include delineation of spring catchment areas and investigation of hydrographs (Hobbs & Smart 1989), simulation of karst water level changes (Csepregi & Lorberer 1989), schematic hydrological diagrams (Gunn 1991; Hardwick & Gunn 1995), flow switching (Bottrell & Gunn 1991), network flow modelling of selective enlargement of competing flowpaths (Groves & Howard 1991; 1994), and simulation of cave formation processes (Shen et al. 1993; Shi 1993).

3 PROPOSED USE OF A SPREADSHEET

A typical computerised spreadsheet is an array of cells containing data, located in named columns and rows. Entries may be data, references or formulae. When a reference is used in a formula, it is treated as an *address*, and the *contents* of that other cell are returned as data to the cell containing the reference. Alternatively, complex formulae may be used to return the results of arithmetical or logical calculations operating on the values in a number of other cells. This completes the basic specification of a spreadsheet, and its typical use is in adding columns of figures, and in tabulating both alphanumeric and numerical data to form a simple database, such as a list of names, addresses and telephone numbers. In this form, spreadsheet software has been available since the 1970s.

However, modern spreadsheets are much more than simple accounting or tabulation aids. Colour graphics may be used, a wide range of interactive charts may be generated, instructions to the user may be listed in text boxes, and buttons placed which cause complex programs to be executed. Complex macros may be written in a high-level language such as Visual Basic, data may be imported from an integrated word processor or database, graphics may be imported from clip art or draw/paint packages, and printouts generated of any area of formulae, data or charts, with cell boundaries removed if desired. Several sheets may be interlinked, and the whole becomes an applications package embodying many features of hypermedia. Finally, a novel use is that spreadsheets may be used for investigating time series, using complex formulae embodying logical relationships which are not evaluated until run time, and it is particularly with time series simulation that this article is concerned.

4 A HYDROLOGICAL SIMULATION OF RAINFALL, A STREAM AND ASSOCIATED WATER STORAGES AND TRANSPORTS

If the spreadsheet is to be used to simulate rainfall, a stream and associated water storages and transports, the data model is represented as a *graph*, i.e. a series of *nodes* connected by *edges*. Both nodes and edges are *objects*, i.e. entities which have both data and intelligence. In the hydrological simulation the nodes are reservoirs which store water, and these are indicated by circles. The edges are processes which distribute water in the percentage proportions (fixed or variable) given on the output edges, with some variable time delay. It is possible to have recursive edges, applicable for the reservoir nodes, where the contents of the reservoir in the next time period is partially a function of the contents of the same reservoir in the previous time period, and partially a function of other inputs. There will be outputs from the reservoirs even if there are no inputs, which explains why a stream continues to flow even if there has been no rainfall, getting most of its low-level flow from the storage reservoirs. The recursive edges are indicated by circular arcs with arrows, and each reservoir effectively has a memory of its own previous state with a delay of one time period in the current time granularity. A spreadsheet is normally used as a combinational machine, where cells are evaluated in a dendritic fashion, the outputs (cell evaluations) are a direct function of the inputs at a point in time (i.e. of other cell values), and no circular linkages are allowed (these produce the error message 'Cannot resolve circular references'). But in the simulation described the spreadsheet has been used in a novel fashion as a sequential machine, with some time relationships. Since there may still be no circular linkages, successive rows of the spreadsheet are used for successive time periods in the current time granularity. Furthermore, the time relationships are not restricted to invariant formulae for the functions of previous variables, but may embody some logical tests which are evaluated continuously. This is just what is required for a simulation where the soil storage may become saturated, for example, new cave streams may be activated as flood levels increase, or the transmission or feedback parameters change with time.

Excel is a registered trademark of the Microsoft Corporation, Redmond, WA, USA, to which full acknowledgement is given.

As an illustration of the type of expressions which have to be entered, consider a possible formula for Infiltration. For simplicity constant transmission or feedback coefficients will be employed, although as explained above these can easily be modelled to change with time. Infiltration is conditional on whether the soil can hold any more water (i.e. has not reached the soil maximum storage). A conditional formula IF is available, followed by two parts, the THEN part for use when the condition is TRUE, and the second ELSE part for use when it is FALSE. IF statements may be nested to give very complex constructions, that is the THEN and ELSE parts may themselves be IF statements:

=IF((H1<=H3),0,(IF((H3+G4*0.65<=H1),G4*0.65,H1–H3)))

Given that soil storages are entered in column H and surface storages in column G, this statement says for row 4 of the sheet (i.e. time period 4):

```
IF the soil maximum (stored in the absolute cell $H$1) is less than or equal
to the desired soil storage from the previous time period (i.e. from a relative
cell in the previous row)
THEN    Return 0 Infiltration
ELSE    IF (the Soil Storage from the previous time period)
        + (Surface Storage for the current period * 0.65) is less
        than or equal to the soil maximum storage
        THEN    Return Infiltration as (Surface Storage for the
                current period * 0.65)
        ELSE    Return Infiltration as (soil maximum storage –
                Soil Storage for the previous time period)
```

Excel is also capable of generating many types of chart without the necessity for programming. It is only necessary to select the range of cells for which a chart is required, and to manipulate the type of chart, axes, legends, titles and labels as required using menus. A typical example of what is achievable is illustrated in the Figure below for the final cave simulation.

5 EXTENSION OF THE MODELLING PRINCIPLES TO THE HYDROLOGICAL SIMULATION OF SINKS, VADOSE PASSAGES, PHREAS AND RESURGENCE FOR A CAVE SYSTEM

The spreadsheet may be used to extend the model to simulate the sinks, vadose passages, phreas and resurgence of a cave system being affected by a flood pulse input from a rainfall event. The data model is generally linear for a simple through cave system.

In the graph of the data model, the nodes (reservoirs) are the initial input rainfall channelled into a surface stream, the input junction (where the input streams join underground), the input to the oxbow complex, the epiphreas, and the resurgence, all represented by circles. The edges (processes) are the input generator (representing sinking streams of different volumes and passage lengths which join underground), the first vadose delay, a passage complexor (representing a pattern of oxbow passages which branch, have different lengths, then rejoin), the second vadose delay, an output complexor (representing a pattern of passages of different lengths which feed the common phreas), and an exponential follower simulating the action of the epiphreas, all represented by edges.

The initial input employed is the stream simulation which has been described above. The next task is to simulate the effect of several sinks with different stream volumes and passage lengths which join at the input junction underground. The effect is to produce a multi-peaked input resulting from the superimposition of several single input peaks of

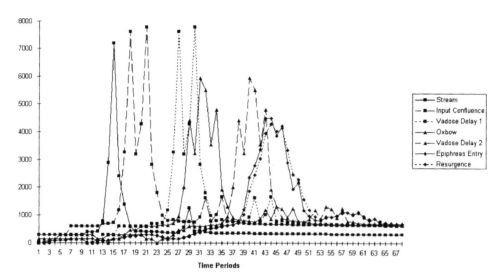

Simulated Cave Flows

Multiple line graph generated on-line for the simulation of a simple cave system involving input generator, first vadose delay, passage complexor, second vadose delay, output complexor and exponential follower.

different heights and timings. This is referred to as the *input generator* in this article.

The first vadose delay representing a single passage from the input junction onwards may be simulated by a simple delay of a number of lines in the spreadsheet.

The cave stream now arrives at a series of passage junctions, oxbows and flood-activated passages where the flood pulse waveform is progressively modified. In some cases the stream will divide into two streams, of different volumes depending on the relative sizes of their respective passages. In other cases the division will only occur if a specific flood level has been reached, when an overflow passage activates. Each of the active passages may further subdivide in a dendritic fashion, and then some passages will rejoin. This division and rejoining after different passage lengths have been traversed is termed the *passage complexor* in this paper, and it may be simulated in terms of logical formulae with passage weights and activation levels which are suitable for later manipulation by an artificial intelligence engine, such as a neural net or genetic algorithm. By examining the components of the resultant bit string some insight into possible passage configurations may be gained.

After the passage complexor there is a second vadose delay representing a single passage between the oxbows and the phreas, which may be simulated

by a simple delay of a number of lines in the spreadsheet.

The cave stream now divides again into several passages which enter the phreas separately. When the flood pulse from one of these component passages reaches the phreas, it will be transmitted immediately to the resurgence, a single pulse at the second vadose delay thus producing multiple pulses at the resurgence. This situation is equivalent to the passage complexor, and is referred to in this paper as the *output complexor*.

Finally, there is the effect of backing-up of flood water at the phreas entrance to be simulated. Initially the phreas cannot transmit the increased flow, so the water backs up until the increased pressure is sufficient for the new volume of water to be forced through the restricted passages of the phreas. This can be simulated by a simple exponential rise in level, the difference between the old and new water levels at the phreas entrance being referred to as the *epiphreas*. As the flood subsides, a similar exponential decay will occur. This mechanism is referred to in this paper as an *exponential follower*.

6 ARTIFICIAL INTELLIGENCE METHODS

It has been mentioned above that the passage complex within the "black box" of the unexplored

446

cave may perhaps be investigated by means of an artificial intelligence engine. There are several possibilities for this, including neural nets, Bayesian belief networks and Genetic Algorithms. The actual algorithm employed is of less interest than the possible solutions derived, i.e. possible passage configurations which would process the observed input from the sinks to produce the observed output at the resurgence.

As an example one such group of algorithms, Genetic Algorithms (GAs), will be described. GAs are general purpose search algorithms, which use methods similar to animal DNA (reproduction, crossover, mutation) to exploit the available population space (Goldberg, 1953; Mitchell, 1996). Reproduction is a complete copy of an individual (parthenogenetic), crossover allows the chromosomes of two parents to be spliced together in any position, which provides rapid search, and mutation is the random changing of bits, which is only occasionally beneficial but which can lead to radical change. In addition inversion may be employed to invert the binary values of a random string of bits. In the search for possible passage configurations, the binary string describing the passage network is subjected to crossover, mutation and inversion operations, and the output evaluated for increasing similarity to the observed resurgence output. An example of GAs being used to control the Usk Reservoir in the Swansea Valley, UK is given by Peggs et al. (1995). The evaluation of GAs for the analysis of passage networks is work in progress by the author.

7 CONCLUSIONS

It has been shown that modern spreadsheet software can satisfy many of the requirements for cave hydrological simulation. This can be achieved with a minimum of programming. The spreadsheet implementation of the model allows on-line updating of the equations and charts during a simulation, and "what if" experiments with different passage configurations and different flood activation levels may be carried out, until the generated output approximates to what is observed in the field.

The ultimate aim is a degree of artificial intelligence, using neural nets, genetic algorithms or other techniques, where the computer will adjust its model of the cave system inside the "black box" automatically until given inputs generate outputs which approximate to reality. Observing the formulae and passage generators/complexors which

then exist in the model may give some insight into possible cave passage configurations in unexplored systems.

REFERENCES

Ashton, K. 1965. Preliminary report on a new hydrological technique. *Cave Research Group of Great Britain Newsletter* 98: 2–5.

Ashton, K. 1966. The analysis of flow data from karst drainage systems. *Trans. CRG* 7(2): 161–203.

Ashton, K. 1967. The University of Leeds Hydrological Survey Expedition to Jamaica 1963. *Trans. CRG* 9(1): 36–51.

Bottrell, S. and J. Gunn 1991. Explorations of Peak and Speedwell Caverns: Flow switching in the Castleton karst aquifer. *Cave Science* 18(1): 47-49.

Csepregi, A. and A. Lorberer 1989. Computer simulation of the karstwater level changes in the Transdanubian Mountain Ranges. *Proc. X Intern. Congress of Speleology, Budapest 1989* (2): 466-469.

Dwerryhouse, A. 1905. The underground waters of North-West Yorkshire. Part II. *Proc. Yorks, Geol. Soc.* 15: 248–292.

Goldberg, D.E. 1953. *Genetic algorithms in search, optimization, and machine learning.* Addison-Wesley Publishing Company, Inc., Wokingham.

Grover, N.C. and Harrington, A.W. 1943. *Stream flow: Measurements, records and their uses.* Dover Publications Inc., New York.

Groves, C.G. and A.D. Howard 1991. Simulation modeling of early karst development. *National Speleological Society Bulletin* 53(2): 118.

Groves, C.G. and A.D. Howard 1994. Network flow modeling of developing karst aquifers: Selective enlargement of competing flowpaths. *National Speleological Society Bulletin* 56(2): 110-111.

Gunn, J. 1991. Explorations of Peak and Speedwell Caverns: Water tracing experiments in the Castleton karst 1950-1990. *Cave Science* 18(1): 43-46.

Hardwick, P. and J. Gunn 1995. Landform-groundwater interactions in the Gwenlais karst, South Wales. In Brown, A.G. (ed.), *Geomorphology and groundwater.* John Wiley & Sons, Chichester, 75-92.

Hobbs, S.L. and P.L. Smart 1989. Delineation of the Banwell spring catchment area and the nature of the spring hydrograph. *Proc. Univ. Bristol Spelaeol. Soc.* 18(3): 359-366.

Howarth, J.H. 1902. The underground waters of North-West Yorkshire. Part I. *Proc. Yorks, Geol. Soc.* 14: 1–48.

de Marsily, G. 1986. *Quantitative hydrogeology: Groundwater hydrology for engineers.* Academic Press Inc., San Diego.

Mitchell, M. 1996. *An introduction to genetic algorithms.* MIT Press, Cambridge, Massachusetts.

Myers, J.O. 1953. Cave Physics. In Cullingford, C.H.D. (ed.), *British Caving*, Routledge and Kegan Paul Limited, London, 226-251.

Oko, U. 1994. Groundwater modeling. In Eslinger, E., U. Oko, J.A. Smith and G.H. Holliday, *Introduction to environmental hydrogeology*, Society for Sedimentary Geology, Tulsa, 11.1-11.9.

Peggs, T., J.C. Miles and C.J. Moore 1995. Genetic based learning of a civil engineering problem. In *Galesia '95, First IEE/IEEE International Conference on Genetic Algorithms in Engineering Systems: Innovations and Applications*, Institute of Electrical Engineers, 394-399.

Shen, J., J. Wang, Q. Yu and W. Gao 1993. Analysis for the origin mechanism of large cave systems in Qingjiang river reaches of Western Hubei, China. *Proc. XI Intern. Congress of Speleology, Beijing 1993*: 107-109.

Shi, M. 1993. Flow action in the formation and development of caves. *Proc. XI Intern. Congress of Speleology, Beijing 1993*: 114.

Wilcock, J.D. 1968. Some developments in pulse-train analysis. *Trans. CRG* 10(2): 73–98.

email: j.d.wilcock@ soc.staffs.ac.uk

Tracer Hydrology 97, Kranjc (ed.) © 1997 Balkema, Rotterdam, ISBN 90 5410 875 4

Author index

Aragno, M. 39
Atalay, I. 199
Audra, P. 203

Barner, W.L. 205
Bat, M. 295
Behrens, H. 305
Belyaev, V.V. 135
Ben Sbih, Y. 431
Benedini, M. 27
Benet, I. 405
Benko, J. 331
Biondić, B. 113
Bricelj, M. 3
Brouyère, S. 11, 397

Calmels, P. 121
Campana, M.E. 389
Carabin, G. 397
Carrera, J. 405
Čenčur Curk, B. 19
Chadha, D.S. 369
Chang Guang-Ye 237
Cucchi, F. 213

d'Alessandro, M. 405
Dassargues, A. 397
Di Fazio, A. 27
Doerfliger, N. 33
Drew, D. 33
Drost, W. 305

Ebhardt, G. 415
Einsiedl, F. 77
Eiswirth, M. 313

Fank, J. 153
Fei Guang-Chan 237
Feldtner, N. 377
Fomovsky, M.A. 135
Formentin, K. 33, 39
Foucher, J.-C. 327
François, O. 121
Freixes, A. 219
Friedrich, K. 55
Fu Rong-An 237

Gangl, G. 227

Garašić, M. 229
García-Gutiérrez, M 405
Gaspar, E. 47, 51
Gaspar, R.D. 47, 51
Giorgetti, F. 213
Gómez, P. 245
Gu Wei-Zu 237
Gudkov, D.I. 127
Guimerá, J. 245, 405
Günay, G. 269

Hadi, S. 55, 69
Harrison, I. 99
Hartmann, A. 339
Hartsch, K. 353
Harum, T. 153
Hildebrand, A.C. 161
Hoehn, E. 63
Höhener, P. 63
Hötzl, H. 189, 269
Hulla, J. 331
Hunkeler, D. 63

Ingraham, N.L. 389
Ivičić, D. 273

Jacobson, R.L. 389
Jakowski, A.E. 415
Jing Zhi-mno 237

Kaglyan, A.E. 135
Kapelj, S. 113
Klenus, V.G. 135
Kloppmann, W. 327
Kogovšek, J. 19, 167, 255
Kollmann, W.F.H. 347
Kostolanský, M. 331
Kovács, T. 331
Kranjc, A. 213
Kuvaev, A. 347
Kuz'menko, M.I. 127

Lachassagne, P. 327
Leader, R.U. 99
Leal, L. 143
Leibundgut, Ch. 55, 69, 161, 181
Lin Zeng-Ping 237

Lindenlaub, M. 161
Liu, H. 255

Malačič, V. 263
Maloszewski, P. 55
Marinetti, E. 213
Martínez, L. 245
Martelat, A. 327
Matičič, B. 321
Merheb, F. 121
Merkl, G. 305
Mesić, S. 113
Mikulla, C. 77
Mišič, M. 3
Monterde, M. 219
Müller, I. 39

Nabivanets, Yu.B. 135
Nagabhushanam, P. 287
Nasvit, O.I. 135
Nativ, R. 269

Obal, M. 19
Ohlenbusch, R. 313
Orth, J.P. 305

Papesch, W. 173
Paunica, I. 47
Pauwels, H. 327
Pavičić, A. 273
Pecly, J. 143
Petrič, M. 255
Plško, J. 331
Polajnar, J. 295
Pristov, J. 295
Ptak, T. 423
Pulido-Bosch, A. 431

Ramoneda, J. 219
Rank, D. 173
Reddy, D.V. 287
Reichert, B. 269
Rentier, C. 11
Roldão, J. 143
Rossi, P. 39

Saaltink, M. 405

Sadler, W.R. 389
Sauter, M. 435
Schlumprecht, Ch. 77
Schnell, K. 313
Sciannamblo, D. 85
Seiler, K.-P. 339
Simon, T. 279
Solomon, K. 269
Spizzico, M. 85
Stanescu, S.P. 51
Stichler, W. 153
Sukhija, B.S. 287
Supper, R. 347

Tadolini, T. 85

Tezcan, L. 269
Trebušak, I. 377
Treskatis, C. 353
Trišič, N. 295

Ufrecht, W. 361
Uggeri, A. 91
Uhlenbrook, S. 181

Vallejos, A. 431
Veselič, M. 19
Vigna, B. 91
Vives, L. 405
Vukovič, A. 263

Ward, R.S. 99, 369
Wilcock, J.D. 443
Williams, A.T. 99, 299, 369
Wohnlich, S. 77
Wolf, M. 305
Wolkersdorfer, C. 377

Ye Gui-Jun 237

Zeyer, J. 63
Zheng Ping-Sheng 237
Zischak, R. 189
Zojer, H. 105